NEUROMETHODS

Series Editor
Wolfgang Walz
University of Saskatchewan,
Saskatoon, SK, Canada

For further volumes:
http://www.springer.com/series/7657

Neuromethods publishes cutting-edge methods and protocols in all areas of neuroscience as well as translational neurological and mental research. Each volume in the series offers tested laboratory protocols, step-by-step methods for reproducible lab experiments and addresses methodological controversies and pitfalls in order to aid neuroscientists in experimentation. Neuromethods focuses on traditional and emerging topics with wide-ranging implications to brain function, such as electrophysiology, neuroimaging, behavioral analysis, genomics, neurodegeneration, translational research and clinical trials. *Neuromethods* provides investigators and trainees with highly useful compendiums of key strategies and approaches for successful research in animal and human brain function including translational "bench to bedside" approaches to mental and neurological diseases.

Translational Research Methods for Major Depressive Disorder

Edited by

Yong-Ku Kim

Department of Psychiatry, Korea University Ansan Hospital, Ansan, Korea (Republic of)

Meysam Amidfar

Deptartment of Speech Therapy, Isfahan University of Medical Sciences, Isfahan, Iran

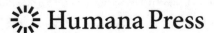 Humana Press

Editors
Yong-Ku Kim
Department of Psychiatry
Korea University Ansan Hospital
Ansan, Korea (Republic of)

Meysam Amidfar
Deptartment of Speech Therapy
Isfahan University of Medical Sciences
Isfahan, Iran

ISSN 0893-2336   ISSN 1940-6045 (electronic)
Neuromethods
ISBN 978-1-0716-2085-4   ISBN 978-1-0716-2083-0 (eBook)
https://doi.org/10.1007/978-1-0716-2083-0

This Humana imprint is published by the registered company Springer Science+Business Media, LLC part of Springer Nature.
The registered company address is: 1 New York Plaza, New York, NY 10004, U.S.A.

Preface to the Series

Experimental life sciences have two basic foundations: concepts and tools. The *Neuromethods* series focuses on the tools and techniques unique to the investigation of the nervous system and excitable cells. It will not, however, shortchange the concept side of things as care has been taken to integrate these tools within the context of the concepts and questions under investigation. In this way, the series is unique in that it not only collects protocols but also includes theoretical background information and critiques which led to the methods and their development. Thus it gives the reader a better understanding of the origin of the techniques and their potential future development. The *Neuromethods* publishing program strikes a balance between recent and exciting developments like those concerning new animal models of disease, imaging, *in vivo* methods, and more established techniques, including, for example, immunocytochemistry and electrophysiological technologies. New trainees in neurosciences still need a sound footing in these older methods in order to apply a critical approach to their results.

Under the guidance of its founders, Alan Boulton and Glen Baker, the *Neuromethods* series has been a success since its first volume published through Humana Press in 1985. The series continues to flourish through many changes over the years. It is now published under the umbrella of Springer Protocols. While methods involving brain research have changed a lot since the series started, the publishing environment and technology have changed even more radically. Neuromethods has the distinct layout and style of the Springer Protocols program, designed specifically for readability and ease of reference in a laboratory setting.

The careful application of methods is potentially the most important step in the process of scientific inquiry. In the past, new methodologies led the way in developing new disciplines in the biological and medical sciences. For example, Physiology emerged out of Anatomy in the nineteenth century by harnessing new methods based on the newly discovered phenomenon of electricity. Nowadays, the relationships between disciplines and methods are more complex. Methods are now widely shared between disciplines and research areas. New developments in electronic publishing make it possible for scientists that encounter new methods to quickly find sources of information electronically. The design of individual volumes and chapters in this series takes this new access technology into account. Springer Protocols makes it possible to download single protocols separately. In addition, Springer makes its print-on-demand technology available globally. A print copy can therefore be acquired quickly and for a competitive price anywhere in the world.

Saskatoon, SK, Canada   *Wolfgang Walz*

Preface

Major depressive disorder (MDD) is a debilitating psychological disorder with a constellation of heterogeneous symptoms characterized by altered emotional, cognitive, and behavioral functions that occur about twice as often in women than in men and affects about 6% of the adult population worldwide each year. Pathophysiology of depression is complex and is hypothesized to involve several biological processes, including neurotransmitter dysfunction, neuronal networks alteration, inadequate neuroendocrine stress response, and chronic inflammation. Despite recent advances in molecular, genetic, and imaging research, the exact pathophysiological mechanisms underlying depression remain elusive. The variability in symptoms, combined with the lack of definitive biomarkers, has presented a major challenge for modeling depression in the laboratory. Improvement of the understanding between clinical practitioners and basic researchers through the development of a common context for depression is a critical step to develop better treatment and prevention of depression. Translational research generally refers to the application of knowledge generated by advances in basic sciences research translated into new approaches for diagnosis, prevention, and treatment of disease. This direction is called bench-to-bedside. Translational research by using of basic science for practical application contributes to identify potential therapeutic targets for disorders, which currently are treated with limited success and consequently improve the health and well-being of society.

This book contains literature that offers a timely review of recent advances in translational research methods discussing the techniques used to study the development of a systems neuroscience paradigm with the goal of understanding brain function and pathophysiology of major depression, and suggestions of new therapeutic approaches for the treatment of major depression. The book is part of the *Neuromethods* series, which focuses on cutting-edge as well as well-established but updated methods and protocols in translational research areas of MDD. This book offers theoretical background interwoven together with tested laboratory protocols and step-by-step methods for reproducible lab experiments to aid neuroscientists and neurobiologists in laboratory testing and experimentation. In this book, we aim to offer the reader an overview of neuroimaging, genetic, electrophysiological, and behavioral research methods in animal models of major depression and application of related findings in clinical management of depressive patients. We have brought together a distinguished cadre of authors with the aim of covering a broad array of translational research methods related to MDD, ranging from behavioral, molecular, and neuroimaging methods to established and innovative pharmacologic and therapeutic neuromodulation techniques. The book aims at bridging these research methods to provide an update of literature relevant to understanding major depression, its neurobiology, and its treatment. Following is a brief overview of the chapters and their content, meant to serve as a guide to navigating the book.

The first part of this book discusses advanced technical approaches to exploring well-established pathophysiological mechanisms and current hypotheses underlying major depression. This part begins with three chapters dedicated to the role of translational research methods in the development of current hypotheses of MDD including hypothalamus-pituitary-adrenal axis (HPA) dysfunction, immune hypothesis and glial dysfunction and altered activity of brain circuits. The fourth and final chapter presents recent ideas, current challenges, and future prospects regarding the discovery of a variety of biomarkers and animal models for depression and how these can help improve diagnosis and treatment.

The second part details the behavioral research methods for major depression. In this part, we will discuss how depression-like behavior can be induced in rodents, and the relationship between these models and depression in humans. This part emphasizes the value of behavioral tests and animal paradigms in modeling at least some aspects of major depression. Use of animal research by offering the experimental control to test neurobehavioral theories with a degree of control and precision that may not always be possible or ethical with human research is incredibly important for continued advancement of the field. Different animal models should be used to study different aspects of depression because finding one valid animal model of depression that reflects all the aspects of the disease is impossible. The translation of paradigms from animal to human will increase generalization of findings and support the development of more effective mental health treatments. Comparing animal findings to human conditions is useful for understanding the complexity of depression and developing clinical diagnoses and animal models in parallel to overcome translational uncertainties. Similarities and differences in findings from analogous animal and human paradigms are discussed, and opportunities are highlighted for future research and paradigm development that will support the clinical utility of these translational paradigms. Chapter 5 assesses the types of rodent models which have been developed in the search for new antidepressants and presents a brief discussion of the biological, primarily biochemical, markers in the blood of depressed patients. Chapter 6 reviews aspects of the neurobiology of suicide through the study of preclinical models with specific emphasis to potential biomarker discovery and treatment targets for mood disorders highly comorbid with suicide. Chapter 7 describes some of the behavioral techniques used to assess cognitive function in the chronic social defeat stress mice model, a well-established model of depression.

The next part illustrates the cellular and molecular research methods for major depression. Specifically, the authors assess the translational value of genetic, optogenetic, and electrophysiologic research methods in order to advance our understanding of the circuit mechanisms of depression and antidepressant treatments. The great promise of this section for depression lies in the power of these technologies to reveal novel mechanistic insight into the disorder to support the development of effective treatments. The development of targeted genetic manipulations and optogenetics, where a light-activated ion channel can be selectively inserted and activated in a population of neurons, makes it possible to study specific circuits and/or signaling pathways in behavioral dimensions relevant to depression. Chapter 8 presents the Principle of Optogenetic and the utility of optogenetics in circuit level analysis of depression in rodents, which led to the identification of specific circuitries involved in the depressive-like behaviors that has great potential for untangling the complex pathophysiology of depression. Chapter 9 presents the main electrophysiological methods currently used to study the effects of antidepressant treatments on neurotransmission. Chapter 10 provides an overview of large-scale transcriptomic profiling on the outset of the two most robustly used high-throughput expression analysis platforms, including RNA microarray and RNA-based next-generation sequencing, and their roles in the context of MDD pathophysiology.

The fourth section of the book describes the major technical advances in noninvasive neuroimaging along with important clinical and translational applications in major depressive disorder. Chapter 11 describes the principles for the positron emission tomography (PET) quantification of radioligand binding and discusses PET method applied to research on the pathophysiology of MDD. Chapter 12 addresses several methods of magnetic resonance imaging (MRI) category such as the task functional MRI (T-FMRI), resting-state functional MRI (Rs-FMRI), diffusion tensor imaging (DTI), diffusion spectrum imaging (DSI), voxel-based morphometry (VBM), and magnetic resonance spectroscopy (MRS) as major tools for the translational research that can help us understand the pathophysiology of MDD. Regional changes in brain structure, functional connectivity, and metabolite concentrations have been reported in depressed patients, giving insight into the networks and brain regions involved, and MRI is a valuable translational tool that can be used to investigate alterations in the structure, function, and molecular makeup of the brain. MRS imaging offers considerable promise for monitoring metabolic alterations associated with MDD. The translation of MRS to clinical practice has been hampered by the lack of technical standardization. There are multiple methods of acquisition, post-processing, and analysis whose details greatly impact the interpretation of the results. Chapter 13 presents a review of the basic principles and important technical issues of MRS followed by a review of the applications of MRS in MDD and a discussion of the implications for the pathophysiology of MDD. Chapter 14 provides practical approaches of multi-modal brain imaging and illustrates how to acquire and analyze the PET and MRI data in multi-modal imaging studies for depressive patients that enables deeper understanding of brain-based pathophysiology for depression.

Treatment options for patients suffering from MDD include psychotherapy, pharmacotherapy, and therapeutic neuromodulation. The fifth and final section of this book focuses on the pharmacological and non-pharmacological interventions including antidepressant agents and their properties such as sexual side effects, neuroimaging biomarkers and neurobiological mechanisms related to antidepressant effect of modulators of serotonin and glutamate receptors as well as neuromodulation and brain stimulation techniques. Neuromodulation techniques are a group of device-based technologies that target specific neural structures via invasive and noninvasive treatments, with the goal of therapeutically modifying pathological patterns of brain activity and circuit connectivity. Invasive treatments require the surgical implantation of stimulating electrodes such as deep brain stimulation (DBS), and noninvasive techniques are able to modulate brain activity transcranially, without surgical intervention, such as transcranial direct current stimulation. These techniques grow from the development of a systems neuroscience paradigm that highlights the role of neural circuits and their processing strategies to understand healthy brain function, neuropsychiatric pathophysiology, and therapeutic mechanisms of action. Chapter 15 addresses the discovery and development of antidepressant agents targeting the neurotransmitter system and their properties. Chapter 16 presents step-by-step instructions on how to perform a supervised tDCS session for the treatment of MDD. Chapter 17 presents a review of preclinical and clinical studies with DBS investigating on different brain targets and animal models of depression that have helped understand the mechanisms of action of DBS as well as candidate biomarkers for treatment response and perspectives for improving DBS's efficacy and safety to treat patients with treatment-resistant depression (TRD). In Chapter 18 the principles of PET are summarized, well-established radiotracers for imaging the serotonergic system are presented, and an overview of serotonergic neurotransmission and its components involved in the pathophysiology of major depression is provided with special focus on molecular neuroimaging studies performed with PET. Chapter 19 presents a review of clinical and preclinical findings related to the antidepressant effect of ketamine and other glutamate receptor modulators in treatment-resistant depression and suicide as well as molecular and neuroimaging mechanisms of the rapid antidepressant effect of ketamine. Chapter 20 provides a translational model for human sexual behavior by presenting methods and models of normal sexual behavior in the male rat to examine the inhibitory and stimulatory effects of antidepressant and other drugs on sexual behavior of male rats that relates well to their known and predicted effects in humans. This section is organized such that each chapter provides an authoritative review of the literature on the translational interventions for MDD.

The field of translational psychiatry is evolving very fast, incorporating new technologies, new indications, and more effective and safer uses of established techniques. These chapters, individually and in their aggregate, provide a broad summary of translational research methods in major depression and a detailed update on its treatment. The broad range of topics covered as well as the methodological depth provided throughout the volume will ensure that the reader is left with a strong appreciation for the progress that has been made in the development and application of translational research methods in major depression as well as the tremendous potential that exists for further major advances in the coming years. We hope this text will prove useful to clinicians and researchers alike and will promote future innovations that advance translational neuroscience of major depression which promise to lead to the development of more potent therapies for the patients who need them.

Ansan, Korea   *Yong-Ku Kim*
Isfahan, Iran   *Meysam Amidfar*

Contents

Contributors

MEYSAM AMIDFAR • *Department of Speech Therapy, School of Rehabilitation, Isfahan University of Medical Sciences, Isfahan, Iran*

FEYZA ARICIOGLU • *Department of Pharmacology and Psychopharmacology Research Unit, Faculty of Pharmacy, Marmara University, Istanbul, Turkey*

AYLA ARSLAN • *School of Advanced Studies, University of Tyumen, Tyumen, Russia*

ORKUN AYDIN • *Department of Psychology, International University of Sarajevo, Sarajevo, Bosnia and Herzegovina*

LUCAS BORRIONE • *Service of Interdisciplinary Neuromodulation, Department and Institute of Psychiatry, University of São Paulo Medical School, São Paulo, Brazil; Laboratory of Neuroscience and National Institute of Biomarkers in Psychiatry, Department and Institute of Psychiatry, University of São Paulo Medical School, São Paulo, Brazil*

MICHEL BOURIN • *Neurobiology of Anxiety and Depression, University of Nantes, Nantes, France*

ANDRÉ R. BRUNONI • *Service of Interdisciplinary Neuromodulation, Department and Institute of Psychiatry, University of São Paulo Medical School, São Paulo, Brazil; Laboratory of Neuroscience and National Institute of Biomarkers in Psychiatry, Department and Institute of Psychiatry, University of São Paulo Medical School, São Paulo, Brazil; Center for Clinical and Epidemiological Research and Interdisciplinary Center for Applied Neuromodulation, University Hospital, University of São Paulo, São Paulo, Brazil*

MU-HONG CHEN • *Department of Psychiatry, Taipei Veterans General Hospital, Taipei, Taiwan; Division of Psychiatry, Faculty of Medicine, National Yang Ming Chiao Tung University, Taipei, Taiwan; Institute of Brain Science, National Yang Ming Chiao Tung University, Taipei, Taiwan*

ANTHONY J. CLEARE • *Centre for Affective Disorders, Department of Psychological Medicine, Institute of Psychiatry, Psychology and Neuroscience, King's College London, London, UK*

VALERIA A. CUELLAR • *Faillace Department of Psychiatry and Behavioral Sciences, Center of Excellence on Mood Disorders, McGovern Medical School, The University of Texas Health Science Center at Houston (UTHealth), Houston, TX, USA*

STEPHEN DE PRÊTRE • *Atlas Pharmaceuticals B.V, Brugge, Belgium*

ALEXANDRE PAIM DIAZ • *Faillace Department of Psychiatry and Behavioral Sciences, Center of Excellence on Mood Disorders, McGovern Medical School, The University of Texas Health Science Center at Houston (UTHealth), Houston, TX, USA; Translational Psychiatry Program, Faillace Department of Psychiatry and Behavioral Sciences, McGovern Medical School, The University of Texas Health Science Center at Houston (UTHealth), Houston, TX, USA*

TANER DOGAN • *Unity Biotechnology, South San Francisco, CA, USA*

THOMAS DRAGO • *Trinity College Institute of Neuroscience, Trinity College Dublin, Dublin, Ireland*

YOGESH DWIVEDI • *Department of Psychiatry and Behavioral Neurobiology, University of Alabama at Birmingham, Birmingham, AL, USA*

ALPER EVRENSEL • *Department of Psychiatry, Uskudar University, Istanbul, Turkey; NP Brain Hospital, Istanbul, Turkey*

MARC FAKHOURY • *Department of Natural Sciences, School of Arts and Sciences, Lebanese American University, Byblos, Lebanon*

ALBERT J. FENOY • *Faillace Department of Psychiatry and Behavioral Sciences, Center of Excellence on Mood Disorders, McGovern Medical School, The University of Texas Health Science Center at Houston (UTHealth), Houston, TX, USA; Vivian L Smith Department of Neurosurgery, McGovern Medical School, The University of Texas Health Science Center at Houston, Houston, TX, USA*

BRISA S. FERNANDES • *Faillace Department of Psychiatry and Behavioral Sciences, Center of Excellence on Mood Disorders, McGovern Medical School, The University of Texas Health Science Center at Houston (UTHealth), Houston, TX, USA; Translational Psychiatry Program, Faillace Department of Psychiatry and Behavioral Sciences, McGovern Medical School, The University of Texas Health Science Center at Houston (UTHealth), Houston, TX, USA*

MICHAEL FRITZ • *Department of Forensic Psychiatry and Psychotherapy, University of Ulm, Ulm, Germany*

MADELINE HAINES • *Trinity College Institute of Neuroscience, Trinity College Dublin, Dublin, Ireland*

PATRICIA A. HANDSCHUH • *Department of Psychiatry and Psychotherapy, Medical University of Vienna, Vienna, Austria*

JOSIEN JANSSEN • *Unit Behavioral Neuroscience, Department of Neurobiology, Groningen Institute for Evolutionary Life Sciences (GELIFES), University of Groningen, Groningen, The Netherlands*

HONG JIN JEON • *Department of Psychiatry, Depression Center, Samsung Medical Center, Sungkyunkwan University School of Medicine, Seoul, South Korea; Department of Health Sciences & Technology, Department of Medical Device Management & Research, and Department of Clinical Research Design & Evaluation, Samsung Advanced Institute for Health Sciences & Technology (SAIHST), Sungkyunkwan University, Seoul, South Korea*

MARIO F. JURUENA • *Centre for Affective Disorders, Department of Psychological Medicine, Institute of Psychiatry, Psychology and Neuroscience, King's College London, London, UK*

JOHN R. KELLY • *Trinity College Institute of Neuroscience, Trinity College Dublin, Dublin, Ireland*

HYEWON KIM • *Department of Psychiatry, Hanyang University Hospital, Seoul, South Korea*

YONG-KU KIM • *Department of Psychiatry, College of Medicine, Korea University, Seoul, South Korea*

MELISANDE E. KONADU • *Department of Psychiatry and Psychotherapy, Medical University of Vienna, Vienna, Austria*

PRAVEEN KORLA • *Department of Psychiatry and Behavioral Neurobiology, University of Alabama at Birmingham, Birmingham, AL, USA*

CHIEN-HAN LAI • *Institute of Biophotonics, National Yang-Ming University, Taipei, Taiwan; PhD Psychiatry & Neuroscience Clinic, Taoyuan, Taiwan*

RUPERT LANZENBERGER • *Department of Psychiatry and Psychotherapy, Medical University of Vienna, Vienna, Austria*

BRIAN E. LEONARD • *Department of Pharmacology, National University of Ireland, Galway, Ireland*

JOHAN LUNDBERG • *Center for Psychiatry Research, Department of Clinical Neuroscience, Region Stockholm and Karolinska Institutet, Stockholm, Sweden*

ADRIANO H. MOFFA • *School of Psychiatry, University of New South Wales, Sydney, NSW, Australia*

Matej Murgas • *Department of Psychiatry and Psychotherapy, Medical University of Vienna, Vienna, Austria*

Erik O'Hanlon • *Trinity College Institute of Neuroscience, Trinity College Dublin, Dublin, Ireland*

Berend Olivier • *Department of Psychopharmacology, Utrecht Institute for Pharmaceutical Sciences, Utrecht University, Utrecht, The Netherlands; Department of Psychiatry, Yale University School of Medicine, New Haven, CT, USA*

Jocelien D. A. Olivier • *Unit Behavioral Neuroscience, Department of Neurobiology, Groningen Institute for Evolutionary Life Sciences (GELIFES), University of Groningen, Groningen, The Netherlands*

Tommy Pattij • *Department of Anatomy and Neurosciences, Center for Neurogenomics and Cognitive Research, VU University Medical Center, Amsterdam, The Netherlands*

Graziano Pinna • *The Psychiatric Institute, Department of Psychiatry, University of Illinois at Chicago, Chicago, IL, USA*

Joao Quevedo • *Faillace Department of Psychiatry and Behavioral Sciences, Center of Excellence on Mood Disorders, McGovern Medical School, The University of Texas Health Science Center at Houston (UTHealth), Houston, TX, USA; Translational Psychiatry Program, Faillace Department of Psychiatry and Behavioral Sciences, McGovern Medical School, The University of Texas Health Science Center at Houston (UTHealth), Houston, TX, USA; Neuroscience Graduate Program, UTHealth Graduate School of Biomedical Sciences, The University of Texas MD Anderson Cancer Center, Houston, TX, USA; Translational Psychiatry Laboratory, Graduate Program in Health Sciences, University of Southern Santa Catarina (UNESC), Criciúma, SC, Brazil*

Kesidha Raajakesary • *Trinity College Institute of Neuroscience, Trinity College Dublin, Dublin, Ireland*

Laís B. Razza • *Service of Interdisciplinary Neuromodulation, Department and Institute of Psychiatry, University of São Paulo Medical School, São Paulo, Brazil; Laboratory of Neuroscience and National Institute of Biomarkers in Psychiatry, Department and Institute of Psychiatry, University of São Paulo Medical School, São Paulo, Brazil*

Darren William Roddy • *Trinity College Institute of Neuroscience, Trinity College Dublin, Dublin, Ireland*

Raquel Romay-Tallon • *The Psychiatric Institute, Department of Psychiatry, University of Illinois at Chicago, Chicago, IL, USA*

Bhaskar Roy • *Department of Psychiatry and Behavioral Neurobiology, University of Alabama at Birmingham, Birmingham, AL, USA*

Marsal Sanches • *Faillace Department of Psychiatry and Behavioral Sciences, Center of Excellence on Mood Disorders, McGovern Medical School, The University of Texas Health Science Center at Houston (UTHealth), Houston, TX, USA*

Leo R. Silberbauer • *Department of Psychiatry and Psychotherapy, Medical University of Vienna, Vienna, Austria*

Sama F. Sleiman • *Department of Natural Sciences, School of Arts and Sciences, Lebanese American University, Byblos, Lebanon*

Jair C. Soares • *Faillace Department of Psychiatry and Behavioral Sciences, Center of Excellence on Mood Disorders, McGovern Medical School, The University of Texas Health Science Center at Houston (UTHealth), Houston, TX, USA; Translational Psychiatry Program, Faillace Department of Psychiatry and Behavioral Sciences, McGovern Medical*

School, The University of Texas Health Science Center at Houston (UTHealth), Houston, TX, USA

BENJAMIN SPURNY-DWORAK • *Department of Psychiatry and Psychotherapy, Medical University of Vienna, Vienna, Austria*

TUNG-PING SU • *Department of Psychiatry, Taipei Veterans General Hospital, Taipei, Taiwan; Division of Psychiatry, Faculty of Medicine, National Yang Ming Chiao Tung University, Taipei, Taiwan; Institute of Brain Science, National Yang Ming Chiao Tung University, Taipei, Taiwan; Department of Psychiatry, Cheng Hsin General Hospital, Taipei, Taiwan*

NEVZAT TARHAN • *Department of Psychiatry, Uskudar University, Istanbul, Turkey; NP Brain Hospital, Istanbul, Turkey*

MIKAEL TIGER • *Center for Psychiatry Research, Department of Clinical Neuroscience, Region Stockholm and Karolinska Institutet, Stockholm, Sweden*

SHIH-JEN TSAI • *Department of Psychiatry, Taipei Veterans General Hospital, Taipei, Taiwan; Division of Psychiatry, Faculty of Medicine, National Yang Ming Chiao Tung University, Taipei, Taiwan; Institute of Brain Science, National Yang Ming Chiao Tung University, Taipei, Taiwan*

PINAR UNAL-AYDIN • *Department of Psychology, International University of Sarajevo, Sarajevo, Bosnia and Herzegovina*

ALLAN H. YOUNG • *Centre for Affective Disorders, Department of Psychological Medicine, Institute of Psychiatry, Psychology and Neuroscience, King's College London, London, UK*

JE-YEON YUN • *Seoul National University Hospital, Seoul, South Korea; Yeongeon Student Support Center, Seoul National University College of Medicine, Seoul, South Korea*

Part I

Translational Research Approach and Current Hypotheses of Major Depressive Disorder

Chapter 1

The Role of Hypothalamic-Pituitary-Adrenal Axis in the Pathophysiology of Major Depression: A Translational Research Perspective of the Prednisolone Suppression Test

Mario F. Juruena, Anthony J. Cleare, and Allan H. Young

Abstract

Glucocorticoids mediate their actions, including feedback regulation of the hypothalamic-pituitary-adrenal (HPA) axis, through two distinct corticosteroid receptor subtypes: mineralocorticoid receptors (MR) and glucocorticoid receptors (GR). Dexamethasone, specifically the dexamethasone suppression test (DST), was the first and most studied glucocorticoid for assessing HPA axis activity; unfortunately, it has pharmacodynamic and pharmacokinetic features that are very distinct from cortisol. We have developed a suppressive test using prednisolone, which is pharmacologically more similar to cortisol than dexamethasone and can therefore obtain a more physiological assessment of HPA axis regulation. We focus here the prednisolone suppression test (PST), summarizing the preclinical work, studies in healthy control subjects, and applications to date in clinical populations. We propose that prednisolone at the 5-mg dosage, together with the assessment of salivary cortisol, can be used to investigate both impaired and enhanced glucocorticoid-mediated negative feedback. The work undertaken to date suggests that, in depression, results using the PST differ from those of the DST, in that patients who are non-suppressors to dexamethasone show normal suppressive responses to prednisolone. Furthermore, preserved suppressive responses to prednisolone predict treatment response in previously difficult-to-treat depression, whereas prednisolone non-suppression is predictive of severe treatment resistance even to the most intensive treatments. We argue that the different results using the PST, which probes both MR and GR, and the DST, which probes only GR, suggest a dissociation between GR and MR function in depressive patients. Thus, we propose that the prednisolone suppression test may offer specific biological and clinical information related to its action at both the GR and the MR. Therefore, the relevance of HPA axis dysfunction in psychiatric disorders and the ability to distinguish "true" from "pseudo-" treatment-resistant depression suggest that the PST holds an important clinical tool.

Key words Hypothalamic-pituitary-adrenal (HPA) axis, Corticosteroid receptor, Mineralocorticoid receptor, Glucocorticoid receptor, Cortisol, Prednisolone, Prednisolone suppression test

Yong-Ku Kim and Meysam Amidfar (eds.), *Translational Research Methods for Major Depressive Disorder*, Neuromethods, vol. 179, https://doi.org/10.1007/978-1-0716-2083-0_1,
© The Author(s), under exclusive license to Springer Science+Business Media, LLC, part of Springer Nature 2022

1 Introduction

The hypothalamic-pituitary-adrenal (HPA) axis constitutes one of the major endocrine systems that maintain homeostasis when the organism is challenged or stressed. Activation of the HPA axis is perhaps the most important endocrine component of the stress response. Abnormal activation of the HPA axis, with increased circulating levels of cortisol, is one potential explanation for many of the features of depression, and many previous studies have described an impaired HPA negative feedback, leading to hyper-cortisolemia, in the more severe forms of depression [1, 2].

Cortisol mediates its action, including feedback regulation of the HPA axis, through two distinct intracellular corticosteroid receptor subtypes referred to as mineralocorticoid receptors (MR) and glucocorticoid receptors (GR) [3, 4]. The type I receptor (MR) has a limited distribution and is found in relatively high density in the hippocampus [5] and in sensory and motor sites outside the hypothalamus [6]. The expression of type II receptors (GR) is more widespread and is found in the hippocampus, amygdala, hypothalamus, and catecholaminergic cell bodies of the brain-stem [7] (see Fig. 1).

There is a theory that suggests a GR defect may mediate the impaired negative feedback thought to cause hypercortisolemia in depression [8]. Under basal levels of cortisol, negative feedback is mediated mainly through the MR in the hippocampus, whereas under stress and high cortisol concentrations, feedback is mediated by the less sensitive GR in the hippocampus, hypothalamus, and pituitary [3]. The balance in these MR- and GR-mediated effects on the stress system is of crucial importance to the set point of the HPA axis activity [3]. It is proposed that the maintenance of corticosteroid homeostasis and the balance in MR-/GR-mediated effects limit vulnerability to stress-related diseases in genetically predisposed individuals [9].

2 From Dexamethasone Suppression Test

The dexamethasone suppression test (DST) was the first, and is to date the most studied, biological marker in research on depressive disorders. In 1968, Bernard Carroll and colleagues showed that depressed patients fail to suppress plasma cortisol to the same extent as non-depressed control subjects [10]. This impaired feedback inhibition by dexamethasone has been demonstrated in depressed patients by a variety of studies, many occurring in the 1970s and 1980s [11, 12]. Early studies in the 1980s proposed the use of the DST to diagnose the melancholic subtype of depression and pointed to the high specificity of the DST in melancholia [13, 14]. However, in the 1990s, several studies found that the

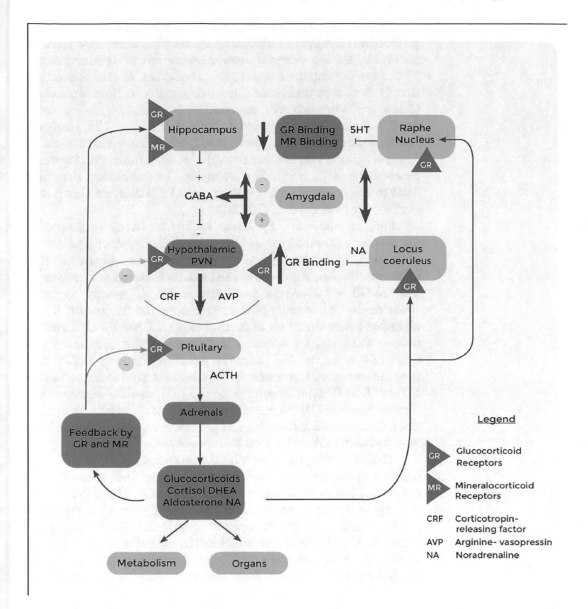

Fig. 1 Schematic diagram of hypothalamic-pituitary-adrenal (HPA) axis negative feedback. It describes HPA axis regulation and negative feedback (−) of cortisol via glucocorticoid receptors (GRs) and mineralocorticoid receptors (MRs), including the hippocampus, amygdala, raphe nucleus, locus coeruleus, and relation via serotonin (5HT) and noradrenaline (NA) with GR/MRs and adrenal hormones. (Adapted from ref. [17])

sensitivity of the DST in the diagnosis of the DSM-III defined melancholic subclass of major depression was only approximately 35–45%, although the specificity was higher at approximately 70–89% [15, 16]. A meta-analysis to determine the significance of differences in rates of non-suppression of cortisol indicated a high probability that a greater rate of cortisol non-suppression occurs in psychotic depression (64% vs 41% in non-psychotic patients) [18].

In summary, studies using the DST have shown that a high proportion of patients with various affective disorders have elevated cortisol levels that escape the suppressive effect of dexamethasone. The use of dexamethasone for a suppressive test has the advantages that it does not cross-react in most cortisol radioimmunoassay (RIAs) and is not subject to reactivation by 11-beta-hydroxysteroid dehydrogenase type 1 in central feedback sites [11, 12]. Unfortunately, dexamethasone has pharmacodynamic and pharmacokinetic features that are very distinct from those of the human endogenous glucocorticoid, cortisol. For example, dexamethasone does not bind to the corticosteroid-binding globulin (CBG) and has a longer half-life than cortisol [19, 20].

Furthermore—and the most important of these distinctive features—dexamethasone and cortisol differ in their abilities to bind and activate the GR and the MR. In fact, the in vitro affinity of dexamethasone for the human GR is 14-fold higher than that of cortisol (Ki = 1.1 nM for dexamethasone vs 15 nM for cortisol), while the in vitro affinity of dexamethasone for the human MR is eightfold lower than that of cortisol (K = 1.1 nM for dexamethasone vs 0.13 nM for cortisol). Furthermore, dexamethasone can fully activate human GR-mediated gene transcription but even at the highest concentrations is unable to fully activate human MR-mediated gene transcription [21], possibly because the dexamethasone-MR complex is much less stable than the dexamethasone-GR complex [22]. Therefore, the DST can only investigate the GR in patients with depression.

These features, together with the low sensitivity of the DST in detecting patients with major depression, have limited the further use of this test in both research and routine clinical practice. While the administration of dexamethasone suppresses endogenous plasma cortisol levels in normal individuals (suppressors) by means of negative feedback inhibition in the majority of published studies, 50% of patients with major depression show an "early escape" (non-suppressors) from dexamethasone suppression [12].

3   Dexamethasone Suppression/CRH Stimulation (DEX/CRH) Test

The most sensitive neuroendocrine function test to detect HPA dysregulation, until now, combines the DST and the corticotrophin-releasing hormone (CRH) stimulation test in the dexamethasone suppression/CRH stimulation (DEX/CRH) test [23–25]. Indeed, Heuser et al. [25] concluded from their studies that the sensitivity of this test is above 80%, depending on age and gender. In this test, patients are pretreated with a single dose (1.5 mg) of dexamethasone at 23.00 h and received intravenously 100 μg CRF at 15.00 h the following day. Although CRH-elicited ACTH response is blunted in depressives, dexamethasone pretreatment produces the opposite effect in the same group and paradoxically enhances ACTH release following CRH. Similarly, CRH-induced cortisol release is much higher in DEX-pretreated patients than in patients treated with CRF challenge alone. The interpretation of the above findings is as follows: DEX due to its low binding to corticosteroid-binding globulin and its decreased access to the brain [25] acts primarily at the pituitary to suppress ACTH. The subsequent decrease of cortisol and the failure of DEX to compensate for the decreased cortisol levels in the nervous tissue create a situation that is sensed by central regulatory elements of the HPA system as a partial and transient adrenalectomy. In response to this situation, the secretion of central neuropeptides that are capable of activating ACTH secretion—mainly CRH and vasopressin—is increased.

Watson et al. [26] compared the use of the DEX/CRH test and the DST in patients with mood disorders and control subjects. They found a close correlation between the cortisol responses on the two tests. The sensitivity of DEX/CRH was 61.9% and the specificity 71.4%, whereas the sensitivity of the DST was 66.6% and the specificity 47.6%. This suggests that the two tests measure common pathology but that the DEX/CRH test is more specific and hence has better diagnostic utility [26].

Nevertheless, the DEX/CRH test remains limited by the pharmacokinetic profile of dexamethasone and the lack of MR receptor activity. Therefore, until recently, there were no tests that could fully assess the contribution of both GR- and MR-mediated negative feedback in the HPA axis overactivity of depression.

4   Prednisolone Suppression Test

Prednisolone is a synthetic glucocorticoid that, like dexamethasone, is widely used as an anti-inflammatory and immunosuppressive drug. Prednisolone mimics cortisol in many ways. Like cortisol, it binds to CBG, and its half-life is similar to that of cortisol [19, 20]. However, the most important of these similarities is that prednisolone and cortisol are similar in their abilities to bind and activate the GR and the MR [19, 27]. Thus, in studies examining rat MR, prednisolone and cortisol have similar affinities for MR (C50 = 20 nM for prednisolone and 16 nM for cortisol [28]) and similar abilities to inhibit sodium excretion in adrenalectomized rats, an index of MR activity [29], whereas dexamethasone has a 3.5-fold lower affinity (C50 = 57 nM) for the MR [28] and shows no activity in the sodium excretion assay [30]. In studies examining human GR, prednisolone has an affinity that is twofold higher than that of cortisol, whereas dexamethasone has an affinity than is sevenfold higher than that of cortisol [31, 32]; and in another study examining mouse GR, prednisolone has a relative potency

Table 1
Characteristics of dexamethasone/prednisolone suppression tests

	Dexamethasone	Prednisolone
Pharmacodynamic	GR	MR/GR
Pharmacokinetic	Long half-life	≅ Cortisol
CBG binding	Does not bind	Binds

CBG corticosteroid binding globulin, *GR* glucocorticoid receptor, *MR* mineralocorticoid receptor
Adapted from ref. [40]

to activate GR function that is the same as cortisol, whereas dexamethasone has a relative potency that is fourfold higher than that of cortisol [31] (see Table 1).

Grossmann et al. [33] contrasted the MR and GR properties of different steroids, regularly used in clinical practice, in the equal in vitro test system complemented by a system to test the steroid-binding affinities (see Fig. 2). They concluded that the potency of a GC is increased by an 11-hydroxy group; both its potency and its selectivity are increased by the D1-dehydro configuration and a hydrophobic residue [33].

Therefore, prednisolone is similar to cortisol in its ability to probe both the GR and the MR. Of course, an important advantage of using prednisolone, rather than cortisol, as a test for both the GR and the MR is that this avoids the confounding effects of the persistence in circulation of the administered cortisol. Indeed, prednisolone is particularly useful in examining suppression of salivary cortisol [19], which represents the bioavailable fraction (5–10%) of plasma cortisol, and therefore reflects more accurately the hormone that reaches and binds the corticosteroid receptors [34].

5   The Prednisolone Suppression Test Method

In these studies, salivary cortisol measurements were used to assess changes in plasma free cortisol levels in response to prednisolone 5 mg. The salivary cortisol was measured using a time-resolved immunofluorescent assay (TR-FIA), as previously described [19, 35–37].

Using the Genesis 100 Robotic Sample Processor (TECAN, Goring on Thames, UK), 50 μL of the sample or standard was added to the wells of microtitration strips (with goat anti-rabbit IgG) and incubated (30 min) with 100 μL of cortisol antibody (1/18000). After further incubation (30 min) with 100 μL of europium-labeled cortisol (1/130) and a washing step, 200 μL of the Enhancement Solution was added, and the strips are shaken for

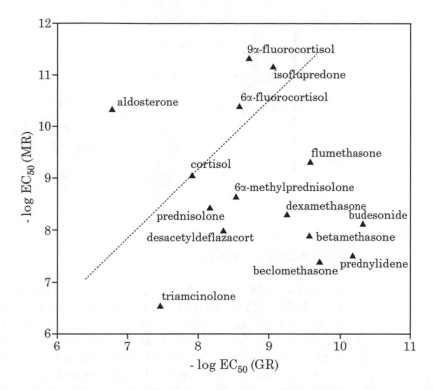

Fig. 2 Selectivity of steroids regarding MR and GR. GR potency increases from left to right, and MR potency increases from bottom to top. The diagonal line separates typical gluco- from mineralocorticoids. Selectivity increases with the perpendicular distance from that line: to the bottom right for glucocorticoids, to the top left for mineralocorticoids. Prednisolone lies close to cortisol in this scheme, unlike dexamethasone. Adapted from ref. [33]

3 min. The fluorescence was then measured in an ARCUS fluorometer linked to Multicalc for RIA data processing and compared with that obtained from a standard curve. The method is a solid-phase TR-FIA (see Fig. 3). Cortisol in saliva (test/standard) and tracer (europium-labeled cortisol) compete for limited quantity of rabbit anticortisol antiserum (polyclonal), which has been raised against cortisol 3-(O-carboxymethyl)oxime (C3-CMO) attached to BSA. The bound cortisol/tracer is separated from the free cortisol by a second antibody (goat anti-rabbit IgG), which is present in excess amounts in the wells of microtiter plates. The bound europium is released by "enhancement reagent" and its fluorescence measured in a fluorometer (Wallac, UK), [36, 37, 40].

With the exception of cortisol antibody batch B930818 (Biogenesis, Poole, Dorset, UK), all other reagents were from PerkinElmer Life Sciences (Cambridge, UK). The intra-assay precision was 8.8% at 0.3 nmol/L, 8.9% at 1.0 nmol/L, and 6.6% at 4.6 nmol/L. The inter-assay precision was 7.7% at 2.1 nmol/L and 5.9% at 9.2 nmol/L. The minimal detection concentration (MDC) was 0.1 nmol/L, and there was no "drifting" evident in

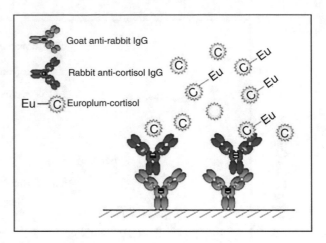

Fig. 3 Principle of the TR-FIA

assays up to 200 wells. Plasma cortisol was quantified using the DSL-2100 "Active Cortisol" Radioimmunoassay (RIA) Kit (Diagnostic System Laboratories). The intra-assay precision was 6% at 74 nmol/L and 5% at 360 nmol/L. The minimum detectable concentration (MDC) was 14 nmol/L. The cross-reactivity of the antiserum was prednisolone (33%), prednisone (1%), and dexamethasone (<1%). Plasma levels of prednisolone were measured by high-performance liquid chromatography (Hewlett Packard UV Detector linked to ChemStation collection system). The calibration graph of the method was in the range of 5 ng/mL–500 ng/mL. The intra-assay precision for prednisolone was 11.2% at 5 ng/mL, 5.2% at 18 ng/mL, and 2.0% at 225 ng/mL. The inter-assay precision was 10.7% at 5 ng/mL, 9.6% at 18 ng/mL, and 3.1% at 225 ng/mL. Plasma prednisolone was measured at the Clinical Trials Laboratory Services Ltd., London UK.

6   Prednisolone Suppression Test Assessing Glucocorticoid and Mineralocorticoid Receptor Function

The evidence summarized above suggests that prednisolone is similar to cortisol in its ability to probe both the GR and the MR and theoretically provides a more naturalistic probe than dexamethasone. Indeed, prednisolone is particularly useful in examining the suppression of salivary cortisol [19], which represents the bioavailable fraction (5–10%) of plasma cortisol, and therefore more accurately reflects the hormone that reaches and binds the corticosteroid receptors [34]. In contrast, salivary cortisol responses to dexamethasone show a large variability [35].

In our studies, saliva samples were collected using Salivettes (Sarstedt, Leicester, UK) and measured using a time-resolved immunofluorescent assay (TR-FIA), as described previously [36, 37]. The cross-reactivity of the antiserum with prednisolone was 28% [37]. Plasma levels of prednisolone (from samples collected at 09.00 the following morning) were measured by high-performance liquid chromatography (Hewlett Packard UV Detector linked to ChemStation collection system), again as previously described in ref. [37].

Jerjes et al. [38] compared capillary gas chromatography, which distinguishes urinary cortisol and prednisolone metabolites, and salivary cortisol immunoassay. Twenty adult volunteers began the study at 21.00 and collected sequential 3-hourly urine samples for a 24-h period (day 1). Prednisolone (5 mg) was taken at 24.00 at the end of day 1. Subjects then began the second 24-h period and continued collecting 3-hourly urine samples until 21.00 on the following day (day 2). Both urinary cortisol metabolites and salivary cortisol were found to have potential for investigation of changed HPA axis negative feedback and for investigation of mild degrees of both glucocorticoid resistance and supersensitivity, based on a convenient pre- and post-dose urinary collection between 09.00 and 18.00 and salivary sampling at 09.00 [38]. Of note was the fact that prednisolone was not detectable in saliva at 09.00, suggesting that cross-reactivity may have limited importance in any event.

7   Prednisolone Suppression Test in Healthy Control Subjects

Because previous studies had shown that 20 mg/day of prednisolone induces complete suppression of cortisol secretion [39], Pariante et al. [19] investigated the capacity for lower doses (2.5 mg, 5 mg, and 10 mg) to detect subtle changes in HPA axis negative feedback. Based on the results, we proposed a prednisolone suppression test (PST) using 5 mg of prednisolone; when taken at 22.00 the night before testing, this produces a mean suppression of salivary cortisol of around 30–40% and is thus an ideal tool to investigate glucocorticoid-mediated negative feedback on the HPA axis.

The same authors compared the effects on plasma and salivary cortisol of the 5-mg prednisolone dose and low-dose dexamethasone (0.5 mg) in a single-blind, repeated-measure design [19]. At all doses, prednisolone caused a larger suppression of salivary cortisol (approximately 20% after 2.5 mg, 30–35% after 5 mg, and 70–75% after 10 mg) than of plasma cortisol (approximately 5% after 2.5 mg, 10% after 5 mg, and 35% after 10 mg). Dexamethasone 0.5 mg gave 80% suppression of plasma cortisol and 90% suppression of salivary cortisol. Plasma and salivary cortisol levels were more consistently correlated in each subject after prednisolone than after dexamethasone. The authors proposed that salivary cortisol responses to 5 mg of prednisolone can be used to investigate both impaired and enhanced glucocorticoid-mediated negative feedback [19].

One application of the PST to date has been to demonstrate an increase in negative feedback induced by acute antidepressant treatment [27]. Healthy individuals were treated with the selective serotonin reuptake inhibitor citalopram (20 mg/day) for 4 days, with the PST administered before and after citalopram treatment. The authors found that citalopram treatment increased the degree of suppression induced by prednisolone from approximately 22% before citalopram to 45% after citalopram. Thus, the antidepressant was able to amplify glucocorticoid-mediated negative feedback on the HPA axis after just 4 days of treatment. These data support the idea that one of the mechanisms by which antidepressants exercise their effects is by normalizing HPA axis hyperactivity due to an increased function of the corticosteroid receptors.

8   Prednisolone Suppression Test in Depression

We have published a series of studies using the PST to understand more about the status of MR and GR in major depression. The first study compared the DST and the PST in 18 patients with severe and treatment-resistant depression and a control group matched for age, body mass index (BMI), and gender distribution [40]. All patients and control subjects underwent both tests, with salivary cortisol output measured between 09.00 and 22.00. The main result of this study was the evident dissociation between the salivary cortisol responses of depressed patients and dexamethasone and their response to prednisolone (see Table 2).

After dexamethasone, depressed patients show less suppression than control subjects; this is consistent with the extensive literature demonstrating dexamethasone non-suppression in depression [11, 12, 41]. However, these same depressed patients who are resistant to dexamethasone show normal suppression after prednisolone, i.e., the same degree as shown in healthy control subjects. In other words, we found a normal suppressive response to prednisolone even though the same individuals demonstrated non-suppression to dexamethasone [40]. Furthermore, in control subjects, there was a correlation between suppression by prednisolone and suppression by dexamethasone, indicating that control subjects are equally sensitive to both glucocorticoids. In contrast, no such correlation was present in depressed patients, confirming the dissociation between sensitivity to prednisolone and resistance to dexamethasone in depression [40].

Our preferred interpretation of these findings is that depressed patients (or, specifically, that sample of depressed patients) show a selective impairment of GR sensitivity, probed by dexamethasone,

Table 2
Salivary cortisol from 09.00 to 22.00, measured as the total cortisol output (area under the curve, AUC) after placebo (AUCPLACEBO), prednisolone (AUCPRED), and dexamethasone (AUCDEX) and percentage of suppression by prednisolone and dexamethasone in patients and control subjects

	Controls ($n = 14$)	Patients ($n = 18$)	F(df) values	P values
$AUC_{PLACEBO}$ (nmol/L/h)	36.6 (2.7)	80.9 (10.3)	13.6 (1, 31)	0.001
AUC_{PRED} (nmol/L/h)	21.7 (2.8)	50.3 (8.9)	7.5 (1, 31)	0.01
AUC_{DEX} (nmol/L/h)	5.6 (1.6)	44.6 (10.9)	9.8 (1, 31)	0.004
Percentage suppression by prednisolone	−41.1% (5.2)	−36.4% (6.7)	0.3 (1, 31)	0.6
Percentage suppression by dexamethasone	−85.1% (3.4)	−45.9% (13.5)	6.3 (1, 31)	0.018

Results given as mean (SEM). Patients show impaired suppression to dexamethasone but normal suppression to prednisolone in comparison with control subjects. Adapted from ref. [40]

whereas MR sensitivity, additionally probed by prednisolone, is retained [40]. This interpretation is consistent with the only study that has so far examined MR function in depression, using the MR antagonist spironolactone [42]. In this study, the authors administered spironolactone in the morning and found that depressed patients showed a larger activation of cortisol secretion compared with control subjects. This suggests that MR activity in depressed patients is preserved, or possibly higher than in control subjects, to compensate for the reduced GR function [42]. We used an MR agonist (prednisolone) to test directly the ability of the MR to suppress the HPA axis, rather than inferring this ability by blocking MR with an antagonist, and it is striking that these two different approaches reach the same conclusion. It is intriguing to speculate that the (hyper)functional MR could represent a protective mechanism that prevents further biological and clinical deterioration in depressed patients.

In summary, as endogenous HPA axis feedback involves both GR and MR and given the above suggestions that in depression MR can compensate for altered GR function, we believe that prednisolone provides a more valid test of the HPA axis in depression and one distinct from dexamethasone.

9 Prednisolone Suppression Test and Treatment Response/Resistance in Depression

A fascinating use for the PST in future research, and potentially in clinical practice, is in helping to delineate subgroups of depression and in particular helping to understand the mechanism of treatment resistance in more severe and chronic forms of depression. To this end, we administered the PST to a group of 45 patients with major depression who were historically defined as resistant to antidepressant therapy; they then underwent a period of intensive inpatient treatment for depression [37]. Forty-six healthy control subjects were also tested. The protocol involved taking either placebo or prednisolone at 22.00 and salivary cortisol at 09.00, 12.00, and 17.00 on the following day. In this study, patients had higher salivary cortisol levels compared with control subjects, both after placebo and after prednisolone, thus confirming the previous findings that patients with severe and treatment-resistant depression have a hyperactive HPA axis [3, 36, 37, 40, 43]. However, patients and healthy control subjects showed similar degrees of suppression to prednisolone, consistent with the data from our previous study [36, 37, 40].

As already reviewed earlier, studies in depression suggest impaired GR function [12, 44] and possibly upregulated MR function [42]. As prednisolone is active at both receptor sites, our results, taken together with these previous studies, are compatible with the notion that, in severe, treatment-resistant depression, there is a change in differential responsiveness of the HPA axis to GR and MR, with increased MR signaling compensating for impaired GR function [37]. Thus, there may be a general resetting of HPA activity with markedly raised basal cortisol levels, suggesting a new set point for HPA function, but with intact negative feedback when this is measured using a more "physiological" challenge able to activate both GR and MR function [45].

A number of prior studies have also investigated whether HPA axis dysfunction is associated with subsequent response to treatment. Our study described above also went on to look at the subsequent response to intensive, inpatient treatment in this group of patients in order to assess whether the PST on admission was a predictor of their subsequent response to treatment [37]. A particularly interesting aspect of these findings was that, although this group of patients with depression as a whole showed preserved negative feedback, this did not apply to all patients [37]. Thus, after intensive treatment, just over half the participants (53%) were classified as treatment responders, with a concomitant improvement in several clinical measures. Those classed as non-responders had been prospectively treated with an intensive, evidence-based treatment package and thus represent a well-defined and truly treatment-resistant population (rather than an insufficiently treated or "pseudo"-resistant population). The PST was able to distinguish these two groups prospectively; thus, there was a higher post-prednisolone cortisol release (representing impaired suppression) in the severe treatment-resistant group compared with the treatment-responsive group. In contrast, no relationship was found with clinical response for basal cortisol levels. Thus, all patients showed HPA axis overactivity, whereas the severe treatment-resistant group also showed non-suppression after

Table 3

Prednisolone suppression rest (PST) summary values in patients with treatment-resistant depression (TRD) and control subjects, calculated as total salivary cortisol output (area under the curve, AUC) after placebo (AUCPLACEBO) and prednisolone 5 mg (AUCPRED)

	A $UC_{PLACEBO}$	AUC_{PRED}	Percentage suppression AUC_{PRED}
Healthy control subjects ($n = 46$)	33.8 (2.5)	16.1 (1.6)	−49.6 (4.0)
Whole TRD sample: PST on admission ($n = 45$)	55.1 (5.1)	32.1 (4.4)	−42.2 (4.8)
P	<0.001	0.001	0.240
TRD with subsequent treatment response: PST on admission ($n = 24$)	53.1 (8.2)	23.5 (4.2)	−52.5 (4.7)
TRD with subsequent treatment non-response: PST on admission ($n = 21$)	57.2 (5.7)	41.9 (7.7)	−30.6 (8.2)
P	0.694	0.046	0.022

Response to treatment is based on a 50% reduction in the Hamilton Depression Rating Scale scores (21 items). Subsequent treatment responders had a higher percentage suppression on admission (52.5%), which was comparable with that of control subjects, whereas subsequent treatment non-responders had impaired suppression (30.6%) (results given as mean and SEM). Adapted from ref. [37]

prednisolone and hence an abnormally impaired negative feedback system (see Table 3).

The implication of this is that there may be a subgroup of patients within those who are severely depressed who have significant neuroendocrine dysfunction, represented by a disturbed HPA axis feedback and an imbalance in the ratio of MR/GR signaling, who are less responsive to the treatments that are currently available and offered in an inpatient affective disorder unit. It may be that the underlying difference in these patients is an inability to compensate for GR resistance by increased MR function.

It is important to underline that we assessed a sample of treatment-resistant depression (TRD) patients with a wide range of axis I comorbidity [37], mainly anxiety symptoms and disorders (generalized anxiety disorder, social phobia, post-traumatic stress disorder, and others). Substantial data suggest that depressed patients with anxiety diagnoses have worse depressive symptoms, a worse clinical course, higher risk of suicide, and possibly different family history [46–50]; however, the influence of comorbid anxiety disorders on the neuroendocrine picture of major depression has not been well studied. In one study, Young and colleagues [51] noted that anxiety disorders occur in approximately 30% of patients with major depressive disorder and concluded that depressed patients with comorbid anxiety disorders show even greater impairment to the negative feedback on the HPA axis than that observed in depressive patients without comorbid anxiety disorders [51]. However, this effect of anxiety was not directly observable in our study [37].

In another prospective study, we have also investigated the changes in the PST before and after an intensive inpatient therapeutic package and assessed any relationship with clinical improvement. Briefly, as in the previously described study, patients took a 5-mg capsule of placebo or prednisolone at 22.00 and gave salivary samples for cortisol analysis at 09.00, 12.00, and 17.00 on the following day. This was then repeated before discharge, after an average of 20 weeks of inpatient treatment. We found no significant change in the response to prednisolone after treatment, even in those who showed a clear improvement in depressive symptoms.

Therefore, the response to prednisolone could be a "trait" feature, able to signal the ability to respond to treatment, but not influenced by the response to treatment itself [37, 45].

This study adds to the mounting evidence that this novel challenge tests different aspects of the HPA axis and offers different biological and clinical information compared with the classic DST and DEX/CRH test [37].

Overall, these data in patients with more severe and treatment-resistant depression suggest that the prednisolone test is able to provide clinically relevant information about the HPA axis, in that those who continue to exhibit a normal level of negative feedback show an increased likelihood of a clinical response to treatment [37]. On the other hand, a continued dysregulation of the HPA system is a strong predictor for negative treatment outcome. This failure to achieve normal cortisol levels might indicate that the underlying limbic dysregulation is not yet resolved. Alternatively, persistent non-suppression may be a trait marker for a more chronic and treatment-resistant group of patients, with different demographic and illness characteristics. Prospective studies, with careful characterization of course of illness and assessment of the same biological marker at different and clearly defined phases of depression, are required to address these questions and to determine whether there may be utility for the PST in routine clinical use [45].

10   Conclusion

Based on the data summarized here, we suggest that the 5-mg PST is a useful tool for investigating the HPA axis. We suggest the 5-mg dose so as to be able to measure changes to the HPA axis in both directions. A 22.00 dosing time is suggested, and assessment of the degree of suppression is most conveniently and sensitively measured using salivary cortisol estimation, ideally at 09.00 and, if possible, other time points during the following day.

We provide evidence that the PST, in contrast to the DST and the DEX/CRH test, probes both the MR and the GR [19, 36, 37, 40] and hence provides a more physiological measure of suppression. Application of the test in clinical populations to date has found non-suppression (and reduced negative feedback) in severe forms of depression.

Furthermore, it is possible that there is a dissociation between GR and MR function present in subgroups of patients with psychiatric disorders, which has not yet been determined using existing tools. We suggest that future studies might usefully examine the PST in larger samples of patients, in patients with other psychiatric disorders (including subtypes of depression such as melancholia and atypical depression), and in non-psychiatric populations of individuals with early- and adult-life stressors.

A particularly intriguing finding is that the PST was related to treatment outcome in severe depression, prospectively distinguishing "true" from "pseudo"-resistant depression. In addition to providing potential insights into the mechanism of treatment resistance, this finding, if replicated, could have clinical utility and deserves further study.

References

1. Nemeroff CB, Evans DL (1984) Correlation between the dexamethasone suppression test in depressed patients and clinical response. Am J Psychiatry 141(2):247–249

2. Gold PW, Goodwin FK, Chrousos GP (1988) Clinical and biochemical manifestations of depression. Relation to the neurobiology of stress. N Engl J Med 319(7):413–420

3. de Kloet ER, Vreugdenhil E, Oitzl MS, Joels M (1998) Brain corticosteroid receptor balance in health and disease. Endocr Rev 19(3):269–301

4. McEwen BS (2000) Allostasis and allostatic load: implications for neuropsychopharmacology. Neuropsychopharmacology 22(2):108–124

5. Reul JM, van den Bosch FR, de Kloet ER (1987) Relative occupation of type-I and type-II corticosteroid receptors in rat brain following stress and dexamethasone treatment: functional implications. J Endocrinol 115(3):459–467

6. Arriza JL, Simerly RB, Swanson LW, Evans RM (1988) The neuronal mineralocorticoid receptor as a mediator of glucocorticoid response. Neuron 1(9):887–900

7. Fuxe K, Harfstrand A, Agnati LF et al (1985) Immunocytochemical studies on the localization of glucocorticoid receptor immunoreactive nerve cells in the lower brain stem and spinal cord of the male rat using a monoclonal antibody against rat liver glucocorticoid receptor. Neurosci Lett 60(1):1–6

8. Modell S, Yassouridis A, Huber J, Holsboer F (1997) Corticosteroid receptor function is decreased in depressed patients. Neuroendocrinology 65(3):216–222

9. Juruena MF, Cleare AJ, Bauer ME, Pariante CM (2003) Molecular mechanisms of glucocorticoid receptor sensitivity and relevance to affective disorders. Acta Neuropsychiatrica 15(6):354–367

10. Carroll BJ, Martin FI, Davies B (1968) Pituitary–adrenal function in depression. Lancet 1(7556):1373–1374

11. Arana GW, Baldessarini RJ, Ornsteen M (1985) The dexamethasone suppression test for diagnosis and prognosis in psychiatry: commentary and review. Arch Gen Psychiatry 42(12):1193–1204

12. Ribeiro SC, Tandon R, Grunhaus L, Greden JF (1993) The DST as a predictor of outcome in depression: a meta-analysis. Am J Psychiatry 150(11):1618–1629

13. Carroll BJ, Feinberg M, Greden JF et al (1981) A specific laboratory test for the diagnosis of melancholia. Standardization, validation, and clinical utility. Arch Gen Psychiatry 38(1):15–22

14. Carroll BJ (1982) The dexamethasone suppression test for melancholia. Br J Psychiatry 140:292–304

15. Rush AJ, Weissenburger JE (1994) Melancholic symptom features and DSM-IV. Am J Psychiatry 151(4):489–498

16. Rush AJ, Giles DE, Schlesser MA et al (1996) The dexamethasone suppression test in patients with mood disorders. J Clin Psychiatry 57(10):470–484

17. Juruena MF, Agustini B, Cleare AJ, Young AH. A translational approach to clinical practice via stress-responsive glucocorticoid receptor signalling. Stem cell investigation. 2017;(16) 4:13

18. Nelson JC, Davis JM (1997) DST studies in psychotic depression: a metaanalysis. Am J Psychiatry 154(11):1497–1503

19. Pariante CM, Papadopoulos AS, Poon L et al (2002) A novel prednisolone suppression test for the hypothalamic-pituitary-adrenal axis. Biol Psychiatry 51(11):922–930

20. Orth D, Kovacs W (1998) The adrenal cortex. In: Wilson J, Foster D, Kronenberg H, Larsen P (eds) Williams textbook of endocrinology, 9th edn. WB Saunders Company, Philadelphia, pp 517–664

21. Rupprecht R, Reul JM, van Steensel B et al (1993) Pharmacological and functional characterization of human mineralocorticoid and glucocorticoid receptor ligands. Eur J Pharmacol 247(2):145–154

22. Reul JM, Gesing A, Droste S et al (2000) The brain mineralocorticoid receptor: greedy for ligand, mysterious in function. Eur J Pharmacol 405(1–3):235–249

23. von Bardeleben U, Holsboer F (1991) Effect of age on the cortisol response to human corticotropin-releasing hormone in depressed patients pretreated with dexamethasone. Biol Psychiatry 29(10):1042–1050

24. von Bardeleben U, Holsboer F, Stalla GK, Muller OA (1985) Combined administration of human corticotropin-releasing factor and lysine vasopressin induces cortisol escape from dexamethasone suppression in healthy subjects. Life Sci 37(17):1613–1618

25. Heuser I, Yassouridis A, Holsboer F (1994) The combined dexamethasone/CRH test: a refined laboratory test for psychiatric disorders. J Psychiatr Res 28(4):341–356

26. Watson S, Gallagher P, Smith MS, Ferrier IN, Young AH (2006) The dex/CRH test—is it better than the DST? Psychoneuroendocrinology 31(7):889–894

27. Pariante CM, Papadopoulos AS, Poon L et al (2004) Four days of citalopram increase suppression of cortisol secretion by prednisolone in healthy volunteers. Psychopharmacology 177(1–2):200–206

28. Lan NC, Matulich DT, Morris JA, Baxter JD (1981) Mineralocorticoid receptor like aldosterone-binding protein in cell culture. Endocrinology 109(6):1963–1970

29. Liddle GW (1958) Aldosterone antagonists. AMA Arch Intern Med 102(6):998–1004

30. Slater JD, Moxham A, Hurter R, Nabarro J (1959) Clinical and metabolic effects of aldosterone antagonism. Lancet 2:931–934

31. Ballard PL, Carter JP, Graham BS, Baxter JD (1975) A radioreceptor assay for evaluation of the plasma glucocorticoid activity of natural and synthetic steroids in man. J Clin Endocrinol Metab 41(2):290–304

32. Lan NC, Graham B, Bartter FC, Baxter JD (1982) Binding of steroids to mineralocorticoid receptors: implications for in vivo occupancy by glucocorticoids. J Clin Endocrinol Metab 54(2):332–342

33. Grossmann C, Scholz T, Rochel M et al (2004) Transactivation via the human glucocorticoid and mineralocorticoid receptor by therapeutically used steroids in CV-1 cells: a comparison of their glucocorticoid and mineralocorticoid properties. Eur J Endocrinol 151(3):397–406

34. Kirschbaum C, Hellhammer DH (1994) Salivary cortisol in psychoneuroendocrine research: recent developments and applications. Psychoneuroendocrinology 19(4): 313–333

35. Reynolds RM, Bendall HE, Whorwood CB, Wood PJ, Walker BR, Phillips DI (1998) Reproducibility of the low dose dexamethasone suppression test: comparison between direct plasma and salivary cortisol assays. Clin Endocrinol 49(3):307–310

36. Juruena MF, Cleare AJ, Papadopoulos AS, Poon L, Lightman S, Pariante CM (2010 Nov) The prednisolone suppression test in depression: dose-response and changes with antidepressant treatment. Psychoneuroendocrinology 35(10):1486–1491

37. Juruena MF, Pariante CM, Papadopoulos AS, Poon L, Lightman S, Cleare AJ (2009) Prednisolone suppression test in depression: prospective study of the role of HPA axis dysfunction in treatment resistance. Br J Psychiatry 194(4):342–349

38. Jerjes WK, Cleare AJ, Wood PJ, Taylor NF (2006) Assessment of subtle changes in glucocorticoid negative feedback using prednisolone: comparison of salivary free cortisol and urinary cortisol metabolites as endpoints. Clin Chim Acta 364(1–2):279–286

39. Seidegard J, Simonsson M, Edsbacker S (2000) Effect of an oral contraceptive on the plasma levels of budesonide and prednisolone and the

influence on plasma cortisol. Clin Pharmacol Ther 67(4):373–381

40. Juruena MF, Cleare AJ, Papadopoulos AS, Poon L, Lightman S, Pariante CM (2006) Different responses to dexamethasone and prednisolone in the same depressed patients. Psychopharmacology 189(2):225–235

41. Baldessarini RJ, Arana GW (1985) Does the dexamethasone suppression test have clinical utility in psychiatry? J Clin Psychiatry 46(2 Pt 2):25–29

42. Young EA, Lopez JF, Murphy-Weinberg V, Watson SJ, Akil H (2003) Mineralocorticoid receptor function in major depression. Arch Gen Psychiatry 60(1):24–28

43. Gold PW, Chrousos GP (2002) Organization of the stress system and its dysregulation in melancholic and atypical depression: high vs low CRH/ NE states. Mol Psychiatry 7(3):254–275

44. Holsboer F (2000) The corticosteroid receptor hypothesis of depression. Neuropsychopharmacology 23(5):477–501

45. Juruena M. The neurobiology of treatment resistant depression: role of the hypothalamic-pituitary-adrenal axis and glucocorticoid and mineralocorticoid receptor function PhD thesis, University of London, Institute of Psychiatry/King's College London; 2007

46. Coryell W, Tsuang D (1992) Hypothalamic–pituitary–adrenal axis hyperactivity and psychosis: recovery during an 8-year follow-up. Am J Psychiatry 149(8):1033–1039

47. Van VC, Winokur G, Behar D, Lowry M (1984) Depressed women with panic attacks. J Clin Psychiatry 45(9):367–369

48. Brown C, Schulberg HC, Shear MK (1996) Phenomenology and severity of major depression and comorbid lifetime anxiety disorders in primary medical care practice. Anxiety 2(5):210–218

49. Joffe RT, Bagby RM, Levitt A (1993) Anxious and nonanxious depression. Am J Psychiatry 150(8):1257–1258

50. Kara S, Yazici KM, Gulec C, Unsal I (2000) Mixed anxiety–depressive disorder and major depressive disorder: comparison of the severity of illness and biological variables. Psychiatry Res 94(1):59–66

51. Young EA, Abelson JL, Cameron OG (2004) Effect of comorbid anxiety disorders on the hypothalamic–pituitary–adrenal axis response to a social stressor in major depression. Biol Psychiatry 56(2):113–120

The Role of Glial Pathology in Pathophysiology and Treatment of Major Depression: Clinical and Preclinical Evidence

Alper Evrensel and Nevzat Tarhan

Abstract

Major depressive disorder is the most common neuropsychiatric disorder affecting millions of people worldwide, with severe consequences and causing the greatest loss of workforce. The monoamine hypothesis is still valid in explaining the etiopathogenesis of depression. Current treatment approaches aim to change the monoamine levels in the synaptic space with various mechanisms of action. However, relapse rates could not be significantly reduced with antidepressant drugs developed and introduced in the last 50 years. The neuroinflammation hypothesis comes to the fore as a solution alternative to these treatment searches stuck in the synaptic gap. One of the important pillars of neuroinflammation is glial dysfunction. Studies investigating cytokine, interleukin, and brain-derived neurotrophic factor (BDNF) levels associated with microglia and astroglia cells are increasing. Antidepressant activity can be obtained, and new antidepressant drug candidates can be determined by means of ligands that agonize and antagonize the glial activity. In this article, clinical and preclinical studies on glial dysfunction in the etiopathogenesis of depression and the treatment approaches recommended on this basis are discussed.

Key words Major depression, Microglia, Astroglia, Neuroinflammation, Dysbiosis

1 Introduction

Depression affects nearly half a billion people worldwide and is the most common neuropsychiatric disorder causing workforce loss [1–3]. Epidemiologists predict that after 10 years, depression will be among the top three diseases that cause disability and death worldwide [2, 3]. Alcohol-substance abuse (approximately 30% of the cases) and death caused by the high risk of suicide are the leading problems that develop secondary to depression [4–7].

Now, depression treatment is based on the monoamine hypothesis, and conventional antidepressant drugs (according to the order of discovery, monoamine oxidase inhibitors, tricyclic antidepressants, selective serotonin reuptake inhibitors, and

Yong-Ku Kim and Meysam Amidfar (eds.), *Translational Research Methods for Major Depressive Disorder*, Neuromethods, vol. 179, https://doi.org/10.1007/978-1-0716-2083-0_2,

serotonin-norepinephrine reuptake inhibitors) aim to increase the levels of serotonin, noradrenaline, and dopamine neurotransmitters in the synaptic cleft [8]. However, these drugs have some disadvantages due to the late onset of action (sometimes up to 3–4 weeks), efficacy levels of around 70%, and side effects (sexual dysfunction, sedation, weight gain, various urinal and gastrointestinal problems) that negatively affect the continuity of treatment [9]. A better understanding of the cellular and molecular mechanisms associated with depression in order to expand this pathophysiological explanation and treatment approaches that are stuck in the synaptic space may offer us more successful treatment opportunities.

Numerous research and evidence suggested that various brain regions and circuits (prefrontal cortex, cingulate cortex, amygdala, hippocampus, thalamus, and striatum) were critically important in the regulation of mood, emotion, and reward systems, and in cases of depression, these regions and circuits had disruptions in their function and blood flow showed reductions [10–12]. In postmortem studies, it was determined that significant structural changes occurred in these brain regions [13].

The immune hypothesis of depression dates back to ancient times [14]. According to this view, while the levels of proinflammatory cytokines and immune mediators in the circulation of many depression patients increase, the number of anti-inflammatory cytokines and mediators decreases. Although this can be described as a coincidence, recent evidence points to the causal role of proinflammatory cytokines in the etiology of depression [15]. In this direction, the emergence of symptoms associated with depression (anxiety, cognitive slowdown, etc.) can be shown in healthy people in a correlation with the proinflammatory cytokines that increase after vaccination [16–18]. Proinflammatory cytokines also cause depressive symptoms in animal models [19–21]. Interestingly, the reduction of these proinflammatory cytokines leads to regression of depression-like behaviors [22–24]. These findings and observations suggest that the relationship between neuroinflammation and depression involves causality rather than an incidental correlation.

2   The Immune Hypothesis of Depression

Depression is common in people with chronic inflammatory diseases such as rheumatoid arthritis, Crohn's disease, and Behcet's disease [25, 26]. Approximately half of the patients who take interferon (an immune mediator that induces proinflammatory cytokines) for hepatitis C virus treatment have the risk of developing iatrogenic depression [27]. The proinflammatory cytokine interleukin-6 (IL-6), which rises after typhoid vaccine in healthy individuals, is thought to cause depression [18]. Additionally, recent meta-analysis studies have provided evidence that levels of proinflammatory cytokines (IL-6, IL-1β, tumor necrosis factor alpha, and C-reactive protein) are increased and decreased with antidepressant treatments in patients with depression [28, 29]. In recent years, opinions emphasizing the role of autoimmune processes in the etiopathogenesis of depression increased significantly [30]. These and similar observations suggest that immune processes and inflammation may play a role in the pathophysiology of depression.

The debate over whether this is the cause or the result of depression has been going on for years. Nevertheless, some recent cohort studies reveal that depression may develop following the rise of proinflammatory cytokines with immune stimulation [31, 32]. However, in order to clarify the role of inflammation in depression, it is necessary to separate and determine the immune effect on specific symptoms, because, for example, it has been determined that depression patients whose physical symptoms (anergy, sleep, and appetite problems) are at the forefront rather than psychological symptoms (unhappiness, pessimism, etc.) have increased CRP levels [33]. A similar situation exists in cancer patients who developed secondary depression due to interferon therapy [34]. It is clear that there is a correlation between depression and immune hyperactivity. However, the evidence obtained in the light of the evaluations made for the establishment of causal relationship suggests that an inflammation is likely to be stimulated for an uncertain reason before the development of depression. Therefore, the validity of the immune hypothesis in depression increases day by day, and immune system hyperactivity comes to the fore as a risk factor for depression.

One of the important areas in the etiopathogenesis of depression is the immune-kynurenine pathway [35]. Serotonin and kynurenine are synthesized from a common precursor: tryptophan [36]. More than 90% of tryptophan is converted to kynurenine by the enzymes indoleamine-2,3-dioxygenase (IDO) found in all somatic cells and tryptophan-2,3-dioxygenase (TDO) found only in hepatocytes [37, 38]. IDO is in two different configurations, IDO1 and IDO2 [39]. IDO1 and TDO activity rate increases when systemic inflammation occurs. Proinflammatory cytokines and glucocorticoids, the molecular precursors of systemic inflammation, strongly stimulate both the IDO1 pathway and the TDO pathway, converting tryptophan into kynurenine [38]. The metabolites of kynurenine (kynurenic acid and quinolinic acid) have a stimulating effect on N-methyl D-aspartate (NMDA) and alpha-7 nicotinic cholinergic receptors [40]. Kynurenic acid shows anti-inflammatory and neuroprotective properties, while quinolinic acid shows excitotoxic properties [41, 42]. In order to place the immune- kynurenine pathway in its place in the depression etiology chain, we should also touch on the gut-brain axis. When the intestinal microbiota composition is disturbed (dysbiosis), serotonin synthesis from tryptophan in the intestines decreases [43]. In addition, lipopolysaccharides that get into the systemic circulation due to leaky gout can trigger low-grade inflammation by stimulating the immune system [44]. Under these conditions, serotonin production from tryptophan is further reduced with the effect of additional inflammation added to the cycle, and tryptophan catabolism shifts from serotonin to kynurenine pathway. This cyclical system may play a role in the etiopathogenesis of depression [35].

3   Glial Dysfunction in Depression

As can be seen, not only synaptic neurotransmitter anomaly is involved in the etiopathogenesis of depression. There is increasing evidence that systemic inflammation crosses the blood-brain barrier and negatively affects neuronal and glial cell functions. Glial cells including microglia, astrocytes, and oligodendrocytes are the supporting elements of the brain tissue. Brain electrophysiology can be performed healthily as a result of the complex interaction of neurons and glial cells.

3.1   Microglial Dysfunction in the Etiopathogenesis of Depression

Microglia are macrophages that function immune in the central nervous system. Macrophages are one of the most important cells of innate immunity in the systemic circulation [45–47]. Microglia are activated in conditions that cause inflammation (infection, injury, toxicity, or neurodegeneration) just like macrophages [45]. In this direction, it also plays a role in the regulation of normal neurogenesis after the apoptosis of neurons [48, 49]. It even functions on synapse maturation in normal embryogenesis [45]. Under any circumstances, long-term activation of microglia may have significant effects on synaptic functions and functions of other glia [50]. The source of these interactions is the proinflammatory cytokines secreted mainly by microglia, and the primary source of cytokines circulating in the central nervous system are microglia [47, 50, 51].

The morphology and functions of microglia differ under normal and pathological conditions. Normal physiological functions often change as a result of exposure to a pathological condition (acute or chronic infection, trauma, toxins, psychological stress, neurodegenerative diseases) [52]. This complex network of interactions may have significant effects on cognitive function, behavior, and mood [52, 53].

Many studies have shown that there is a correlation between depressive symptoms and immune reaction [43]. This systemic immune reaction triggered by bacterial or viral infections is thought to activate microglia in the central nervous system [54, 55]. There are studies suggesting that even the microglia morphology is different in depressed patients who have attempted suicide [56, 57]. It is claimed that the factor that activates systemic immunity is endotoxins in the structure of intestinal lipopolysaccharide (LPS) [43]. It is possible to induce depressive symptoms in healthy people by the administration of endotoxins [18, 58]. Similar results have been obtained in animal experiments [21, 59]. Interestingly, this interaction is reversible. Depression-like behavior is alleviated if microglial activation is inhibited by genetic mutations or pharmacological tools (e.g., microglial inhibitor minocycline) [21, 60, 61].

Many preclinical and clinical studies examining microglial activity in depression showed increased microglial biomarkers [62, 63]. It was shown that ionized calcium-binding adapter molecule 1 (Iba-1), one of the microglia cytoskeletal structural elements, increased during microglial activation [64]. Interestingly, it was found that Iba-1 expression increased in the dorsal anterior cingulate cortex of depressed patients who attempted suicide [65]. Similar results were obtained in stress-induced depression models [66]. Also increased CD11b, another microglial biomarker, was found in the mouse brain (especially in the amygdala, striatum, and hippocampus) after LPS administration [67]. The increase of these microglial biomarkers indicated the presence of microglial dysfunction in the pathophysiology of depression as well as of therapeutic importance.

3.2 The Role of Inflammatory Cytokines, Transcription Factors, and BDNF

The immune system performs its complex functions through mediators synthesized and secreted by leukocytes and lymphocytes, the most important of which are cytokines. These mediators in the protein structure allow immune cells to communicate with each other [68]. Various cytokines have different effects that enable immune cells to be activated (proinflammatory) or inhibited (anti-inflammatory) [69]. Proinflammatory activity is the primary function of cytokines, and they initiate inflammation in body tissues. Cytokines with major proinflammatory activity include interleukin (IL)-6, IL-1β, IL-15, IL-17, and IL-18, tumor necrosis factor (TNF)-α, and interferon-γ (IFN-γ) [69]. The main anti-inflammatory cytokines that function to calm the immune system are IL-4, IL-10, and IL-13 [69, 70]. Imbalance between proinflammatory cytokines and anti-inflammatory cytokines may pave the way for various immune system-related disorders that negatively affect all body functions. Previous studies showed an increase in the level of proinflammatory cytokines and a decrease in the level of anti-inflammatory cytokines in depression [15, 43]. Serum IL-1β and TNF-α levels of patients with depression are high, and these two cytokines are major proinflammatory cytokines produced by microglia [71–73]. It is observed that these cytokines increase after LPS administration in healthy human volunteers, and this increase correlates with depressive symptoms [18, 58].

According to a well-known expression, "correlation does not prove causality." For this reason, many animal and human studies were conducted to test whether there was a causal relationship between proinflammatory cytokines and depressive symptoms. In one of these studies, IL-1β and TNF-α knockout mice exhibited fewer depressive symptoms than healthy ones in models of depression [19]. It was observed that the immune reaction initiated by the administration of IL-1β and TNF-α into the cerebrospinal fluid (CSF) induced depression-like behavior [20]. One of the study designs that provide important evidence is prospective follow-up studies. In a study in which 4415 children were followed up from the age of 9 to the age of 18, a correlation was found between IL-6 levels and depressive symptoms [31]. In another 5-year prospective study conducted on an elderly sample (n = 656), a correlation was found between IL-6 and CRP levels and depressive symptoms [32]. Two recent meta-analyses examining the results of these studies, in summary, point to a causal relationship between neuroinflammation and depression [28, 29]. In this respect, approaches reducing the production of proinflammatory cytokines may have beneficial results in the treatment of depression.

Transcription factors play an important role in the regulation of neuroinflammation. Some of the studies examining the relationship between inflammation and depression have focused on nuclear factor-kappa B (NF-kB) functions [74]. Under normal conditions, NF-kB is found in the cytoplasm of all cells, bound to the IkB protein and inactive. In the inflammatory process, the IkB kinase enzyme is activated, and then NF-Kb and IkB are separated. As a result, NF-KB is released and activated and moves toward the cell nucleus. Reaching the cell nucleus, NF-kB leads to the transcription of many genes that alter cell functions [74, 75]. In addition, NF-kB is activated in peripheral immune cells of healthy people exposed to stress [76]. Activation of NF-kB stimulates the production of proinflammatory cytokines. In addition, it plays an important role in carrying peripheral immune signals to the CNS by crossing the blood-brain barrier [77, 78]. For example, behavioral response obtained by peripheral IL-1β administration can be prevented by inhibition of NF-kB [79].

One of the mediators playing an anti-inflammatory role in the cell is peroxisome proliferator-activated receptor type-α (PPAR-α) [80]. PPAR-α shows its anti-inflammatory effect by inhibiting NF-kB activity [81]. PPAR-α knockout mice exhibit more depression-like behavior than normal mice [82]. Moreover, pharmacological activation of PPAR-α (via WY-14643, a selective agonist) causes a decrease in LPS-induced depression-like behavior in mice [83]. Although there is not enough evidence yet, treatment options that allow obtaining anti-inflammatory efficacy through PPAR-α and NF-kB pathways in the treatment of depression may be promising.

One of the important links in the chain between the immune system and depression is brain-derived neurotrophic factor (BDNF). BDNF activity disorder is thought to be involved in the etiopathogenesis of neuroinflammatory depression [84]. BDNF is a proinflammatory mediator and is also secreted by microglia [84]. Microglial BDNF increases neuronal excitability by causing a decrease in the concentration of chlorine in the neuron [85]. This situation can be reversed by γ-aminobutyric acid (GABA), resulting in an antidepressant activity [86]. BDNF expression increases in the hippocampus of the mice that exhibit LPS-induced depression-like behavior [87]. BDNF leads to NF-kB activation via tyrosine receptor kinase B (Trk-B) [88]. That is why pharmacological blockade of Trk-B causes regression in LPS-induced depression-like behaviors [89]. In summary, approaches that reduce the increase in microglial BDNF and the inflammatory processes that occur through it may be important in the treatment of depression and may be new therapeutic options.

3.3 The Role of Microglial Nicotinic Acetyl Cholinergic Mechanisms

Research on the role of nicotinic acetylcholine receptors (nAChRs) in addiction disorders and also in other psychiatric disorders such as depression are increasing in recent years [5, 90]. nAChRs consist of nine alpha (α) and three beta (β) protein subunits [91]. α7 nAChRs are also secreted by microglial cells in the prefrontal cortex and limbic system and play an anti-inflammatory role [92, 93]. This result is likely due to inhibition of microglial activity and reduced proinflammatory cytokine production [94, 95]. There is also evidence that this neuroimmune dysfunction may contribute to the pathogenesis of depression [93, 96].

It is thought that there are various problems in synaptic cholinergic pathways in the etiology of many neurodegenerative disorders, and these problems may trigger neuroinflammation by reducing the acetylcholine levels in microglia [97]. According to preclinical studies, the α7 subunit of nAChRs plays a key role in triggering neuroinflammation and is therefore shown as one of the potential treatment grounds for depression [91, 92]. However, there are some differences between neuronal α7 nAChRs and microglial α7 nAChRs. For example, following stimulation of microglial α7 nAChRs, intracellular signaling pathways become activated and cause calcium influx from the interstitial cleft into the cell [98], inhibition of NF-kB-mediated activity, and reduced production of proinflammatory cytokines [95, 98, 99]. In an experimental study of this, after stimulation of NF-kB with LPS by using GTS-21 (full agonist of α7 nAChRs), proinflammatory cytokine production could be reduced [100]. Hence, it is thought that new therapeutic potentials for depression can be achieved through various pharmacological interventions that modulate α7 nAChR activity. It will be useful to take a brief look at the evidence obtained in this direction.

In one of these experimental studies, $\alpha7$ nAChR knockout mice scored higher in depression tests compared to normal mice [101]. Conversely, with the $\alpha7$ nAChR agonist effect (via PNU 282987 and GTS-21), a regression in depression-like behavior was observed, and this result was interpreted as an antidepressant effect [102, 103]. Ligands that directly affect the $\alpha7$ nAChR receptor as well as those that produce positive allosteric effects have been studied in depression models. For example, cotinine (an $\alpha7$ nAChR-positive allosteric modulator ligand) caused reduced depression-like behavior in mice [104]. In another mouse experiment, a different $\alpha7$ nAChR-positive allosteric modulator ligand (3-furan-2-yl-N-p-tolyl-acrylamide) was found to have antidepressant activity [105].

These preclinical studies are interesting as they show that by inhibiting microglial activity, an important step has been taken in the neuroinflammatory etiopathogenesis of depression and that antidepressant treatments can be developed toward these goals. The fact that the LPS-induced depression model was used in these studies also supports the hypothesis that intestinal leakage and LPS infiltration into the systemic circulation are the sources of neuroinflammation [14, 43]. These preclinical studies support that by inhibiting microglial activity, proinflammatory cytokine production can be reduced, and as a result, anxiolytic and antidepressant-like efficacy can be achieved.

3.4 Astroglial Dysfunction in the Etiopathogenesis of Depression

Astroglia (astrocytes) provide support to neurons in the central nervous system in many ways. These functions ensure metabolic balance, synaptic growth, neuron maturation, and continuity of the blood-brain barrier [46, 106, 107]. In addition to these functions, it has been determined that astrocytes have another important function in recent years: activation and orientation of microglia [108]. Astrocytes stimulated by cytokines play a key role in neuroinflammation when the inflammatory reaction begins with NF-kB activation [108]. This activation increases the production of proinflammatory cytokines through positive feedback, and there is growing evidence that this increase leads to neuroinflammation and subsequent depression [73, 78].

Synaptic growth and neuron maturation, one of the basic functions of astrocytes, disappears in case of inflammation, and the neuroprotective effect turns into a neurotoxic effect [70]. Depression-like behaviors obtained by LPS stimulation in mouse experiments can be prevented by ketamine (a chemical secreted from astrocytes for neuroprotective effect) [109]. In light of this result, it can be said that during neuroinflammation, astrocytes stop ketamine release, and as a natural reflection of this, the inhibited neurotoxic effect is released.

Astrocyte dysfunctions in patients with depression are concentrated in some parts of the brain. Accordingly, it was determined that the astrocyte nuclei in the prefrontal cortex of patients with depression who underwent postmortem brain examination after committing suicide were larger [110], as were the cytoplasmic dimensions of astrocytes in the anterior cingulate cortex [111]. According to these observations, it can be said that patients with depression have astroglial cell hypertrophy, which is a finding in favor of neuroinflammation.

Astrocyte activity can be determined by many biomarkers. For example, a biomarker called S100 calcium-binding protein B (S100B) is accepted as an indicator of astrocyte activity and increases in the CSF [112] and plasma [113, 114] of patients with depression. Another biomarker, glial fibrillary acidic protein (GFAP), which plays a role in the formation of the astrocyte cytoskeleton and synthesized by astrocytes, is higher in elderly patients with depression, especially in the orbitofrontal cortex [115, 116]. Similar findings were obtained in inflammatory depression models of experimental animals [67, 117]. As a result, increased astrocyte activity, which is determined both experimentally and by biomarker measurements, can be interpreted as an indicator of neuroinflammation in depression. In this direction, treatment approaches that regulate astrocyte functions and reduce pathology can offer therapeutic solutions for depression.

4 Conclusions

Approximately three decades have passed since the immune hypothesis of depression was introduced, and all the evidence accumulated reveals that immune mechanisms play an important role in the etiopathogenesis of depression. There are opinions that the source that activates these immune mechanisms is intestinal leakage and systemic LPS infiltration, and in the central nervous system, it is the glial cell activation. Evidence is growing regarding the mediator role of systemic immune system parameters, proinflammatory cytokines, transcription factors, and BDNF in the etiology of depression. It is thought that microglia that gained activity and started to produce proinflammatory cytokines are associated with depression in the prefrontal cortex and hippocampus. Nicotinic cholinergic mechanisms seem to play an important role in microglial activation. In this context, research on α7 nAChRs has been increasing in recent years. According to the evidence obtained from animal experiments, agonizing α7 nAChRs (e.g., via PNU 282987 and GTS-21) cause decreased production of proinflammatory cytokines and regression of neuroinflammation. These findings are interpreted as obtaining an antidepressant-like effect in LPS-induced depression models. Similarly, it has been shown that ligands that are α7 nAChR-positive allosteric modulators (e.g., PNU120596) also reduce microglia activity and exhibit anti-inflammatory activity in the brain. Although these evidences lead to important insights in understanding the pathogenesis of depression, neuroinflammation, and neuron-glia interaction, it is clear that more research is needed for this information to enter into clinical use as new options in the treatment of depression. However, these seminal and inspiring developments raise our hope that the missing links in depression treatment will be found over time.

References

1. Kessler RC, Bromet EJ (2013) The epidemiology of depression across cultures. Annu Rev Public Health 34:119–138

2. World Health Organization (WHO) (2017) Depression and other common mental disorders Global Health estimates. World Health Organization, Geneva

3. Vos T, Abajobir AA, Abate KH et al (2017) Global, regional, and national incidence, prevalence, and years lived with disability for 328 diseases and injuries for 195 countries, 1990-2016: a systematic analysis for the global burden of disease study 2016. Lancet 390:1211–1259

4. Davis L, Uezato A, Newell JM, Frazier E (2008) Major depression and comorbid substance use disorders. Curr Opin Psychiatry 21:14–18

5. Rahman S (2015) Targeting brain nicotinic acetylcholine receptors to treat major depression and co-morbid alcohol or nicotine addiction. CNS Neurol Disord Drug Targets 14:647–653

6. Morozova M, Rabin RA, George TP (2015) Co-morbid tobacco use disorder and depression: a re-evaluation of smoking cessation therapy in depressed smokers. Am J Addict 24:687–694

7. Fassberg MM, Cheung G, Canetto SS et al (2016) A systematic review of physical illness, functional disability, and suicidal behaviour among older adults. Aging Ment Health 20:166–194

8. Ferrari F, Villa RF (2017) The neurobiology of depression: an integrated overview from biological theories to clinical evidence. Mol Neurobiol 54:4847–4865

9. Bleakley S (2013) Review of the choice and use of antidepressant drugs. Prog Neurol Psychiatry 17:18–26

10. Krishnan V, Nestler EJ (2008) The molecular neurobiology of depression. Nature 455(7215):894–902

11. Nestler EJ (2015) Role of the brain's reward circuitry in depression: transcriptional mechanisms. Int Rev Neurobiol 124:151–170

12. Pandya M, Altinay M, Malone DA Jr, Anand A (2012) Where in the brain is depression? CurrPsychiatry Rep 14:634–642

13. Drevets WC, Price JL, Furey ML (2008) Brain structural and functional abnormalities in mood disorders: implications for neurocircuitry models of depression. Brain Struct Funct 213:93–118

14. Maes M (1995) Evidence for an immune response in major depression: a review and hypothesis. Prog Neuro-Psychopharmacol Biol Psychiatry 19:11–38

15. Leonard BE (2018) Inflammation and depression: a causal or coincidental link to the pathophysiology? Acta Neuropsychiatr 30(1):1–16

16. Brydon L, Harrison NA, Walker C, Steptoe A, Critchley HD (2008) Peripheral inflammation is associated with altered substantia nigra activity and psychomotor slowing in humans. Biol Psychiatry 63:1022–1029

17. Eisenberger NI, Berkman ET, Inagaki TK et al (2010) Inflammation-induced anhedonia: endotoxin reduces ventral striatum responses to reward. Biol Psychiatry 68:748–754

18. Harrison NA, Brydon L, Walker C et al (2009) Inflammation causes mood changes through alterations in subgenual cingulate activity and mesolimbic connectivity. Biol Psychiatry 66:407–414

19. Goshen I, Kreisel T, Ben-Menachem-Zidon O et al (2008) Brain interleukin-1 mediates chronic stress-induced depression in mice via adrenocortical activation and hippocampal neurogenesis suppression. Mol Psychiatry 13:717–728

20. Kaster MP, Gadotti VM, Calixto JB, Santos AR, Rodrigues AL (2012) Depressive-like behavior induced by tumor necrosis factor-alpha in mice. Neuropharmacology 62:419–426

21. O'Connor JC, Andre C, Wang Y et al (2009) Interferon-gamma and tumor necrosis factor-alpha mediate the upregulation of indoleamine 2,3-dioxygenase and the induction of depressive-like behavior in mice in response to bacillus Calmette-Guerin. J Neurosci 29: 4200–4209

22. Koo JW, Duman RS (2008) IL-1beta is an essential mediator of the antineurogenic and anhedonic effects of stress. Proc Natl Acad Sci U S A 105:751–756

23. Murray CL, Obiang P, Bannerman D, Cunningham C (2013) Endogenous IL-1 in cognitive function and anxiety: a study in IL-1RI/mice. PLoS One 8(10):e78385

24. Naude PJ, Dobos N, van der Meer D et al (2014) Analysis of cognition, motor performance and anxiety in young and aged tumor necrosis factor alpha receptor 1 and 2 deficient mice. Behav Brain Res 258:43–51

25. Dickens C, Creed F (2001) The burden of depression in patients with rheumatoid arthritis. Rheumatology (Oxford) 40(12): 1327–1330

26. Kilinçarslan S, Evrensel A (2020) The effect of fecal microbiota transplantation on psychiatric symptoms among patients with inflammatory bowel disease: an experimental study. Actas Esp Psiquiatr 48(1):1–7

27. Bonaccorso S, Marino V, Biondi M (2002) Depression induced by treatment with interferon-alpha in patients affected by hepatitis C virus. J Affect Disord 72(3):237–241

28. Haapakoski R, Mathieu J, Ebmeier KP, Alenius H, Kivimäki M (2015) Cumulative meta-analysis of interleukins 6 and 1β, tumour necrosis factor α and C-reactive protein in patients with major depressive disorder. Brain Behav Immun 49:206–215

29. Goldsmith DR, Rapaport MH, Miller BJ (2016) A meta-analysis of blood cytokine network alterations in psychiatric patients: comparisons between schizophrenia, bipolar disorder and depression. Mol Psychiatry 21(12):1696–1709

30. Shin C, Kim YK (2019) Autoimmunity in microbiome-mediated diseases and novel therapeutic approaches. Curr Opin Pharmacol 49:34–42

31. Khandaker GM, Pearson RM, Zammit S, Lewis G, Jones PB (2014) Association of serum interleukin 6 and C-reactive protein in childhood with depression and psychosis in young adult life: a population-based longitudinal study. JAMA Psychiat 71(10): 1121–1128

32. Zalli A, Jovanova O, Hoogendijk WJ, Tiemeier H, Carvalho LA (2016) Low-grade inflammation predicts persistence of depressive symptoms. Psychopharmacology 233(9): 1669–1678

33. Jokela M, Virtanen M, Batty GD, Kivimäki M (2016) Inflammation and specific symptoms of depression. JAMA Psychiat 73(1):87–88

34. Capuron L, Gumnick JF, Musselman DL et al (2002) Neurobehavioral effects of interferon-alpha in cancer patients: phenomenology and paroxetine responsiveness of symptom dimensions. Neuropsychopharmacology 26(5): 643–652

35. Evrensel A, Ünsalver BÖ, Ceylan ME (2020) Immune-kynurenine pathways and the gut microbiota-brain axis in anxiety disorders. Adv Exp Med Biol 1191:155–167

36. Palego L, Betti L, Rossi A, Giannaccini G (2016) Tryptophan biochemistry: structural, nutritional, metabolic, and medical aspects in humans. J Amino Acids 2016:8952520

37. Clarke G, McKernan DP, Gaszner G (2012) A distinct profile of tryptophan metabolism along the kynurenine pathway downstream of toll-like receptor activation in irritable bowel syndrome. Front Pharmacol 3:90

38. O'Mahony SM, Clarke G, Borre YE, Dinan TG, Cryan JF (2015) Serotonin, tryptophan metabolism and the brain-gut-microbiome axis. Behav Brain Res 277:32–48

39. Fatokun AA, Hunt NH, Ball HJ (2013) Indoleamine 2,3-dioxygenase 2 (IDO2) and the kynurenine pathway: characteristics and potential roles in health and disease. Amino Acids 45(6):1319–1329

40. Forrest CM, Youd P, Kennedy A et al (2002) Purine, kynurenine, neopterin and lipid peroxidation levels in inflammatory bowel disease. J Biomed Sci 9(5):436–442

41. Kaszaki J, Erces D, Varga G (2012) Kynurenines and intestinal neurotransmission: the role of N-methyl-D-aspartate receptors. J Neural Transm 119(2):211–223

42. Stone TW, Darlington LG (2013) The kynurenine pathway as a therapeutic target in cognitive and neurodegenerative disorders. Br J Pharmacol 169(6):1211–1227

43. Evrensel A, Ceylan ME (2015) The gut-brain Axis: the missing link in depression. Clin Psychopharmacol Neurosci 13(3): 239–244

44. Evrensel A, Ceylan ME (2016) Fecal microbiota transplantation and its usage in neuropsychiatric disorders. Clin Psychopharmacol Neurosci 14(3):231–237

45. Paolicelli RC, Bolasco G, Pagani F et al (2011) Synaptic pruning by microglia is necessary for normal brain development. Science 333(6048):1456–1458

46. Lacagnina MJ, Rivera PD, Bilbo SD (2017) Glial and neuroimmune mechanisms as critical modulators of drug use and abuse. Neuropsychopharmacology 42:156–177

47. Bachtell RK, Jones JD, Heinzerling KG, Beardsley PM, Comer SD (2017) Glial and neuroinflammatory targets for treating substance use disorders. Drug Alcohol Depend 180:156–170

48. Schafer DP, Lehrman EK, Kautzman AG et al (2012) Microglia sculpt postnatal neural circuits in an activity and complement-dependent manner. Neuron 74:691–705

49. Sierra A, Encinas JM, Deudero JJ et al (2010) Microglia shape adult hippocampal neurogenesis through apoptosis-coupled phagocytosis. Cell Stem Cell 7:483–495

50. Kierdorf K, Prinz M (2013) Factors regulating microglia activation. Front Cell Neurosci 7:44

51. Kim YS, Joh TH (2006) Microglia, major player in the brain inflammation: their roles in the pathogenesis of Parkinson's disease. Exp Mol Med 38:333–347

52. Prinz M, Priller J (2014) Microglia and brain macrophages in the molecular age: from origin to neuropsychiatric disease. Nat Rev Neurosci 15:300–312

53. Eyre H, Baune BT (2012) Neuroplastic changes in depression: a role for the immune system. Psychoneuroendocrinology 37:1397–1416

54. Yirmiya R, Rimmerman N, Reshef R (2015) Depression as a microglial disease. Trends Neurosci 38:637–658

55. Rock RB, Gekker G, Hu S et al (2004) Role of microglia in central nervous system infections. Clin Microbiol Rev 17:942–964

56. Bayer TA, Buslei R, Havas L, Falkai P (1999) Evidence for activation of microglia in patients with psychiatric illnesses. Neurosci Lett 271:126–128

57. Steiner J, Bielau H, Brisch R et al (2008) Immunological aspects in the neurobiology of suicide: elevated microglial density in schizophrenia and depression is associated with suicide. J Psychiatr Res 42:151–157

58. Krabbe KS, Reichenberg A, Yirmiya R (2005) Low-dose endotoxemia and human neuropsychological functions. Brain Behav Immun 19:453–460

59. Remus JL, Dantzer R (2016) Inflammation models of depression in rodents: relevance to psychotropic drug discovery. Int J Neuropsychopharmacol 19(9):pyw028

60. Henry CJ, Huang Y, Wynne A et al (2008) Minocycline attenuates lipopolysaccharide (LPS)- induced neuroinflammation, sickness behavior, and anhedonia. J Neuroinflammation 5:15

61. Bhattacharya A, Lord B, Grigoleit JS et al (2018) Neuropsychopharmacology of JNJ-55308942: evaluation of a clinical candidate targeting P2X7 ion channels in animal models of neuroinflammation and anhedonia. Neuropsychopharmacology 43:2586–2596

62. Kettenmann H, Hanisch UK, Noda M, Verkhratsky A (2011) Physiology of microglia. Physiol Rev 91:461–553

63. Frick LR, Williams K, Pittenger C (2013) Microglial dysregulation in psychiatric disease. Clin Dev Immunol 2013:608654

64. Ohsawa K, Imai Y, Sasaki Y, Kohsaka S (2004) Microglia/macrophage-specific protein Iba1 binds to fimbrin and enhances its actin-bundling activity. J Neurochem 88:844–856

65. Torres-Platas SG, Cruceanu C, Chen GG, Turecki G, Mechawar N (2014) Evidence for increased microglial priming and macrophage recruitment in the dorsal anterior cingulate white matter of depressed suicides. Brain Behav Immun 42:50–59

66. Kreisel T, Frank MG, Licht T et al (2014) Dynamic microglial alterations underlie stress-induced depressive-like behavior and suppressed neurogenesis. Mol Psychiatry 19:699–709

67. Parrott JM, Redus L, O'Connor JC (2016) Kynurenine metabolic balance is disrupted in the hippocampus following peripheral lipopolysaccharide challenge. J Neuroinflammation 13(1):124

68. Reichenberg A, Yirmiya R, Schuld A et al (2001) Cytokine-associated emotional and cognitive disturbances in humans. Arch Gen Psychiatry 58:445–452

69. Alcami A (2003) Viral mimicry of cytokines, chemokines and their receptors. Nat Rev Immunol 3:36–50

70. Mechawar N, Savitz J (2016) Neuropathology of mood disorders: do we see the stigmata of inflammation? Transl Psychiatry 6(11):e946

71. Dowlati Y, Herrmann N, Swardfager W et al (2010) A meta-analysis of cytokines in major depression. Biol Psychiatry 67:446–457

72. Hannestad J, DellaGioia N, Bloch M (2011) The effect of antidepressant medication treatment on serum levels of inflammatory

cytokines: a meta-analysis. Neuropsychopharmacology 36:2452–2459

73. Dantzer R, O'Connor JC, Freund GG, Johnson RW, Kelley KW (2008) From inflammation to sickness and depression: when the immune system subjugates the brain. Nat Rev Neurosci 9:46–56

74. Caviedes A, Lafourcade C, Soto C, Wyneken U (2017) BDNF/NF-κB Signaling in the neurobiology of depression. Curr Pharm Des 23(21):3154–3163

75. Li Q, Verma IM (2002) NF-kappa B regulation in the immune system. Nat Rev Immunol 2:725–734

76. Bierhaus A, Wolf J, Andrassy M et al (2003) A mechanism converting psychosocial stress into mononuclear cell activation. Proc Natl Acad Sci U S A 100:1920–1925

77. Miller AH, Raison CL (2016) The role of inflammation in depression: from evolutionary imperative to modern treatment target. Nat Rev Immunol 16:22–34

78. Liu T, Zhang L, Joo D, Sun SC (2017) NF-kappa B signaling in inflammation. Signal Transduct Target Ther 2:17023

79. Godbout JP, Berg BM, Krzyszton C, Johnson RW (2005) Alpha-tocopherol attenuates NFkappaB activation and pro-inflammatory cytokine production in brain and improves recovery from lipopolysaccharide induced sickness behavior. J Neuroimmunol 169:97–105

80. Pistis M, Melis M (2010) From surface to nuclear receptors: the endocannabinoid family extends its assets. Curr Med Chem 17:1450–1467

81. Genolet R, Wahli W, Michalik L (2004) PPARs as drug targets to modulate inflammatory responses? Curr Drug Targets Inflamm Allergy 3:361–375

82. Song L, Wang H, Wang Y-J et al (2018) Hippocampal PPARα is a novel therapeutic target for depression and mediates the antidepressant actions of fluoxetine in mice. Br J Pharmacol 175:2968–2987

83. Yang R, Wang P, Chen Z et al (2017) WY-14643, a selective agonist of peroxisome proliferator-activated receptor-alpha, ameliorates lipopolysaccharide-induced depressive-like behaviors by preventing neuroinflammation and oxido-nitrosative stress in mice. Pharmacol Biochem Behav 153:97–104

84. Lima Giacobbo B, Doorduin J, Klein HC et al (2019) Brain-derived neurotrophic factor in brain disorders: focus on neuroinflammation. Mol Neurobiol 56:3295–3312

85. Ferrini F, De Koninck Y (2013) Microglia control neuronal network excitability via BDNF signalling. Neural Plast 2013:429815

86. Luscher B, Shen Q, Sahir N (2011) The GABAergic deficit hypothesis of major depressive disorder. Mol Psychiatry 16:383–406

87. Tomaz VS, Cordeiro RC, Costa AM et al (2014) Antidepressant-like effect of nitric oxide synthase inhibitors and sildenafil against lipopolysaccharide-induced depressive-like behavior in mice. Neuroscience 268:236–246

88. Marini AM, Jiang X, Wu X et al (2004) Role of brain-derived neurotrophic factor and NF-kappaB in neuronal plasticity and survival: from genes to phenotype. Restor Neurol Neurosci 22:121–130

89. Zhang JC, Wu J, Fujita Y et al (2015) Antidepressant effects of TrkB ligands on depression-like behavior and dendritic changes in mice after inflammation. Int J Neuropsychopharmacol 18:pyu077

90. Dineley KT, Pandya AA, Yakel JL (2015) Nicotinic ACh receptors as therapeutic targets in CNS disorders. Trends Pharmacol Sci 36:96–108

91. Gotti C, Zoli M, Clementi F (2006) Brain nicotinic acetylcholine receptors: native subtypes and their relevance. Trends Pharmacol Sci 27:482–491

92. de Jonge WJ, Ulloa L (2007) The alpha7 nicotinic acetylcholine receptor as a pharmacological target for inflammation. Br J Pharmacol 151:915–929

93. King JR, Gillevet TC, Kabbani N (2017) A G protein-coupled α7 nicotinic receptor regulates signaling and TNF-α release in microglia. FEBS Open Bio 7:1350–1361

94. Shytle RD, Mori T, Townsend K et al (2004) Cholinergic modulation of microglial activation by alpha 7 nicotinic receptors. J Neurochem 89:337–343

95. Abbas M, Alzarea S, Papke RL, Rahman S (2017) The alpha7 nicotinic acetylcholine receptor positive allosteric modulator attenuates lipopolysaccharide-induced activation of hippocampal IkappaB and CD11b gene expression in mice. Drug Discov Ther 11(4):206–211

96. Patel H, McIntire J, Ryan S, Dunah A, Loring R (2017) Anti-inflammatory effects of astroglial α7 nicotinic acetylcholine receptors are mediated by inhibition of the NF-κB pathway and activation of the Nrf2 pathway. J Neuroinflammation 14:192

97. Carnevale D, De Simone R, Minghetti L (2007) Microglia-neuron interaction in

inflammatory and degenerative diseases: role of cholinergic and noradrenergic systems. CNS Neurol Disord Drug Targets 6:388–397

98. Suzuki T, Hide I, Matsubara A et al (2006) Microglial alpha7 nicotinic acetylcholine receptors drive a phospholipase C/IP3 pathway and modulate the cell activation toward a neuroprotective role. J Neurosci Res 83:1461–1470

99. Corradi J, Bouzat C (2016) Understanding the bases of function and modulation of alpha7 nicotinic receptors: implications for drug discovery. Mol Pharmacol 90:288–299

100. Yue Y, Liu R, Cheng W et al (2015) GTS 21 attenuates lipopolysaccharide-induced inflammatory cytokine production in vitro by modulating the Akt and NF-kappaB signaling pathway through the alpha7 nicotinic acetylcholine receptor. Int Immunopharmacol 29:504–512

101. Zhang JC, Yao W, Ren Q et al (2016) Depression-like phenotype by deletion of alpha7 nicotinic acetylcholine receptor: role of BDNF-TrkB in nucleus accumbens. Sci Rep 6:36705

102. Melis M, Scheggi S, Carta G et al (2013) PPARalpha regulates cholinergic-driven activity of midbrain dopamine neurons via a novel mechanism involving alpha7 nicotinic acetylcholine receptors. J Neurosci 33:6203–6211

103. Zhao D, Xu X, Pan L et al (2017) Pharmacologic activation of cholinergic alpha7 nicotinic receptors mitigates depressive-like behavior in a mouse model of chronic stress. J Neuroinflammation 14:234

104. Grizzell JA, Iarkov A, Holmes R, Mori T, Echeverria V (2014) Cotinine reduces depressive-like behavior, working memory deficits, and synaptic loss associated with chronic stress in mice. Behav Brain Res 268:55–65

105. Targowska-Duda KM, Feuerbach D, Biala G, Jozwiak K, Arias HR (2014) Antidepressant activity in mice elicited by 3-furan-2-yl-N-p-tolyl-acrylamide, a positive allosteric modulator of the alpha7 nicotinic acetylcholine receptor. Neurosci Lett 569:126–130

106. Chaboub LS, Deneen B (2013) Astrocyte form and function in the developing central nervous system. Semin Pediatr Neurol 20:230–235

107. Linker KE, Cross SJ, Leslie FM (2019) Glial mechanisms underlying substance use disorders. Eur J Neurosci 50(3):2574–2589

108. Colombo E, Farina C (2016) Astrocytes: key regulators of neuroinflammation. Trends Immunol 37:608–620

109. Walker AK, Budac DP, Bisulco S et al (2013) NMDA receptor blockade by ketamine abrogates lipopolysaccharide-induced depressive-like behavior in C57BL/6J mice. Neuropsychopharmacology 38:1609–1616

110. Rajkowska G, Stockmeier CA (2013) Astrocyte pathology in major depressive disorder: insights from human postmortem brain tissue. Curr Drug Targets 14(11):1225–1236

111. Torres-Platas SG, Hercher C, Davoli MA et al (2011) Astrocytic hypertrophy in anterior cingulate white matter of depressed suicides. Neuropsychopharmacology 36:2650–2658

112. Grabe HJ, Ahrens N, Rose HJ, Kessler C, Freyberger HJ (2001) Neurotrophic factor S100 beta in major depression. Neuropsychobiology 44:88–90

113. Goncalves CA, Leite MC, Nardin P (2008) Biological and methodological features of the measurement of S100B, a putative marker of brain injury. Clin Biochem 41:755–763

114. Rothermundt M, Arolt V, Wiesmann M et al (2001) S-100B is increased in melancholic but not in non-melancholic major depression. J Affect Disord 66:89–93

115. Miguel-Hidalgo JJ, Baucom C, Dilley G et al (2000) Glial fibrillary acidic protein immunoreactivity in the prefrontal cortex distinguishes younger from older adults in major depressive disorder. Biol Psychiatry 48:861–873

116. Miguel-Hidalgo JJ, Waltzer R, Whittom AA et al (2010) Glial and glutamatergic markers in depression, alcoholism, and their comorbidity. J Affect Disord 127:230–240

117. Laumet G, Edralin JD, Chiang AC et al (2018) Resolution of inflammation-induced depression requires T lymphocytes and endogenous brain interleukin-10 signaling. Neuropsychopharmacology 43:2597–2605

Chapter 3

Neural Circuits Underlying the Pathophysiology of Major Depression

Meysam Amidfar and Yong-Ku Kim

Abstract

Major depressive disorder (MDD) is a complex neuropsychiatric syndrome characterized by pervasive disturbances in mood regulation, reward sensitivity, cognitive control, and neurovegetative functioning that is associated with disrupted functional connectivity across brain networks. Probe in vivo function of specific circuits using optogenetic tools combined with integration of chemogenetic tools and recent advances in vivo imaging and electrophysiological techniques will improve our understanding of the circuit mechanisms of depression. Recent research in patients and preclinical animal models has focused on identifying the neural circuits that mediate separable characteristics of depression, and connections between the prefrontal cortex (PFC) and subcortical regions such as the dorsal raphe nucleus (DRN) and the ventral tegmental area (VTA) have emerged as candidate targets. Combination of animal models of major depression together with the development of novel and sophisticated technologies to study neural circuit changes provides a good possibility for the development of newer and better therapeutics for the treatment of MDD in the near future. Understanding the exact nature of the causally important abnormalities in these circuits will contribute to identifying the neural circuits underlying MDD and to modulate circuit and behavioral dysfunction with accurate and individual experimental interventions. Here, we review recent technological advances designed to precisely monitor and manipulate neural circuit activity.

Key words Neural circuits, Optogenetic, Chemogenetics, Major depression, Prefrontal cortex circuit

1 Introduction

Major depressive disorder (MDD) as one of the most common psychiatric disorders constitutes a seriously leading cause of disability and public health problem of very high prevalence [1, 2]. MDD is characterized by the presence of a wide range of symptoms, such as depressed mood, anhedonia, fatigue or decreased energy, irritability, impaired concentration or decision-making, psychomotor agitation or retardation, insomnia or hypersomnia, weight loss, or weight gain [3]. Cognitive and emotional disturbances in MDD implicate brain systems involved in the mood regulation, attention, social interaction, reward processing, neurovegetative function

Yong-Ku Kim and Meysam Amidfar (eds.), *Translational Research Methods for Major Depressive Disorder*, Neuromethods, vol. 179, https://doi.org/10.1007/978-1-0716-2083-0_3,
© The Author(s), under exclusive license to Springer Science+Business Media, LLC, part of Springer Nature 2022

(i.e., sleep, appetite, energy, weight, libido), motivation, and stress responses [4]. Brain systems involved in stress responses, social interaction, the regulation of mood, anxiety, fear, reward processing, neurovegetative function (i.e., sleep, appetite, energy, weight, libido), motivation, and attention are responsible for the clinical phenomenology of major depression [3]. A regulatory system at four levels within the brain has been suggested as a depression model that implicated in control of behavioral output, including (1) the cerebral cortex (isocortex, limbic cortex, corticoid amygdala, and hippocampal complex) as the highest level; (2) the subcortical forebrain (dorsal striatum, ventral striatum, and extended amygdala); (3) the midbrain (monoaminergic regulation centers); and (4) the habenula, which connects the cerebral cortex and midbrain systems [5]. Although the development of neuroimaging technologies has enabled significant advances toward elucidating the neural circuits mediating separable features of depression in patients and preclinical animal models, the underlying etiology and pathophysiology of MDD are still not entirely understood [6]. The heterogeneous clinical picture of major depression is due to diverse constellation of symptoms crossing multiple domains including mood and cognition (affect, attention, and concentration), motor and vegetative function (psychomotor retardation or agitation, sleep, and appetite), reward (anhedonia) systems, and perception (guilt, biases, and judgments) that is probably governed by reciprocal projections between typically frontal cortical structures, governing cognitive functions, striatal and basal ganglia structures influencing reward/motivation, and limbic/autonomic projections influencing vegetative and autonomic features [7]. MDD involves a multilayered dysfunction of anatomic structures, circuits, and neurotransmitter systems that makes it as a multifaceted illness with diversity of symptoms [7]. Structural and functional changes potentially induced by altered glutamatergic and gamma-aminobutyric acid (GABA) transmission in the prefrontal cortex (PFC) and hippocampus have been suggested as main brain circuitry abnormalities underlying depression, and reversal of these abnormalities by manipulation of neuronal activity provides hope for treatment of refractory depression [8].

Dynamic interactions between discrete regions of the brain produce complex animal behavior, and characterizing functional connections between brain regions lead to a full understanding of how the brain generates behavior [9]. The human brain comprises approximately 100 billion heterogeneous neurons that anatomically and neurochemically construct 100 trillion brain connections leading to a complex network of interconnected activity that regulates behavior [10]. The major circuits of the central nervous system are created by projection neurons through their interconnections [11]. Brain regions typically comprise various types of projection neurons, and different projection classes, such as those that interconnect locally or receive inputs from upstream regions ("long-range inputs"), lead to a variety of projection neuron circuits [11]. Unique properties of neural circuits shaped by various subtypes of neurons with diverse molecular and electrical properties are responsible for the complex and different functions of the brain and elucidating the distinct roles of each neural circuit critical to our understanding of both normal and abnormal brain function [12]. Understanding the architecture and function of neural circuits that consist of a complex network of varying neural subtypes is one of the main goals of systems neuroscience [13]. Because of the complexity of neural circuitry, understanding of CNS circuit function in MDD with traditional methods has been challenged, and genetically and temporally precise manipulations to probe detailed underlying mechanisms have been developed [14]. The development of methods permitting recording and manipulation of neural circuits defined by connectivity has enabled describing the neural circuits that mediate separable features of depression in patients and preclinical animal models [6]. Refinement and evolution of the technological possibilities to investigate brain connectivity on a functional level emerged depression as a complex network disorder rather than a dysfunction of a few specific brain regions [15].

Here, we present a review literature of what is established knowledge about the animal circuitry underlying depressive symptoms that highlight the role of different brain areas and functional connectivity between them in the pathophysiology of MDD. In this chapter, we will review the advanced techniques that are being applied for studying neural circuit function, such as optogenetic techniques alone and in combination with other in vivo measurement strategies that systematically characterize the connectivity and function of precise neural circuits involved in depression and make possible translation of the animal brain circuitry to the human brain.

2   New Methodologies for Defining Neural Circuits in Depression

Low spatiotemporal resolution of available techniques in humans, including functional connectivity analysis and resting-state fMRI studies, diffusion tensor imaging (DTI), EEG, and MEG, results in nonspecific findings related to neural circuits underlying neuropsychiatric disease [16]. Then, animal research is currently needed to understand disease at the level of the circuit [16]. The techniques required to investigate neural activity in animal models of disease have contributed to current understanding of circuits within the brain as well as establishing a causal relationship between a specific neurocircuit abnormality and disease [16]. These novel techniques by selectively targeting mutations to specific brain circuits in the mouse rather than mutating all brain cells and targeting a gene known to be highly penetrant for human disease to an individual, relatively conserved, circuit element allow the researcher to "home in" on how a gene affects a single brain circuit, which contributes to determining whether that circuit is involved in generating an abnormal behavioral phenotype [16]. A main long-lasting purpose of systems neuroscience is to characterize the different roles of neural subtypes in brain circuit function [13]. One critical method for meeting this goal is the causal manipulation of selective cell types [13]. This strategy employs techniques for in vivo optical stimulation of specific populations of excitatory neurons and inhibitory interneurons combined with electrophysiology [13]. Analysis of animal models of depression with electrophysiological, optogenetics, chemogenetics, and molecular techniques will contribute to revealing the neural circuit-specific mechanisms of depression [14, 17–21]. The selective targeting of specific neural circuits has been made possible with the emergence of genetic techniques to selectively manipulate cellular activity, including optogenetics, which uses channels that are activated by light, and chemogenetics, which uses engineered G protein-coupled receptors that are activated by otherwise inert drug-like small molecules [22–24]. Optogenetics via light-activated channels and chemogenetics via engineered G protein-coupled receptors activated by otherwise inert molecules confer unprecedented control of cellular activity, and both tools powerfully and precisely manipulate neuronal activity [25]. Enabling cell-type and pathway-specific activation or silencing of neurons is the common feature of both optogenetic and chemogenetic methodologies [26, 27]. Optogenetic and chemogenetic techniques have allowed previously unfeasible levels of functional circuit mapping in neuroscience [28]. In addition, deconstruction of the complex whole-brain networks that are fundamental to behavioral states has been made possible by monitoring neural activity in freely moving animals using the coupling of chemogenetics with imaging techniques [24]. Chemogenetic and optogenetic techniques have been employed in nonhuman primates and may eventually have therapeutic worth in humans, especially in the treatment of behavioral disorders, and then may have considerable translational potential [24, 27]. It has demonstrated that optogenetic and chemogenetic tools can produce sustained changes in behavior when they are utilized in a way that is somewhat similar to clinical treatments for depression, including TMS, DBS, and ketamine [29]. While the application of optogenetic and chemogenetic tools in clinical circumstances remains to be achieved, these techniques may characterize specific synaptic targets that reverse depressive phenotypes in a manner that is not possible with the indiscriminate circuit modulatory methods existing for human use [29].

New investigative methods such as optogenetics for circuit-level study of animal models of depression, in which genes of interest are knocked out or mutated, enable the researcher to chemically modify neuronal activity, visualize long-range projections in intact brains, and control neuronal activity with light [16]. High reliability of optogenetic and chemogenetic tools in control of neuronal activity in preclinical models has contributed to illuminate the role of the PFC and its circuitry to depression-like behavior [29]. Probing synaptic connections that is traditionally performed by methodologies, such as electrical stimulation, is not faithful for axons originating from different cell types, because with enhancement of distance between neurons, the recording networks between labeled neuron pairs become ineffective [23, 30–32]. However, channelrhodopsin-2 (ChR2) as an alternative approach can be genetically targeted and activated using photostimulation instead of electrical stimulation, making possible the activation of neurons and their axons and consequently mapping the connections of molecularly defined cell populations [23, 30–32]. These techniques promise to reach a refined view of the circuit mechanisms underlying depression and potential mechanistic targets for development and reversal of depression-associated circuit abnormalities by providing unique access to specific circuits and neuronal subpopulations [29]. Translational studies in rodents using optogenetic and chemogenetic tools contribute to identifying nodes and connections within the networks of the limbic system that involved motivated behavior and affective phenotypes [33]. While humans have much greater degree of cortical elaboration compared to the mouse and other animal models, some basic circuit components of the mouse brain could be employed as beneficial models of similar human circuits [16]. Although humans have many more circuit elements, mice exhibit largely similar basic neuron subtypes and connections among them in several brain areas [16]. Behavioral phenotypes induced by mutations in such highly conserved neurocircuitries in the mouse are likely to be similar to those in humans [16]. Moreover, the incorporation of chemogenetics and optogenetics into animal models could prominently improve our understanding of neural circuits [24]. Although optogenetics has been more widely employed in depression studies, chemogenetics suggests distinctive benefits for probing circuit function, remarkably making it possible to ideally probe sustained cellular alterations through targeted, chronic inhibition or activation [25]. Optogenetic tools by mapping the circuitry underlying depression-like behavior in animal models have elucidated the role of the medial prefrontal cortex, ventral hippocampus, nucleus accumbens, ventral tegmental area, and limbic areas in stress susceptibility [25]. Combination of optogenetics or chemogenetics with in vivo imaging techniques, such as fMRI, fiber photometry,

or single-cell calcium imaging, will contribute to precise understanding of the neural circuit mechanisms of depression [25].

2.1 Optogenetics

Optogenetics, or the optical control of neural activity through
artificial incorporation of light-sensitive proteins into cell membranes and through regulation of the activity of specific cell types,
contributes to a better understanding of their role in the local
network activity and overall brain function in vivo [13]. Optogenetic technologies employ light to control biological processes
within targeted cells in vivo with high temporal precision
[34]. Optogenetic tools employ light-activated optogenetic probes
for identification and manipulation of specific cell types in the
nervous system and their neural circuits in awake, behaving animals
[35]. Optogenetic technology combines optical imaging and
genetic targeting to visualize genetically targeted neurons in living
animals and to track and control electrical and biochemical events
within targeted cell types [36]. Optical manipulations via light-
activated channels and pumps provide an appropriate method
with high reliability for stimulating and inhibiting the firing of
specific neural populations in vivo [13, 37–41]. The use of the
light-driven chloride pump halorhodopsin (NpHR) and the light-
sensitive, cation permeable channelrhodopsin-2 (ChR2) led to the
development of optogenetic approaches for high-speed, light-
induced activation or silencing of neurons [12, 23, 31, 42,
43]. Optical manipulation of activity in neural circuits with light-
sensitive   rhodopsins,   such   as   the   *Chlamydomonas*
channelrhodopsin-2 (ChR2), may help in revealing both the normal and pathologic circuit functions [44]. A variety of expression
targeting strategies, such as the transgenesis and viral transduction
methods, are employed for genetic targeting of cell types and
selective expression of opsin genes in defined subsets of neurons
in the brain. These methods are critical for dissecting the roles of
specific cell types in neural circuits, including mapping connectivity,
measuring activity, and inactivating and activating specific neurons
[44–46]. Viral transduction is the most commonly used strategy to
date for the expression of ChR2 in brain tissue [13].

It has demonstrated that transgenic mice that express ChR2 in
subsets of neurons contribute to in vivo light-induced activation
and mapping of neural circuits [32, 47]. Genetically modified
rodents, when combined with optogenetics using target opsins
to genetically defined cell types in animal models, enable selective
control of neuromodulatory function with unique temporal precision in genetically defined neuronal subpopulations and their
projections, providing new approaches for circuit analysis [34, 41,
48–51].

2.2   Chemogenetics   Chemogenetics is synonymous with designer receptors exclusively activated by designer drugs (DREADDs) that are used ubiquitously to modulate GPCR activity in vivo and have been widely applied in the field of behavioral neuroscience [24, 25]. DREADDs contribute to monitor the electrophysiological, biochemical, and behavioral outputs of specific neuronal types and better understanding of the links between brain activity and behavior [24]. The chemogenetic approach utilizes genetically modified G protein-coupled receptors (GPCRs) commonly referred to as "designer receptors exclusively activated by designer drugs" or DREADDs, including the first-generation DREADDs, (hM4Di, hM3Dq, Gs-D) which developed from the human muscarinic receptor, and the more recently developed "kappa opioid receptor DREADDs" (KORD), which was formed from the kappa opioid receptor [28]. Focal or systemic delivery of the DREADD ligand results in neuronal silencing or excitation [28]. Since longer-term chemogenetic approaches in contrast to the properties of optogenetic manipulators can generate continued hyperpolarization and depolarization effects [52–55], they might be a more useful choice for reproducing common physiological features of certain behavioral/disease conditions that are related to unrelenting hyper- or hypo-activated neural circuits [56].

2.3   DREAMM   DREADD-assisted metabolic mapping (DREAMM) has been developed as a new biobehavioral imaging methodology that combines DREADD chemogenetic technology, which allows remote in vivo control of cell-specific firing [57], with positron emission tomography (PET) and [18F]fluoro-2-deoxyglucose (FDG), which is a direct marker of brain function, to generate concurrently quantitative, dynamic, whole-brain, and regionally unbiased, cell type-specific functional circuit mapping during the awake, freely moving state [10, 58]. DREAMM provides the distinctive flexibility of combining various techniques, including molecular, viral, transgenic, and pharmacological, in order to probe whole-brain functional networks and concomitant behaviors using stimulation or inhibition of a region-specific and neurochemically specific cell type in a mechanistic and unbiased manner [10]. A typical DREAMM experiment first includes regional or global expression of stimulatory or inhibitory DREADDs in targeted cell types induced by a genetic engineering manipulation (e.g., viral vector-based or transgenic model) and then undergoes chemogenetic DREADD modulation combined with an FDG-PET behavioral imaging strategy [10]. In addition to FDG, DREAMM also can be employed and accompanied by an array of other PET probes (even in the same subject) that target alternative/complementary cellular processes, such as enzymatic processes, neurotransmitter dynamics, and receptor signaling, thus providing investigation on both the global long-range functional networks involved in cell type-specific behaviors and the underlying neurochemistry [10]. DREAMM is now being used in experimental rodents and nonhuman primates but has the potential to be applied in the future in humans [10]. DREAMM would have the ability to be applied together with advanced cellular genetic engineering techniques, such as combination of DREADD and cell replacement therapeutics [59], to evaluate the effectiveness of functional integration of such transplanted cells within neuronal networks [10]. In sum, DREAMM holds potential capacity as an effective, reverse engineering approach to visualize in vivo-specific neuronal assembly networks correlated with normal and pathologic behavior and to consequently develop knowledge about different neuronal circuits related to psychiatric disorders [10].

2.4 Integrated Optogenetic and Electrophysiology Approaches

The integration of optical tools with traditional electrophysiological techniques provides a powerful strategy for investigation on the local cortical circuit activity in vivo comprising a complex and flexible series of interactions between excitatory and inhibitory neurons [37]. The use of light activation for assessing neural network dynamics requires combination of optical stimulation with neural recording techniques [13]. Different projections from specific neuronal subpopulations to different postsynaptic target regions often result in unique firing patterns, indicating that the way of integration of particular neurons within a circuit defines their functional role [56]. Therefore, computational role of participant cells within a complex circuit can be described by phototagging of genetically marked neurons in a projection-specific fashion [60]. Genetically guided electrophysiological approaches enable identifying extracellular recordings within brain tissue originating from an extensive range of various cell types and provide an avenue for elucidating activity patterns of select neuronal subpopulations [56]. The extracellular recordings could be used to recognize information encoding, neuronal communication, processing, propagation, and computation of neuronal circuits [61]. Thus, extracellular recordings as a noninvasive, long-term, multi-cell method that measure (with sub-millisecond time resolution) the extracellular field potential generated by an action potential discharge from either a single neuron (single-unit recording) or neuronal population (multi-unit recording) have been developed for studying neuronal circuit connectivity, physiology, and pathology [62]. The extracellular recordings enable simultaneous and long-term recordings of local field potentials (LFPs) and extracellular action potentials (EAPs) from a population of neurons at millisecond time scale [61].

A new era of optogenetic-electrophysiological experiments that are capable of probing local and long-range connectivity is performed by the preserved photoexcitability of ChR2-expressing axons in brain slices [31]. Channelrhodopsin-2 (ChR2)-assisted circuit mapping (CRACM) has been developed as an efficient method to dissect local and long-range synaptic circuits in brain slices [31]. CRACM technique includes delivering light-gated channel channelrhodopsin-2 (ChR2) to neuronal axons in vivo that photostimulation of ChR2-positive axons can be transduced reliably into single action potentials in brain slices [31]. CRACM maps connections between presynaptic ChR2-positive neurons, and postsynaptic neurons could be targeted by whole-cell recordings [31]. Progresses in electrophysiological phototagging methods have been facilitated by current advancements in vivo extracellular recording tools accompanied by the developments in light sources [56]. This method is accomplished by activation of ChR2 that has been localized in axons and presynaptic terminals by light delivery through implanted optical fibers. Correspondingly, implantation of multielectrode recording devices is performed neighboring the somas of these ChR2-targeted neurons [56]. The optical fiber and single extracellular electrodes or multielectrode arrays may be placed several hundred micrometers from the original virus injection site, where neural activity evoked in vivo by activation of cell-type-specific expression of ChR2 [13]. 30–45 min is necessary for injection of viral vectors, and 1–4 h is necessary for in vivo electrophysiology with optogenetic stimulation [13]. Combining photostimulation with whole-cell recordings of synaptic currents makes it possible to map circuits between presynaptic neurons, defined by ChR2 expression, and postsynaptic neurons, defined by targeted patching [31]. Introducing a Cre recombinase-dependent viral vector encoding the light-activated cation channel, channelrhodopsin-2 (ChR2), to genetically distinct neuronal populations in various Cre driver transgenic mouse lines can be used for distinguishing the firing profiles of genetically defined neuronal populations [30, 44, 51, 53]. Through localized delivery of a ubiquitous viral vector, region-specific targeting of ChR2 provides anatomical specificity [56]. ChR2 can be a beneficial physiological tag or marker when extracellular recordings are performed and as a brief pulse of blue light provokes a short-latency action potential in cells expressing ChR2 that is reliably emerged across multiple light presentations [13, 37, 56]. Consequently, during in vivo extracellular recordings, cells expressing ChR2 are distinguishable from "ChR2-negative" neurons based on their electrical responses to light [56]. One approach for channelrhodopsin-2 (ChR2)-based synaptic circuit analysis combined photostimulation of virally transfected presynaptic neurons' axons with whole-cell electrophysiological recordings from retrogradely labeled postsynaptic neurons [11]. This approach utilized the preserved photoexcitability of ChR2-expressing axons in brain slices and could be used to assess either local or long-range functional connections. Stereotaxic injections are employed both to express ChR2 selectively in presynaptic axons of interest (using rabies or adeno-associated viruses) and to label two types of post-synaptic projection neurons of interest using fluorescent retrograde tracers [11]. Whole-cell electrophysiological recordings of targeted tracer-labeled postsynaptic neurons and sampling voltage or current responses to blue light-emitting diode (LED) photostimulation of ChR2-expressing axons for assessing of synaptic connections are performed in brain slices [11]. This method enables rapid, quantitative characterization of synaptic connectivity between defined pre- and postsynaptic classes of neurons [11].

Two major advantages have been suggested for the integration of optical and electrophysiological methods in order to in vivo explore network dynamics including rigorously identifying the subpopulation of recorded neurons belonging to the targeted cell class and the ability to directly manipulate the activity level of a specific class of cell concurrent with observing ongoing network activity or behavior [37]. However, this simultaneous application can produce a variety of electrical artifacts during recordings as a result of different causes, including possible photoelectric effects or temperature-dependent effects on electrode conduction properties [13].

2.5 Optogenetic fMRI (ofMRI)

Functional magnetic resonance imaging (fMRI) is a specialist form of magnetic resonance imaging associated with imaging changes in blood flow and oxygenation that is correlated with regional variation in metabolic activity in the brain [63]. fMRI is a widely used imaging method that is sensitive to changes in the blood oxygenation level-dependent (BOLD) signal and detects regions of the brain that change their level of neuronal activation in response to specific experimental conditions [64]. The BOLD signal is a manifestation of the change in contrast at the individual voxel level that is generated by distortion of the MRI scanner's magnetic field induced by blood flow responses such as the ratio of oxygenated to deoxygenated hemoglobin [65]. fMRI studies produce activation maps of different regions in the brain, which typically depict the average level of engagement in response to a specific stimulus or during a specific task [64]. The location of the BOLD signal response is delineated using this contrast that is co-registered to a structural brain template [65]. The correlation of the change in signal with the stimulus or event presentation is employed for identifying a specific brain region with any potential functional involvement [65].

The approach of integration of optogenetics and rodent fMRI, termed optogenetic fMRI (ofMRI), entails acquisition of whole-brain fMRI data while using light to selectively activate or deactivate genetically defined populations of neurons expressing light-sensitive opsins [66]. ofMRI as a novel approach that combines optogenetic control of neural circuits with high-field fMRI selectively stimulates brain regions while simultaneously recording

BOLD fMRI signals during optical stimulation, allowing for detection of activation in the stimulated site as well as in downstream functional targets [67–69]. The ofMRI method has been employed to identify and examine functional connections between several brain regions including the hippocampus, basal ganglia, cerebellum, motor cortex, and prefrontal cortex [9, 66, 70–74]. The advantages of ofMRI approach are elucidating the functional activation of various brain regions in response to output from a distinct brain region and defining potential functional pathways by which one area can manipulate activity across downstream regions [9]. To assess the relationship between the BOLD signals and actual neural activity in one brain region, the neural responses of the brain region to the optogenetic stimulation could be examined with in vivo electrophysiology in un-anesthetized mice, and the single-cell and spike time resolution action potential firing rates of neurons were recorded while simultaneously modulating the downstream brain regions via light stimulation [9]. The combined ofMRI and in vivo electrophysiology methods will contribute to exploring the downstream changes in BOLD signal and electrophysiological activity in downstream brain circuits and pathways [9].

3 Depression as a Neural Circuit Disorder

Many brain areas may underlie different symptom clusters of depression such as negative emotions, impaired cognition, sad or dysphoric affect, and anxiety-related symptoms [75, 76]. Reductions in gray matter volume and glial density in brain regions mediating the cognitive aspects of depression, such as feelings of worthlessness and guilt, have been confirmed by a large body of postmortem and neuroimaging evidence [77–79]. A large body of neuroanatomical techniques including functional and structural imaging, as well as analysis of lesions and histological methods, indicates that anatomical networks link the medial prefrontal cortex (mPFC) and a few related cortical areas to the ventral striatum and pallidum, the hypothalamus, the amygdala, the medial thalamus, the periaqueductal gray, and other parts of the brainstem centrally involved in the pathophysiology of depression [4]. Thus, translation of clinical subdomains of depressive symptomatology into dysfunctional neurocircuitry might be relevant for the development of a direct modulation of human brain structures in vivo [15]. Translational evidence provided by preclinical and clinical studies revealed more consistent antidepressant response following deep brain stimulation (DBS) applied to several neuromodulatory targets, such as the ventromedial prefrontal cortex, anterior limb of internal capsule (ALIC), medial forebrain bundle (MFB), nucleus accumbens (NAc), lateral habenula (LHb), subthalamic nucleus, and subcallosal cingulate gyrus (SCG) [80]. Influence of neuroanatomical locations on DBS-related antidepressant effects and modulatory effects of DBS on monoamine neurotransmitters in target regions or interconnected brain networks suggests that depression is a neural circuit disorder [80].

3.1 Prefrontal Cortex Circuit

Optogenetic techniques have been used to target cell bodies in the prefrontal cortex of mice in order to better understand the underlying mechanisms of DBS treatments for depression [81]. Several brain nuclei downstreams of the mPFC regulate depression-related behavior, including the periaqueductal gray [82], basolateral amygdala [83], dorsal raphe and lateral habenula [84], and nucleus accumbens [85]. Disrupted coherence between mPFC neural activity and multiple downstream regions in normal animals, such as the amygdala and ventral tegmental area, has been suggested as a central brain mechanism underlying MDD [86]. Clinical evidence supported the model derived from rodent studies that suggest chronically increased mPFC activity typical of the depressed state disrupts normally synchronous activity within and between subcortical regions [86, 87]. The PFC through the PFC-DRN pathway plays a crucial role in modulating emotional responses [84]. Activity of two of these subcortical regions including the dorsal raphe nucleus (DRN) and the ventral tegmental area (VTA) as sources of the majority of forebrain serotonin (5-hydroxytryptamine; 5-HT) and DA neurons, respectively, is controlled by the mPFC [6]. By combining optogenetics with fMRI and in vivo electrophysiological recordings to probe interactions between cortical and subcortical brain regions, it has found that dopamine neuron stimulation drives striatal activity, whereas locally increased mPFC excitability reduces this striatal response and inhibits the behavioral drive for dopaminergic stimulation [70]. These findings suggest a mechanism by which mPFC regulates the dynamic interactions between specific distant subcortical regions governing dopamine-driven reward behavior, with important implications for anhedonia [70]. Moreover, elevated excitability of the mPFC could reduce striatal BOLD responses to the stimulation of dopamine neurons [70]. One chemogenetic study showed that activation of the mPFC to medial dorsal thalamus (MDT) circuit via hM3Dq DREADDs is sufficient to elicit a decrease in depression-like behavior in mice [88]. The MDT is the major thalamic input to the mPFC that integrates information from DRN, substantia nigra, the amygdala, VTA, and the ventral pallidum [88].

3.2 Hippocampus

Optogenetic inhibition of the ventral hippocampus (vHIP)-mPFC pathway could abolish the antidepressant-like effect of ketamine in mice, demonstrating that activity in the vHIP-mPFC pathway is necessary for the antidepressant-like effect of ketamine [89]. In addition, chemogenetic activation of vHIP-PFC but not vHIP-NAc activation mimicked the antidepressant-like response to ketamine [89]. Chemogenetic-based stimulation of the entorhinal cortex-hippocampal dentate gyrus circuitry, including administration of the designer ligand clozapine-N-oxide (CNO) in mice expressing excitatory DREADDs in excitatory neurons of entorhinal cortex for 5 weeks, showed that chronic stimulation of glutamatergic entorhinal cortical afferents to the dentate gyrus ameliorates behavioral symptoms of depression under both acute and chronic stressful conditions [90]. Based on this evidence, antidepressant-like behaviors induced by chronic – but not acute – stimulation of the entorhinal cortex-hippocampal dentate gyrus circuit by enhancing hippocampal function from mood to memory and targeting of this novel circuit for MDD treatment may be targeted for depression treatment and support modern framing of depression as a "circuitopathy" [7, 90]. Targeting of this novel circuit for treatment of MDD may preclude side effects of current brain stimulation therapies, such as memory loss and cognitive impairment, and may address the great request for novel possibilities of treatment of MDD [90]. In vivo, low-frequency stimulation of the vHIP-NAc pathway induces a pro-resilient effect in defeated mice and increases social interaction, whereas acute enhancement of this pathway induces susceptibility in defeated mice, suppressing social interaction, suggesting that vHIP via interaction with NAc bidirectionally modulates susceptibility to depression-like behavior [91].

3.3 Ventral Tegmental Area (VTA)

The mesolimbic dopaminergic pathway, including dopaminergic (DA) neurons in VTA and their projections to the nucleus accumbens (NAc), plays an essential role in the recognition of emotionally salient stimuli such as aversion and reward [92, 93]. NAc as a region critical for reward and motivation is implicated in the pathophysiology of depression [91]. NAc is innervated by limbic structures including excitatory afferents from the medial PFC, the basolateral amygdala, and the ventral hippocampus [94]. The NAc integrates cortico-limbic afferents with dopaminergic modulation from the ventral tegmental area [91, 95]. It has suggested that DA neurons of the VTA-NAc circuit (mesolimbic) play a crucial role in modulating depression-related behaviors and rapid antidepressant effects [14, 17, 96–98]. Anatomical studies reveals a sparse projection from mPFC to the VTA [99–102], and this innervation of the VTA by mPFC comprises relatively equivalent proportions of synapses onto VTA dopaminergic and GABAergic cells [99, 103].

Optogenetic and chemogenetic approaches have provided crucial insight related to the role of the VTA as a predominantly dopaminergic structure in the mesolimbic reward circuit in depression-like behaviors [14, 17]. Optogenetic phasic stimulation of VTA neurons projecting to the NAc, but not to the medial prefrontal cortex (mPFC), induced susceptibility to social defeat stress, whereas optogenetic inhibition of the VTA-NAc projection induced resilience, and inhibition of the VTA-mPFC projection promoted susceptibility in freely behaving mice, suggesting neural circuit-specific mechanisms of depression [17]. Pharmacological inhibition of the mPFC could enhance DA phasic firing [104]. Pharmacological inhibition of the infralimbic prefrontal cortex (ILPFC) in rats exposed to chronic mild stress (CMS) could restore inhibited DA firing in the medial VTA induced by pharmacological excitation of the ILPFC [105]. Thus, inhibition of ILPFC in the CMS-exposed rats or inhibition of BA25 in patients with depression via DBS probably by engaging distant downstream brain networks can restore normal dopamine-mediated hedonic functions [105]. Therefore, translation of this finding to MDD in humans would offer a target site within the ILPFC/BA25 in order to normalize dopamine neuron drive as a therapeutic intervention [105].

3.4 Dorsal Raphe Nucleus (DRN)

It has suggested that DRN serotonergic neurons integrate different information that are encoded by various upstream brain areas through the functional synaptic inputs [106]. Examining the functional synaptic connection from the upstream brain areas to the 5-HT neurons as well as GABAergic neurons in the DRN using channelrhodopsin-2 (ChR2)-assisted circuit mapping (CRACM) revealed that lateral hypothalamus, preoptic area, substantia nigra, and amygdala send both glutamatergic and GABAergic projections to the DRN, with less connection to the contralateral wing of DRN, but the PFC and LHb bilaterally send mostly glutamatergic projections to the DRN [106]. The LHb-DRN pathway probably mediates the role of lateral habenula (LHb) in emotional regulation [107]. Anatomically medial prefrontal cortex (mPFC) neurons project to DRN, and the activity of DRN 5-HT neurons is strongly regulated by the mPFC [108]. The development of optogenetic tools for neural circuit intervention has enabled the direct investigation on the functional role of the projection from the mPFC to the DRN in behaviors relevant to depression in rodents [6, 109, 110]. A quantitative method for continuous assessment and control of active response to behavioral challenge, synchronized with single-unit electrophysiology and optogenetics in freely moving rats for recording millisecond-precision neural and behavioral data alongside optogenetic control during the forced swim test (FST), revealed that acute optogenetic stimulation of those mPFC cells projecting to the DRN increased escape-related behaviors without increasing generalized locomotor activity, indicating a specific role of this projection in selection of an effortful motivated behavioral pattern during challenging circumstances [84]. Chemogenetic stimulation of the serotoninergic projection from DRN to the nucleus accumbens (NAc), using DREADD technology promoted antidepressant-like behaviors, confirming a role for DRN-NAc pathway in depression-like behavior [111]. DREAMM mapping of brain-wide metabolic changes revealed that chronic activation of DRN neurons decreases metabolic activity in cortical areas, habenula, hippocampus, and central and medial thalamic nuclei, and acute DRN activation decreases metabolic activity in the periaqueductal gray and central and medial thalamic nuclei and increases metabolic activity in the motor cortices, cingulate cortex, and primary somatosensory [25].

3.5 Habenula

The lateral habenula (LHb), a small brain region in the dorsal diencephalon, projects to brainstem nuclei including VTA and DRN and receives projections from the limbic forebrain [112]. It has hypothesized that reduced motor activity in major depression might be a result of the inhibitory effect of the LHb on dopamine neurons and hyperactivity of the LHb might lead to hypoactivity of dopamine neurons [113]. It has demonstrated that activation of the LHb-DA neuron pathway is aversive [114–116] and that hyperactivation of this pathway plays a role in the etiology of depression [117, 118]. The habenula nucleus in the epithalamus is one of the few forebrain structures projecting to the dorsal raphe and habenular-raphe interactions, which play an important role in mood regulation and the etiology of depression [119–121]. Disturbed activity in the habenula-raphe pathway has been suggested as common mechanism for abnormalities of mood and cognition in depression [120]. The phasic activation of LHb could transiently repress the activity of VTA DA neurons [122] via connection with GABAergic neurons in the rostromedial tegmental nucleus (RMTg) that project to dopamine (DA) neurons in the VTA [123, 124]. GABAergic rostromedial tegmental nucleus (RMTg, also known as tail VTA), serotonergic (5-HT) dorsal and median raphe (DRN, MRN), and midbrain aminergic centers, which include the dopaminergic (DA) VTA and substantia nigra pars compacta (SNc), are the primary output regions from the lateral habenula [125]. Using DREADD, it has found that inhibition of the LHb decreases depression-like behavior possibly by enhancement of serotonergic activity [112]. Increased activation of the LHb leads to the downregulation of the serotonergic, noradrenergic, dopaminergic systems and functional inhibition of the LHb via DBS generated antidepressive effects [126]. Chronic DBS of the LHb in a rat model of depression could effectively improve depressive symptoms likely as a result of elevated monoamine levels [127].

4 Conclusion

Understanding functional organization of the brain at the circuit level will make possible translation of basic neuroscience evidence to clinical settings. Innovative technologies in systems neuroscience can synergistically use to dissect neural circuit function [56]. These collaborative approaches will help generate a holistic understanding of circuit-wide function that underlies behavioral processes [56]. This chapter presents a paradigm shift in pathophysiology of major depression, suggesting circuit-level abnormalities impacting function across multiple brain regions and neurotransmitter systems but not dysregulation of single neurotransmitter systems. In this review, we addressed mapping brain activity and neural connectivity in rodents using optogenetics in conjunction with either functional magnetic resonance imaging or in vivo electrophysiology. The recently developed and strong novel experimental methods such as DREADD and optogenetics and many others have finally opened the door to circuit-level investigations of psychiatric disease in the animal model. Opsins and DREADDs compared to traditional pharmacological techniques allow significantly increased spatial, cell-type, and temporal specificity, permitting for exact division of the neural circuits underlying depression-like behavior. We selected to focus on these techniques because they are capable of assessing neural network dynamics most closely linked to human studies [128]. The power of optogenetics and chemogenetics to reveal novel mechanistic insight and development of effective treatments for depression is the great promise of these neural control technologies [25]. By probing the specific brain regions, pathways, or cell types that mediate behavioral dysregulation, we can precisely characterize the locus of disease-associated dysfunction [25]. Structural changes within the prefrontal cortex (PFC) and functional connectivity abnormalities between PFC and distal brain regions demonstrate circuit-level abnormalities in depression [30]. Dysfunction in an extended network involving the mPFC and anatomically related limbic, striatal, thalamic, and basal forebrain structures underlies the pathophysiology of depression. Specific neural circuits within the limbic-cortical system mediate mood and emotional regulation, and MDD represents a brain-based disorder that leads to dysregulation of these circuits. Translational imaging and molecular and behavioral research methods converge in developing the understanding of these disrupted circuits and providing exciting and powerful future treatments. Finally, neurocircuitry-based approaches can facilitate innovations in therapeutic interventions, such as deep brain stimulation, by locating nodes as targets for effective treatments and identifying the patients who are likely to benefit from such interventions.

References

1. Kessler RC, Berglund P, Demler O, Jin R, Koretz D, Merikangas KR et al (2003) The epidemiology of major depressive disorder: results from the National Comorbidity Survey Replication (NCS-R). JAMA 289(23): 3095–3105

2. Murphy JM, Laird NM, Monson RR, Sobol AM, Leighton AH (2000) A 40-year perspective on the prevalence of depression: the Stirling County study. Arch Gen Psychiatry 57(3):209–215

3. Association D-AP (2013) Diagnostic and statistical manual of mental disorders. American Psychiatric Publishing, Arlington

4. Price JL, Drevets WC (2012) Neural circuits underlying the pathophysiology of mood disorders. Trends Cogn Sci 16(1):61–71

5. Loonen AJ, Ivanova SA (2016) Circuits regulating pleasure and happiness—mechanisms of depression. Front Hum Neurosci 10:571

6. Post RJ, Warden MR (2018) Melancholy, anhedonia, apathy: the search for separable behaviors and neural circuits in depression. Curr Opin Neurobiol 49:192–200

7. Lozano AM, Lipsman N (2013) Probing and regulating dysfunctional circuits using deep brain stimulation. Neuron 77(3):406–424

8. Duman RS, Sanacora G, Krystal JH (2019) Altered connectivity in depression: GABA and glutamate neurotransmitter deficits and reversal by novel treatments. Neuron 102(1): 75–90

9. Choe KY, Sanchez CF, Harris NG, Otis TS, Mathews PJ (2018) Optogenetic fMRI and electrophysiological identification of region-specific connectivity between the cerebellar cortex and forebrain. NeuroImage 173: 370–383

10. Michaelides M, Hurd YL (2015) DREAMM: a biobehavioral imaging methodology for dynamic in vivo whole-brain mapping of cell type-specific functional networks. Neuropsychopharmacology 40(1):239

11. Yamawaki N, Suter BA, Wickersham IR, Shepherd GM (2016) Combining optogenetics and electrophysiology to analyze projection neuron circuits. Cold Spring Harb Protoc 10:pdb.prot090084

12. Zhao S, Cunha C, Zhang F, Liu Q, Gloss B, Deisseroth K et al (2008) Improved expression of halorhodopsin for light-induced silencing of neuronal activity. Brain Cell Biol 36(1–4):141–154

13. Cardin JA, Carlén M, Meletis K, Knoblich U, Zhang F, Deisseroth K et al (2010) Targeted optogenetic stimulation and recording of neurons in vivo using cell-type-specific expression of Channelrhodopsin-2. Nat Protoc 5(2):247

14. Tye KM, Mirzabekov JJ, Warden MR, Ferenczi EA, Tsai H-C, Finkelstein J et al (2013) Dopamine neurons modulate neural encoding and expression of depression-related behaviour. Nature 493(7433): 537–541

15. Höflich A, Michenthaler P, Kasper S, Lanzenberger R (2019) Circuit mechanisms of reward, anhedonia, and depression. Int J Neuropsychopharmacol 22(2):105–118

16. Theyel B (2018) Animal models in psychiatric disease: a circuit-search approach. Harv Rev Psychiatry 26(5):298

17. Chaudhury D, Walsh JJ, Friedman AK, Juarez B, Ku SM, Koo JW et al (2013) Rapid regulation of depression-related behaviours by control of midbrain dopamine neurons. Nature 493(7433):532–536

18. Francis TC, Chandra R, Friend DM, Finkel E, Dayrit G, Miranda J et al (2015) Nucleus accumbens medium spiny neuron subtypes mediate depression-related outcomes to social defeat stress. Biol Psychiatry 77(3):212–222

19. Francis TC, Chaudhury D, Lobo MK (2014) Optogenetics: illuminating the neural basis of rodent behavior. Open Access Anim Physiol 6:33–51

20. Han M-H, Friedman AK (2012) Virogenetic and optogenetic mechanisms to define potential therapeutic targets in psychiatric disorders. Neuropharmacology 62(1):89–100

21. Russo SJ, Nestler EJ (2013) The brain reward circuitry in mood disorders. Nat Rev Neurosci 14(9):609–625

22. BN A, Li X, Pausch MH, Herlitze S, Roth RL (2007) Evolving the lock to fit the key to create a family of G protein-coupled receptors potently activated by an inert ligand. Proc Natl Acad Sci U S A 104:5163–5168

23. Boyden ES, Zhang F, Bamberg E, Nagel G, Deisseroth K (2005) Millisecond-timescale, genetically targeted optical control of neural activity. Nat Neurosci 8(9):1263–1268

24. Whissell PD, Tohyama S, Martin LJ (2016) The use of DREADDs to deconstruct behavior. Front Genet 7:70

25. Muir J, Lopez J, Bagot RC (2019) Wiring the depressed brain: optogenetic and chemogenetic circuit interrogation in animal models of depression. Neuropsychopharmacology 44(6):1013–1026

26. Bernstein JG, Boyden ES (2011) Optogenetic tools for analyzing the neural circuits of behavior. Trends Cogn Sci 15(12):592–600

27. Urban DJ, Roth BL (2015) DREADDs (designer receptors exclusively activated by designer drugs): chemogenetic tools with therapeutic utility. Annu Rev Pharmacol Toxicol 55:399–417

28. Forcelli PA (2017) Applications of optogenetic and chemogenetic methods to seizure circuits: where to go next? J Neurosci Res 95(12):2345–2356

29. Hare BD, Duman RS (2020) Prefrontal cortex circuits in depression and anxiety: contribution of discrete neuronal populations and target regions. Mol Psychiatry:1–17

30. Atasoy D, Aponte Y, Su HH, Sternson SM (2008) A FLEX switch targets Channelrhodopsin-2 to multiple cell types for imaging and long-range circuit mapping. J Neurosci 28(28):7025–7030

31. Petreanu L, Huber D, Sobczyk A, Svoboda K (2007) Channelrhodopsin-2–assisted circuit mapping of long-range callosal projections. Nat Neurosci 10(5):663–668

32. Wang H, Peca J, Matsuzaki M, Matsuzaki K, Noguchi J, Qiu L et al (2007) High-speed mapping of synaptic connectivity using photostimulation in Channelrhodopsin-2-transgenic mice. Proc Natl Acad Sci U S A 104(19):8143–8148

33. Spellman T, Liston C (2020) Toward circuit mechanisms of pathophysiology in depression. Am J Psychiatr 177(5):381–390

34. Gradinaru V, Zhang F, Ramakrishnan C, Mattis J, Prakash R, Diester I et al (2010) Molecular and cellular approaches for diversifying and extending optogenetics. Cell 141(1):154–165

35. Carter ME, de Lecea L (2011) Optogenetic investigation of neural circuits in vivo. Trends Mol Med 17(4):197–206

36. Deisseroth K, Feng G, Majewska AK, Miesenböck G, Ting A, Schnitzer MJ (2006) Next-generation optical technologies for illuminating genetically targeted brain circuits. J Neurosci 26(41):10380–10386

37. Cardin JA (2012) Dissecting local circuits in vivo: integrated optogenetic and electrophysiology approaches for exploring inhibitory regulation of cortical activity. J Physiol Paris 106(3–4):104–111

38. Gradinaru V, Mogri M, Thompson KR, Henderson JM, Deisseroth K (2009) Optical deconstruction of parkinsonian neural circuitry. Science 324(5925):354–359

39. Han X, Qian X, Bernstein JG, H-h Z, Franzesi GT, Stern P et al (2009) Millisecond-timescale optical control of neural dynamics in the nonhuman primate brain. Neuron 62(2):191–198

40. Huber D, Petreanu L, Ghitani N, Ranade S, Hromádka T, Mainen Z et al (2008) Sparse optical microstimulation in barrel cortex drives learned behaviour in freely moving mice. Nature 451(7174):61–64

41. Tsai H-C, Zhang F, Adamantidis A, Stuber GD, Bonci A, De Lecea L et al (2009) Phasic firing in dopaminergic neurons is sufficient for behavioral conditioning. Science 324(5930):1080–1084

42. Ernst OP, Murcia PAS, Daldrop P, Tsunoda SP, Kateriya S, Hegemann P (2008) Photoactivation of channel rhodopsin. J Biol Chem 283(3):1637–1643

43. Zhang F, Prigge M, Beyrière F, Tsunoda SP, Mattis J, Yizhar O et al (2008) Red-shifted optogenetic excitation: a tool for fast neural control derived from Volvox carteri. Nat Neurosci 11(6):631

44. Zhang F, Gradinaru V, Adamantidis AR, Durand R, Airan RD, De Lecea L et al (2010) Optogenetic interrogation of neural circuits: technology for probing mammalian brain structures. Nat Protoc 5(3):439–456

45. Luan H, White BH (2007) Combinatorial methods for refined neuronal gene targeting. Curr Opin Neurobiol 17(5):572–580

46. Luo L, Callaway EM, Svoboda K (2008) Genetic dissection of neural circuits. Neuron 57(5):634–660

47. Arenkiel BR, Peca J, Davison IG, Feliciano C, Deisseroth K, Augustine GJ et al (2007) In vivo light-induced activation of neural circuitry in transgenic mice expressing channelrhodopsin-2. Neuron 54(2):205–218

48. Kravitz AV, Freeze BS, Parker PR, Kay K, Thwin MT, Deisseroth K et al (2010) Regulation of parkinsonian motor behaviours by optogenetic control of basal ganglia circuitry. Nature 466(7306):622–626

49. Tye KM, Prakash R, Kim S-Y, Fenno LE, Grosenick L, Zarabi H et al (2011) Amygdala circuitry mediating reversible and bidirectional control of anxiety. Nature 471(7338):358–362

50. Witten IB, Lin S-C, Brodsky M, Prakash R, Diester I, Anikeeva P et al (2010) Cholinergic interneurons control local circuit activity and cocaine conditioning. Science 330(6011):1677–1681

51. Witten IB, Steinberg EE, Lee SY, Davidson TJ, Zalocusky KA, Brodsky M et al (2011) Recombinase-driver rat lines: tools, techniques, and optogenetic application to dopamine-mediated reinforcement. Neuron 72(5):721–733

52. Alexander GM, Rogan SC, Abbas AI, Armbruster BN, Pei Y, Allen JA et al (2009) Remote control of neuronal activity in

transgenic mice expressing evolved G protein-coupled receptors. Neuron 63(1):27–39

53. Atasoy D, Betley JN, Su HH, Sternson SM (2012) Deconstruction of a neural circuit for hunger. Nature 488(7410):172–177

54. Krashes MJ, Koda S, Ye C, Rogan SC, Adams AC, Cusher DS et al (2011) Rapid, reversible activation of AgRP neurons drives feeding behavior in mice. J Clin Invest 121(4): 1424–1428

55. Sasaki K, Suzuki M, Mieda M, Tsujino N, Roth B, Sakurai T (2011) Pharmacogenetic modulation of orexin neurons alters sleep/wakefulness states in mice. PLoS One 6(5): e20360

56. Jennings JH, Stuber GD (2014) Tools for resolving functional activity and connectivity within intact neural circuits. Curr Biol 24(1): R41–R50

57. Rogan SC, Roth BL (2011) Remote control of neuronal signaling. Pharmacol Rev 63(2): 291–315

58. Michaelides M, Anderson SAR, Ananth M, Smirnov D, Thanos PK, Neumaier JF et al (2013) Whole-brain circuit dissection in free-moving animals reveals cell-specific mesocorticolimbic networks. J Clin Invest 123(12):5342–5350

59. Dell'Anno MT, Caiazzo M, Leo D, Dvoretskova E, Medrihan L, Colasante G et al (2014) Remote control of induced dopaminergic neurons in parkinsonian rats. J Clin Invest 124(7):3215–3229

60. Jennings JH, Sparta DR, Stamatakis AM, Ung RL, Pleil KE, Kash TL et al (2013) Distinct extended amygdala circuits for divergent motivational states. Nature 496(7444): 224–228

61. Obien MEJ, Deligkaris K, Bullmann T, Bakkum DJ, Frey U (2015) Revealing neuronal function through microelectrode array recordings. Front Neurosci, vol 62, p 423

62. Spira ME, Hai A (2013) Multi-electrode array technologies for neuroscience and cardiology. Nat Nanotechnol 8(2):83

63. Christie IN, Wells JA, Southern P, Marina N, Kasparov S, Gourine AV et al (2013) fMRI response to blue light delivery in the naive brain: implications for combined optogenetic fMRI studies. NeuroImage 66:634–641

64. Rogers BP, Morgan VL, Newton AT, Gore JC (2007) Assessing functional connectivity in the human brain by fMRI. Magn Reson Imaging 25(10):1347–1357

65. Rajagopalan P, Krishnan KR, Passe TJ, Macfall JR (1995) Magnetic resonance imaging using deoxyhemoglobin contrast versus positron emission tomography in the assessment of brain function. Prog Neuro-Psychopharmacol Biol Psychiatry 19(3): 351–366

66. Lee HJ, Weitz AJ, Bernal-Casas D, Duffy BA, Choy M, Kravitz AV et al (2016) Activation of direct and indirect pathway medium spiny neurons drives distinct brain-wide responses. Neuron 91(2):412–424

67. Kahn I, Desai M, Knoblich U, Bernstein J, Henninger M, Graybiel AM et al (2011) Characterization of the functional MRI response temporal linearity via optical control of neocortical pyramidal neurons. J Neurosci 31(42):15086–15091

68. Lee JH, Durand R, Gradinaru V, Zhang F, Goshen I, Kim D-S et al (2010) Global and local fMRI signals driven by neurons defined optogenetically by type and wiring. Nature 465(7299):788–792

69. Shih Y-YI, Chen Y-Y, Lai H-Y, Kao Y-CJ, Shyu B-C, Duong TQ (2013) Ultra high-resolution fMRI and electrophysiology of the rat primary somatosensory cortex. NeuroImage 73:113–120

70. Ferenczi EA, Zalocusky KA, Liston C, Grosenick L, Warden MR, Amatya D et al (2016) Prefrontal cortical regulation of brain-wide circuit dynamics and reward-related behavior. Science 351(6268)

71. Lee JH (2012) Informing brain connectivity with optogenetic functional magnetic resonance imaging. NeuroImage 62(4): 2244–2249

72. Leong AT, Chan RW, Gao PP, Chan Y-S, Tsia KK, Yung W-H et al (2016) Long-range projections coordinate distributed brain-wide neural activity with a specific spatiotemporal profile. Proc Natl Acad Sci 113(51): E8306–E8E15

73. Ryali S, Shih Y-YI, Chen T, Kochalka J, Albaugh D, Fang Z et al (2016) Combining optogenetic stimulation and fMRI to validate a multivariate dynamical systems model for estimating causal brain interactions. NeuroImage 75:398–405

74. Weitz AJ, Fang Z, Lee HJ, Fisher RS, Smith WC, Choy M et al (2015) Optogenetic fMRI reveals distinct, frequency-dependent networks recruited by dorsal and intermediate hippocampus stimulations. NeuroImage 107:229–241

75. Ressler KJ, Mayberg HS (2007) Targeting abnormal neural circuits in mood and anxiety disorders: from the laboratory to the clinic. Nat Neurosci 10(9):1116–1124

76. Ressler KJ, Nemeroff CB (2000) Role of sero-tonergic and noradrenergic systems in the pathophysiology of depression and anxiety disorders. Depress Anxiety 12(S1):2–19

77. Drevets WC (2001) Neuroimaging and neuropathological studies of depression: implications for the cognitive-emotional features of mood disorders. Curr Opin Neurobiol 11(2): 240–249

78. Harrison PJ (2002) The neuropathology of primary mood disorder. Brain 125(7): 1428–1449

79. Sheline YI (2003) Neuroimaging studies of mood disorder effects on the brain. Biol Psychiatry 54(3):338–352

80. Dandekar M, Fenoy A, Carvalho A, Soares J, Quevedo J (2018) Deep brain stimulation for treatment-resistant depression: an integrative review of preclinical and clinical findings and translational implications. Mol Psychiatry 23(5):1094–1112

81. Covington HE, Lobo MK, Maze I, Vialou V, Hyman JM, Zaman S et al (2010) Antidepressant effect of optogenetic stimulation of the medial prefrontal cortex. J Neurosci 30(48): 16082–16090

82. Franklin TB, Silva BA, Perova Z, Marrone L, Masferrer ME, Zhan Y et al (2017) Prefrontal cortical control of a brainstem social behavior circuit. Nat Neurosci 20(2):260–270

83. Vialou V, Robison AJ, LaPlant QC, Covington HE III, Dietz DM, Ohnishi YN et al (2010) ΔFosB in brain reward circuits mediates resilience to stress and antidepressant responses. Nat Neurosci 13(6):745

84. Warden MR, Selimbeyoglu A, Mirzabekov JJ, Lo M, Thompson KR, Kim S-Y et al (2012) A prefrontal cortex–brainstem neuronal projection that controls response to behavioural challenge. Nature 492(7429):428–432

85. Liu Z, Wang Y, Cai L, Li Y, Chen B, Dong Y et al (2016) Prefrontal cortex to accumbens projections in sleep regulation of reward. J Neurosci 36(30):7897–7910

86. Hultman R, Mague SD, Li Q, Katz BM, Michel N, Lin L et al (2016) Dysregulation of prefrontal cortex-mediated slow-evolving limbic dynamics drives stress-induced emotional pathology. Neuron 91(2):439–452

87. Drysdale AT, Grosenick L, Downar J, Dunlop K, Mansouri F, Meng Y et al (2017) Resting-state connectivity biomarkers define neurophysiological subtypes of depression. Nat Med 23(1):28–38

88. Miller OH, Bruns A, Ammar IB, Mueggler T, Hall BJ (2017) Synaptic regulation of a thalamocortical circuit controls depression-related behavior. Cell Rep 20(8):1867–1880

89. Carreno F, Donegan J, Boley A, Shah A, DeGuzman M, Frazer A et al (2016) Activation of a ventral hippocampus–medial prefrontal cortex pathway is both necessary and sufficient for an antidepressant response to ketamine. Mol Psychiatry 21(9):1298–1308

90. Yun S, Reynolds RP, Petrof I, White A, Rivera PD, Segev A et al (2018) Stimulation of entorhinal cortex–dentate gyrus circuitry is antidepressive. Nat Med 24(5):658–666

91. Bagot RC, Parise EM, Pena CJ, Zhang H-X, Maze I, Chaudhury D et al (2015) Ventral hippocampal afferents to the nucleus accumbens regulate susceptibility to depression. Nat Commun 6(1):1–9

92. Koob GF (2008) A role for brain stress systems in addiction. Neuron 59(1):11–34

93. Wenzel JM, Rauscher NA, Cheer JF, Oleson EB (2015) A role for phasic dopamine release within the nucleus accumbens in encoding aversion: a review of the neurochemical literature. ACS Chem Neurosci 6(1):16–26

94. Groenewegen HJ, Wright CI, Beijer AV, Voorn P (1999) Convergence and segregation of ventral striatal inputs and outputs. Ann N Y Acad Sci 877(1):49–63

95. Goto Y, Grace AA (2008) Limbic and cortical information processing in the nucleus accumbens. Trends Neurosci 31(11):552–558

96. Cao J-L, Covington HE, Friedman AK, Wilkinson MB, Walsh JJ, Cooper DC et al (2010) Mesolimbic dopamine neurons in the brain reward circuit mediate susceptibility to social defeat and antidepressant action. J Neurosci 30(49):16453–16458

97. Friedman AK, Walsh JJ, Juarez B, Ku SM, Chaudhury D, Wang J et al (2014) Enhancing depression mechanisms in midbrain dopamine neurons achieves homeostatic resilience. Science 344(6181):313–319

98. Willner P, Hale AS, Argyropoulos S (2005) Dopaminergic mechanism of antidepressant action in depressed patients. J Affect Disord 86(1):37–45

99. Carr DB, Sesack SR (2000) Projections from the rat prefrontal cortex to the ventral tegmental area: target specificity in the synaptic associations with mesoaccumbens and mesocortical neurons. J Neurosci 20(10): 3864–3873

100. Gabbott PL, Warner TA, Jays PR, Salway P, Busby SJ (2005) Prefrontal cortex in the rat: projections to subcortical autonomic, motor, and limbic centers. J Comp Neurol 492(2): 145–177

101. Sesack SR, Carr DB (2002) Selective prefrontal cortex inputs to dopamine cells: implications for schizophrenia. Physiol Behav 77(4–5):513–517

102. Watabe-Uchida M, Zhu L, Ogawa SK, Vamanrao A, Uchida N (2012) Whole-brain mapping of direct inputs to midbrain dopamine neurons. Neuron 74(5):858–873

103. Beier KT, Steinberg EE, DeLoach KE, Xie S, Miyamichi K, Schwarz L et al (2015) Circuit architecture of VTA dopamine neurons revealed by systematic input-output mapping. Cell 162(3):622–634

104. Jo YS, Mizumori SJ (2016) Prefrontal regulation of neuronal activity in the ventral tegmental area. Cereb Cortex 26(10): 4057–4068

105. Moreines JL, Owrutsky ZL, Grace AA (2017) Involvement of infralimbic prefrontal cortex but not lateral habenula in dopamine attenuation after chronic mild stress. Neuropsychopharmacology 42(4):904–913

106. Zhou L, Liu M-Z, Li Q, Deng J, Mu D, Sun Y-G (2017) Organization of functional long-range circuits controlling the activity of serotonergic neurons in the dorsal raphe nucleus. Cell Rep 18(12):3018–3032

107. Amo R, Fredes F, Kinoshita M, Aoki R, Aizawa H, Agetsuma M et al (2014) The habenulo-raphe serotonergic circuit encodes an aversive expectation value essential for adaptive active avoidance of danger. Neuron 84(5):1034–1048

108. Celada P, Puig MV, Casanovas JM, Guillazo G, Artigas F (2001) Control of dorsal raphe serotonergic neurons by the medial prefrontal cortex: involvement of serotonin-1A, GABAA, and glutamate receptors. J Neurosci 21(24):9917–9929

109. Tye KM, Deisseroth K (2012) Optogenetic investigation of neural circuits underlying brain disease in animal models. Nat Rev Neurosci 13(4):251–266

110. Yizhar O, Fenno LE, Davidson TJ, Mogri M, Deisseroth K (2011) Optogenetics in neural systems. Neuron 71(1):9–34

111. You I-J, Wright SR, Garcia-Garcia AL, Tapper AR, Gardner PD, Koob GF et al (2016) 5-HT 1A autoreceptors in the dorsal raphe nucleus convey vulnerability to compulsive cocaine seeking. Neuropsychopharmacology 41(5): 1210–1222

112. Nair SG, Strand NS, Neumaier JF (2013) DREADDing the lateral habenula: a review of methodological approaches for studying lateral habenula function. Brain Res 1511: 93–101

113. Hikosaka O (2010) The habenula: from stress evasion to value-based decision-making. Nat Rev Neurosci 11(7):503–513

114. Lammel S, Lim BK, Ran C, Huang KW, Betley MJ, Tye KM et al (2012) Input-specific control of reward and aversion in the ventral tegmental area. Nature 491(7423):212–217

115. Shabel SJ, Proulx CD, Trias A, Murphy RT, Malinow R (2012) Input to the lateral habenula from the basal ganglia is excitatory, aversive, and suppressed by serotonin. Neuron 74(3):475–481

116. Stamatakis AM, Stuber GD (2012) Activation of lateral habenula inputs to the ventral midbrain promotes behavioral avoidance. Nat Neurosci 15(8):1105–1107

117. Li B, Piriz J, Mirrione M, Chung C, Proulx CD, Schulz D et al (2011) Synaptic potentiation onto habenula neurons in the learned helplessness model of depression. Nature 470(7335):535–539

118. Li K, Zhou T, Liao L, Yang Z, Wong C, Henn F et al (2013) βCaMKII in lateral habenula mediates core symptoms of depression. Science 341(6149):1016–1020

119. Aghajanian G, Wang RY (1977) Habenular and other midbrain raphe afferents demonstrated by a modified retrograde tracing technique. Brain Res 122(2):229–242

120. Morris J, Smith K, Cowen P, Friston K, Dolan RJ (1999) Covariation of activity in habenula and dorsal raphe nuclei following tryptophan depletion. NeuroImage 10(2):163–172

121. Sakai K, Salvert D, Touret M, Jouvet M (1977) Afferent connections of the nucleus raphe dorsalis in the cat as visualized by the horseradish peroxidase technique. Brain Res 137(1):11–35

122. Matsumoto M, Hikosaka O (2007) Lateral habenula as a source of negative reward signals in dopamine neurons. Nature 447(7148):1111–1115

123. Jhou TC, Geisler S, Marinelli M, Degarmo BA, Zahm DS (2009) The mesopontine rostromedial tegmental nucleus: a structure targeted by the lateral habenula that projects to the ventral tegmental area of Tsai and substantia nigra compacta. J Comp Neurol 513(6):566–596

124. Kaufling J, Veinante P, Pawlowski SA, Freund-Mercier MJ, Barrot M (2009) Afferents to the GABAergic tail of the ventral tegmental area in the rat. J Comp Neurol 513(6): 597–621

125. Yang Y, Wang H, Hu J, Hu H (2018) Lateral habenula in the pathophysiology of depression. Curr Opin Neurobiol 48:90–96

126. Sartorius A, Henn FA (2007) Deep brain stimulation of the lateral habenula in treatment resistant major depression. Med Hypotheses 69(6):1305–1308

127. Meng H, Wang Y, Huang M, Lin W, Wang S, Zhang B (2011) Chronic deep brain stimulation of the lateral habenula nucleus in a rat model of depression. Brain Res 1422:32–38

128. Snyder AZ, Bauer AQ (2019) Mapping structure-function relationships in the brain. Biol Psychiatry Cogn Neurosci Neuroimaging 4(6):510–521

Chapter 4

Translational Research Approach to Neurobiology and Treatment of Major Depression: From Animal Models to Clinical Treatment

Michel Bourin

Abstract

The literature indicates that about two-thirds of patients with depression do not achieve remission on initial treatment and that the likelihood of non-response increases with the number of treatments tested. Providing ineffective therapies has significant consequences on individual and societal costs, including persistent distress and poor well-being, risk of suicide, loss of productivity, and waste of healthcare resources. The vast literature on depression indicates a large number of biomarkers that may improve the treatment of people with depression. In addition to the neurotransmitter and neuroendocrine markers that have been studied extensively for many decades, recent data points to the inflammatory response (and more generally the immune system) and metabolic and growth factors involved in depression. The combination of these biological biomarkers found in animals has led to the creation of behavioral models in rodents and zebrafish capable of better understanding depression and contributing to the discovery of new antidepressants.

   Key words Depression, Biomarkers, Inflammation, Behavioral models, Zebrafish

---

1   Introduction

The rise of neuroscience has provided many avenues for exploring and understanding the mechanisms involved in the development of major depression. Genetic vulnerability and the neurodevelopmental hypothesis are supported by numerous observations, against the backdrop of neurophysiological, neurocircuitry, and neurocognition abnormalities, which brain imaging and various complementary examinations highlight. However, do the biomarkers and endophenotypes studied allow us to identify more precisely the subjects at risk and to offer more effective treatment to prevent the transition to depression? Through a review of recent studies and the latest meta-analyses on the subject, we will address the limits of

Yong-Ku Kim and Meysam Amidfar (eds.), *Translational Research Methods for Major Depressive Disorder*, Neuromethods, vol. 179, https://doi.org/10.1007/978-1-0716-2083-0_4,
© The Author(s), under exclusive license to Springer Science+Business Media, LLC, part of Springer Nature 2022

divinatory art based on neuroscience and cohort studies, as well as the hopes raised by the development of personalized medicine [1].

According to the World Health Organization, depression is the leading cause of disability, affecting an estimated 350 million people worldwide. Although the treatments currently available are safe, there is significant variability in the results of antidepressant treatments depending on the patient. To date, there is no clinical assessment to predict with a high degree of certainty whether a particular patient will respond to a given antidepressant.

Moreover, 5–15% of the population is at risk of having a unipolar or bipolar depressive episode during their lifetime. The patients express intense fatigue, deep sadness, loss of pleasure and interest, or a strong feeling of worthlessness, which is accompanied by great difficulties of concentration. Unlike a temporary episode of sadness, the major depressive episode lasts beyond 15 days without mood remission despite stimulation from the outside. It can lead to the person's isolation or even to suicide. Research has been able to highlight the existence of a genetic vulnerability, but it is the combination of risk factors such as life situations or events (death, sentimental breakdown, loss of work, painful childhood, etc.) and the occurrence of recurrent depressive episodes (relapses) that can determine the degree of severity of the disease.

Today, many effective treatments exist to treat this pathology: chemical treatments, (antidepressants) which make it possible to relieve depressive symptoms, often associated with medical and psychotherapeutic care, and physical treatments such as deep electrical stimulation or electroconvulsivotherapy, therapeutic means used in the most resistant forms.

However, progress should be made in the individual management of depressed patients, whether they are bipolar or not. To meet this challenge, scientists have developed, among other things, new experimental approaches in animals, focused on extreme phenotypes in response to antidepressant treatment. In the future, this specific approach could serve as a model for the discovery of improved treatments adapted to patients suffering from depression.

Current research suggests that depression results from a combination of genetic, biological, ecological, and psychological aspects. Depression is a major psychiatric disorder worldwide, with significant economic and psychological pressure on society [2]. Fortunately, depression, even in the most severe cases, can be treated. The sooner the treatment can start, the more effective it is. Therefore, robust biomarkers are needed to improve diagnosis in order to speed up the process of drug and/or drug discovery for every patient with the disorder. These are objective and peripheral physiological indicators making it possible to predict the probability of the onset or existence of depression, to stratify them according to severity or symptomatology, to indicate predictions and prognoses, or to follow therapeutic interventions. The purpose of this chapter is to show recent ideas, current challenges, and future prospects regarding the discovery of a variety of biomarkers and animal models for depression and how these can help improve diagnosis and treatment.

2 Validation of Animal Models

No animal model can claim to account for what a psychiatric illness is and the different syndromes it claims to represent [3]. Therefore, the validation of the criteria that each model is supposed to meet to demonstrate its validity is, for practical purposes, largely determined by the purpose of the model and its intended use. Yet, it seems difficult to create and use animals to model diseases as complex as depression [4]. For a psychiatrist, it is difficult to conceptualize that rodents or zebrafish can be helpful in understanding the diverse set of symptoms of depression. Predicting response on animal models to antidepressant drugs also appears highly speculative [5].

The great heterogeneity of depressive symptoms most often associated with anxiety symptoms, especially in bipolar disorder, leads to great difficulty in diagnosis. Clinical studies of antidepressants in depression point to all these difficulties. Moreover, human biomarkers, as we will see below, are hardly compatible with the understanding of animal models [6]. In addition, studies of postmortem human brain samples are unreliable as storage and sample collection suffer from many biases. In contrast, more and more animal models often offer advantages, such as a high level of standardization [7]. Working with standardized animal cohorts can help to minimize bias and to process larger sample sizes, for example, when dealing with small and inexpensive species such as zebrafish and also allowing unlimited access to the central nervous system.

Ideally, an animal model in psychopharmacology should be similar to the human psychiatric disorder in terms of induced behaviors, i.e., creative validity or face validity; the etiology and neurobiological mechanisms underlying the disorder, i.e., theoretical validity or construct validity; and only the treatment response that has been clinically effective, i.e., predictive validity [8].

Construct validity must be present in MDD animal models if depressive behavior and associated characteristics can be clearly and unambiguously seen and interpreted in the model, which is not the case [9]. It is difficult, for example, to talk about anhedonia in mice even in the chronic mild test [10]. The face validity criterion is fulfilled if the model has similar or comparable elements in terms of anatomical, biochemical, neuropathological, or behavioral characteristics between animals and humans [11]. Predictive validity focuses on the ability of an animal model to serve as a tool for pharmacological research: antidepressant agents, which induce antidepressant-like effects in animals, are also expected to show similar or comparable effects in humans [12]. Based on these criteria, the strength of an animal model system can be estimated. Behavioral aspects of MDD-related phenotypes as well as behavioral tests to treat the effects of antidepressant agents have been characterized in various animal experimental approaches: to model depression-like phenotypes, a number of different strategies have been used, for example, stress selection during distinct windows of vulnerability in animal life to induce lasting behavioral and neurobiological changes [13]. For excellent and recent in-depth reviews of animal models of depression-like conditions and more recent attempts to model circuit-based symptomatic dimensions in MDD, see ref. [14]. Given the plethora of different attempts to model MDD-like phenotypes over the past several decades, concluding that we need to fundamentally rethink animal models for depressive disorders may seem pretentious. However, how can we overcome current limitations and advance the field to finally translate fundamental advancements into better care for depressed patients? In the context of antidepressant research, the majority of animal models and related publications traditionally analyze and discuss an average effect of treatment or manipulation with respect to the respective control condition. There is only very limited insight into why so many patients do not show an answer, despite the fact that antidepressants have been shown to be effective in general. Unfortunately, the enigma of the heterogeneity of the response to antidepressants has not been systematically addressed to date, although it has long been recognized as one of the critical factors hindering the discovery of antidepressant drugs and the clinical evaluation and approval of potentially new compounds [15].

3   Clinical Profile of Depression

Depression is a recent concept, since it was built over the course of the twentieth century with as many acceptances as there are views on it: psychological and psychiatric, of course, but also philosophical, sociological, even political, and economic [16]. Since most of the symptoms of depression are more or less pronounced variations of normal reactions, each individual represents them personally, and an infinite number of models can be developed to define the depression entity, its contours, and its mechanisms. It was therefore necessary to set a perimeter and major benchmarks to agree on the object itself. This is only from the 1980s, when the DSM-3 established the concept of "major depressive episode," with diagnostic criteria that have since changed very little. The clinicians use very littles standardized tools in their daily practice; therefore, the existence of a common frame of reference, taught during studies and constantly mentioned during continuing education and conferences, clearly improves the relevance of the diagnoses made and the opportunities for fruitful exchanges between practitioners.

Nosography, as defined by international classifications, is admittedly relatively poor compared to the great diversity and complexity of the psychopathology of affective disorders. But the concepts of major depressive episode, bipolarity, and endogenous or melancholic depression are now relatively consensual and operational. While these are a priori complex notions, the pathological threshold between normality and depressive episode seems rather well established by current criteria. Indeed, all the explorations of possible "sub-syndromic" depressive states, that is to say below the symptomatic level of the major depressive episode, showed that they were few in number and not very significant. However, at least two important concepts remain insufficiently mastered: psychopathological heterogeneity in depression and the link between diagnoses and optimal therapeutic choices [17].

The second point (who to deal with and in what way?) follows quite clearly from the first, and somewhat beyond our purpose, we will only address the first here. Indeed, these are the persistent difficulties in "typing" depressions, in the absence of sufficiently established etiopathogenic models. Even if all clinicians agree quite easily on the existence of different forms of depression, any attempt to construct a typology comes up against the absence of external markers and the polymorphous and intricate aspect of the semiological presentations. Simple distinctions are now proposed, for example, between recurrent and non-recurrent forms (necessarily fragile and temporary concept), or "endogenous" forms strongly marked by "biological" symptoms) and non-endogenous, but these categorizations are relatively poor and lead to few practical recommendations. A few other markers are proposed but remain fairly general and probably cover various pathophysiological realities: seasonality, postpartum occurrence, anxiety, or mixed symptoms associated (characteristics newly introduced in the DSM-5).

Another example of this nosographic problem was given by the debates around mourning during the development of the DSM-5 [18]. The authors' intention was not to systematically exclude grieving situations from the scope of depression, as was the case in previous versions. This choice is based on epidemiological data confirming that depressions occurring after bereavement can have the same symptomatic and prognostic severity as non-bereavement depressions and respond similarly to the recommended therapeutic strategies.

This proposal sparked strong opposition from many clinicians, considering that the grieving reaction is a normal and necessary adaptation phenomenon and that there is no justification for making it a pathological entity. In the last version of DSM-5, bereavement is no longer an exclusion criterion for making a diagnosis of EDM, but it is clearly mentioned that it is up to the practitioner, on a case-by-case basis, to judge whether the patient's condition is pathological or not and that symptoms of "classic" bereavement (sadness, nightmares, guilt, etc.) are not sufficient to justify a diagnosis. It is also proposed, in the appendix, a new category requiring additional studies, qualified as "complex and pathological bereavement." This topic shows to what extent the clinical tools available today remain insufficient to carry out a qualitative assessment of depressive states and that most of the observations and analyzes come down to the know-how (which remains an art) of clinicians. However, there are still symptoms and management rules that remain relevant:

1. If the diagnosis of a depressive episode is confirmed by a set of clinical arguments: self-deprecation, guilt, psychomotor slowing down, anhedonia, anxiety, weight loss, sleep disorder, etc.

2. If patient cooperation can be obtained for an adequate dosage, evaluation should continue for the 4–8 weeks necessary for the development of full efficacy.

3. Whether the patient is regularly monitored to assess the therapeutic benefit and the interest in continuing treatment or not during the consolidation phase (4 months after remission of symptoms).

4   Neurobiological Substrates of Depression

Three neurotransmitters are particularly important for the proper functioning of the brain and for the stability of morale: serotonin, norepinephrine, and dopamine [19]. This hypothesis has been put forward since the 1960s and the discovery of the first antidepressants whose primary mode of action is to inhibit serotonin reuptake, i.e., to increase his brain count. More recently, data from neuroimaging, neurophysiology, and cell and molecular biology have made it possible to associate depression with various neurobiological phenomena, including dysfunction of neuroplasticity. Neuroplasticity corresponds to changes in the organization and structure of certain neuronal elements, producing a modification or modulation of their function (neurogenesis, apoptosis, synaptic plasticity, reorganization of the composition of neural networks, etc.) [20].

The changes would depend on factors in the environment in which the subject lives (e.g., stress, alcohol.) and the environment in which the nervous system is located (e.g., glucocorticoid levels). Knowing that serotonin (5-HT) regulates several aspects of brain plasticity and that most effective antidepressants facilitate 5-HT transmission, it has been proposed that these psychotropic drugs also exert their therapeutic effect by promoting neuroplasticity [21].

The brain-derived neurotrophic factor (BDNF) controls the survival and differentiation of specific populations of neurons. In the developing cerebral cortex, BDNF regulates the growth and complexity of the dendrites of pyramidal neurons [22]. Given the fundamental role played by dendrites in the transmission of neural information, changes in the growth of dendrites in cortical neurons by BDNF can have major consequences on the functioning of the brain and in particular the cerebral cortex. Despite these observations, the mechanisms underlying these effects are still poorly understood.

5   Identify Biomarkers of Depression

Two important concepts, at least, remain insufficiently mastered: psychopathological heterogeneity within depression and the link between diagnoses and optimal therapeutic choices. Who treats and in what way, arises quite clearly from the first and is somewhat related to our subject; we will only address the first here. These are, in fact, persistent difficulties in "typing" depressions, in the absence of sufficiently established etiopathogenic models [23].

Even if all clinicians agree quite easily on the existence of different forms of depression, any attempt to construct a typology comes up against the absence of external markers and the polymorphous and intricate aspect of semiological presentations. Simple distinctions are proposed today, for example, between recurrent and non-recurrent forms (notion necessarily fragile and temporary), or "endogenous" forms (strongly marked by "biological" symptoms) and non-endogenous, but these categorizations are relatively poor and result in few practical recommendations. Some other markers are proposed but remain fairly general and probably cover various physiopathological realities: seasonality, postpartum onset, anxious, or mixed symptoms associated (characteristics newly introduced in DSM-5).

The challenge is to define biological markers and brain imaging to facilitate the diagnosis of depression [24]. To identify these markers, brain activity (through functional magnetic resonance imaging and EEG) and biological activity (markers of inflammation in the blood) of depressed patients should be measured. The results should thus make it possible to develop new therapeutic strategies as close as possible to the patients' difficulties. The definition of biological markers should allow more precise monitoring of the effects of these therapeutic interventions [25]. The results of this project will therefore contribute to the development of personalized medicine adapted to the specific needs of the depressed patient.

The predisposition to depression, in particular bipolar, is of genetic origin; however, for example, following intense stress (loss of a loved one, divorce) or continuous stress (e.g., at the workplace), the subject can decompensate. In some people, it is only after another stressful episode (even if not very intense) that depression can develop. Thus, the first stress would leave a mark in the brain, modifying neural networks in a lasting way. These individuals are considered at risk, that is, they have a high probability of developing depression as a result of other stress.

Achieving recognition of these at-risk populations requires characterizing vulnerability to depression. To study it, a research team relied on a model reproducing intense social stress in rats [26]. This protocol induces a modification of the structure of neurons in certain regions of the brain, in particular in the hippocampus, an area involved in many learning and memorization processes. At the same time, the level of BDNF, a molecule involved in cell growth, was greatly reduced in this region but also in the blood [27]. After a few weeks, half of the stressed animals returned to their normal state, while the other half retained the neural changes and low BDNF levels. Following further stress of lower intensity, depressive symptoms only appeared in this second group, identifying them as a vulnerable population. The researchers then characterized the measurement of the level of BDNF in the blood as a biological marker of predisposition to depression. This study opens up new perspectives aimed at identifying subjects predisposed to developing depression in an at-risk population. The aim is to enable early pharmacological and/or behavioral therapy aimed at preventing the development of the disease.

6   The Term "Biomarker" Is Used in Different Ways

In its simplest form, it is a characteristic of an organism that can be objectively measured over time. This definition is supplemented by at least two other characteristics: the expected utility of the marker and the technique necessary for its measurement.

– Distinguishing the absence or presence of a disease is the most frequent utility (diagnostic biomarker). In the field of pathology, we note the presence of measures which monitor its appearance (predictive biomarker) and its evolution (prognostic biomarker). Finally, in the field of care, there are also markers that can predict the efficacy or adverse effects of therapy (treatment biomarker) [28].

– Schematically, we can group biomarkers into large families: clinical, behavioral, laboratory, electrophysiological, and imaging [29].

- The clinical markers are those taken from the history or somatic examination of an individual. The cardiovascular risk factors that constitute the Framingham score are an illustration of this [30]. In psychiatry, the presence of an anxious symptomatology, for example, predicts a poorer response to antidepressants [31].
- Behavioral approaches use all kinds of experimental markers such as exposure to stimuli, cognitive demands, or physical exertion [32].
- Laboratory biomarkers include genetic and epigenetic markers, hormones, cytokines, neuropeptides, enzymes, and other unique measures. They can also reflect more complex biological systems such as proteomes or metabolomes by a combination of measurement [33].
- Electrophysiological biomarkers include electroencephalogram (simple or evoked potentials), electrocardiogram (especially analysis of rhythm variation), skin conductance, and electromyogram [34].
- Imaging techniques use structural (scanner and especially magnetic resonance), functional (fMRI) imaging, positron emission tomography (PET), or single-photon emission computed tomography (SPECT) to the study of connectivity between brain regions [35].

The field of study of biomarkers is an extension of the extensive literature of "trait" markers and "state" markers of diseases of which it is a form of renewal [36]. By emphasizing the dynamic nature of measurements linked to interindividual and intra-individual variations over time, biomarkers attempt to refine these concepts. This renewal of markers is particularly useful in psychiatric pathologies as their heterogeneity seems great and the stability of the diagnosis uncertain over time. However, it is medicine as a whole that is concerned, with definite benefits for some specialties. This is particularly the case with oncology, whose diagnosis and prescriptions have been upset by the discovery of new measures opening the way to the so-called "personalized" medicine.

Depression is on a case-by-case basis. One study developed a mouse model to identify blood signatures associated with response to antidepressant therapy [37]. It has also shown the importance of the stress-related glucocorticoid receptor in recovery from depression [38]. This model simulated the clinical situation, identifying the good and bad responders to antidepressant treatment. Ultimately, the identification of a set of predictive biomarkers of the response to antidepressants, carried out in mice, would considerably improve the quality of care and treatment for depressed patients, since the second treatment tried is systematically less effective than the first. In the future, this specific approach could serve as a model for the discovery of improved and tailored treatments for patients with depression. The variation in the activity of all genes in the blood has been studied, based on several clinical studies. The results revealed that the GPR56 receptor gene was among the most significantly activated genes, only in patients for whom the antidepressant provides a real therapeutic response (patients called antidepressant responders) and not in those who do not – responders or patients receiving placebo [39]. The interest of this discovery lies in particular in the fact that GPR56 is an easy-to-measure blood biomarker. GPR56, also called ADGRG1, is a particularly complex and still poorly understood receptor [40] that is involved in many biological processes including neurogenesis (formation of neurons) and brain maturation, astrocyte maturation, and also the activation of the immune system.

7   Variability of Biomarkers

The variation of biomarkers over time and according to different situations concerns certain types, for example, proteomics, than others such as genomics [41]. Standards do not exist for most biomarkers or are not accepted most of the time. Indeed, the influence of environmental factors on markers often depends on genetics as well as physiological differences between people. This makes it difficult to assess the activity of biomarkers and identify biological abnormalities. Due to the number of potential biomarkers, it is difficult to measure them on a large scale as well as in a full panel of relevant markers [42].

Many factors have been reported to alter protein levels in biological systems in patients with affective disorders. In addition to research-related factors, length and conditions of storage may lead to the degradation of certain compounds, as well as time of day, ethnicity, exercise, diet, usage of tobacco, or other addictive substances. This is also true for certain pathologies such as concomitant inflammatory diseases, cardiovascular diseases, and even other somatic diseases. For example, although inflammation is more often seen in depressed but healthy individuals compared to non-depressive groups, depressed people who also have a concomitant disease associated with immunity often have higher levels of cytokines that can therefore be considered as biomarkers in unipolar or even bipolar depression [43].

Endocrine and immune responses play a well-known role in the stress response, and transient stress at the time of biological specimen collection is rarely measured in studies despite the variability of this factor among individuals with symptoms of depression. Acute and chronic psychological stressors accentuate short- and long-term inflammatory responses [44]. These findings are made with respect to early childhood and childhood stress, as there is less interference with the origin of the inflammation. Thus, during childhood traumatic events, increased inflammation has also been reported only in depressed children [45]. Conversely, people with depression and with a history of childhood trauma may have no trauma reactions early in life. Stress-induced alterations in the HPA axis appear to be related to cognitive function, as well as to the depression subtype or variation in HPA-related genes. Stress also has short- and long-term effects on mechanisms. It is not clear exactly how childhood trauma affects biomarkers in adult depression, but it is possible that stress early in life predisposes some people to experiencing stress reactions in adulthood, which are amplified psychologically and/or biologically [46].

Neurocognitive dysfunction is common in people with affective disorders, especially in patients with MDD [47]. Cognitive deficits appear to be linked to resistance to treatment. Norepinephrine and dopamine are important for cognitions such as learning processes and memory. There are generally associated elevated inflammatory responses probably affecting cognitions during depressive episodes. Researchers have proposed that CRP is more closely related to cognitive performance than to the main symptoms of depression [48].

The difference in hormone activity between men and women interferes with the sensitivity of genders to depression. Age and sex do not cause differences in inflammatory cytokines in patients. The differences between patients and controls are greater in studies evaluating samples from younger patients, while gender, body mass index (BMI), and clinical factors did not influence these meta-analyses [49]. The increase in adipose tissue stimulates cytokine production and is closely related to associated metabolic markers. Resistance to antidepressants has been shown in patients with increased weight gain and higher BMI [50]. A significant portion of subjects with unipolar depression and obesity have the same activation of the innate immune system, which causes an inflammatory reaction [51].The biomarkers of this inflammatory state, markedly increased in these depressed patients, include tumor necrosis factor or TNF-alpha and interleukins IL-1 and IL-6 (pro-inflammatory cytokines), C-reactive protein and fibrinogen (proteins secreted by the liver in the acute phase of inflammation), and chemokines [52]. For example, in a large proportion of obese people, the excessive accumulation of lipids in the cells of the fatty tissue of the abdomen compromises the supply of oxygen to the fat cells. This oxygen deficiency generates cellular stress, which in turn triggers a hypersecretion of proteins, including pro-inflammatory cytokines. In the case of depression, a great deal of work in depressed patients shows that a significant subset of these have a low-noise inflammatory state, caused by activation of the immune system without underlying medical causes [53]. Although the exact trigger for this inflammatory process is unknown, adipose tissue hypoxia, cellular stress (at the level of the endoplasmic reticulum), and especially the activation induced by saturated fatty acids have been identified as important processes in the onset of a chronic low-grade inflammatory state. Thus, there is again a significant increase in pro-inflammatory cytokines, which will influence the functioning of the brain and modify mood and behavior. These findings have led to the now well-established hypothesis that the brain interprets immune activation as a stressor [54].

In a model of mice rendered depressive, the enriched environment makes it possible to attenuate the increase in the levels of mRNA of pro-inflammatory cytokines (IL-1β, IL-6, and TNF-α) in the hypothalamus and the hippocampus, two areas of the brain closely related to depression. An orientation of the activation profile of the microglia toward an anti-inflammatory type M2 profile is stimulated in mice in an enriched environment, compared to mice housed under standard conditions, which in turn exhibit a pro-inflammatory phenotype. The anti-inflammatory effects on microglia and the behavioral antidepressant effects of the enriched environment are based, at least in part, on the ApN. The IV injection of gApN indeed mimics the anti-inflammatory effects of the enriched environment on the activation profile of the microglia with the induction of a decrease in the expression and production of pro-inflammatory cytokines while increasing the levels of the mRNAs of anti-inflammatory markers such as Arg1 (Arginase 1) and IL-10 and improves anxiety-depressive behaviors in mice [55]. Thus, it has been shown that AdipoRon, an adiponectin receptor agonist, acts as an antidepressant and metabolic regulator in a mouse model of depression [56].

Numerous biomarker studies of depression have been performed by collecting samples from non-drug subjects in order to reduce heterogeneity. However, many of these assessments are taken after a longer or shorter drug withdrawal period, leading to potentially significant uncertainty about residual changes in disease, compounded by the range of treatments that may have effects on the patient's inflammation. Some studies have excluded the use of psychotropic drugs, but not other drugs: in particular, the oral contraceptive pill is frequently allowed in research participants and not controlled in analyses, which has been observed to increase the levels of certain hormones and cytokines [57]. However, the many drug treatments for depression have distinct and complex pharmacological properties, suggesting that the different treatment options may have varying degrees of effects on the inflammatory process [58]. In addition to the effects on monoamines, certain drugs specifically targeting serotonin (i.e., SSRIs) would cause changes in inflammation, and noradrenergic antidepressants (e.g., SNRIs) would produce slight changes [59]. The effects of drugs alone or in combination with others on biomarkers are often difficult to assess. These are likely influenced by other factors, including duration of treatment (few trials assess long-term drug use), sample heterogeneity, and non-stratification of participants based on response to medication treatment [60].

8 Depression Subtypes

To date, no homogeneous subgroup of depressive episodes or disorders has been able to reliably distinguish between patients on the basis of symptom presentation or responsiveness to treatment. The existence of a subgroup in which biological changes would be more pronounced could pave the way for a particular treatment [61]. Several subtypes of depression based on neurobiological differences corresponding to clinical subtypes of depression have been proposed: hypercortisolism with melancholic depression, hypocortisolism reflecting an atypical subtype with anhedonia, and a subtype characterized by high inflammation [62]. The literature has shown an inflammatory subtype in depression. The clinical correlates between high inflammation and depression are still uncertain; it is not known which subjects are truly likely to participate in a possible cohort. However, it has been proposed that people with atypical depression may have higher levels of inflammation than the melancholic subtype [24]. This is hardly consistent with the data regarding the HPA axis in the melancholic and atypical subtypes. Resistant depression or depression with significant somatic symptoms has also been suggested as a potential inflammatory subtype. Other potential subtypes for an inflammatory subtype involve symptoms resembling metabolic syndrome [63].

The propensity for hypomania can biologically distinguish patients with depression. Bipolar disorder is a multifaceted group of mood disorders; in particular, it appears that subsyndromal bipolar disorder is more prevalent than previously recognized [64]. The inaccurate and/or delayed diagnosis of bipolar disorder has recently been identified; it takes an average time of 10 years to make the diagnosis. This delay leads to a deterioration of the patient's condition, resulting in a higher cost of managing the disease [65]. The majority of patients with bipolar disorder initially have one or more depressive episodes, and most often unipolar depression is the usual diagnosis. It is generally accepted that a subject who has experienced two or more marked depressive episodes should be considered to have bipolar disorder [66]. This underlines the need for reliable markers as the stakes are high. Some leads have been developed; however, these data are scarce; what seems most relevant is that bipolar patients no longer respond to antidepressants after 2 years of treatment or even become worse [67, 68]. Resistance to treatment and bipolar disorder are not independent of reasoning; there is probably a continuum of constructs which increases the difficulty in identifying subtypes [69]. In addition to subtyping, it should be noted that many laboratory abnormalities seen in depression are found similar in patients with other diagnoses. Thus, transdiagnostic examinations are also potentially important [70]. Finally, real animal models as well as real biomarkers are yet to be discovered as regards bipolarity [71, 72]. The heritability of bipolar disorder is explained by several groups of genes involved in particular in the fusion of synaptic vesicles (SNAP25), the physiology of calcium channels (CANB2, CACNI, CACNAIC, CACNAl0,etc.), or the synaptic pathways (NCAM; Neurexin, NRXl; Neuroligin, NLG). These genes are associated with other pathologies such as schizophrenia or autism [73].

9 Biomarker Precision Challenges

A large number of potentially useful biomarkers are under study, posing a challenge for psychiatry in determining markers that are truly specific for a particular pathology. Relatively few biomarkers have been adequately researched for depression, and for the most part, their precise role in healthy and clinical populations is not well understood. Despite this, a number of attempts have been made to come up with panels of promising biomarkers. These are essentially markers of oxidative stress capable of improving the response to treatment: BDNF, cortisol, soluble TNFα type II receptor, antitrypsin alpha28, apolipoprotein CIII, epidermal growth factor, myeloperoxidase, prolactin, and resistin. Unfortunately, most of these parameters are found in pathologies as diverse as cancer or diabetes [74], which is not surprising because these pathologies can lead to real depressive episodes.

Due to technological advances, it is now possible to evaluate a large number of biomarkers simultaneously at a lower cost and with higher sensitivity than in the past. Today, this ability to measure many compounds is advancing faster than our ability to efficiently analyze and interpret data [75]. Rather, it is metabolomics, which is the systematic study of the unique chemical imprint left by biological processes during metabolism. Thus, the analysis of metabolites present in the body or released with natural secretions would make it possible to constitute a metabolomic signature that evolves over the course of life, at the rate of mutations, changes in the body, and associated diseases [76]. Associated with the study of genes and proteins, it will complete the rectangle of the four "-omics" within the circle represented by the body [77]. With the ultimate goal of the study of the metabolites being to obtain a concordance between the profile analyzed in the biofluids and the biochemistry of the pathological tissue considered, it is necessary to determine how a tissue metabolomic profile is reflected in the metabolomic signature of blood or urinary samples.

The development of tools for analysis and detection of biological samples, such as chromatography and mass spectrometry, and the discovery of the crucial impact for understanding the pathologies of some of the metabolites identified in these samples (e.g., microRNAs) now allow metabolomics to develop and find its applications in medicine [78]. Big data using new analytical approaches and standards will help to solve this problem, and new methodologies are proposed; one example is the development of a statistical approach based on flow analysis to discover potential new metabolic markers based on their reactions between networks and to integrate gene expression into metabolite data. Machines learning techniques are used to predict treatment outcomes in studies with big data.

The simultaneous examination of a set of biomarkers is an alternative to the inspection of isolated markers which could provide a more precise view of the complex web of biological systems or networks [79]. Also, after helping to disentangle contrasting data and interactions are well understood, biomarker data can then be aggregated or indexed. One of the challenges is to identify the optimal method to achieve this, and this may require technological improvements and/or new analytical techniques. Historically, ratios between two distinct biomarkers have given interesting results. Few attempts have been made to aggregate biomarker data on a larger scale, such as those using principal component analysis of tissue sources of proinflammatory cytokines. Composite biomarker panels are both a challenge and an opportunity for future research to identify meaningful and reliable endpoints that can be used to improve treatment elements [60]. The use of big data is likely necessary to address the current challenges of heterogeneity, biomarker variability, and identifying optimal markers for the development of translational research applied to depression. This raises technological and scientific challenges; the health sciences have recently started to use big data analysis much later than the corporate sector. Following further regression analyses, three biomarkers were selected in association with the most important depressive symptoms (highly variable size of red blood cells and serum glucose and bilirubin levels). The authors conclude that big data can be used effectively to generate hypotheses [80]. Larger biomarker phenotyping projects are currently underway and will help advance our future journey in the neurobiology of depression [81].

10   Future Prospects

The research results to date require replication in large-scale studies. This is especially true for new biomarkers, such as the thymus chemokine and activating-regulating chemokine and growth factor tyrosine kinase 2, which to our knowledge have not been studied in clinically depressed and healthy control samples [82]. Big data studies should analyze complete sets of biomarkers, using increasingly sophisticated analytical techniques, to fully determine the relationships between markers and the factors that may modify them in pathological and non-pathological populations.

Regarding the selection of biomarkers, multiple panels may be needed for different potential pathways that the research might involve. Taken together, the current evidence indicates that biomarker profiles are definitely altered in a subpopulation of individuals currently suffering from depression. This can be established within or between categories of diagnoses, which would explain a certain inconsistency in the results observed in this literature. Quantification of a biological subgroup (or subgroups) can be more efficiently facilitated by large-scale cluster analysis of panels of biomarker networks in depression [83]. This would illustrate the intra-population variability; latent class analyses might show distinct clinical features based, for example, on inflammation.

10.1 Specific Effects of Treatment on Inflammation

All commonly prescribed treatments for depression should be thoroughly evaluated to determine their specific biological effects, which also explains the effectiveness of treatment trials [84]. This may allow more personalized design of concepts related to biomarkers and symptom presentations and may be possible in the context of unipolar and bipolar depression. This will likely be useful for potential new treatments as well as currently indicated treatments.

10.2 Prospective Determination of Response to Treatment

The use of the above techniques will likely result in an improvement in the ability to predict treatment resistance prospectively. More authentic and persistent (long-term EEG) measures of response to treatment may help. Evaluating other valid measures of patient well-being (such as quality of life and daily functioning) might provide a more holistic assessment of treatment outcomes, which might be more closely associated with biomarkers. Although biological activity alone is not able to distinguish treatment responders from non-responders, a concomitant measurement of biomarkers with psychosocial or demographic variables could be integrated with information on biomarkers to develop a predictive model of insufficient response to the disease treatment [85]. If a reliable model is developed to predict response (for the depressed population or a subpopulation) and is validated retrospectively, a translational design can establish its applicability in a large controlled trial.

10.3 Toward Stratified Treatments

Currently, patients with depression are not routinely referred to an optimized intervention program [86]. If validated, a stratified trial design could be used to test a model to predict nonresponse and/or to determine where a patient should be sorted in a staged care model. This could be useful in both standardized and naturalistic treatment settings for different types of intervention. Ultimately, a clinically viable model could be developed to provide patients with the most appropriate treatment, to recognize those at risk of developing refractory depression, and to provide improved care and monitoring for these patients. Patients identified as at risk of resistance to treatment may be prescribed concomitant psychological and pharmacological treatment or combination pharmacotherapy [87]. As a speculative example, participants without elevated proinflammatory cytokines might be indicated to receive psychological rather than pharmacologic treatment, while a subset of patients with particularly elevated inflammation might receive an anti-inflammatory agent during treatment, a standard treatment. Similar to stratification, personalized treatment selection strategies may be possible in the future. For example, a particular depressed individual might have elevated TNFα levels, but no other biological abnormalities, and might benefit from short-term treatment with a TNFα antagonist [88]. Personalized treatment may also include monitoring expression for the duration of required continuation treatment or to detect early markers of relapse.

10.4 New Treatment Targets

There are a large number of potential treatments that may be effective for depression that have not been adequately examined, including new or reused interventions from other medical disciplines [89]. Some of the more popular targets have been anti-inflammatory drugs such as celecoxib (and other cyclooxygenase-2 inhibitors), TNFα antagonists, etanercept and infliximab, minocycline, or aspirin [90]. These antiglucocorticoid compounds that appear promising, including ketoconazole and metyrapone, have been studied for depression [38, 91], but both have drawbacks with their side effect profile, and the clinical potential of metyrapone is uncertain. Mifepristone and the corticosteroids fludrocortisone and spironolactone and dexamethasone and hydrocortisone may also be effective in treating depression in the short term [92]. Targeting glutamate N-methyl-d-aspartate receptor antagonists, including ketamine, may represent effective treatments for depression [93]. Omega-3 polyunsaturated fatty acids influence inflammatory and metabolic activity and appear to have some efficiency and have antidepressant effects through relevant neurobiological pathways [94].

In this way, the biochemical effects of antidepressants have been used for clinical benefits in other disciplines: in particular gastroenterological, neurological, and nonspecific diseases. The anti-inflammatory effects of antidepressants may be part of the mechanism of these drugs [95]. Lithium has also been suggested to reduce inflammation through glycogen synthase kinase pathways [96]. Focusing on these effects could prove informative for the signature of a depressive biomarker, and, in turn, the biomarkers could represent novel markers.

11   Toward New Animal Model Concepts

Much of the current understanding of the pathogenesis of mood disorders at large has come from animal models [13]. However, due to the unique and complex characteristics of mental disorder in humans, rather than seeking to model the depressive syndrome as a whole, we seek to model some of its more well-known symptoms (from a dimensional perspective), such as cognitive disorders, psychomotor disorders, impulsivity, or even disorders of the response to the environment (stress), to pleasant or unpleasant events. Molecular, genetic, and epigenetic factors are also clearly used to build animal models of depression in which we will try to replicate the transcriptional and translational changes observed in depressed patients [97].

Beyond reproduction of the etiological factors and/or neurobiological mechanisms of depression, animal models must also be sensitive to treatments (pharmacological or not) which are recognized as the most effective in humans [98, 99]. Human depression, like other psychiatric disorders, is often triggered or precipitated by stressful events in vulnerable individuals. This is why the majority of animal models of depression are based on exposure to different types of stress. Each stressor is characterized by its intensity, duration, and frequency of exposure, as well as the period of cerebral maturation during which it is applied [100]. Three types of stressors are currently known: acute stress, repeated acute stress, and chronic stress.

If we know the mechanisms of action of antidepressants, no one today knows why they are effective. Scientists wonder why it takes several weeks to get an effect. If the compounds act directly on the communication between the neurons, the effectiveness should appear from the first shots.

It is in particular to answer this question of delay of action that American scientists carried out a study on mice. After placing them in anxiety-inducing conditions, they administered the two kinds of antidepressants, an SSRI and a tricyclic. Then, they X-rayed some mice, targeting a particular region of the brain: the hippocampus. This "irradiation" specifically destroys dividing cells. Rodents subjected to X-rays did not see their condition improve, unlike their little comrades who were also treated with antidepressants [101]. There is only one possible explanation: to be effective, antidepressants promote the formation of new neurons. This discovery is capital. It provides an answer not only to the question of the effectiveness of antidepressants but also to better understand this disease. Indeed, making new cells does not happen overnight. This is why the treatment gives results only when these new neurons are functional, after a few weeks. The Japonese practitioners considered for a long time that the only valid depression was post-stroke depression, which was cured by antidepressants that were found to promote neuronal regrowth in these subjects [102, 103]. Neurons have the ability to make new connections at synapses more easily, thus facilitating communication between neurons in the form of chemical messages (neurotransmitters or neurotransmitters). The brain would therefore be better able to adapt to stress, overcome depression, and form new neurons.

What are its links with mood? But this also opens the way to new treatments: it could be interesting to search for molecules that promote the production of new cells in the hippocampus. It should be noted that this formation of new neurons is certainly only one of the many mechanisms of action of antidepressants in the brain. And it remains difficult to extrapolate the results obtained in mice to humans. Antidepressants still keep a large part of their mystery. But we now know that these drugs can have one or more mechanisms of action.

12   Why Should We Use Animals to Model Complex Diseases Like MDD?

What could be the strengths of an animal model and what are its limits? From a psychiatrist's perspective, it is difficult to agree that rodents or even species such as zebrafish could be useful in investigating a complex mental disorder characterized by a diverse set of symptoms such as MDD. The same goes for the question of response to psychopharmacological treatment. Great heterogeneity in the symptomatology of MDD and a close association with other comorbid psychiatric disorders in a large proportion of patients with MDD are major drawbacks and confounding factors for clinical studies [12]. The exclusive use of peripheral tissues, such as blood, may have only limited value in deciphering the neurobiology of depression, as the brain is only accessible indirectly, for example, through neuroimaging approaches [104]. In addition, postmortem human brain samples suffer from many confounding variables such as pH change, molecule degradation, age bias, and a bias in favor of suicide victims. In contrast, animal models offer unique advantages, such as a high level of standardization. Working with standardized animal cohorts can help to minimize bias; process larger sample sizes, for example, when it comes to small and profitable species such as zebrafish; and, finally, allow an unrestricted access to the organ of interest, i.e., the central nervous system.

The power of an animal model can be described in terms of three key elements: construction, face, and predictive validity. Constructive validity is present in MDD animal models, though depressive behavior and associated characteristics can be clearly and unambiguously seen and interpreted in the model. The face validity criterion is met if the model has similar or comparable elements in terms of "anatomical, biochemical, neuropathological or behavioral characteristics" between animal and man [105].

Predictive validity focuses on the ability of an animal model to serve as a tool for pharmacological research: antidepressant agents, which induce antidepressant-like effects in animals, are also expected to show similar or comparable effects in humans. [106, 107]. Based on these criteria, the strength of an animal model system can be estimated. Behavioral aspects of MDD-related phenotypes as well as behavioral tests to treat the effects of antidepressant agents have been characterized in various animal experimental approaches: to model depression-like phenotypes, a number of different strategies have been used, for example, stress selection during distinct windows of vulnerability in animal life to induce lasting behavioral and neurobiological changes [108]. For excellent and recent in-depth reviews of animal models of depression-like conditions and more recent attempts to model the circuit-based symptomatic dimensions in MDD, see refs. [109, 110]. Considering the plethora of different attempts to model MDD-like phenotypes over the past decades, concluding that we need to fundamentally rethink animal models for depressive disorders may seem pretentious. However, how can we overcome current limitations and advance the field to finally translate fundamental advancements into better care for the patients?

In the context of antidepressant research, the majority of animal models and related publications traditionally analyze and discuss an average effect of treatment or manipulation with respect to the respective control condition [111]. There is only very limited insight into why so many patients do not show an answer, despite the fact that antidepressants have been shown to be effective in general. Unfortunately, the enigma of the heterogeneity of response to antidepressants has not been systematically addressed to date, although it has long been recognized as one of the critical factors hampering the discovery of antidepressant drugs and the clinical evaluation and approval of potentially new compounds [112]. Therefore, to pave the way for so-called precision psychiatry, we wish to propose a framework for translational studies in individualized medicine in psychiatry.

Individuality – generally defined as the collection of divergent behavioral and physiological traits among individuals – develops when unique environmental influences act on the genome, in complex pathways, to produce phenotypic diversity. Individuality is considered essential for the development of several neuropsychiatric disorders. Focusing on individuality rather than average results has gained increasing attention in both rodents and zebrafish [113–115].

Approaches to focus on heterogeneity and individuality within a cohort of mice have been used successfully in the context of stress research to identify putative neurobiological pathways modulating individual sensitivity and resilience: in 2007, the Nestler group published the results of a groundbreaking study, where they did not analyze the mere effect of some manipulation (in this case, a paradigm of chronic social defeat stress), but stratified each mouse based on its performance in a defined behavioral test of social interaction as an outcome [116]. Stratification of animals based on their performance on the social interaction test allowed a focus on differences within the experimental group, accompanied by the advantage that the two new "extreme" subgroups (above or below a certain threshold) become more homogeneous, which could facilitate the discovery of true candidates.

In resilience research, this stratification approach has been shown to be successful in a number of excellent publications in recent years. In order to identify the neurobiological mechanisms underlying the response to antidepressant treatment, it was recently set up an animal experimental approach using stratification into extreme subpopulations among a considerably high number of inbred mice, genetically homogeneous in response to antidepressant treatment [117].

In addition to the significant mean group effect between the groups treated with antidepressants and vehicles, we continued to select, in the cohort of animals treated with paroxetine, subpopulations of good and bad responders based on their results in one of the main behavioral tests evaluating antidepressant efficacy in rodents [118]. Indeed, it was possible to identify specific transcriptomic signatures associated with the state of response in murine blood and to translate and successfully validate the results of our animal model in a cohort of depressed patients [37]. Enfin, nous pourrions révéler un rôle particulier du récepteur des glucocorticoïdes (GR) dans la formation de la réponse aux antidépresseurs, ce qui est d'autant plus intéressant que ces données ont été générées par un modèle animal utilisant une approche sans hypothèse. The putative role of glucocorticoid receptor (GR) in modulating antidepressant-like effects had previously been suggested through basic and clinical research on depression based on hypotheses supporting the validity of this model.

We believe this is the first step toward a more in-depth and dimensional analysis of the different and more complex behavioral signatures of the response to antidepressant therapy. Future studies should implement cluster analyses of phenotypic results, for example, by automated behavioral analysis in an animal's domestic cage.

To develop an approach to identify stratification into different subpopulations among a large number of responders using a low-cost animal model, it was recently established an animal experimental paradigm where we analyzed the responses and the behavior of a group of zebrafish subjected to stress exposure. As a vertebrate, the zebrafish show a strong homology to the main neuromodulatory circuits involved in stress and emotional regulation [119]. In addition, they exhibit behavioral phenotypes to identify "depression-like" cues and are sensitive to different mind-altering drugs. However, studies so far have focused on the average the effect of drug treatments on population and have not carefully examined heterogeneity and individuality. Results suggest the existence of a clear stratification in behavioral outcomes following stress exposure in zebrafish. Since zebrafish are cheap to maintain in large numbers, and genetic manipulations of their genome are fairly easy, they provide a powerful complementary animal model for rodents to test for heterogeneity in antidepressant responses [120]. Thus, individual results and significant stratification of the experimental group should be considered instead of the average group effects in animal models of depression and response prediction to improve translation between preclinical research and clinical trials. As recent examples show, this strategy could help increase the success rate when extrapolating results from bench to bedside and back.

Modelling an anxiety-depressive phenotype in female mice, by adapting a neuroendocrine model of depression developed in males, is based on the chronic administration of corticosterone. On the other hand, there is a study of the comparison of neurogenesis between responding and non-responding mice to chronic fluoxetine treatment or resistant to two successive treatment strategies with a different mechanism of action (fluoxetine and then imipramine). Furthermore, data from the clinical literature suggests that a peripheral marker, the β-arrestin 1 protein, may be a marker of depression and response to treatment. Variations in this potential clinical biomarker were therefore measured in this model. This work has shown the complexity of inducing an anxiety/depressive phenotype in female mice in a stable and robust manner via chronic administration of corticosterone. In males, resistance to antidepressant treatment was modeled in the CORT model. Neurogenic processes appear to play an essential role in the response to treatment, as a lack of response is associated with impaired adult hippocampal neurogenesis. If, in this model, the peripheral expression of β-arrestin 1 is not reduced in mice exhibiting an anxiety-depressive phenotype, it nevertheless makes it possible to distinguish the responding mice from the mice resistant to treatment, which validates its interest in as a biomarker of the antidepressant response [121].

13   Conclusion

However, overwhelming contrasting data shows that a number of challenges must be addressed before biomarker research can be applied to improve the management and care of people with depression. Due to the sheer complexity of biological systems, the simultaneous examination of a full range of markers in large sample sizes presents a considerable advantage in uncovering interactions between biological and psychological states in individuals. Optimizing the measurement of neurobiological parameters and clinical measures of depression will likely facilitate a better understanding. This review also emphasizes the importance of examining potentially modifying factors (such as disease, age, cognition, and medications) in a coherent understanding of the biology of depression and the mechanisms of resistance to treatment. It is likely that certain markers will be the most promising for predicting response to treatment or resistance to specific treatments in a subset of patients, and simultaneous measurement of biological and psychological data could improve the ability to prospectively identify those with risk. Establishing a panel of biomarkers has implications for improving diagnostic accuracy and prognosis, as well as for individualizing treatments at the earliest possible stage of depressive illness and developing new effective therapeutic targets. These implications may be limited to subgroups of depressed patients. Pathways leading to these possibilities complement recent research strategies aimed at more closely linking clinical syndromes to underlying neurobiological substrates. In addition to reducing heterogeneity, it can facilitate a move toward parity between physical and mental health. Clearly, while much work is needed, establishing the relationship between relevant biomarkers and depressive disorders has substantial implications for reducing the burden of depression at the individual and societal level. The big challenge of current research is to combine human biomarkers of depression compatible with animal models that have included human biomarkers compatible with animal neurobiology.

References

1. Thibaut F, Bourin M (2015) Precision medicine Editorial. Int J Emerg Ment Health 17: 367

2. Jacob KS (2012) Depression: a major public health problem in need of a multi-sectoral response. Indian J Med Res 136:537–539

3. Gass P, Wotjak C (2013) Rodent models of psychiatric disorders--practical considerations. Cell Tissue Res 354:1–7

4. Krishnan V, Nestler EJ (2011) Animal models of depression: molecular perspectives. Curr Top Behav Neurosci 7:121–147

5. Bourin M, Redrobe JP, Hascoet M, Baker GB, Colombel MC (1996) A schematic representation of the psychopharmacological profile of antidepressants. Prog Neuro-Psychopharmacol Biol Psychiatry 20: 1389–1402

6. Demin KA, Sysoev M, Chernysh MV, Savva AK, Koshiba M, Wappler-Guzzetta EA, Song C, De Abreu MS, Leonard B, Parker MO, Harvey BH, Tian L, Vasar E, Strekalova T, Amstislavskaya TG, Volgin AD, Alpyshov ET, Wang D, Kalueff AV (2019) Animal models of major depressive disorder and the implications for drug discovery and development. Expert Opin Drug Discov 14: 365–378

7. Institute of Medicine (2013) Improving the utility and translation of animal models for nervous system disorders: workshop summary. The National Academies Press, Washington, DC. https://doi.org/10. 17226/13530

8. Maximino C, van der Staay FJ. Behavioral models in psychopathology: epistemic and

semantic considerations. Behav Brain Funct. 2019 1;15(1):1

9. Belzung C, Lemoine M (2011) Criteria of validity for animal models of psychiatric disorders: focus on anxiety disorders and depression. Biol Mood Anxiety Disord 1(1):9

10. Scheggi S, De Montis MG, Gambarana C (2018) Making sense of rodent models of anhedonia. Int J Neuropsychopharmacol 21: 1049–1065

11. Varga OE, Hansen AK, Sandøe P, Olsson IA (2010) Validating animal models for preclinical research: a scientific and ethical discussion. Altern Lab Anim 38:245–248

12. Herzog DP, Beckmann H, Lieb K, Ryu S, Müller MB (2018) Understanding and predicting antidepressant response: using animal models to move toward precision psychiatry. Front Psych 9:512

13. Wang Q, Timberlake MA 2nd, Prall K, Dwivedi Y (2017) The recent progress in animal models of depression. Prog Neuro-Psychopharmacol Biol Psychiatry 77:99–109

14. Planchez B, Surget A, Belzung C (2019) Animal models of major depression: drawbacks and challenges. J Neural Transm (Vienna) 126:1383–1408

15. Penn E, Tracy DK (2012) The drugs don't work? Antidepressants and the current and future pharmacological management of depression. Ther Adv Psychopharmacol 2: 179–188

16. Bourin M (2020) History of depression through the ages. Arch Depress Anxiety 6: 010–018

17. Goldberg D (2011) The heterogeneity of "major depression". World Psychiatry 10: 226–228

18. Pies RW (2014) The bereavement exclusion and DSM-5: an update and commentary. Innov Clin Neurosci 11:19–22

19. Krishnan V, Nestler EJ (2008) The molecular neurobiology of depression. Nature 455: 894–902

20. Sharma A (2013) N, Classen J, Cohen LG. Neural plasticity and its contribution to functional recovery. Handb Clin Neurol 110: 3–12

21. Yohn CN, Gergues MM, Samuels BA (2017) The role of 5-HT receptors in depression. Mol Brain 10:28

22. Cohen-Cory S, Kidane AH, Shirkey NJ, Marshak S (2010) Brain-derived neurotrophic factor and the development of structural neuronal connectivity. Dev Neurobiol 70: 271–288

23. Labermaier C, Masana M, Müller MB (2013) Biomarkers predicting antidepressant treatment response: how can we advance the field? Dis Markers 35:23–31

24. Strawbridge R, Young AH, Cleare AJ (2017) Biomarkers for depression: recent insights, current challenges and future prospects. Neuropsychiatr Dis Treat 13:1245–1262

25. Calvo MS, Eyre DR, Gundberg CM (1996) Molecular basis and clinical application of biological markers of bone turnover. Endocr Rev 17:333–368

26. Blugeot A, Rivat C, Bouvier E, Molet J, Mouchard A, Zeau B, Bernard C, Benoliel JJ, Becker C (2011) Vulnerability to depression: from brain neuroplasticity to identification of biomarkers. J Neurosci 31(36): 12889–12899

27. Miranda M, Morici JF, Zanoni MB, Bekinschtein P (2019) Brain-derived neurotrophic factor: a key molecule for memory in the healthy and the pathological brain. Front Cell Neurosci 13:363

28. Califf RM (2018) Biomarker definitions and their applications. Exp Biol Med (Maywood) 243:213–221

29. Mackey S, Greely HT, Martucci KT (2019) Neuroimaging-based pain biomarkers: definitions, clinical and research applications, and evaluation frameworks to achieve personalized pain medicine. Pain Rep 4(4):e762

30. Grundy SM, Balady GJ, Criqui MH, Fletcher G, Greenland P, Hiratzka LF, Houston-Miller N, Kris-Etherton P, Krumholz HM, LaRosa J, Ockene IS, Pearson TA, Reed J, Washington R, Smith SC Jr (1998) Primary prevention of coronary heart disease: guidance from Framingham: a statement for healthcare professionals from the AHA task force on risk reduction. American Heart AssociationCirculation 97:1876–1887

31. Bagby RM, Ryder AG, Cristi C (2002) Psychosocial and clinical predictors of response to pharmacotherapy for depression. J Psychiatry Neurosci 27:250–257

32. Parr LA (2001) Cognitive and physiological markers of emotional awareness in chimpanzees (Pan troglodytes). Anim Cogn 4: 223–229

33. Hung L, Wu H, Hsieh K, Lee G (2014) Microfluidic platforms for discovery and detection of molecular biomarkers. Microfluid Nanofluid 16:941–963

34. Pourmohammadi S, Maleki A (2020) Stress detection using ECG and EMG signals: a comprehensive study. Comput Methods Prog Biomed 193:105482

35. Holzschneider K, Mulert C (2011) Neuroimaging in anxiety disorders. Dialogues Clin Neurosci 13:453–461

36. Khoury R, Nasrallah HA (2018) Inflammatory biomarkers in individuals at clinical high risk for psychosis (CHR-P): state or trait? Schizophr Res 199:31–38

37. Carrillo-Roa T, Labermaier C, Weber P, Herzog DP, Lareau C, Santarelli S, Wagner KV, Rex-Haffner M, Harbich D, Scharf SH, Nemeroff CB, Dunlop BW, Craighead WE, Mayberg HS, Schmidt MV, Uhr M, Holsboer F, Sillaber I, Binder EB, Müller MB (2017) Common genes associated with antidepressant response in mouse and man identify key role of glucocorticoid receptor sensitivity. PLoS Biol 15(12):e2002690

38. Kling MA, Coleman VH, Schulkin J (2009) Glucocorticoid inhibition in the treatment of depression: can we think outside the endocrine hypothalamus? Depress Anxiety 26: 641–649

39. Belzeaux R, Gorgievski V, Fiori LM, Lopez JP, Grenier J, Lin R, Nagy C, Ibrahim EC, Gascon E, Courtet P, Richard-Devantoy S, Berlim M, Chachamovich E, Théroux JF, Dumas S, Giros B, Rotzinger S, Soares CN, Foster JA, Mechawar N, Tall GG, Tzavara ET, Kennedy SH (2020) Turecki G.GPR56/ADGRG1 is associated with response to antidepressant treatment. Nat Commun 11(1): 1635

40. Salzman GS, Ackerman SD, Ding C et al (2016) Structural basis for regulation of GPR56/ADGRG1 by its alternatively spliced extracellular domains. Neuron 91: 1292–1304

41. Frantzi M, Bhat A, Latosinska A (2014) Clinical proteomic biomarkers: relevant issues on study design and technical considerations in biomarker development. Clin Transl Med 3:7

42. Strimbu K, Tavel JA (2010) What are biomarkers? Curr Opin HIV AIDS 5:463–466

43. Berk M, Williams LJ, Jacka FN, O'Neil A, Pasco JA, Moylan S, Allen NB, Stuart AL, Hayley AC, Byrne ML, Maes M (2013) So, depression is an inflammatory disease, but where does the inflammation come from? BMC Med 11:200

44. Liu YZ, Wang YX, Jiang CL (2017) Inflammation: the common pathway of stress-related diseases. Front Hum Neurosci 11:316

45. De Bellis MD, Zisk A (2014) The biological effects of childhood trauma. Child Adolesc Psychiatr Clin N Am 23:185–vii

46. Sherin JE, Nemeroff CB (2011) Posttraumatic stress disorder: the neurobiological impact of psychological trauma. Dialogues Clin Neurosci 13:263–278

47. Lam RW, Kennedy SH, Mclntyre RS, Khullar A (2014) Cognitive dysfunction in major depressive disorder: effects on psychosocial functioning and implications for treatment. Can J Psychiatr 59:649–654

48. Krogh J, Benros ME, Jørgensen MB, Vesterager L, Elfving B, Nordentoft M (2014) The association between depressive symptoms, cognitive function, and inflammation in major depression. Brain Behav Immun 35:70–76

49. Rainville JR, Hodes GE (2019) Inflaming sex differences in mood disorders. Neuropsychopharmacology 44:184–199

50. Capuron L, Lasselin J, Castanon N (2017) Role of adiposity-driven inflammation in depressive morbidity. Neuropsychopharmacology 42:115–128

51. Shelton RC, Miller AH (2011) Inflammation in depression: is adiposity a cause? Dialogues Clin Neurosci 13:41–53

52. Kany S, Vollrath JT, Relja B (2019) Cytokines in inflammatory disease. Int J Mol Sci 20: 6008

53. Felger JC, Lotrich FE (2013) Inflammatory cytokines in depression: neurobiological mechanisms and therapeutic implications. Neuroscience 246:199–229

54. Anisman H (2002) Stress, immunity, cytokines and depression. Acta Neuropsychiatrica 14:251–261

55. Chabry J, Nicolas S, Cazareth J, Murris E, Guyon A, Glaichenhaus N, Heurteaux C, Petit-Paitel A (2015) Enriched environment decreases microglia and brain macrophages inflammatory phenotypes through adiponectin-dependent mechanisms: relevance to depressive-like behavior. Brain Behav Immun 50:275–287

56. Nicolas S, Debayle D, Béchade C, Maroteaux L, Gay AS, Bayer P, Heurteaux C, Guyon A, Chabry J (2018) Adiporon, an adiponectin receptor agonist acts as an antidepressant and metabolic regulator in a mouse model of depression. Transl Psychiatry 16(8):159

57. Zettermark S, Perez Vicente R, Merlo J (2018) Hormonal contraception increases the risk of psychotropic drug use in adolescent girls but not in adults: a pharmacoepidemiological study on 800 000 Swedish women. PLoS One 13(3):e0194773

58. Adzic M, Brkic Z, Mitic M, Francija E, Jovicic MJ, Radulovic J, Maric NP (2018) Therapeutic strategies for treatment of inflammation-

related depression. Curr Neuropharmacol 16: 176–209

59. Szałach ŁP, Lisowska KA, Cubała WJ (2019) The influence of antidepressants on the immune system. Arch Immunol Ther Exp 67:143–151

60. Davis KD, Aghaeepour N, Ahn AH, Angst MS, Borsook D, Brenton A, Burczynski ME, Crean C, Edwards R, Gaudilliere B, Hergenroeder GW, Iadarola MJ, Iyengar S, Jiang Y, Kong JT, Mackey S, Saab CY, Sang CN, Scholz J, Segerdahl M, Tracey I, Veasley C, Wang J, Wager TD, Wasan AD, Pelleymounter MA (2020) Discovery and validation of biomarkers to aid the development of safe and effective pain therapeutics: challenges and opportunities. Nat Rev Neurol 16: 381–400

61. Brand SJ, Moller M, Harvey BH (2015) A review of biomarkers in mood and psychotic disorders: a dissection of clinical vs preclinical correlates. Curr Neuropharmacol 13: 324–368

62. Juruena MF, Bocharova M, Agustini B, Young AH (2018) Atypical depression and non-atypical depression: is HPA axis function a biomarker? A systematic review. J Affect Disord 233:45–67

63. Penninx BW, Milaneschi Y, Lamers F, Vogelzangs N (2013) Understanding the somatic consequences of depression: biological mechanisms and the role of depression symptom profile. BMC Med 11:129

64. Phillips ML, Kupfer DJ (2013) Bipolar disorder diagnosis: challenges and future directions. Lancet 381:1663–1671

65. Singh T, Rajput M (2006) Misdiagnosis of bipolar disorder. Psychiatry (Edgmont) 3(10):57–63

66. Cuellar AK, Johnson SL, Winters R (2005) Distinctions between bipolar and unipolar depression. Clin Psychol Rev 25:307–339

67. Geddes JR, Miklowitz DJ (2013) Treatment of bipolar disorder. Lancet 381:1672–1682

68. Bourin M (2017) Are antidepressants useful in bipolar disease? Arch Depress Anxiety 3(2): 058–059

69. Nusslock R, Frank E (2011) Subthreshold bipolarity: diagnostic issues and challenges. Bipolar Disord 13:587–603

70. Bystritsky A, Khalsa SS, Cameron ME, Schiffman J (2013) Current diagnosis and treatment of anxiety disorders. P T 38:30–57

71. Nestler EJ, Hyman SE (2010) Animal models of neuropsychiatric disorders. Nat Neurosci 13:1161–1169

72. Muneer A (2020) The discovery of clinically applicable biomarkers for bipolar disorder: a review of candidate and proteomic approaches. Chonnam Med J 56:166–179

73. Heyes S, Pratt WS, Rees E et al (2015) Genetic disruption of voltage-gated calcium channels in psychiatric and neurological disorders. Prog Neurobiol 134:36–54

74. Heidari F, Rabizadeh S, Mansournia MA, Mirmiranpoor H, Salehi SS, Akhavan S, Esteghamati A, Nakhjavani M (2019) Inflammatory, oxidative stress and anti-oxidative markers in patients with endometrial carcinoma and diabetes. Cytokine 120:186–190

75. Hughes JP, Rees S, Kalindjian SB, Philpott KL (2011) Principles of early drug discovery. Br J Pharmacol 162:1239–1249

76. Johnson CH, Ivanisevic J, Siuzdak G (2016) Metabolomics: beyond biomarkers and towards mechanisms. Nat Rev Mol Cell Biol 17:451–459

77. Hasin Y, Seldin M, Lusis A (2017) Multiomics approaches to disease. Genome Biol 18(1):83

78. Gowda GA, Zhang S, Gu H, Asiago V, Shanaiah N, Raftery D (2008) Metabolomics-based methods for early disease diagnostics. Expert Rev Mol Diagn 8: 617–633

79. McDermott JE, Wang J, Mitchell H et al (2013) Challenges in biomarker discovery: combining expert insights with statistical analysis of complex omics data. Expert Opin Med Diagn 7:37–51

80. Dipnall JF, Pasco JA, Berk M, Williams LJ, Dodd S, Jacka FN, Meyer D. Fusing data mining, machine learning and traditional statistics to detect biomarkers associated with depression. PLoS One. 2016 5;11(2): e0148195

81. Leuchter AF, Hunter AM, Krantz DE, Cook IA (2014) Intermediate phenotypes and biomarkers of treatment outcome in major depressive disorder. Dialogues Clin Neurosci 16:525–537

82. Leighton SP, Nerurkar L, Krishnadas R, Johnman C, Graham GJ, Cavanagh J (2018) Chemokines in depression in health and in inflammatory illness: a systematic review and meta-analysis. Mol Psychiatry 23:48–58

83. Lynch CJ, Gunning FM, Liston C (2020) Causes and consequences of diagnostic heterogeneity in depression: paths to discovering novel biological depression subtypes. Biol Psychiatry 88:83–94

84. Kirsch I (2014) Antidepressants and the placebo effect. Z Psychol 222:128–134

85. Perlman K, Benrimoh D, Israel S, Rollins C, Brown E, Tunteng JF, You R, You E, Tanguay-Sela M, Snook E, Miresco M, Berlim MT. A systematic meta-review of predictors of antidepressant treatment outcome in major depressive disorder. J Affect Disord. 2019 15;243:503–515

86. Unützer J, Park M (2012) Strategies to improve the management of depression in primary care. Prim Care 39:415–431

87. Voineskos D, Daskalakis ZJ, Blumberger DM (2020) Management of Treatment-Resistant Depression: challenges and strategies. Neuropsychiatr Dis Treat 16:221–234

88. Brymer KJ, Romay-Tallon R, Allen J, Caruncho HJ, Kalynchuk LE (2019) Exploring the potential antidepressant mechanisms of TNFα antagonists. Front Neurosis 11(13):98

89. Kraus C, Kadriu B, Lanzenberger R, Zarate CA Jr, Kasper S (2019) Prognosis and improved outcomes in major depression: a review. Transl Psychiatry 3(9):127

90. Kohler O, Krogh J, Mors O, Benros ME (2016) Inflammation in depression and the potential for anti-inflammatory treatment. Curr Neuropharmacol 14:732–742

91. Sigalas PD, Garg H, Watson S, McAllister-Williams RH, Ferrier IN (2012) Metyrapone in treatment-resistant depression. Ther Adv Psychopharmacol. 2:139–149

92. Ninomiya EM, Martynhak BJ, Zanoveli JM, Correia D, da Cunha C, Andreatini R (2010) Spironolactone and low-dose dexamethasone enhance extinction of contextual fear conditioning. Prog Neuro-Psychopharmacol Biol Psychiatry 34:1229–1235

93. Mathews DC, Henter ID, Zarate CA (2012) Targeting the glutamatergic system to treat major depressive disorder: rationale and progress to date. Drugs 72:1313–1333

94. Levant B (2013) N-3 (omega-3) polyunsaturated fatty acids in the pathophysiology and treatment of depression: pre-clinical evidence. CNS Neurol Disord Drug Targets 12:450–459

95. Kopschina Feltes P, Doorduin J, Klein HC, Juárez-Orozco LE, Dierckx RA, Moriguchi-Jeckel CM, de Vries EF (2017) Anti-inflammatory treatment for major depressive disorder: implications for patients with an elevated immune profile and non-responders to standard antidepressant therapy. J Psychopharmacol 31:1149–1165

96. Nassar A, Azab AN (2014) Effects of lithium on inflammation. ACS Chem Neurosci 18(5):451–458

97. Menke A, Binder EB (2014) Epigenetic alterations in depression and antidepressant treatment. Dialogues Clin Neurosci 16:395–404

98. Söderlund J, Lindskog M (2018) Relevance of rodent models of depression in clinical practice: can we overcome the obstacles in translational neuropsychiatry? Int J Neuropsychopharmacol 21:668–676

99. Planchez B, Surget A, Belzung C (2019) Animal models of major depression: drawbacks and challenges. J Neural Transm 126:1383–1408

100. Schneiderman N, Ironson G, Siegel SD (2005) Stress and health: psychological, behavioral, and biological determinants. Annu Rev Clin Psychol 1:607–628

101. Surget A, Tanti A, Leonardo ED, Laugeray A, Rainer Q, Touma C, Palme R, Griebel G, Ibarguen-Vargas Y, Hen R, Belzung C (2011) Antidepressants recruit new neurons to improve stress response regulation. Mol Psychiatry 16:1177–1188

102. Bourin M (2018) Post-stroke depression and changes in behavior and personality. Arch Depress Anxiety 4(1):031–033

103. Elzib H, Pawloski J, Ding Y, Asmaro K (2019) Antidepressant pharmacotherapy and poststroke motor rehabilitation: a review of neurophysiologic mechanisms and clinical relevance. Brain Circ 5:62–67

104. Kaltenboeck A, Harmer C (2018 Oct 8) The neuroscience of depressive disorders: a brief review of the past and some considerations about the future. Brain Neurosci Adv 2:2398212818799269

105. Bourin M, Fiocco AJ, Clenet F (2001) How valuable are animal models in defining antidepressant activity? Hum Psychopharmacol 16:9–21

106. David DJP, Nic Dhonnchadha B.A., Jolliet P., Hascoet M., Bourin M. The use of animal models in defining antidepressant response. Brain Pharmacol, 2001, 1, 11–35

107. Gardier AM, Bourin M (2001) Appropriate use of "knockout" mice as models of depression or models of testing the efficacy of antidepressants. Psychopharmacology 153:393–394

108. Stepanichev M, Dygalo NN, Grigoryan G, Shishkina GT, Gulyaeva N (2014) Rodent models of depression: neurotrophic and neuroinflammatory biomarkers. Biomed Res Int 2014:932757

109. Abelaira HM, Réus GZ, Quevedo J (2013) Animal models as tools to study the

pathophysiology of depression. Braz J Psychiatry 35(Suppl 2):S112–S120

110. Reus GZ, Abelaira HM, Leffa DD, Quevedo J (2014) Cognitive dysfunction in depression: lessons learned from animal models. CNS Neurol Disord Drug Targets 13:1860–1870

111. Bourin M (2018) In: Kim YK (ed) The use of animal models in defining antidepressant response: a translational approach in understanding depression. Springer Nature, Singapore, pp 233–242

112. Hillhouse TM, Porter JH (2015) A brief history of the development of antidepressant drugs: from monoamines to glutamate. Exp Clin Psychopharmacol 23:1–21

113. Langova V, Vales K, Horka P, Horacek J (2020) The role of zebrafish and laboratory rodents in schizophrenia research. Front Psych 27(11):703

114. Pantoja C, Hoagland A, Carroll EC, Karalis V, Conner A, Isacoff EY Neuromodulatory regulation of Behavioral individuality in zebrafish. Neuron. 2016 3; 91:587–601

115. Feyissa DD, Aher YD, Engidawork E, Höger H, Lubec G, Korz V. Individual differences in male rats in a Behavioral test battery: a multivariate statistical approach. Front Behav Neurosci. 2017 17; 11:26

116. Krishnan V, Han MH, Graham DL, Berton O, Renthal W, Russo SJ, Laplant Q, Graham A, Lutter M, Lagace DC, Ghose S, Reister R, Tannous P, Green TA, Neve RL, Chakravarty S, Kumar A, Eisch AJ, Self DW, Lee FS, Tamminga CA, Cooper DC, Gershenfeld HK, Nestler EJ. Molecular adaptations underlying susceptibility and resistance to social defeat in brain reward regions. Cell. 2007 19; 131:391–404

117. Italia M, Forastieri C, Longaretti A, Battaglioli E, Rusconi F (2020) Rationale, relevance, and limits of stress-induced psychopathology in rodents as models for psychiatry research: an introductory overview. Int J Mol Sci 21:7455

118. Glover ME, Pugh PC, Jackson NL et al (2015) Early-life exposure to the SSRI paroxetine exacerbates depression-like behavior in anxiety/depression-prone rats. Neuroscience 284:775–797

119. Cheng RK, Jesuthasan SJ, Penney TB (2014) Zebrafish forebrain and temporal conditioning. Philos Trans R Soc Lond Ser B Biol Sci 369(1637):20120462

120. Stewart AM, Braubach O, Spitsbergen J, Gerlai R, Kalueff AV (2014) Zebrafish models for translational neuroscience research: from tank to bedside. Trends Neurosci 37: 264–278

121. Mekiri M, Gardier AM, David DJ, Guilloux JP (2017) Chronic corticosterone administration effects on behavioral emotionality in female c57bl6 mice. Exp Clin Psychopharmacol 25:94–104

Part II

Behavioral Research Methods for Major Depression

Chapter 5

Translational Strategies for Developing Biomarkers for Major Depression: Lessons Learned from Animal Models

Feyza Aricioglu and Brian E. Leonard

Abstract

This review critically assesses the types of rodent models which have been developed in the search for new antidepressants. In order to determine the validity upon which the models are based, a brief discussion of the biological, primarily biochemical, markers in the blood of depressed patients is presented. The rodent models that have been developed are mainly based on simulating the behavioral changes, particularly stress-induced changes, seen in depressed patients. Despite the numerous different acute behavioral tests and more complex chronic models which have been developed, advances in the discovery of novel types of antidepressant have been disappointing. Possible reasons for this are ascribed to the mechanistic interpretation and unnatural environment that rodents are placed together with the restrictive application of compounds developed for their presumed selectivity for the biogenic amine neurotransmitters. It is proposed that a new generation of rodent models based on chronic drug administration to rodents in biologically meaningful environmental situations is essential if this area of drug development is to be successful.

Key words Clinical biomarkers, Rodent models, Acute and chronic models, Model limitations

1 Is the Search for Animal Models of Major Depression a Lost Cause?

Serendipity played a major role in the discovery of the first clinically useful psychotropic drugs that revolutionized the treatment of major depression, schizophrenia, and anxiety disorders. The success of these discoveries was ascribed to their effects on brain biogenic amines, the monoamine neurotransmitters which had been identified at that time. Retrospectively, it is now realized that this was an oversimplification of the situation and has contributed to the lack of success in the development of any novel psychotropic drugs in the past three decades. It is becoming increasingly recognized that major psychiatric disorders are not due to a single deficit in brain monoamines but more likely to result from the interactions between the genetic predisposition and adverse environmental factors, such as stress [1]. For example, major depressive disorder

Yong-Ku Kim and Meysam Amidfar (eds.), *Translational Research Methods for Major Depressive Disorder*, Neuromethods, vol. 179, https://doi.org/10.1007/978-1-0716-2083-0_5,

(MDD) can arise from traumatic environmental experience(s) and/ or negative life events as the important risk factors that trigger MDD particularly in those who are genetically prone to depression.

It is against this background that the problems of developing relevant animal models of MDD, or indeed any other model of major psychiatric disorder, must be considered. As the pharmaceutical industry has always led the development of psychotropic drugs, it is understandable that valid animal models are essential for the development of better psychotropic drugs. However, as few novel psychotropic drugs have been development in recent years, investment has switched from psychotropic drug development to more immediately successful and lucrative targets [2]. This presentation will consider which animal models have been developed and critically assess their limitations. For practical reasons, rodents are the animals of choice for models of MDD, and therefore this presentation will be restricted to rodent models.

The first part of this presentation will provide a summary of the clinical biochemical markers, which may give an insight into the pathophysiology of depression.

2   Are There Any Meaningful Biomarkers for Depression?

MDD is a prevalent, heterogeneous disorder that is characterized by depressed mood, anhedonia, and, frequently, defective cognitive function. Despite the worldwide occurrence of MDD, with a lifetime prevalence of 17% [3], the biological basis for the heterogeneity of the disorder is poorly defined, and its effective treatment is still limited. Indeed, it has been estimated that only 50% of patients adequately respond to antidepressant drug treatment. Thus, in order to improve the targeting of treatments and to identify subtypes of the disorder which may respond to specific treatments, the development of biomarkers for the accurate diagnosis and treatment of the disorder is important, thereby enabling MDD patients to be separated into clinically homogeneous populations. To date, the search for qualitative and quantitative tests for MDD and its treatment has been unsuccessful [4, 5].

The importance of discovering relevant biomarkers is widely recognized. Multiple dysregulated neuronal circuits result in multiple changes in neurotransmitter, immune, endocrine, genetic, and metabolic processes, leading to the conclusion that no single biomarker is likely. So how can a biomarker be defined? There are many definitions, but the one which encapsulates the concept both accurately and simply is that suggested by the Clinical Pharmacology and Therapeutics Working Group (2001): "A property that can be both objectively measured and considered as an indicator of a normal biological process or a response to a therapeutic intervention."

Biomarkers are divided into diagnostic markers, which identify the presence or absence of the disorder, and treatment biomarkers, which are intended to predict the response to the treatment administered. For practical purposes, biomarkers should have a selectivity and specificity of at least 80%. Three main groups have been recognized:

1. Trait markers persist throughout the disorder.

2. State markers are transient and reflect the presence of the disorder.

3. Endophenotype markers which recognize the genetic basis of the disorder in the individual and family relatives.

Despite the increase in knowledge of experimental and clinical neuroscience of MDD, no biomarkers have been found that fulfill these criteria. There are many reasons for this failure, but the most important is the diverse nature and clinical classification of MDD. The aim of the following presentation is to discuss some of the attempts being made to improve the success in discovering relevant biomarkers. As a caveat, it must be stressed that MDD seldom exists without other psychiatric and medical comorbidities which complicate the clinical picture of the disorder. In addition to the differences in genotypic and phenotypic expression and the complexity of MDD, this presentation will be confined to blood biomarkers which are most likely to be of practical value. Wider and detailed discussions of the pathophysiology of MDD and the identification of potential biomarkers have been published [5–8]. A comprehensive list of putative biomarkers is shown in Tables 1 and 2 at the end of this section of the review.

2.1 Potential Preclinical Biomarkers for MDD

It is likely that if there had been a suitable animal model of MDD, preferably a rodent model, the search for a useful biomarker would have been achieved. But, after decades of research into suitable models, the success has been limited. Even if it was possible to develop a model which demonstrated qualitatively relevant behavioral and biological changes (neurochemical, endocrine, immune, pathophysiological changes in organ function, etc.) to those seen in MDD patients, the rodent phenotype could not possibly simulate the distinctly human characteristics of the disorder. Feelings of guilt, suicidal thoughts, psychological states of happiness, euphoria, pleasure, and remorse are uniquely human and have evolved into a structurally complex brain over millions of years. A more detailed discussion of animal models and there application in the search for biomarkers will be considered in another section of this chapter. In summary, many rodent models involve subjecting the animal to external stressors in an attempt to replicate the effects of stress seen in MDD patients. The chronic mild stress model is frequently used to demonstrate anhedonia and cognitive dysfunction, which is

Table 1
Biomarkers of depression (modified from ref. [9])

Biomarker	Sub-marker	Changes in biomarker	Reference

Neurotransmitters	Glutamate
	Gamma-aminobutyric acid
	  (GABA)
	Serotonin (5-HT)
	Noradrenaline (NA)
	Dopamine (DA)
	Metabolites of 5-HT
	  (5-HIAA) and NA
	  (MHPG)
		↑Glutamate
		↓NA
		↓DA
		↓5-HIAA
		↓MHPG
			[9–20]

Neuroendocrine	Cortisol
	Dexamethasone suppression
	  test (DST)
	Thyroid hormones
		Dysregulation in circadian/
		  sleep-wake cycles
		↑Saliva cortisol
		↑HPA-axis activation
		Thyroid dysfunction
		↑TSH
			[9, 20–27]

Metabolic	High-density lipoprotein
	  (HDL)
	Low-density lipoprotein
	  (LDL)
	Polyunsaturated fatty acids
	  (PUFAs)
	Sphingomyelin
	Adipokines leptin and ghrelin
	Insulin
		Abnormal PUFAs
		↓ HDL
		↑ LDL
		↓ Adipokine
		↑, ↓, or ↔ leptin and ghrelin
		↑ Insulin
			[28–32]

Kynurenine pathway	Kynurenic acid (KYNA) 3-hydroxykynurenine (3-HK) Quinolinic acid (QA)	↓Tryptophan ↑Kynurenine ↑QA ↓KYNA	[9, 20, 33–36]

Oxidative and
nitrosative
stress
	Malondialdehyde (MDA)
	8-hydroxy-2-deoxiguanosine
	  (8-OHdG)
	Total antioxidant capacity
	  (TAC)
	Reactive oxygen/nitrogen
	  species (ROS/RNS)
	Superoxide dismutase
	  (SOD)
	Catalase (CAT)
	Myeloperoxidase (MPO)
	Antibodies against
	  neopitopes
		↑SOD
		Lipid peroxidation
		↓SOD and catalase
			[9, 20, 37, 38]

Inflammatory/ immune factors	Interleukin-1 (IL-1) IL-2 IL-4 IL-6 IL-8 IL-10 Tumor necrosis factor α (TNF-α) Interferon-γ (INF-γ)	↑IL-1 ↑IL-6 ↑TNF-α ↑IFN-γ	[9, 20, 33, 39–41]

(continued)

Table 1
(continued)

Biomarker	Sub-marker	Changes in biomarker	Reference
Growth factors	Brain-derived neurotrophic factor (BDNF) fibroblast (FGF-2) Insulin-like growth factor-1 (IGF-1) Nerve growth factor (NGF) Growth hormone (GH) Vascular endothelial growth factor (VEGF) Glial cell line-derived neurotrophic factor (GDNF)	↓BDNF ↑BDNF ↓IGF-1 ↑VEGF ↓ or ↔ VEGF	[9, 20, 42–49]
Neuroanatomy neurocircuitry	Volume changes, blood flow changes Glucose metabolism changes	↓ Volume in hippocampal, prefrontal cortex, orbitofrontal cortex, and basal ganglia ↑Blood flow in the amygdala, orbital cortex, and medial thalamus ↓Blood flow in the prefrontal cortex and anterior cingulate cortex ↑glucose metabolism in the amygdala, orbital cortex, and medial thalamus ↓NAA in frontal cortex and Subcortical regions	[9, 20, 50–56]
Intracellular signaling molecules	cAMP MAPK/ERK Glycogen synthase kinase-3 (GSK-3) Adenosine triphosphate (ATP)	↓ cAMP and MAPK/ERK pathway activity ↑expression of MKP GSK-3 ATP	[9, 57–61]
Genetic markers		Polymorphisms in 5-HT Transporter, 5-HT receptor-2A, BDNF, and tryptophan hydroxylase Val/Met polymorphism ↓Neuropeptide Y expression	[9, 20, 62–67]
Proteomic markers	Insulin secretion Glyoxalase-1 Dihydropyrimidinase-related protein-2 leptin IL-1, BDNF Micro-RNAs (miRNAs)	Abnormal insulin secretion ↓Glyoxalase-1 and dihydropyrimidinase-related protein-2 ↑Leptin, IL-1, BDNF proteins 28 miRNAs upregulated and 2 miRNAs downregulated in Treatment	[9, 20, 41, 68–71]

Table 2
Experimental depression models symptoms and biological markers

Model	Suitable strain	Symptoms	Biomarkers	Reference
Chronic unpredictable mild stress	Sprague-Dawley Wistar-Kyoto C57BL/6 (J) and ICR Mice Demonstrates good predictive validity	Induces changes in HPA axis despair, anhedonia, apathy, sleep disturbance, psychomotor abnormalities, body weight loss. Reduced sucrose preference, immobility in FST/TST, decreased grooming. Induces both anxiety- and depression-like behaviors in FST, EPM, and sucrose preference tests	Neurotrophin alterations: • Prefrontal cortex and hippocampus and hypothalamus (BDNF ↓). • Amygdala (BDNF and VEGF unchanged). • Hippocampus (VEGF unchanged) inflammation: • Blood (IL-1 ↑, TNF- ↑, IL-6 ↑). • Prefrontal cortex (IFN- ↑, TNF-. ↑, IL-6 ↑, L-1 ↑, TNF- ↑, IL-10 ↓) • Hippocampus (IL-1 ↑, TNF- ↑, IL-18 ↑, IL-4 ↑ TGF-. ↓ IL-6 mRNA ↑) Hypercortisolemia, cortico-limbic alterations, neurotransmission Alterations	[72–77]
Learned helplessness	Rats and mice Holtzman CLH zebrafish Demonstrates good predictive validity	Induces changes in HPA axis anhedonia, despair, social withdrawal, eating behavior abnormalities. Failures to escape from escapable, signaled shock. Induces anxiety- and depression-like behaviors with good similarity to the symptoms of depression with cognitive and neuroendocrine Impairments	Neurotrophin alterations: • Prefrontal cortex and hippocampus (BDNF ↓ or no changes). • Hippocampus (BDNF mRNA ↓, NGF ↑). Hypercortisolemia Increased 5-HT neurotransmission and 5-HT1A receptor desensitization	[73–77]
Early-life stress 1. Early maternal separation. 2. Maternal deprivation.	Sprague-Dawley Demonstrates good predictive validity	Induces changes in HPA axis induces long-lasting behavioral changes until adulthood. Despair but not anhedonia, sleep disturbances, psychomotor abnormalities. Induces learning and memory	Neurotrophin alterations: • Hippocampus (BDNF ↑, NGF ↑, NT-3 ↑). • Cerebellum (IGF-1 and IGF-1R mRNAs ↑). Inflammation:	[75–79]

(continued)

Table 2
(continued)

Model	Suitable strain	Symptoms	Biomarkers	Reference
		deficits; depressive- and anxious-like behavior in open field and EPM Suitable for studying the interaction between genes and environment in a newborn animal	• Blood (IL-1 ↑, IL-6 ↑). • Brain (chemokine ligand 7 ↓, chemokine receptor 4 ↓, IL-10. ↓, IL-1 ↓, IL-5 ↓) Hypercortisolemia, Increased activity in the amygdala, altered noradrenergic neurotransmission	
Prenatal stress	Demonstrates good predictive validity	Induces changes in HPA axis induces depression- and anxiety- like phenotype: Prolonged immobility (despair-like behavior) in FST and TST tests, anhedonia- reduced preference to sucrose solution and anxiety in EPM Useful in the research and elucidation of the epigenetic mechanisms underlying the consequences of gestational stress or early-life stress to later depression and anxiety	Neurotrophin alterations: • Amygdala and hippocampus (BDNF/BDNF expression ↓). • Hippocampus (methylation of BDNF exon IV ↑ m-BDNF/pro-BDNF ↓). Inflammation: Hippocampus (IL-1 mRNA ↑ IL-10 ↓).	[76, 77]
Social isolation	Sprague-Dawley Demonstrates good predictive validity	Induces changes in HPA axis Social withdrawal, anhedonia, body weight loss, eating behavior abnormalities, psychomotor abnormalities Induces both anxiety- and depression-like behaviors in FST, TST EPM, and sucrose Preference test	Neurotrophin alterations, • Hippocampus (BDNF mRNA ↓) İnflammation. • Blood (IL-6 ↓, IL-4 ↓, TNF- ↑, IFN ↑).	[76, 77, 80, 81]
Social deficit	Sprague-Dawley Wistar-Kyoto Demonstrates good predictive validity	Social avoidance, reduced center time in open field, immobility in FST/TST, reduced sucrose preference, circadian and metabolic changes Induces both anxiety- and depression-like behaviors in OF, EPM, sucrose	Neurotrophin alterations, • Hippocampus (BDNF ↓, BDNF mRNA ↓, NGF ↑, FGF2, FGFR1 mRNAs ↓). • Nucleus accumbens (BDNF ↑). • Blood (NGF ↑).	[76, 77, 81, 82]

(continued)

Table 2
(continued)

Model	Suitable strain	Symptoms	Biomarkers	Reference
		preference, social interaction tests, and memory dysfunction Induces changes in altered function of the noradrenergic, serotonergic, cholinergic, GABAergic, and glutamatergic neurotransmitter systems	Inflammation • Hippocampus (IL-6 ↑). • Brain (IL-6 mRNA ↑, IL-1 unchanged, TNF- mRNAs unchanged).	
		Useful in investigation of the chronic psychomotor agitated depression		
Olfactory bulbectomy	Sprague-Dawley Demonstrates good predictive validity	Transient anhedonia, despair	Changes in cortico-limbic circuits, hippocampus (BDNF ↓, BDNF mRNA ↑, NGF mRNA ↓) Inflammation Blood and brain (IL-1 ↑, TNF- ↑) increased corticosterone, changes In neurotransmission	[76, 77, 83, 84]

usually attenuated by chronic antidepressant treatment. The direct activation of the immune system by the administration of pro-inflammatory cytokines has been used as a relevant acute model of depression, as this results in changes in the immune, endocrine, and central neurotransmitters, which are accompanied by behavioral changes. These, and related models, are discussed in details elsewhere [85–87].

2.2 A Brief Survey of the Potential Monoamine Biomarkers of MDD

Monoamine biomarkers have been a major interest for several decades following the widespread acceptance of the biogenic amine theory of depression in the 1950s and 1960s [88]. While monoamine targets have formed the foundation of pharmacotherapy for MDD since the 1990s, monoamine targets have failed to provide a breakthrough in the discovery of novel treatments. Indeed, the most promising novel targets now center on immune and endocrine changes, which are a common feature of MDD, bipolar disorder and schizophrenia. These will be considered later in this presentation.

With regard to the search for potential biomarkers for MDD based on the genetic abnormalities of the monoamines, attention has been focused on the changes in the synthesis, transport, and metabolism of noradrenaline and serotonin. For example, the serotonin transporter 5HTTLPR and the postsynaptic receptor 5HTR2A are of considerable interest because the genes encoding for these factors have several allelic forms. Thus, the gene for the serotonin transporter has two main alleles of interest, long and short forms, which modulate serotonin transport and have possible predictive value in identifying those patients most likely to respond to selective serotonin reuptake inhibitor (SSRI) type of antidepressants [89]; patients with the l-allele were more likely to respond to treatment than those with the s-allele. As a caution, it must be added that these findings were based on a Caucasian population where the frequency of the l-allele was up to 43%, whereas in an Asian population, the frequency is less than 12% and the s-allele is predominant [90]. However, there is no clinical evidence that the response of Asian patients with MDD is any less responsive to SSRIs than Caucasian patients!

The results for the biomarker studies of the 5HT2A gene in the antidepressant response are more substantial. Thus, the Munich Antidepressant Response Signature study established that the gene marker predicted the response of patients to citalopram, a result that has been widely replicated by more than 20 other research groups for citalopram and for other antidepressants [8]. While this research looks promising, it is important to emphasize the limitation in relying on the genetic basis of serotonin transmission to define useful markers for treatment response. The genes involved in catecholamine metabolism have also been investigated. Thus, the monoamine oxidase A and catechol-O-methyltransferase are of particular interest: the two isoforms of tryptophan hydroxylase have also received attention. While there is some evidence that the genes for these enzymes are changed in some patients with MDD, there is no convincing evidence that effective antidepressant treatment normalizes the aberrant activity [8]. A similar conclusion has been drawn for genetic studies of the ionotropic glutamate kainate 4 receptor gene.

2.3 Growth Factor Biomarkers

Brain-derived neurotrophic factor (BDNF) has been the most widely studied growth factor, particularly in depression, because of its vital role in neuroplasticity, cell migration, and cell survival. BDNF is not only released from neurons but also from endothelial cells, platelets, and leucocytes. There is evidence that BDNF is decreased in the serum of patients with MDD and that it plays a role in ameliorating the response of the brain to stress, an effect which could be linked to its involvement in neuronal repair and synaptogenesis [91, 92] This is plausible as stress and depression are known to decrease BDNF expression and reduce neurogenesis in the hippocampus [93]. These changes have been associated with the activation of glucocorticoid release which reduced BDNF gene expression [94]. The reduction in serum BDNF is reported to be reversed by effective antidepressant treatment [91]. While these results suggest that BDNF could act as a useful biomarker for MDD, it is still unclear to what extent serum BDNF reflects changes in the CSF and brain. Furthermore, changes in serum BDNF are not unique for MDD, and it is still uncertain if the changes are a cause or a consequence of the disorder.

There are many other growth factors which are being explored both for their physiological functions and for their connection to the pathophysiology of depression (e.g., vascular endothelial growth factor (VEGF)), but so far there is no evidence that they are convincing biomarkers of MDD.

2.4 Cytokines as Biomarkers

One of the first publications to consider the importance of immune dysregulation in MDD was by Smith in his publication on the macrophage hypothesis of depression [95]. Both the activation (increase in pro-inflammatory mediators) and suppression (reduction in natural killer cells and in lymphocyte proliferation) occur in MDD. Unlike the acute changes in the inflammatory response following stress, the chronic changes seen in MDD are maladaptive and contribute to the symptoms of the disorder [96]. The causal relationship between chronic inflammation and MDD is provided by the results of clinical studies, which demonstrate the link between the rise in serum pro-inflammatory cytokines, such as the IL-1, IL-6, TNF-alpha, IFN-gamma, and CRP, and the cardinal symptoms of the disorder. These inflammatory changes are mainly reversed by effective antidepressant treatment [33, 97]. The changes in neuronal function in the brain, which are caused by the pro-inflammatory cytokines, contribute to neurodegenerative changes associated with the long-term effects of MDD. These neurodegenerative changes are caused by an increase in neuronal excitotoxicity combined with a reduction in BDNF and related neuroprotective factors [98, 99]. Such changes can be partly explained by the activation of the pro-inflammatory cytokines of the tryptophan-kynurenine pathway in the microglia, which leads to the synthesis of the neurotoxin quinolinic acid; this acts as an agonist at NMDA glutamate receptors [100]. Quinolinic acid, together with the increase in the reactive oxygen and nitrogen species which arise as a consequence of the oxidative stress, is responsible for the neuronal and mitochondrial destruction. This contributes to the neurodegenerative changes which underlie dementia in some elderly depressed patients [101, 102]. While it is necessary to be cautious in concluding that the pro-inflammatory cytokines are causative rather than co-incidental factors in MDD, Khandaker et al. [102] concluded from a Mendelian randomization analysis in patients from the UK Biobank sample that IL-6 and CRP are likely to be causally linked to depression [103]. Against this somewhat optimistic conclusion, Berk and co-workers concluded that the poor sensitivity and selectivity of inflammatory markers have failed to benefit to the development of novel treatments for MDD [5]. Despite these differences in views, so far the search for meaningful biomarkers from the immune and endocrine systems seems to be the most promising for the future. From the immune markers, it appears that IL-6 and CRP, together with TNF-alpha, appear to be most promising, but many more large controlled studies need to be undertaken to substantiate this conclusion.

2.5 Neuroendocrine Biomarkers of MDD

Extensive clinical research has established that neuroendocrine and metabolic functions are deranged in MDD. The changes are linked to an increased secretion of corticotropin-releasing factor (CRF) and the consequent rise in serum cortisol. This reflects the increase in CRF expressing neurons in limbic regions coupled with a reduction in CRF receptors in the frontal cortex [104, 105]. While these studies were undertaken in depressed patients who committed suicide, they do support the view that the hypothalamic-pituitary-adrenal (HPA) axis is malfunctional in MDD.

Whereas a negative feedback mechanism usually operates in non-depressed individuals in response to a stressful challenge, thereby resulting in the rapid reduction in the serum cortisol level once the stressful stimulus ceases, in most cases of MDD, the enhanced responsiveness of the HPA axis to stress is prolonged, resulting in an increase in duration of the action of circulating cortisol on glucocorticoid receptors. As a consequence, the peripheral and central glucocorticoid receptors are desensitized and hypercortisolemia is maintained [42]. In most treatment-responsive patients, the glucocorticoid receptors are re-sensitized, and the serum cortisol levels return to normal. The dexamethasone suppression test (DST) was developed as a marker for melancholic depression [106]. In this test, melancholic, psychotic, and elderly patients with MDD fail to show a reduction in serum cortisol following the administration of the potent glucocorticoid dexamethasone, thereby demonstrating that the glucocorticoid receptors remain in a desensitized state. Effective antidepressant treatment usually normalizes the cortisol response. More recently, the sensitivity of the DST test has been increased by combining the DST with a CRF stimulation following the administration of dexamethasone 15 hours earlier. Depressed patients, particularly melancholic patients, show an exaggerated cortisol response which is normalized by effective treatment [107, 108]. Changes in the serum concentration of cortisol and CRF have been proposed as biomarkers of MDD, particularly when they are added to a broad biomarker panel [109].

Pro-inflammatory cytokines activate the HPA axis and thereby contribute to hypercortisolemia in depressed patients. Pro-inflammatory cytokines impair the negative feedback mechanism by increasing the relatively inert isoform and decreasing the active isoform of the glucocorticoid receptors [110]. This illustrates the interaction which not only contributes to the psychopathology of immune-endocrine interactions but also demonstrates the complexity of the identifying biomarker changes.

2.6 Metabolic Biomarkers of MDD

Circulating leptin and ghrelin are important biomarkers of energy homeostasis. Leptin levels decrease in patients with chronic depression, an effect which has also been detected in those exposed to chronic stress [111]. Studies in rodents show that leptin has antidepressant-like effects which could be linked to an increase in the synthesis of BDNF expression in the hippocampus [112]. Ghrelin shows the opposite changes to leptin in depressed patients [113]. These observations are particularly important because the metabolic syndrome which frequently accompanies MDD is associated with obesity. As leptin and ghrelin are closely involved in brain energy homeostasis, they could be useful components of a biomarker panel. However, more extensive studies are required to substantiate the value of leptin and ghrelin as biomarkers.

A discussion of potential metabolic biomarkers would not be complete without consideration of the markers of oxidative stress. Superoxide dismutase (SOD) activity has been shown to increase in erythrocytes from patients with MDD and to decrease following effective antidepressant treatment [114, 115]. A more robust marker of oxidative metabolism is malondialdehyde (MDA), which increases in the serum of MDD patients but returns to normal following effective antidepressant therapy [116, 117]. To date, this study has not been replicated in large, adequately controlled trials. From this brief discussion of biomarkers for MDD and response to treatment, it is clear that none of the biomarkers discussed fulfill the criteria for successful application [5, 118]. Nevertheless, the importance of discovering useful biomarkers is widely recognized, but it is evident that no single biomarker is likely to be of practical, or theoretical, importance. The heterogeneous nature of depression with its numerous subtypes and comorbid medical disorders adds to the difficulty in identifying relevant biomarkers. One possible solution will be to develop biomarkers that target specific symptoms or subtypes. As an example of this approach, Kappelmann and colleagues reported that the results of a large GWAS demonstrated the co-heritability between serum CRP levels and symptoms associated with psychological and metabolic dysregulation (anhedonia, changes in appetite, tiredness, and feelings of inadequacy). Increased IL-6 signaling was found to be associated with suicidality [119].

To conclude, the best option will be to establish a panel of biomarkers which will identify different facets of the depressive phenotype. At present, it seems as though we are a long way from achieving this goal!

Generally accepted biomarkers for depression are summarized in Table 1. Experimental depression models and related biological markers are given also in Table 2.

3　Criteria for a Valid Animal Model

The search for new animal models in the 1960s was stimulated by the lack of success of the widely used rodent models such as the reserpine reversal model, which, while it had been successful in demonstrating the potential antidepressant actions of tricyclic anti-depressants and monoamine oxidase inhibitors, as these drugs could reverse the reserpine-induced hypothermia, failed to register the activity of novel antidepressants like mianserin which failed to inhibit monoamine reuptake. The demand for better (rodent) models thus became evident and resulted in criteria being defined for a suitable model. McKinney and Bunney [71] proposed five criteria which a suitable model should fulfill [120]. These were as follows:

- The similarity between the changes in the model and those seen in patients with MDD. These changes should be readily observed and measurable.
- There should be inter-rater reliability.
- There should be an observable response to the positive effects of antidepressants.

Abrahamson and Seligman [120] added the need for the etiological changes in the model with those occurring in MDD [121]. Willner [121] reevaluated the essential criteria and suggested that there were three essential criteria for an animal model, namely, face validity (similarity to MDD), predictive validity (similar changes seen in the model and MDD following treatment), and construct validity (the causes of the symptoms in the model and MDD are similar) [122, 123].

Before discussing animal models, it is essential to consider the nature and the validity of the biological, primarily biochemical, markers of major depression. Knowing how much validity and reliability criteria are met in the development of experimental models determines the translational value of the model (Fig. 1). This will be followed by details of the established rodent models.

To what extent can these criteria of face, predictive, and construct validity apply? For example, face validity is not possible in a simple rodent model of MDD as it is impossible to recreate all of

Fig. 1 Knowing how much validity and reliability criteria are met in the development of experimental models determines the translational value of the model

the complex behavioral features of MDD. For the predictive validity, the behavioral tests in rodents which are based on known antidepressants fail to detect potential antidepressants which act on different neural processes. The most difficult aspect to simulate in a model is construct validity. As the neurobiological mechanisms underlying depression are not understood, how can an animal model be constructed that fulfills this criterion? In addition to these fundamental problems, how is it possible to translate the behavioral changes in rodents due to the heterogeneous nature of depression, lack of appropriate biomarkers, and the subjective changes in the MDD patient which do not translate to animals? Such critical views of the criteria for an appropriate animal model have been further discussed by Cryan and Holmes , McArthur and Borsini , and Berton et al. [123–125]. Robinson [126] added a useful addendum by adding that an animal model should:

- Enable an endpoint to be measured which can be related to a similar endpoint in the patient. How can a rodent exhibit guilt, remorse, euphoria, and suicidal thoughts?
- Be sensitive to factors linked to the development of the disease being treated.
- Exhibit similar underlying neurobiology/neuropathology.
- Reliably predict the efficacy of treatment which can be translated to the clinical situation.

Clearly none of the models discussed in this presentation fulfills these criteria! Any model, apart from primate models, is bound to be limited by the complexity/structure of the brain. Is it therefore reasonable to expect a rat with a brain weight of about 1.4 gm to simulate the emotional and behavioral output from a human brain of about 1.4 kg?

The behavioral models widely used by the pharmaceutical industry and research workers are screening tests and will not be considered in any detail in the presentation. These tests are covered by the researchers in ref. [127, 128]. Chronic rodent models of MDD will form the basis for further discussion. It is reasonable to expect a valid model of MDD to be an intact animal: neurons do not exist in isolation but are part of an interacting network; neurotransmitters, receptors, growth factors, enzymes, etc. do not function in isolation but are modulated by other concurrent physiological/biochemical processes. Furthermore, antidepressants and most psychotropic drugs only show optimal therapeutic activity after chronic administration. These are some of the reasons why only chronic rodent models will be considered.

4   Chronic Stress-Based Rodent Models

4.1   The Need for an Ethologically Based Assessment of Rodent Behavior

Rodent behavioral models have been linked to the development of apparatuses which the experimenters assumed simulated an aspect of human behavior. This has resulted in the standardization of apparatuses, simple test protocols, and time-efficient and behavioral tests, which are reliable and easy to use. What is often overlooked relates to the biology of the animal before an attempt is made to understand the behavior and neurobiology which implemented the behavior under experimental conditions. To what extent has the limited success in developing good rodent models been due to the mechanistic interpretation of the living animal? Ethologically valid behaviors can be defined as the physiological and environmental stimuli and response that are within the experience that the rodent would encounter and to which it would naturally react [129]. To what extent do the routine experimental stress tests assess ethologically valid behaviors? Some of the problems which are relevant to rodent models will now be considered.

4.2   The Use of Selective Strains (Usually Mouse Strains) to Reduce Variation and Improve Reproducibility

Most rodent strains are inbred for many generations to produce genetically identical animals. This interferes with natural adaptive selection including adaptive changes to an artificial laboratory environment. As the laboratory is standardized to minimize variation, it requires housing to be in standardized plastic cages with sawdust bedding. Such a barren environment is potentially stressful and adversely affects brain development [130, 131]. Rodents are naturally burrowing animals, so the addition of a thickness of appropriate bedding would help to make the environment more ethologically relevant. Over 200 different strains of mice have been created, many of them to exhibit behaviors of human physical or behavioral disorder. The use of different strains by different institutions and environmental differences in the housing and handling of the animals confound the difficulties in comparing experimental results. Such problems are often not considered when the results are published. In short, both genotype and phenotype need to be considered in establishing valid rodent models.

4.3 Housing and Its Importance

Standard laboratory conditions require inbred strains of rodents that are restricted to plastic or metal cages with standard food and water, with or without a thin layer of sawdust or wood shavings for bedding. Research in the 1960s and 1970s demonstrated that not only was the behavior compromised but also important neural and neurochemical parameters were affected [132, 133]. Behavioral changes resulting from such housing include increased aggression, increased fearfulness, and learning deficits [134]. There is now also evidence of changes in cellular immune function. With the recent interest by neuroscientists in psychoimmunology, this observation is particularly relevant [135]. More ethologically based housing models require nesting materials, shelter, space to move freely, social contact with other animals, and, in general, more environmental stimulation. Increasing the complexity of the housing does not necessarily lead to a larger variation in the experimental results [136].

Although rodents have the same five senses as humans, the relative importance of the senses may differ. Olfaction is an obvious example. Rodents use urine deposits to mark territory and recognition. Standard laboratory practice is to clean the cages regularly, which affects group behavior. Thus, transferring nesting material from the dirty to the clean cage would at least be a compromise [137]. Differences between vision in rodents and humans are also important. Rodents are nocturnal burrowing animals, yet many experimental procedures (maze learning, open field, Morris water maze, etc.) are undertaken in daylight [138]. From a biological perspective, all these caveats seem obvious but are often ignored in the standardized approach to the use of rodent models. Similar problems arise when considering models based on taste, sound, and touch, but these aspects will not consider here.

4.4 An Ethologically Based Approach to the Development of a Rodent Model of Depression

The primary emphasis of the standardized laboratory tests is their reproducibility. The relatively simple behavioral tests are often undertaken by junior staff and more recently with the use of automated procedures. Many experimental procedures have been developed in rats (such as spatial learning tasks, Morris water maze) and uncritically applied to mice where the behavioral responses differ [139]. Tests of exploration and spatial learning are adversely influenced by fear and the environment, in which such tests undertaken are usually novel for the rodent which can induce anxious behavior and confounding variability into the interpretation of the result.

Perhaps it is now time to adopt more ethologically based approaches into rodent models of MDD. If the current standardized battery approach was successful, then there would be little reason for changing the status quo. However, the limitations of the current approach may be one of the main reasons why there has been only modest success in the discovery of new psychotropic drugs. Olsson and colleagues [140] in their review of the ethological approach to animal psychology and behavior identified three major areas requiring attention.

These are as follows:

- Behavioral genetics in considering the impact of inbreeding on social communication. The use of transgenic rodents also raises issues regarding the effects on normal behavior.

- The effect of housing on the behavior and physiological/biochemical changes which have already been referred to above.

- The need to apply ethological research methods more extensively to the routine methods which are widely used.

In summary, rodents used in behavioral studies, and particularly as models of MDD, are not inanimate objects whose behavioral responses are predictable reflexes which can be measured by machines. Perhaps it is time to get back to an understanding of the biology of rodents and their behavior before assumptions are made about the neuroscience consequences!

5 Screening Tests and Behavioral Batteries Used to Establish Depression Models

In depression research, it is important to differentiate between screening test and models. Screening tests are based on acute behavioral observations, usually of mice, in which known antidepressants show predictable activity [121]. Models of depression are usually based on observations of more complex forms of behavior and usually following the chronic administration of the drug [141, 142].

The forced swim test (FST) and the tail suspension test (TST) are acute screening tests which are widely used in the pharmaceutical industry and research laboratories. Both these tests meet predictive validity at least for detecting compounds which have a similar profile to known antidepressants. Both tests have reproducibility and are easy to use and have been valuable for the development of a generation of "me-too" antidepressants.

The FST was first developed by Porsolt, Le Pichon, and Jalfre [143]. Initially the test was developed using mice, but more recently rats have also been used with similar results. The FST involves placing the rodent in a container of warm water for 5 min from which it cannot escape. This is a highly stressful environment. Initially the rodent swims vigorously trying to escape. This is followed by short periods of inactivity in which the animal makes minimal movements to avoid drowning. The acute administration of most types of antidepressants shortens the period of immobility. The immobility is considered to be an indication of learned helplessness and therefore an indication of clinical antidepressant action. While most antidepressants do not appreciably alter the locomotor activity of rodents which would limit the usefulness of the immobility time criterion, stimulants and some anticholinergic drugs can cause false-positive results.

The TST was developed as an acute antidepressant test in mice. In this test, mice are suspended at a height by the tail, and the time in which they struggle to escape and the period in which they hang without struggling is recorded over a 6-min period. Acute antidepressant administration shortens the time of immobility [144].

Despite the assumption that both these acute tests are equally effective in detecting potential antidepressant activity, it has been demonstrated that novel GABA-B receptor antagonists, and knockout models of depression, only show activity in the FST and not the TST [145]. The FST and the TST are acute screening tests which do not meet the criteria for structural or face validity.

FST and TST also assesses despair based on how a rodent reacts to an unpleasant environment, whereas elevated plus maze (EPM) and marble burying tests are used to assess anxiety. Open field test (OFT) is often used to test both anxiety and locomotor activity. Anhedonia, a loss of interest in things that were once pleasurable, can be assessed by means of the sucrose preference test, grooming, and body splash test. Morris's water maze and passive avoidance tests can be used to evaluate cognitive functions/deficits due to depression. Rodents are highly social animals, and when placed into an area in which territory has not been established, they socially engage with one another displaying a number of behavioral acts that can be quantitatively measured. Symptoms such as asociality are usually evaluated by social aversion and social interaction tests. Behavioral batteries used to establish depression models are summarized in Fig. 2.

Animal models based on stress exposure such as depression show dramatic changes in behavior, physiology, endocrine function, and neuronal activity in rodents. Loss in body weight, reduced locomotor activity, change in sleeping patterns and increasing number of early-morning waking episodes, increased levels of the adrenocortical hormone cortisol, increased adrenal weight, increased concentration of noradrenaline, and reduced gonadal function help to understand the underlying pathophysiology of major depression [146].

6 Chronic Models of Depression

Stress is an important environmental risk factor that plays an important role in the development of depression, and therefore various types of stressors are used to create chronic models. These include models of learned helplessness, social stress, maternal deprivation, and chronic mild stress models. These will be briefly discussed.

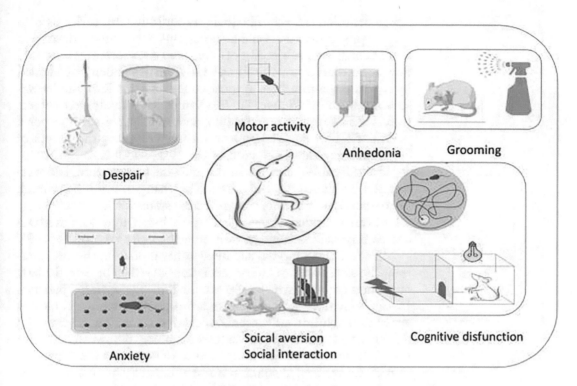

Despair

Motor activity

Anhedonia

Grooming

Anxiety

Soical aversion
Social interaction

Cognitive disfunction

Fig. 2 Behavioral batteries used to establish depression models

Experimental depression models and related biological markers are given also in Table 2.

6.1 Learned Helplessness Model

In this model, the rodent is repeatedly subject to foot shocks which are uncontrolled and repeated and from which the animal is unable to escape. When the rodent is subsequently placed in an environment from which it can escape the shocks, it fails to show attempts to escape. This is considered to be learned helplessness [78, 82, 147].

Rodents subject to learned helplessness behavior exhibit changes in motor activity, loss of weight, disrupted sleep pattern, decreased libido, and memory deficits, which simulate some of the changes seen in major depression. In addition to these physical changes, a decrease in brain noradrenaline and serotonin, increased HPA activity, loss of synapses in the hippocampus, and a decrease in BDNF have also been reported. However, the effects of the stress which causes these changes are of short duration and can be reversed by acute antidepressant administration [82, 144].

6.2 Chronic Unpredictable Mild Stress (CUMS)

The CUMS model is widely used because it meets most of the validity and reliability criteria. In the model first developed by Katz and co-workers [72, 73], rodents were subject to severe unavoidable foot shock daily for 3 weeks. This resulted in hypercortisolemia and anhedonia as indicated by a decrease in sucrose preference. It should be noted that in the model developed by Katz et al., the subjection of animals to such severe stress would now be considered unethical and a violation of standards demanded by regulatory authorities in most countries. The Katz model was later modified by Willner [79, 148], in which rodents were subject to a series of mild, unpredictable, and inescapable stressors for a period of 2–6 weeks. The stressors included reversing the periods of light/dark exposure, agitation of the cage, social isolation, exposure to loud noise, changes in the ambient temperature, and wetting the bedding [79, 146, 148]. The changes in the behavioral, neurochemical, immune, and endocrine systems which accompany CUMS are considered to simulate the major changes seen in MDD and an important advantage over other stress-based models [149].

In the CUMS model, and many other models in this presentation, the assessment of anhedonia is considered to be a particularly important criterion which reflects anhedonia in depressed patients. In experimental studies, rodents are presented with a choice of water or a solution of sucrose/saccharine water. To determine if the stressful experience has changed to choice, the sweetened solution is also presented without a choice of water. As the consumption of the sweetened solution is not affected by hunger or satiety, the change in

choice is considered to reflect the psychological and not the physiological/metabolic state of the animal [148–150].

6.3 Early-Life Stress Model

It is well established that negative and traumatic experiences in the early period of life increase the susceptibility of stress-related pathologies, such as anxiety in adulthood. Thus, exposure to stress in early development is an important risk factor, which plays a role in the later development of depression. In clinical studies, parental loss in childhood and physical or emotional neglect are important factors in the development of mood disorders in adult life [151, 152].

The early postnatal period, in which the nervous system develops rapidly, shows a high degree of plasticity and is particularly vulnerable to stress. In the early-life stress model, the offspring are removed from the nest and kept in a warm environment away from the mother and other siblings for a short time period. The offspring are removed from the mother every day for the first 2 weeks of postnatal life. These animals develop anxiety and depression-like behavior on reaching adulthood, changes associated with an increase in HPA activity, and reduced brain BDNF. The depression-like activity is reversed by chronic antidepressant treatments, and neurogenesis is restored [153, 154].

While this model has many attractive features, there is a lack of standardization of the stressors, the duration and severity of the stress used, and the behavioral endpoints determined. These problems are common to all the chronic stress models!

6.4 Social Defeat Stress

Stress associated with social interaction can be an environmental factor which plays a role in the development of depression. In humans, anxiety, feeling lonely, the need to socially withdrawal, and decreased self-esteem are frequent feature of depression [155]. In rodents, social defeat models simulate some phenotypes of depression. This model involves male rodents in which the aggressive and dominant animal, the invader, is introduced into the cage of the non-dominant resident animal. This interaction may involve direct or indirect contact, in which case a transparent screen is used to separate the animals. Following the exposure, the non-dominant animal develops anhedonia, social avoidance, decreased libido, decreased locomotor activity, anxiety-like behavior, and changes in sleep pattern and circadian rhythm, changes which simulate the depressive phenotype. The anhedonia and social avoidance behavior persist for a period following the defeat exposure. Chronic antidepressant treatments largely reverse these symptoms. The endocrine, neurotransmitter, and BDNF changes are also reversed by chronic antidepressant treatments [105, 156, 157].

6.5 Olfactory Bulbectomy

Olfactory bulbectomy is an important chronic model of depression-like behavior in rodents. Neuroimaging, brain lesions, and neuropathological studies implicate the involvement of limbic structures in MDD [158]. Of the chronic rodent models which have been developed over the past 40 years, the bilateral olfactory bulbectomy rat model (OBX) is particularly pertinent in this respect. Not only does the OBX model exhibit depressive-like behavior which is selectively attenuated by chronic, but not acute, antidepressant treatments, but it also shows hyperactive behavior and irritability in stressful situations, anhedonia, changes in brain monoamine neurotransmitters, neuroinflammation, and changes in the endocrine system which are qualitatively similar to those see in patients with MDD [83, 84, 159, 160]. The OBX model also has a facility to explore other physiological systems which become disrupted in MDD. For example, the loss of olfaction is a frequent feature of severe depression particularly in elderly patients showing cognitive decline. In general, however, depressed patients have poor olfactory discrimination and deficits in anatomical regions sub-serving olfactory function [161]. These regions include the dorsal amygdala [158, 162, 163]. The associated changes in emotional and cognitive function are a reflection of the neuronal loss of dendritic spines and synapses; the increased activity of the microglia, resulting in the increased release of pro-inflammatory cytokines and other inflammatory mediators; and an increase in the glutamatergic system with the release of neurotoxins such as quinolinic acid [84, 98, 159, 164].

In comparison to the frequently used chronic stress models of depression, it is apparent that the OBX model offers many advantages not only for understanding the complexity of the interactions between the neuroanatomical and neurophysiological areas which maybe dysfunctional in depression but also as a model for the detection of novel antidepressants. However, it must be emphasized that the OBX model is not a rodent model for screening compounds for their potential antidepressant activity. Careful brain surgery is required to remove the olfactory bulbs without damaging the prefrontal cortex. A period of at least 2 weeks is required for the animals to recover, and to obtain stable and reproducible behavior responses, the animals must be handled daily by the same experimenter. In most cases, effective antidepressants must be administered for at least 2 weeks to ensure optimal activity. Details of the procedures necessary for the establishment of the OBX model are covered by Song and Leonard and Kelly et al. [83, 84]. In summary, the OBX rat is a model of depression and not a screening method!

6.6 Predator Stress Model

Predator stress can be initiated in rodents by exposing them to the odor or to the sight of the predator. This potent stressful stimulus has been shown to induce depressive-like behavior [165, 166]. Evidence of anhedonia, anxiety-like behavior, and a disruption in social interaction behavior have been reported; there is also evidence of a reduction in neurogenesis as a result of exposure to an aversive odor. Thus, the exposure of a rodent to the sight of a cat, or exposure to cat litter, results in increased aggression, freezing behavior, increased heart rate, and other signs of activation of the autonomic system. In addition, some animals show a decrease in locomotor activity, grooming behavior, and libido. This model should be considered as a model of post-traumatic stress rather than depression. Thus, antidepressants, particularly SSRI antidepressants such as fluoxetine, actively reverse the effects of trauma-induced behaviors, while benzodiazepines preferentially inhibit the avoidance behavior [167].

6.7 Genetic Models

As there is a strong genetic component to MDD, many transgenic models have been created in mice particularly involving targeting the genes involved in serotonergic, noradrenergic transmission and genes regulating the HPA axis. Another experimental approach involves gene mutations and knockout mice. These approaches result in models that exhibit resistance to the effects of antidepressants together with an increased susceptibility to stress. However, depression is a multigenetic condition in which genes interact with one another and with the environment. It is therefore unlikely that models carrying single mutations in a particular pathway can fully reflect MDD in a meaningful way. These limitations have been discussed by several researchers in recent years [168–171]. Genetically altered mice models for depression are summarized in Table 3.

Table 3
Genetically altered mice models for depression (modified from ref. [172])

System	Model	Reference
Serotonergic	5-HT1A receptor knockout 5-HT1B receptor knockout Serotonin transporter knockout 5-HT7 receptor knockout	[172–185]
Noradrenergic	Dopamine-b-hydroxylase knockout a2A-Adrenoceptor knockout a2C-Adrenoceptor knockout a2C-Adrenoceptor overexpressing Noradrenaline transporter knockout	[186–188]
Monoamine oxidase	Monoamine oxidase A knockout monoamine oxidase B knockout	[189–192]
Opioid	Mu opioid receptor knockout Delta opioid receptor knockout	[193]
GABAergic	Glutamic acid decarboxylase (65-kDa isoform) knockout GABAB (1) receptor knockout	[191, 194]
Glutamate	NMDA receptor e4 subunit mGluR7 knockout	[172, 195]
Substance P-related	Tachykinin NK_1 receptor knockout Tac1 gene	[196–198]
HPA axis	Glucocorticoid receptor-impaired transgenic CRF-overexpressing mice CRF2 receptor knockout	[199–202]
Immunological	Tumor necrosis factor a knockout Interleukin-6 knockout	[203, 204]
Other receptors	Adenosine A_{2A} receptor knockout nicotinic b2 knockout CB1 Nociceptin/orphanin FQ receptor NPY-2 Dopamine D5 receptor knockout	[182, 183, 205–208]

6.8 Models Based on Optogenetic Manipulation

This is a new and novel approach to inducing depression-like behavior in rodents in which opsins, light-sensitive channel markers, are used to inhibit or excite specific neurons. Designer receptors exclusively activated by designer drugs (DREADD)-based chemogenetic tools are used to identify the circuitry and cellular signals that specify behavior. Opsins can be specifically activated by DREADDs, for chemogenetics, in specific neurons, thereby enhancing the specific targeting and specificity of cell types in comparison to traditional pharmacological approaches [209]. This permits the precise dissection of neural circuits which underlie depression-like behavior. Muir and colleagues [210] have discussed the importance of DREADD-based chemogenetic strategies for dissecting the functional consequences of activation of complex neural pathways. The potential advantage of the optogenetic approach is that it enables the profiling of genes in regions in which they are primarily involved. Given the increasing knowledge of the complexity of molecular events mediating depression-like behavior, the targeting of specific signaling proteins using the optogenetic approach could become an important method to define the molecular basis of depression in the future [210].

6.9 Zebrafish as Models of Depression, their Usefulness, and Limitations

Both human patients and animal models show individual differences which are recognized as critical traits embodying individuality [211]. Human populations involve patients that are genetically and environmentally heterogeneous and therefore not subject to a standardization procedure applicable to inbred laboratory rodents. As already discussed, individual behavioral differences in rodent models are affected by social status and adaptive strategies, but it is unclear whether this variability is due to consistent individual traits or to random variation [212]. Individual traits and their variation can be assessed by varying the environmental conditions by enrichment, for example. Another strategy is to assess variation in a wide range of species that not only include mammals but also fish. In this respect, zebrafish (*Danio rerio*) are emerging as a popular model for studying CNS disorders [213].

Zebrafish possess all the classical neurotransmitters of humans and have been shown to respond to drugs which are widely used therapeutically [214]. In addition, they display high genetic and physiological homology to humans [215]. Zebrafish may therefore be of interest for their behavioral traits and their variation.

Some experimental methods for exploring individuality in animal models are based on exposing them to novel objects (shy-boldness tests) or to conspecifics (aggression tests). Individual differences have been demonstrated both in rodents [216] and in zebrafish [217]. In fish, boldness can be assessed by specific behavioral endpoints such as their swim level, horizontal position, feeding latency, latency to enter a novel environment, and their reaction to threatening or benign novel objects [217, 218]. Social enrichment affects rodent behavior [219] but also in fish. Rainbow trout have been shown to reduce their boldness after they have observed conspecifics losing a fight [220]. Demin and co-workers [221] have critically assessed intraspecific behavior in rodents and fish and summarized their similarity in terms of boldness, aggressiveness, dominance, and sociability. Responses to anxiety-provoking situations and to pain are also similar [222, 223].

In summary, depression-like states can be induced in zebrafish by stress, genetic, and pharmacological manipulations similar to rodents [224]. For example, acute or chronic stress exposure leads to the disruption of endocrine responses. Chronic psychostimulant treatment provokes behavioral sensitization, while chronic reserpine treatment results in depression-like behavior. It has also been shown in zebrafish that knockout of the glucocorticoid receptor gene results in an increase in cortisol and changes in exploration and habituation behavior similar to those changes seen in rodents.

Clearly there are many aspects of the zebrafish biology and behavior which demand closer investigation to supplement the widespread use of rodents as models for MDD. However, the development of the zebrafish model is recent and requires more research to become applicable as a useful model.

7 Conclusions

The purpose of this presentation was to critically assess the current position of animal models of depression. At present, all of the available rodent models lack criteria to enable them to provide information which would enable a new generation of antidepressants to be identified. From the rodent models available, it is unlikely that a major breakthrough in the development of novel antidepressants is the foreseeable future. As discussed in this presentation, until more emphasis is placed on understanding the biology of the rodent, particularly on the complexity of the behavior, it is unlikely that better models, as distinct from screening tests, will be developed. This will require the (re)introduction of behavioral pharmacology into the drug development process, which will require additional time and added costs. Until there is a major change in the research culture of the pharmaceutical industry which seems to be fixated on computer-driven automated procedures wherever they can be applied, the future looks bleak!

References

1. Caspi A, Moffitt TE (2006) Gene-environmental interactions in psychiatry: joining forces with the neurosciences. Nat Rev Neurosci 7:583–590

2. Hyman SE (2014) Revitalising psychiatric therapeutics. Neuropsychopharmacology 39:220–229

3. Kessler RC, Chiu WT, Demler D et al (2005) Prevalence, severity and co-morbidity of 12 month DSM 4 disorders in the National Comorbidity Survey Replication. Arch Gen Psychiat 62:617–629

4. Lakhan SE, Vieira K, Hamlat E (2010) Biomarkers in psychiatry: drawbacks and potentials for misuse. Int Arch Med 3:1–8

5. Berk M, Walker AJ, Nierenberg AA (2019) Biomarker-guided anti-inflammatory therapies. JAMA Psychiat 76:779–780

6. Kennis M, Gerritsen L, van Dalen M et al (2020) Prospective biomarkers of major depressive disorder: from promises to reality check. Mol Psychiatry 215:321–338

7. Li M, Soczynska JK, Kennedy SH (2011) Inflammatory biomarkers in depression: an opportunity for novel therapeutic interventions. Curr Psychiat Rep 13(5):316–320

8. Breitenstein B, Scheuer S, Holsboer F (2014) Are there meaningful biomarkers of treatment response for depression? Drug Discov Today 19:539–561

9. Yadid G, Nakash R, Deri I, Tamar G, Kinor N, Gispan I et al (2000) Elucidation of the neurobiology of depression: insights from a novel genetic animal model. Prog Neurobiol 62:353–378

10. Powell SB, Geyer MA, Preece MA, Pitcher LK, Reynolds GP, Swerdlow NR (2003) Dopamine depletion of the nucleus accumbens reverses isolation-induced deficits in prepulse inhibition in rats. Neuroscience 119: 233–240

11. Lidberg L, Belfrage H, Bertilsson L, Evenden MM, Åsberg M (2000) Suicide attempts and impulse control disorder are related to low cerebrospinal fluid 5-HIAA in mentally disordered violent offenders. Acta Psychiatr Scand 101:395–402

12. Ruhé HG, Mason NS, Schene AH (2007) Mood is indirectly related to serotonin, norepinephrine and dopamine levels in humans: a meta-analysis of monoamine depletion studies. Mol Psychiatry 12:331–359

13. Hughes JW, Watkins L, Blumenthal JA, Kuhn C, Sherwood A (2004) Depression and anxiety symptoms are related to increased 24-hour urinary norepinephrine excretion among healthy middle-aged women. J Psychosom Res 57:353–358

14. Mooney JJ, Samson JA, Hennen J, Pappalardo K, McHale N, Alpert J, Koutsos M, Schildkraut JJ (2008) Enhanced norepinephrine output during long-term desipramine treatment: a possible role for the extraneuronal monoamine transporter (SLC22A3). J Psychiatr Res 42:605–611

15. Mauri MC, Ferrara A, Boscati L, Bravin S, Zamberlan F, Alecci M, Invernizzi G (1998) Plasma and platelet amino acid concentrations in patients affected by major depression and under fluvoxamine treatment. Neuropsychobiology 37:124–129

16. Sanacora G, Gueorguieva R, Epperson CN, Wu YT, Appel M, Rothman DL, Krystal JH, Mason GF (2004) Subtype-specific alterations of γ-aminobutyric acid and glutamate in patients with major depression. Arch General Psychiatry 61:705–713

17. Wegener G, Harvey BH, Bonefeld B, Müller HK, Volke V, Overstreet DH, Elfving B (2010) Increased stress-evoked nitric oxide signalling in the Flinders sensitive line (FSL) rat: a genetic animal model of depression. Int J Neuropsychopharmacol 13: 461–473

18. Nobis A, Zalewski D, Waszkiewicz N (2020) Peripheral markers of depression. J Clin Med 9:3793

19. Wulff K, Gatti S, Wettstein JG, Foster RG (2010) Sleep and circadian rhythm disruption in psychiatric and neurodegenerative disease. Nat Rev Neurosci 11:589–599

20. Karatsoreos IN (2014) Links between circadian rhythms and psychiatric disease. Front Behav Neurosci 8:162

21. Pariante CM, Lightman SL (2008) The HPA axis in major depression: classical theories and new developments. Trends Neurosci 31: 464–468

22. Pariante CM (2009) Risk factors for development of depression and psychosis: glucocorticoid receptors and pituitary implications for treatment with antidepressants and glucocorticoids. Ann N Y Acad Sci 1179:144–152

23. Trzepacz PT, McCue M, Klein I, Levey GS, Greenhouse J (1988) A psychiatric and neuropsychological study of patients with untreated Graves' disease. Gen Hosp Psychiatry 10:49–55

24. Wysokinski A, Kloszewska I (2014) Level of thyroid-stimulating hormone (TSH) in patients with acute schizophrenia, unipolar depression or bipolar disorder. Neurochem Res 39(7):1245–1253

25. Solberg LC, Olson SL, Turek FW, Redei E (2001) Altered hormone levels and circadian rhythm of activity in the WKY rat, a putative animal model of depression. Am J Physiol Regul Integr Comp Physiol 281:R786–R794

26. Edwards R, Peet M, Shay J, Horrobin D (1998) Omega-3-polyunsaturated fatty acid levels in the diet and in red blood cell membranes of depressed patients. J Affect Disord 48:149–155

27. Maes M, Smith R, Christophe A, Cosyns P, Desnyder R, Meltzer H (1996) Fatty acid composition in major depression: decreased $\omega3$ fractions in cholesteryl esters and increased $C20:4\omega6C20:5\omega3$ ratio in cholesteryl esters and phospholipids. J Affect Disord 38:35–46

28. Lamers F, Bot M, Jansen R, Chan MK, Cooper JD, Bahn S, Penninx BWJH (2016) Serum proteomic profiles of depressive subtypes. Transl Psychiatry 6:e851

29. Domenici E (2010) Early-life stress and antidepressants modulate peripheral biomarkers in a gene- environment rat model of depression. Prog Neuro-Psychopharmacol Biol Psychiatry 34:1037–1048

30. Domenici E, Willé DR, Tozzi F, Prokopenko I, Miller S, McKeown A, Brittain C, Rujescu D, Giegling I, Turck CW et al (2010) Plasma protein biomarkers for depression and schizophrenia by multi analyte

profiling of case-control collections. PLoS One 5:e9166

31. Capuron L, Neurauter G, Musselman DL, Lawson DH, Nemeroff CB, Fuchs D, Miller AH (2003) Interferon-alpha-induced changes in tryptophan metabolism: relationship to depression and paroxetine treatment. Biol Psychiatry 54:906–914

32. Bonaccorso S, Marino V, Puzella A, Pasquini M, Biondi M, Artini M, Almerighi C, Verkerk R, Meltzer H, Maes M (2002) Increased depressive ratings in patients with hepatitis C receiving interferon-a-based immunotherapy are related to interferon-a- induced changes in the serotonergic system. J Clin Psychopharmacol 22:86–90

33. Dowlati Y, Herrmann N, Swardfager W, Liu H, Sham L, Reim EK, Lanctôt KLA (2010) A meta-analysis of cytokines in major depression. Biol Psychiatry 67:446–457

34. Leonard BE, Myint A (2006) Changes in the immune system in depression and dementia: causal or coincidental effects? Dialogues Clin Neurosci 8(2):163–174

35. Myint AM, Steinbusch HWM, Goeghegan L et al (2007) The effect of the COX-2 inhibitor celecoxib on behavioural and immune changes in an olfactory bulbectoimised rat model of depression. Neuroimmunoimmodulation 14:65–71

36. Khanzode SD, Dakhale GN, Khanzode SS, Saoji A, Palasodkar R (2003) Oxidative damage and major depression: the potential antioxidant action of selective serotonin-reuptake inhibitors. Redox Rep 8:365–370

37. Della FP, Abelaira HM, Réus GZ, Ribeiro KF, Antunes AR, Scaini G, Jeremias IC, dos Santos LMM, Jeremias GC, Streck EL, Quevedo J (2012) Tianeptine treatment induces antidepressive-like effects and alters BDNF and energy metabolism in the brain of rats. Behav Brain Res 233:526–535

38. Gimeno D, Marmot MG, Singh-Manoux A (2008) Inflammatory markers and cognitive function in middle-aged adults: the Whitehall II study. Psychoneuroendocrinology 33:1322–1334

39. Gibney SM, Fagan EM, Waldron AM, O'Byrne J, Connor TJ, Harkin A (2014) Inhibition of stress-induced hepatic tryptophan 2,3- dioxygenase exhibits antidepressant activity in an animal model of depressive behaviour. Int J Neuropsychopharmacol 17:917–928

40. Carboni L, Becchi S, Piubelli C, Mallei A, Giambelli R, Razzoli M, Mathé AA, Popoli M, Domenici E (2010) Early-life stress and antidepressants modulate peripheral biomarkers in a gene-environment rat model of depression. Prog Neuro-Psychopharmacol Biol Psychiatry 34:1037–1048

41. Duman CH (2010) Model of depression. Vitam Horm 82:1–21

42. Carroll BJ (1981) A specific laboratory test for the diagnosis of melancholia. Arch Gen Psychiatry 38(1):15

43. Blugeot A, Rivat C, Bouvier E, Molet J, Mouchard A, Zeau B, Bernard C, Benoliel JJ, Becker C (2011) Vulnerability to depression: from brain neuroplasticity to identification of biomarkers. J Neurosci 31:12889–12899

44. Harvey BH, Hamerm M, Louw R, Van Der Westhuizen FH, Malan L (2013) Metabolic and glutathione redox markers associated with brain-derived neurotrophic factor in depressed African men and women: evidence for counterregulation? Neuropsychobiology 67:33–40

45. Roceri M, Hendriks W, Racagni G, Ellenbroek BA, Riva MA (2002) Early maternal deprivation reduces the expression of BDNF and NMDA receptor subunits in rat hippocampus. Mol Psychiatry 7:609–616

46. Duman CH, Schlesinger L, Terwilliger R, Russell DS, Newton SS, Duman RS (2009) Peripheral insulin- like growth factor-1 produces antidepressant-like behavior and contributes to the effect of exercise. Behav Brain Res 198:366–371

47. Khawaja X, Xu J, Liang JJ, Barrett JE (2004) Proteomic analysis of protein changes developing in rat hippocampus after chronic antidepressant treatment: implications for depressive disorders and future therapies. J Neurosci Res 75:451–460

48. Iga JI, Ueno SI, Yamauchi K, Numata S, Tayoshi-Shibuya S, Kinouchi S, Nakataki M, Song H, Hokoishi K, Tanabe H, Sano A, Ohmori T (2007) The Val66Met polymorphism of the brain-derived neurotrophic factor gene is associated with psychotic feature and suicidal behavior in Japanese major depressive patients. Am J Med Genet Part B 144B:1003–1006

49. Elfving B, Plougmann PH, Wegener G (2010) Differential brain, but not serum VEGF levels in a genetic rat model of depression. Neurosci Lett 474:13–16

50. Campbell S, McQueen G (2004) The role of the hippocampus in the pathophysiology of major depression. J Psychiatry Neurosci 29:417–426

51. Sheline YI (2003) Neuroimaging studies of mood disorder effects on the brain. Biol Psychiatry 54:338–352

52. Koolschijn PCMP, van Haren NEM, Lensvelt-Mulders GJLM, Hulshoff Pol HE, Kahn RS (2009) Brain volume abnormalities in major depressive disorder: a meta-analysis of magnetic resonance imaging studies. Hum Brain Mapp 30:3719–3735

53. Lorenzetti V, Allen NB, Fornito A, Yücel M (2009) Structural brain abnormalities in major depressive disorder: a selective review of recent MRI studies. J Affect Disord 117: 1–17

54. Neumeister A, Hu XZ, Luckenbaugh DA, Schwarz M, Nugent AC, Bonne O, Herscovitch P, Goldman D, Drevets WC, Charney DS (2006) Differential effects of 5-HTTLPR genotypes on the behavioral and neural responses to tryptophan depletion in patients with major depression and controls. Arch Gen Psychiatry 63:978–986

55. Brambilla P, Stanley JA, Nicoletti MA, Sassi RB, Mallinger AG, Frank E, Kupfer D, Keshavan MS, Soares JC (2005) 1H magnetic resonance spectroscopy investigation of the dorsolateral prefrontal cortex in bipolar disorder patients. J Affect Disord 86:61–67

56. Gruber S, Frey R, Mlynárik V, Stadlbauer A, Heiden A, Kasper S, Kemp GJ, Moser E (2003) Quantification of metabolic differences in the frontal brain of depressive patients and controls obtained by 1H-MRS at 3 tesla. Investig Radiol 38:403–408

57. Hu LW, Kawamoto EM, Brietzke E, Scavone C, Lafer B (2011) The role of Wnt signaling and its interaction with diverse mechanisms of cellular apoptosis in the pathophysiology of bipolar disorder. Prog Neuro-Psychopharmacol Biol Psychiatry 35:11–17

58. Fone KCF, Porkess MV (2008) Behavioural and neurochemical effects of post-weaning social isolation in rodents-relevance to developmental neuropsychiatric disorders. Neurosci Biobehav Rev 32:1087–1102

59. Maes M, Fišar Z, Medina M, Scapagnini G, Nowak G, Berk M (2012) New drug targets in depression: inflammatory, cell-mediated immune, oxidative and nitrosative stress, mitochondrial, antioxidant, and neuroprogressive pathways. And new drug candidates-Nrf2 activators and GSK-3 inhibitors. Inflammopharmacol 20:127–150

60. Martins-de-Souza D, Guest PC, Harris LW, Vanattou-Saifoudine N, Webster MJ, Rahmoune H, Bahn S (2012) Identification of proteomic signatures associated with depression and psychotic depression in post-mortem brains from major depression patients. Transl Psychiatry 2:e87

61. Cao X, Li LP, Wang Q, Wu Q, Hu HH, Zhang M, Fang YY, Zhang J, Li SJ, Xiong WC (2013) Astrocyte- derived ATP modulates depressive-like behaviors. Nat Med 19: 773–777

62. Lohoff FW (2010) Overview of the genetics of major depressive disorder. Curr Psychiatry Rep 12:539–546

63. Serova L, Sabban EL, Zangen A, Overstreet DH, Yadid G (1998) Altered gene expression for catecholamine biosynthetic enzymes and stress response in rat genetic model of depression. Mol Brain Res 63:133–138

64. Craddock N, Owen MJ, O'Donovan MC (2006) The catechol-O- methyl transferase (COMT) gene as a candidate for psychiatric phenotypes: evidence and lessons. Mol Psychiatry 11:446–458

65. Caberlotto L, Fuxe K, Overstreet DH, Gerrard P, Hurd YL (1998) Alterations in neuropeptide Y and Y1 receptor mRNA expression in brains from an animal model of depression: region specific adaptation after fluoxetine treatment. Mol Brain Res 59: 58–65

66. Melas PA, Mannervik M, Mathé AA, Lavebratt C (2012) Neuropeptide Y: identification of a novel rat mRNA splice-variant that is downregulated in the hippocampus and the prefrontal cortex of a depression-like model. Peptides 35:49–55

67. Tashiro A, Hongo M, Ota R, Utsumi A, Imai T (1997) Hyperinsulin response in a patient with depression. Changes in insulin resistance during recovery from depression. Diabetes Care 20:1924–1925

68. Katon WJ (2008) The comorbidity of diabetes mellitus and depression. Am J Med 121: S8–S15

69. Yang Y, Yang D, Tang G, Zhou C, Cheng K, Zhou J, Wu B, Peng Y, Liu C, Zhan Y, Chen J, Chen G, Xie P (2013) Proteomics reveals energy and glutathione metabolic dysregulation in the prefrontal cortex of a rat model of depression. Neuroscience 247:191–200

70. Bocchio-Chiavetto L, Maffioletti E, Bettinsoli P, Giovannini C, Bignotti S, Tardito D, Corrada D, Milanesi L, Gennarelli M (2013) Blood microRNA changes in depressed patients during antidepressant treatment. Eur Neuropsychopharmacol 23: 602–611

71. McKinney WT, Jr Bunney WE, Jr (1969). Animal model of depression. Review of evidence: implications for research. Arch Gen Psychiatry 21(2): 240–248

72. Katz RJ, Roth KA, Carroll BJ (1981) Acute and chronic stress effects on open field activity in the rat: implications for a model of depression. Neurosci Biobehav Rev 5(2):247–251

73. Katz RJ, Roth KA, Schmaltz K (1981) Amphetamine and tranylcypromine in an animal model of depression: pharmacological specificity of the reversal effect. Neurosci Biobehav Rev 5(2):259–264

74. Willner P, Towell A, Sampson D et al (1987) Reductiuon in sucrose preference by chronic unpredictable mild stress and its restoration by a tricyclic antidepressant. Psychopharmacol 93:358–364

75. Maier SF, Seligman HEP (2016) Learned helplessness at fifty: insights from neuroscience. Psychol Rev 123:1–19

76. Stepanichev M, Dygalo NN, Grigoryan G, Shishkina GT, Gulyaeva N (2014) Rodent models of depression: neurotrophic and neuroinflammatory biomarkers. BioMed Research International Biomed Res Int 2014:932757

77. Becker M, Pinhasov A, Ornoy A (2021) Animal models of depression: what can they teach us about the human disease? Diagnostics 11: 123

78. Seligman ME, Maier SF, Geer JH (1968) Alleviation of learned helplessness in the dog. J Abnorm Psychol 73(3):256–262

79. Willner P, Belzung C (2015) Treatment-resistant depression: are animal models of depression fit for purpose? Psychopharmacology 232:3473–3495

80. Bartolomucci A, Palanza P, Sacerdote P et al (2003) Individual housing induces altered immuno-endocrine responses to psychological stress in male mice. Psychoneuroendocrinology 28(4):540–558

81. Golden AS, Covington HE, Berton O, Russo SJ (2011) A standardized protocol for repeated social defeat stress in mice. Nat Protoc 6(8):1183–1191

82. Krishnan V, Nestler EJ (2011) Animal models of depression: molecular perspectives. Curr Top Behav Neurosci 7:121–147

83. Kelly JP, Wrynn AS, Leonard BE (1997) The olfactory bulbectomised rat as a model of depression: an uypdate. Pharmacol Ther 74: 299–316

84. Song C, Leonard BE (2005) The olfactory bulbectomised rat as a model of depression. Neurosci Biobehav Rev 29(4–5):627–647

85. Osima EE, Pillinger T, Rodriguewz JM et al (2020) Inflammatory markers in depression: a meta-analysis of mean differences and variability in 5166 patients and 5088 controls. Brain Behav Immun 87:901–909

86. Miller GE, Rohleder N, Stekler Kirchbaum C (2005) Clinical depression and regulation of inflammatory response during acute stress. Psychol Med 67:679–687

87. Sluzewska A, Rybakowski J, Bosmans E et al (1996) Indicators of immune activation in major depression. Psychiatric Res 64: 161–167

88. Bunney WE (1975) The current status of research in the catecholamine theories of affective disorder. Psychopharmacol Commun 1:599–609

89. Lesch KP (1997) The 5-HT transporter gene-linked polymorphic region (5HTTLPR) in evolutionary perspective: alternative biallelic variation in rhesus monkeys. J Neural Trans 104:1259–1266

90. Goldman N (2010) The serotonin transporter polymorphism (5HTTLPR): allelic variation and links with depressive symptoms. Depress Anxiety 27:160–169

91. Polyakova M, Stuka K, Shyemburg K et al (2015) BDNF as a biomarker for successful treatment of mood disorder: a systematic and quantitative meta-analysis. J Affect Dis 174: 432–440

92. Aydemir O, Deveri AS, Taneli F (2005) The effect of chronic antidepressant treatment on serum BDNF levels in depressed patients: a preliminary study. Biol Psychiatry 29: 261–265

93. Taliaz D, Loya A, Gersner R, Haramati S, Chen A, Zangen A (2011) Resilience to chronic stress is mediated by hippocampal BDNF. J Neurosci 23;31(12):4475-4483

94. Schaaf MJ, De Kloet ER Vreugdenhi. Corticosterone effects on BDNF expression in the hippocampus. Implications for memory formation. Stress 3(3):201–208

95. Smith RS (1991) The macrophage theory of depression. Med Hypotheses 35:298–306

96. Blume J, Douglas SD, Evans DL (2010) Immune suppression and immune activation in depression. Brain Behav Immun 25(2): 221–229

97. Howren MB, Lamkin DM, Suls J et al (2009) Association of depression with C-reactive protein, IL-1 and IL-6: ameta-analysis. Psychosom Med 1:171–186

98. Myint AM, Kim YK, Verkerk R, Scharpé S, Steinbusch H, Leonard B (2007) Kynurenine pathway in major depression: evidence of impaired neuroprotection. J Affect Disord 98:143–151

99. Guillemin GJ, Smythe GA, Takikawa O, Brew BJ (2005) Expression of indoleamine 2,3 dioxygenase and production of quinolinic

acid by human microglia, astrocytes and neurons. Glia 49:15–23

100. Myint AM, Kim YK (2003) Cytokine-serotonin interaction through IDO: a neuro-degeneration hypothesis. Med Hypoth 61:519–525

101. Capuron L, Miller A (2011) Immune system to brain signalling: neuropharmacological implications. Pharmac Ther 130:226–238

102. Khandaker GM, Zuber V, Rees JMB et al (2020) Shared mechanism between depression and coronary heart disease: findings from mendelian randomization analysis of a large UK population based cohort. Mol Psychiatry 25:1477–1486

103. Raadsheer FC, Oorschot DE, Verwer RWH, Tilders FJH, Swaab DF (1994) Age-related increase in the total number of corticotropin-releasing hormone neurons in the human paraventricular nucleus in controls and alzheimer's disease: comparison of the disector with an unfolding method. The Journal of Comparative Neurol 339:447–457

104. Nemeroff CB, Owens MJ, Bissette G, Andorn AC, Stanley M (1988) Reduced corticotropin releasing factor binding sites in the frontal cortex of suicide victims. Arch Gen Psychiatry 45(6):577–579

105. Schmidt HD, Duman RS (2011) Peripheral BDNF produces antidepressant-like effects in cellular and behavioral models. Neuropsychopharmacology 35:2378–2391

106. Heuser I, Vassouridis A, Holsboer F (1994) The combined dexamethasone/CRH test: a refined lab test for psychiatricdisorders. J Psychiat Res 28:341–356

107. Aubry JM, Gervbason N, Osiek C, Perret G, Rossier MF, Bertschy G, Bondolfi G (2007) The DEX/CRH neuroendocrine test and the prediction of remitted depressive outpatients. J Psychiat Res 41:290–296

108. Pace TW, Hu F, Miller AH (2007) Cytokine effects on glucocorticoid receptor function: relevance to glucocorticoid resistance and pathophysiological treatment of major depression. Brain Behav Immun 21:9–19

109. Lu XY (2007) The leptin hypothesis of depression: a potential link between mood disorders and obesity. Curr Opin Pharmacol 7:648–652

110. Yamada N, Katsuuva O, Ochi Y et al (2011) Impaired CNS leptin action is implicated in depression associated with obesity. Endocrinol 152:2634–2643

111. Lutter M, SakataI Osborne-Lawrence S et al (2008) The oroxigenic hormone ghrelin defends against depressive symptoms in chronic stress. Nat Neurosci 11:752–753

112. Bilici M, Efe H, Koroglu MA, Uydu HA, Bekaroğlu M, Deger O (2001) Antioxidative enzyme activities and lipid peroxidation in major depression: alteration by antidepressant treatments. J Affect Dis 64:43–51

113. Kotan VD, Sarandol E, Kirham E et al (2011) Effects of long-term antidepressant treatment on oxidative status in major depressive disorder: a 24 week follow-up study. Prog Neuropsychopharmacol Biol Psychiat 35:1284–1290

114. Islam MR, Islam MR, Imtiaz A et al (2018) Elevated serum levels of malondialdehyde and cortisol are associated with major depression: a case controlled study. Sage Open Med 6:2050312118773953

115. Camkurt MA, Findikli E, Izci F et al (2016) Evaluation of malondialdehyde, superoxide dismutase and catalase activity and their diagnostic value in drug naive, first episode, non-smoker major depression patients and healthy controls. Psychol Res 238:81–85

116. Kappelmann N, Adoth JA, Georgakis DC et al (2020) Dissecting the association between inflammation, metabolic dysregulation and specific depressive symptoms: a genetic correlation and 2-sample mendelian randomization study. JAMA Psychiat 78(2):161–170

117. Brand SJ, Möller M, Harvey BH (2015) A review of biomarkers in mood and psychotic disorders: a dissection of clinical vs. preclinical correlates. Curr Neuropharmacol 13:324–368

118. Krishnan V, Nestler EJ (2008) The molecular neurobiology of depression. Nature 455:894–902

119. Papp M, Klimek V, Willner P (1994) Parallel changes in dopamine D2 receptor binding in limbic forebrain associated with chronic mild stress-induced anhedonia and its reversal by imipramine. Psychopharmacol (Berl) 115:441–446

120. Abrahamson I, Seligman MEP (1977) Modeling psychopathology in the laboratory: history and rationale. In: Maseri J, Seligman MEP (eds) Psychopathology experimental models. Freemen, San Francisco

121. Willner P (1984) The validity of animal models of depression. Psychopharmacology 83:1–16

122. Willner P (1997) Validity, reliability and utility of the chronic mild stress model of depression: a 10-year review and evaluation. Psychopharmacology 134(4):319–329

123. Cryan JF, Holmes A (2005) The ascent of the mouse and advances in modelling human depression. Nat Dev Drug Discov 4:775–790

124. McArthur R, Borsini E (2006) Animal models of depression in drug discovery: a

historical perspective. Pharmacol Biochem Behav 84(3):436–452

125. Berton O, Hahn CG, Thase ME (2012) Are we getting closer to valid translational models for major depression? Science 338:75–79

126. Robinson ESJ (2016) Improving the translational validity of methods used to study depression in animals. Psychopath Dev 3: 41–63

127. Cryan JF, Slattery DA (2007) Animal models of mood disorders: recent developments. Curr Opin Psychiatry 20:1–7

128. Frazer A, Marilak DA (2005) What should animal models of depression model? Neurosci Biobehav Rev 29:515–523

129. Juavinett AL, Erlich JC, Churchland AK (2018) Decision-making behaviors: weighing ethology, complexity and sensorimotor compatibility. Curr Opin Neurobiol 49:42–50

130. Renner MJ, Rosenzweig MR (1987) Enriched and impoverished environments. Springer, New York

131. Benefiel AC, Greenough WT (1998) Effects of experience and environment on the developing and mature brain: implications for animal laboratory housing. Int Lab Animal Res J 39:1–8

132. Perez C, Canal JR, Damingues E et al (1997) Individual housing influences certain biochemical parameters in the rat. Lab Anim 31:357–361

133. Cummins RA, Livesey PJ, Evans JGM (1977) A developmental theory of environmental enrichment. Science 197:692–694

134. Joseph R, Gallagher RE (1980) Gender and early environment influences on activity, over responsiveness and exploration. Dev Psychobiol 13:527–544

135. Moberg GP, Mench JA (2000) The biology of animal stress, basic principles and implications for animal welfare. CABI Publishing, Wallingford

136. Van de Weerd HA, Aarsen EL, Mulder A et al (2002) Effects of environmental enrichment for mice on variation in experimental results. J Appl Anim Welf Sci 5:87–108

137. Van Loo PLP, Kruitwanger CLJJ, Van Zutphen LFM et al (2000) Modulation of aggression of male mice: influence of cage cleaning regime and cent marks. Anim Welf 9:281–295

138. Kelliher P, O'Connor TJ, Harkin A et al (2000) Varying responses in the rat forced swim test under diurnal and nocturnal conditions. Physiol Behav 69:531–539

139. Gerlai R, Clayton NS (1999) Analysing hippocampal function in transgenic mice in ethological perspective. Trends Neurosci 22: 47–51

140. Olsson R, Livesey CM, Patterson-Kane EG et al (2003) Understanding behaviour; the relevance of ethological approaches in laboratory animal science. Appl Anim Behav Sci 81: 245–264

141. Abelaira HM, Reus GZ, Quevedo J (2013) Animal models as tools to study the pathophysiology of depression. Rev Bras Psiquiatr 35(Suppl 2):S112–S120

142. Deussing JM (2006) Animal models of depression. Drug Discov Today Dis Models 3(4):375–383

143. Porsolt RD, Le Pichon M, Jalfre M (1977) Depression: a new animal model sensitive to antidepressant treatments. Nature 266(5604):730–732

144. Yan HC, Cao X, Das M, Zhu XH, Gao TM (2010) Behavioral animal models of depression. Neurosci Bull 26(4):327–337

145. Slattery DA, Cryan JF (2012) Using the rat forced swim test to assess antidepressant-like activity in rodents. Nat Protoc 7:1009–1014

146. Czeh B, Fuchs E, Wiborg O, Simon M (2016) Animal models of major depression and their clinical implications. Prog Neuro-Psychopharmacol Biol Psychiatry 64: 293–310

147. Pryce CR, Azzinnari D, Spinelli S, Seifritz E, Tegethoff M, Meinlschmidt G (2011) Helplessness: a systematic translational review of theory and evidence for its relevance to understanding and treating depression. Pharmacol Ther 132(3):242–267

148. Willner P (2005) Chronic mild stress (CMS) revisited: consistency and behavioural-neurobiological concordance in the effects of CMS. Neuropsychobiology 52(2):90–110

149. Wiborg O (2013) Chronic mild stress for modeling anhedonia. Cell Tissue Res 354(1):155–169

150. Willner P, Muscat R, Papp M (1992) Chronic mild stress-induced anhedonia: a realistic animal model of depression. Neurosci Biobehav Rev 16(4):525–534

151. Heim C, Binder EB (2012) Current research trends in early life stress and depression: review of human studies on sensitive periods, gene-environment interactions, and epigenetics. Exp Neurol 233(1):102–111

152. Heim C, Nemeroff CB (2001) The role of childhood trauma in the neurobiology of mood and anxiety disorders: preclinical and clinical studies. Biol Psychiatry 49(12): 1023–1039

153. Reus GZ, Stringari RB, Ribeiro KF, Cipriano AL, Panizzutti BS, Stertz L et al (2011) Maternal deprivation induces depressive-like behaviour and alters neurotrophin levels in the rat brain. Neurochem Res 36(3):460–466

154. Ruedi-Bettschen D, Zhang W, Russig H, Ferger B, Weston A, Pedersen EM et al (2006) Early deprivation leads to altered behavioural, autonomic and endocrine responses to environmental challenge in adult Fischer rats. Eur J Neurosci 24(10): 2879–2893

155. Bjorkvist K (2001) Social defeat as a stressor in humans. Physiol Behav 73(3):435–442

156. Blanchard RJ, McKittrick CR, Blanchard DC (2001) Animal models of social stress: effects on behavior and brain neurochemical systems. Physiol Behav 73(3):261–271

157. Rygula R, Abumaria N, Flugge G, Fuchs E, Ruther E, Havemann-Reinecke U (2005) Anhedonia and motivational deficits in rats: impact of chronic social stress. Behav Brain Res 162(1):127–134

158. Fuchs E, Czech B, Kale MH et al (2004) Alterations in neuroplasticity in depression: the hippocampus and beyond. Eur Neuropsychopharmacol 14:S481–S490

159. Song C, Leonard BE (1995) The effect of olfactory bulbectomy in the rat, alone and in combination with antidepressants and endogenous factors, on immune function. Hum Psychopharmacol 10:7–18

160. Jesberger JA, Richardson JS (1988) Brain output dysregulation induced by olfactory bulbectomy: an approximation in the rat of major depressive disorder in humans? Int J Neurosci 38(3–4):241–265

161. Croy J, Symmank A, Schellong J et al (2014) Olfaction as a model of depression. J Affect Dis 160:80–86

162. Mucignata-Caretta C, Bondi M, Caretta A (2006) Time course of alterations after olfactory bulbectomy in mice. Physiol Behav 89: 637–643

163. Mucignata-Caretta C, Caretta S (2004) Animal models of depression: olfactory lesions affect amygdala, subventricular zone and aggression. Neurobiol Dis 16:386–395

164. Yuan TF, Skolnick BM (2014) Role of olfactory system dysfunction in depression. Proc Neuropsychopharmacol Biol Psychiat 54: 26–30

165. Apfelbach R, Blanchard CD, Blanchard RJ, Hayes RA, McGregor IS (2005) The effects of predator odors in mammalian prey species: a review of field and laboratory studies. Neurosci Biobehav Rev 29:1123–1144

166. Burgado J, Harrell CS, Eacret D et al (2014) Two weeks of predatory stress induces anxiety-like behavior with co-morbid depressive-like behavior in adult male mice. Behav Brain Res 275:120–125

167. Wu YP, Gao HY, Ouyang SH et al (2019) Predator stress-induced depression is associated with inhibition of hippocampal neurogenesis in adult male mice. Neural Regen Res 14:298

168. Howard DM, Adams MJ, Clarke TK et al (2019) Genome-wide meta-analysis of depression identifies 102 independent variants and highlights the importance of the prefrontal brain regions. Nat Neurosci 22: 343–352

169. Nam H, Clinton SM, Jackson NL, Kerman IA (2014) Learned helplessness and social avoidance in the Wistar-Kyoto rat. Front Behav Neurosci 8:109

170. Winter C, Vollmayr B, Djodari-Irani A et al (2011) Pharmacological inhibition of the lateral habenula improves depressive-like behavior in an animal model of treatment resistant depression. Behav Brain Res 216:463–465

171. Planchez B, Surget A, Belzung C (2019) Animal models of major depression: drawbacks and challenges. J Neural Transm 126: 1383–1408

172. Cryan JF, Mombereau C (2004) In search of a depressed mouse: utility of models for studying depression-related behavior in genetically modified mice. Mol Psychiatry 9: 326–357

173. Ramboz S, Oosting R, Amara DA, Kung HF, Blier P, Mendelsohn M et al (1998) Serotonin receptor 1A knockout: an animal model of anxiety-related disorder. Proc Natl Acad Sci U S A 95:14476–14481

174. Parks CL, Robinson PS, Sibille E, Shenk T, Toth M (1998) Increased anxiety of mice lacking the serotonin1A receptor. Proc Natl Acad Sci U S A 95:10734–10739

175. Heisler LK, Chu HM, Brennan TJ, Danao JA, Bajwa P, Parsons LH et al (1998) Elevated anxiety and antidepressant-like responses in serotonin 5-HT1A receptor mutant mice. Proc Natl Acad Sci U S A 95:15049–15054

176. Mayorga AJ, Dalvi A, Page ME, Zimov-Levinson S, Hen R, Lucki I (2001) Antidepressant-like behavioral effects in 5-hydroxytryptamine (1A) and 5-hydroxytryptamine(1B) receptor mutant mice. J Pharmacol Exp Ther 298:1101–1107

177. Knobelman DA, Hen R, Blendy JA, Lucki I (2001) Regional patterns of compensation following genetic deletion of either

5-hydroxy- tryptamine (1A) or 5-hydroxytryptamine(1B) receptor in the mouse. J Pharmacol Exp Ther 298: 1092–1100

178. Knobelman DA, Hen R, Lucki I (2001) Genetic regulation of extra- cellular serotonin by 5-hydroxytryptamine (1A) and 5-hydroxytryptamine(1B) autoreceptors in different brain regions of the mouse. J Pharmacol Exp Ther 298:1083–1091

179. Boutrel B, Monaca C, Hen R, Hamon M, Adrien J (2002) Involvement of 5-HT1A receptors in homeostatic and stress-induced adaptive regulations of paradoxical sleep: studies in 5-HT1A knock-out mice. J Neurosci 22:4686–4692

180. De Groote L, Olivier B, Westenberg HG (2002) The effects of selective serotonin reuptake inhibitors on extracellular 5-HT levels in the hippocampus of 5-HT(1B) receptor knockout mice. Eur J Pharmacol 439: 93–100

181. Malagie I, David DJ, Jolliet P, Hen R, Bourin M, Gardier AM (2002) Improved efficacy of fluoxetine in increasing hippocampal 5-hydroxytryptamine outflow in 5-HT(1B) receptor knock-out mice. Eur J Pharmacol 443:99–104

182. Holmes A, Hollon TR, Gleason TC, Liu Z, Dreiling J, Sibley DR et al (2001) Behavioral characterization of dopamine D5 receptor null mutant mice. Behav Neurosci 115: 1129–1144

183. Holmes A, Yang RJ, Murphy DL, Crawley JN (2002) Evaluation of antidepressant-related behavioral responses in mice lacking the serotonin transporter. Neuropsychopharmacology 27:914–923

184. Li Q, Wichems C, Heils A, Van De Kar LD, Lesch KP, Murphy DL (1999) Reduction of 5-hydroxytryptamine (5-HT)(1A)-mediated temperature and neuroendocrine responses and 5-HT(1A) binding sites in 5-HT transporter knockout mice. J Pharmacol Exp Ther 291:999–1007

185. Gobbi G, Murphy DL, Lesch K, Blier P (2001) Modifications of the serotonergic system in mice lacking serotonin transporters: an in vivo electrophysiological study. J Pharmacol Exp Ther 296:987–995

186. Schramm NL, McDonald MP, Limbird LE (2001) The alpha (2a)-adrenergic receptor plays a protective role in mouse behavioral models of depression and anxiety. J Neurosci 21:4875–4882

187. Sallinen J, Haapalinna A, MacDonald E, Viitamaa T, Lahdesmaki J, Rybnikova E et al

(1999) Genetic alteration of the alpha2-adrenoceptor subtype c in mice affects the development of behavioral despair and stress-induced increases in plasma corticosterone levels. Mol Psychiatry 4:443–452

188. Haller J, Bakos N, Rodriguiz RM, Caron MG, Wetsel WC, Liposits Z (2002) Behavioral responses to social stress in noradrenaline transporter knockout mice: effects on social behavior and depression. Brain Res Bull 58:279–284

189. Cases O, Seif I, Grimsby J, Gaspar P, Chen K, Pournin S et al (1995) Aggressive behavior and altered amounts of brain serotonin and norepinephrine in mice lacking MAOA. Science 268:1763–1766

190. Evrard A, Malagie I, Laporte AM, Boni C, Hanoun N, Trillat AC et al (2002) Altered regulation of the 5-HT system in the brain of MAO- a knock-out mice. Eur J Neurosci 15: 841–851

191. Mombereau C, Kaupmann K, Sansig S, van der Putten H, Cryan JF (2003) GABAB receptors play a key role in the modulation of anxiety, depression-related behaviours. Behav Pharmacol 14(Suppl 1):24

192. Grimsby J, Toth M, Chen K, Kumazawa T, Klaidman L, Adams JD et al (1997) Increased stress response and beta-phenylethylamine in MAOB-deficient mice. Nat Genet 17: 206–210

193. Filliol D, Ghozland S, Chluba J, Martin M, Matthes HW, Simonin F et al (2000) Mice deficient for delta- and mu-opioid receptors exhibit opposing alterations of emotional responses. Nat Genet 25:195–200

194. Stork O, Ji FY, Kaneko K, Stork S, Yoshinobu Y, Moriya T et al (2000) Postnatal development of a GABA deficit and disturbance of neural functions in mice lacking GAD65. Brain Res 865:45–58

195. Miyamoto Y, Yamada K, Noda Y, Mori H, Mishina M (2002) Nabeshima T (2002) lower sensitivity to stress and altered mono-aminergic neuronal function in mice lacking the NMDA receptor epsilon 4 subunit. J Neurosci 22:2335–2342

196. Santarelli L, Gobbi G, Blier P, Hen R (2002) Behavioral, physiologic effects of genetic or pharmacologic inactivation of the substance P receptor (NK1). J Clin Psychiatry 63 (Suppl 11):11–17

197. Froger N, Gardier AM, Moratalla R, Alberti I, Lena I, Boni C et al (2001) 5-Hydroxytryptamine (5-HT)1A autoreceptor adaptive changes in substance P (neurokinin 1) receptor knock-out mice mimic

antidepressant-induced desensitization. J Neurosci 21:8188–8197

198. Bilkei-Gorzo A, Racz I, Michel K, Zimmer A (2002) Diminished anxiety-, depression-related behaviors in mice with selective deletion of the Tac1 gene. J Neurosci 22: 10046–10052

199. Montkowski A, Barden N, Wotjak C, Stec I, Ganster J, Meaney M et al (1995) Long-term antidepressant treatment reduces behavioural deficits in transgenic mice with impaired glucocorticoid receptor function. J Neuroendocrinol 7:841–845

200. Groenink L, Dirks A, Verdouw PM, Schipholt M, Veening JG, van der Gugten J et al (2002) HPA axis dysregulation in mice overexpressing corticotropin releasing hormone. Biol Psychiatry 51:875–881

201. Bale TL, Vale WW (2003) Increased depression-like behaviors in corticotropin-releasing factor receptor-2-deficient mice: sexually dichotomous responses. J Neurosci 23:5295–5301

202. van Gaalen MM, Stenzel-Poore MP, Holsboer F, Steckler T (2002) Effects of transgenic overproduction of CRH on anxiety-like behaviour. Eur J Neurosci 15: 2007–2015

203. Yamada K, Lida R, Miyamoto Y, Saito K, Sekikawa K, Seishima M, Nabeshima T (2000) Neurobehavioral alterations in mice with a targeted deletion of the tumor necrosis factor-alpha gene: implications for emotional behavior. J Neuroimmunol 111(1–2): 131–138

204. Calapai G, Crupi A, Firenzuoli F, Inferrera G, Ciliberto G, Parisi A et al (2001) Interleukin-6 involvement in antidepressant action of Hypericum perforatum. Pharmacopsychiatry 34(Suppl 1):S8–S10

205. El Yacoubi M, Ledent C, Parmentier M, Costentin J, Vaugeois JM (2001) Adenosine A2A receptor knockout mice are partially protected against drug-induced catalepsy. Neuroreport 12:983–986

206. Martin M, Ledent C, Parmentier M, Maldonado R, Valverde O (2002) Involvement of CB1 cannabinoid receptors in emotional behaviour. Psychopharmacology 159: 379–387

207. Gavioli EC, Marzola G, Guerrini R, Bertorelli R, Zucchini S, De Lima TC et al (2003) Blockade of nociceptin/orphanin FQ-NOP receptor signalling produces antidepressant-like effects: pharmacological and genetic evidences from the mouse forced swimming test. Eur J Neurosci 17: 1987–1990

208. Tschenett A, Singewald N, Carli M, Balducci C, Salchner P, Vezzani A et al (2003) Reduced anxiety and improved stress coping ability in mice lacking NPY-Y2 receptors. Eur J Neurosci 18:143–148

209. Lobo MK, Nestler EJ, Covington HE (2012) Potential utility of optogenetics in the study of depression. Biol Psychiatry 71:1068–1074

210. Muir J, Lopez J, Bagot RC (2019) Wiring the depressed brain: optogenetic and chemogenetic circuit interrogation in animal models of depression. Neuropsychopharmacology 44:1013–1026

211. Koolhaas JM, Kortre SM, DeBoer SF et al (1999) Coping styles in animals: current status in behaviour and stress-physiology. Neurosci Biobehav Rev 23:925–935

212. Dopfel D, Perez PD, Verbitsky A et al (2019) Individual variability in behaviour: functional networks predict vulnerability using an animal model of PTSD. Nat Commun 10:23–27

213. Khan KM, Collier ADS, Meshalkina DA et al (2017) Zebrafish models in neuropsychopharmacology and CNS discovery. Brit J Pharmacol 174:1925–1944

214. Kalueff AV, Stewart AM, Gerlai R (2014) Zebrafish are an emerging model for studying complex brain disorders. Trends Pharmacol Sci 35:63–75

215. Lieschke GJ, Currie PD (2007) Animal models of human disease: zebrafish swim into view. Nat Rev Genet 8:353–355

216. Dawson C, Skyner LJ, Ryan CP et al (2014) Shyness-boldness but not exploration, predicts glucocorticoid-stress response in Richardson ground squirrels. Ethology 120: 1101–1107

217. Oswald ME, Singer M, Robinson BD (2013) The quantitative genetic architecture of the bold-shy continuum in zebrafish (Danio rerio). PLoS One:e68628

218. White JR, Meeken MG, McCormick MJ, Ferran MCO (2013) Comparison of measures of boldness and their relationships to survival in young fish. PLoS One 8:e68828

219. Haemisch A, Voss T, Gartner K (1994) Effects of environmental enrichment on aggressive behaviour, dominance hierarchies and endocrine status in DBA/2J mice. Physiol Behav 56:1041–1048

220. Frost AI, Winrow-Giffen A, Ashley PJ, Sneddon LV (2007) Plasticity of animal personality traits: does prior experience alter the degree of boldness? Prog Roy Soc Biol Sci 274:333–339

221. Demin KA, Lakstygal AM, Volgin AD et al (2020) Cross species analysis of intraspecies

behavioural differences in mammals and fish. Neurosci Rev 429:33–45

222. Costa FV, Canzian J, Stefanello FV, Kalueff AV, Rosemberg DB (2019) Naloxone prolongs abdominal constriction in a zebrafish based pain model. Neurosci Lett 708: 1343–1346

223. Egan RJ, Bergner CL, Hart PC et al (2009) Understanding behavioural and physiological phenotypes of stress and anxiety in zebrafish. Brain Behav Res 205:38–44

224. De Abreu M, Genario R, Giacomini A et al (2018) Zebrafish as a model of neurodevelopmental disorders. Neuroscience 445:3–11

Chapter 6

Animal Model Approaches to Understanding the Neurobiology of Suicidal Behavior

Raquel Romay-Tallon and Graziano Pinna

Abstract

Severe psychiatric conditions, including major depressive disorder (MDD) and post-traumatic stress disorder (PTSD), are characterized by a strong comorbidity with suicide ideations and attempts. Predictors of suicide risk in humans include impulsivity and aggressiveness, which can be reproduced in rodents by aggressive behavior using the resident-intruder test and other behavioral tests that measure impulsivity, irritability, and hopelessness. Currently, biomarkers or therapies to treat suicide behaviors are lacking. Animal models of behavioral traits of suicide in humans provide means to better understand the neurobiology of suicidal behavior, as well as help to understand neural mechanisms and circuits underlying suicide. The search for biomarker-tailored therapies against suicide is urgently needed. Neurosteroid biosynthesis modulates emotional state and stress response, and evidence shows that its downregulation during pathophysiological conditions underlays rapid changes in mood and may result in affective disorders. Similarly, several lines of preclinical and clinical evidence suggest that these neuromodulators may underlay suicidal risk and thereby may offer valuable biomarkers. Hereinafter, we analyze animal model approaches that reproduce suicidal-like behaviors in rodents and highlight findings on the role of neurosteroid biosynthesis in suicidal behavior observed in stress models of PTSD/suicide in humans.

Key words Social isolation, PTSD, Suicide, Aggressive behavior, Neurosteroids, Endocannabinoid system

1 Introduction

During the ongoing COVID-19 pandemic, suicide attempts have increased dramatically. This adds to already grim suicide death rates that have been increasing for the past several years. Indeed, the latest report from the Center for Disease Control and Prevention has placed suicide among the 10 main causes of death (https://www.cdc.gov/nchs/fastats/leading-causes-of-death.htm). Suicide and suicidal ideation show a crescent incidence within the general population with an increment from 10.5% in 100,000 habitants in 1999 to 14% in 100,000 in 2018. Worldwide, about one million people die for suicide every year, and suicide is the tenth leading

Yong-Ku Kim and Meysam Amidfar (eds.), *Translational Research Methods for Major Depressive Disorder*, Neuromethods, vol. 179, https://doi.org/10.1007/978-1-0716-2083-0_6,

cause of death and the second leading cause of death in young individuals in the age range of 15–29 years (World Health Organization, 2012). It is reported as the second cause of death among the American population within the age range of 10–34 years old (https://www.cdc.gov/media/releases/2018/p0607-suicide-prevention.html). The suicide rate, including suicide attempt and ideation, is threefold higher in specific groups of populations, such as veteran communities and victims of child abuse and violence [1–3].

Suicide is commonly comorbid with mental health conditions, for instance, with post-traumatic stress disorder (PTSD) and/or major depressive disorder (MDD), exacerbated by intrapersonal traits, life stressors, poor coping skills, substance abuse, and access to weapons [4]. Interestingly, among contributing factors to suicide, social isolation is probably one of the most severe [5]. In recent years, during the COVID-19 pandemic, social isolation and lack of physical and social contacts have played a toll both on the prevalence of mood disorders and in the incidence of suicide attempts [6].

Importantly, the field of psychiatry remains backward in establishing efficacious treatments based on diagnostic biomarkers and predictors of treatment efficacy. The first-line treatment for mental disorders, especially PTSD and depression, are the selective serotonin reuptake inhibitors (SSRIs); however, only 50% of the patients improve, and at least 30% show remission of symptoms upon treatment discontinuation [7]. Paradoxically, SSRIs appeared to induce suicide ideation in patients under 18 years old (www.fda.gov/downloads/Drugs/DrugSafety). This evidence leaves a scenario with hardly any available treatment to offer in support of the youngest psychiatric patients. Progress in the development of more efficient treatments and biomarkers for psychiatric disorders, including suicide, offer a slow advance due to the complexity of neurobiological mechanisms of these disorders. In addition, affected individuals present comorbidities with other psychopathologies, such as depression and drug abuse, which makes it even more complex the diagnosis and improvement of symptoms following pharmacological therapies [8]. Therefore, the development of effective pharmacological treatments and the identification of biomarkers for mental disorders, particularly for MDD, or PTSD aggravated by suicide constitute a key priority in the scientific psychiatric community. Biomarkers for MDD, PTSD, and suicide relate to the assessment of neurobiological values easily accessible in plasma, serum, or cerebrospinal fluid [8].

Hereafter, we will review aspects of the neurobiology of suicide through the study of preclinical models with specific emphasis to potential biomarker discovery and treatment targets for mood disorders highly comorbid with suicide.

2 Animal Model Approaches to Understanding the Neurobiology of Suicidal Behavior

Animal models constitute an indispensable tool to study the pathophysiology of psychiatric disorders by mimicking underlying behavioral endophenotypes. Importantly, animal models allow the validation of neurobiological markers useful for diagnosis as well as the development of novel pharmacological treatments [9, 10].

To model suicidal behavior in the laboratory, it is necessary to put the focus on certain behavioral features that resemble suicidal attempts and suicidal ideation in humans, as animal models do not show face validity for suicide. Clinical studies indicate suicidal ideation can be associated with four personality features: aggressiveness, impulsivity, irritability, and hopelessness [11–13].

2.1 Aggressiveness

Numerous investigations evidenced a strong correlation of suicide and suicide attempts in young individuals with a history of aggressiveness and impulsivity. These results seem to be consistent even after taking into consideration underlying psychopathologies [14]. Aggressive behavior can be reproduced in the laboratory in preclinical models by using the resident-intruder paradigm [15] (see Fig. 1, panel 1a–d). The resident and the intruder should be sex-, age- and weight-matched. They are placed together in an environment familiar for the resident; usually, the test is conducted in the resident cage where the resident mouse has been housed for a period of time that varies from days to weeks prior to the test. The aggressive behavior is usually assessed for 10 min after placing the intruder together with the resident. Generally, the resident mouse will be the first to show an offensive aggression to intimidate the opponent, in this case the intruder mouse. During resident-intruder interactions, five main dissuading behavioral and violent traits are evaluated: latency to attack, time spent for the resident to perform the first attack (this measure can also be included as impulsivity), tail rattling (the animal exerts rapid lateral tail movements before or after an attack) (see Fig. 1, panel 1b), wrestling (the two animals stand in an upright posture and violently shove and spar) (see Fig. 1, panel 1c), chasing (the resident pursuits the intruder across the cage), and bite attacks (the resident violently attacks the intruder in vulnerable areas of the body such as throat and/or neck) (see Fig. 1, panel 1d). During the resident-intruder test, the animals should be monitored to avoid life-threatening harm, which would result in an anticipated endpoint of the test [11, 15].

These forms of aggressive behavior can be reduced by acute treatment with the conventional antidepressants, i.e., SSRIs, tricyclic antidepressants (TCA), or monoamine oxidase inhibitors (MAOIs). Conversely, their chronic administration may exacerbate

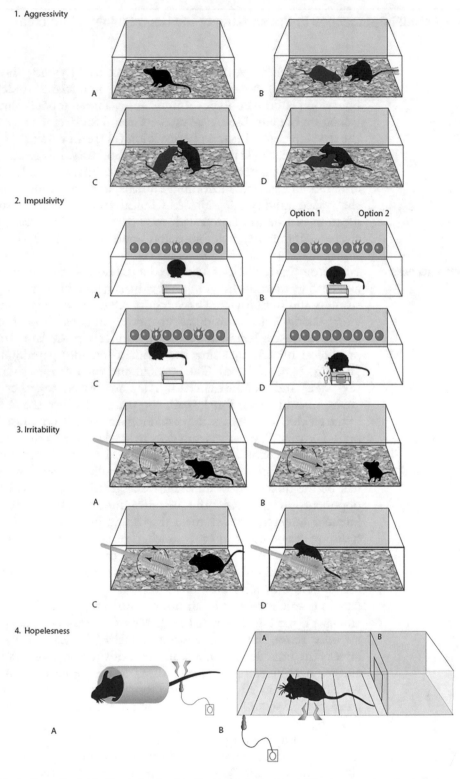

Fig. 1 Evaluation of suicide-like behavioral traits. Panel 1: Resident-intruder paradigm to evaluate aggressivity. (**a**) The test is performed in the resident's cage, where a sex-, age-, and weight-matched intruder is placed. (**b**) Tail rattling: the resident intimidates the intruder shaking its tail. (**c**) Wrestling: the two animals

aggressive behavior in rats [16]. Similarly, suicide ideation and attempts have been shown to be exacerbated in children after treatment with high dosage of SSRIs [17]. However, new studies in the field also suggest that using low SSRI doses may provide a strategy to treat suicidal teens more safely [18].

2.2 Impulsivity

The role of impulsivity in suicidal behavior is still controversial; however, studies support its implication in suicide [19]. It is defined as incapacity of inhibition of behavior and lack of reflection of the consequences of particularly dangerous behaviors [20]. This feature is highly influenced by the age of the patient, given that studies have demonstrated that suicide attempts committed by younger individuals are the consequence of an impulsive act. Conversely, adult individuals that commit suicide are less violent and exhibit fewer suicidal attempts. Aggressive behavior in these individuals is significantly correlated with impulsivity. This behavior can be studied in the laboratory by measuring impulsive decision-making and impulsive action in rodents.

Impulsive decision-making is mainly assessed with operant paradigms (see Fig. 1, panel 2a–d). The most widely used is probably the operant delayed reinforcement tasks. In this test, the rodents must perform a series of tasks associated with different delayed reward periods. The operant tests evaluate whether the rodent selects a more immediate reward or a smaller or larger but delayed one [21]. The animals are placed in a box containing nine windows where the rewards are presented. Prior to the test, the animals undergo water deprivation for 20 h during the first 4 days and 22 h the next 10 days until weight stabilizes. Following this period, the animals are habituated to 10% condensed milk. Before

---

Fig. 1 (continued) increment their physical interaction. (**d**) Bite attack: the resident attacks the intruder by biting its throat, back, and head. Panel 2: Operant paradigm to evaluate impulsivity. (**a**) The tested rodent is presented to nine windows. A light stimulus starts and the rodent starts the test by nose-poking the window. (**b**) Two light stimuli turn on presenting two options: option 1, a small reward, water, with short delay; option 2, a large reward, 10% condensed milk, with a larger delayed. (**c**) The rodent displays impulsive behavior by choosing option 1 and receiving water. (**d**) A light turns on in the window where the reward will be presented. The animal must nose-poke the window to receive the reward. Panel 3: Bottle brush test to measure irritability. (**a**) A white bottle brush (rotating) is presented to the rodent in the opposite corner of the cage. (**b**) The bottle brush (rotating) approaches the animal and touches it in the whiskers, mimicking an attack. (**c**) The animal can respond to the "attack" by approaching the object displaying aggressive behavior. The bottle brush returns to the initial position (rotating). (**d**) The rodent manifests irritable behavior by attacking the bottle brush. The test ends with the bottle brush immobile in the initial position. Panel 4: Learned helplessness paradigm to measure hopelessness. (**a**) The animal is previously exposed to 120 electroshocks in the tail for 60 min. During the training sessions, the animal is placed in a restrainer, a different scenario than that where the test will be performed to avoid the "contextual fear" variable. (**b**) The animal is placed in a shuttle box with grid floors and will receive 30 electroshocks (right panel, a). The box is then open so the animal can escape the electroshocks by entering the compartment (right panel, b)

the test is performed, the animals are presented for 4 consecutive days to four sessions of 10 min each when they have free access to water and condensed milk. Once they are habituated to the rewards, they are trained to nose-poke the center hole as a response for a light stimulus to start the trial (see panel 2a). A small reward, water, will be presented in the left side with no delay, and a large reward, 10% condensed milk, will be presented in the right side with an increased delay (see panel 2b). The side of the cage where the reward is presented is counterbalanced between the experimental animals. The animal needs to nose-poke the window with the chosen reward (see panel 2c). Finally, the animal will receive the reward in a separated window after the light stimulus is off (see panel 2d). Thus, impulsiveness is measured by preference for a small but immediate reward versus a larger but delayed reward [22].

Impulsive action measures the active mechanism of inhibiting a pre-learned response toward a stimulus. The go/no-go, stop-change, and stop-signal reaction time are the most used paradigms to assess an impulsive action. During these tests, the experimental animal learns the go response toward a cued signal, and after subset of trials, the no-go signal is presented, so the animal must inhibit its response. In the stop-change time, the no-go signal is presented following the go signal. The rodents need to inhibit their behavior to receive the reward [23]. Interestingly, the stop-reaction time test measures the latency of the animal to inhibit their response. The stop-reaction time is paired with the go/no-go or stop-change paradigm.

The impulsive behavior can be as well measured with the five-choice serial reaction time task, which also measures behavioral aspects, such as attention and motivation. In this test, the animals will respond in one of the five response apertures after seeing a light stimulus. There is a 5-s interval between trials during which the rodents must refrain from responding. Responses during the 5 s intertrial window are considered a premature response, an act of impulsivity, and it is punished [21].

2.3 Irritability

There is a strong association between irritability and suicide ideation and attempts, and impulsivity, and age plays an important role in modulating it. Irritability is defined as a reduced control over the tempter and irascible behavior [24]. Patients with suicidal ideation present almost 1.5-fold more irritable behavior than those without suicidal thoughts. Also, adolescents with a more irritable behavior are more likely to consider suicide [25]. Interestingly, irritability is also considered a subsyndromal symptom of depression, as patients, particularly children, with psychiatric conditions display irritable behaviors.

The assessment of irritability in preclinical models is complicated as, unlike the beforementioned characteristics, irritability is not purely behavioral and often intermingles with aspects of aggressivity. The most common models are pharmacological paradigms, such as opiate or ethanol withdrawal, although some models establish a hostile environment that induces such behavior. Irritability is commonly assessed by measuring the resistance and aggressiveness to tactile and auditory stimuli. Irritability can be measured as the struggle behavior to be picked up, restrained in supine position, and, in mice, held by the tail [26]. This behavioral feature can also be studied using the bottle brush method [27] (see Fig. 1, panel 3a–d). A bottle brush is placed in the cage of the subject and simulating attacks. Typically, the subject will be attacked 20 times a day, with 15-s interval between attacks. Each attack is composed by five different stages, and each stage lasts 2 s. In a first stage, the brush is rotated toward the subject from the opposite side of the cage, which is established as the starting point (see panel 3a). Following this, the brush in rotation touches the rodent whiskers and returns to the starting point [27] (see panel 3b, c). The brush stays rotating in the starting point. Finally, it remains immobile in the starting point (see panel 3d). In a variant of the bottle brush test, the object returns to the starting point of the cage and stays immobile in vertical position. Different aspects of behavior are analyzed, including escape, biting, digging, jumping out of the cage, boxing (drums the wall of the cage with its front paws), washing its face, climbing on the edge of the cage, following and exploring the brush, and tail rattling [28].

2.4 Hopelessness

Hopelessness is the most persistent behavior in suicidal patients. In fact, hopelessness and low self-esteem are highly correlated with suicidal ideation and attempts. Furthermore, studies in patients with psychiatric disorders have reported that feelings of hopelessness are significantly higher in suicide completers. Like irritability, hopelessness is a subjective feature which makes it difficult to model. The most accepted method to assess this behavioral feature is the learned helplessness paradigm where rodents are exposed to unpredictable and inescapable stressors, such as foot shock. This test is also commonly used to evaluate depressive-like behavior in rodents [29]. The learned helplessness paradigm commonly uses mild electric foot shocks as stressor, and it measures the failure to escape from this stressor. Often, the experimental animals try to avoid electroshocking by touching the grid with their fur; therefore, electrodes can also be placed in the tail [30].

The animal is placed in automated shuttle boxes with grid floors to deliver foot shocks. The animals have the capacity to escape. The animals need to be first exposed to the stressor in the training sessions, often two sessions, followed by the actual test when failure to escape is evaluated [31]. During training sessions, the tested animal is exposed to 120 electroshocks of 0.15 mA of amplitude for 5 s, with intertrial interval of 25–35 s. Each training session lasts 60 min. Often, training and test are performed in the same context by adding the "contextual fear" variable. This implies a confounding effect as the failure to escape responds to fear conditioning instead of helplessness, which is promoted by different neuronal circuitries. Thus, it is recommended to use two different contexts for training and testing. Therefore, during training sessions, the animals can be placed in restrainers; 120 electroshocks are delivered to the tail for 60 min, with a crescent amplitude starting at 0.25 mA–0.60 mA, incrementing 0.05 mA every 15 shocks (see Fig. 1, panel 4a). There intertrial interval is also 25–35 s. During the testing day, the rodents are moved to the behavioral room using a different route than that used during the training session, and they are placed in the shuttle box where they will be allowed to habituate for 1 min, followed by 30 electroshocks with 0.10 mA amplitude for 30 s (see Fig. 1, panel 4b) [30].

Unlike for aggressiveness, impulsivity, and irritability, preclinical models of hopelessness present a good predictive validity, which is further supported by the finding that this behavior is improved with the administration of SSRIs, TCA, and MAOIs.

Taking all together, the development of preclinical model for suicide is challenging as it is not possible to assess suicidal ideation in rodents. However, the ability to model the main factors involved in the mental process of suicide offers a remarkable tool to understand the neurobiology of suicide and, more importantly, to validate biomarkers and develop more efficient pharmacological therapies.

3   Modelling Suicidal Behavior in Socially Isolated Rodents

Individual lifetime history of impulsivity and aggressiveness predicts suicide ideations and attempts. A lifetime history of aggression and impulsivity discriminates young individuals who committed suicide later in life [32, 33]. Regrettably, the neurobiological underpinnings underlying impulsivity and aggressive behavior and how they translate into suicidal behavior remain obscure.

Among several stressors and traumatic events, social isolation stress, defined as an inadequate or nonexistent interpersonal contact, has been suggested as a key risk factor for premature death, worsening of mental disorders, and suicide [34–36]. Studies have shown an increased risk of suicide in participants with limited social interaction [37]. Further, a recent study suggests that prisoners present sevenfold increment for risk of suicide compared with the general population. Isolation is worsened by the limited capacity of control over themselves and the living conditions in custody [38]. The impact of social isolation on increasing the rates of suicide has become a relevant topic due to the lockdown and the social restrictions during the COVID-19 pandemic [6].

Several neuroendocrinological mechanisms are unleashed in response to perceived social isolation or loneliness [39]. Scientific evidence shows salivary cortisol concentrations increase in subjects that have been isolated for extended periods of time [40, 41]. In lymphocytes and monocytes, alteration in neutrophil/lymphocytes and in neutrophil/monocyte ratios was observed in socially isolated individuals that suggest glucocorticoid signaling dysfunction [42]. During brain development, exaggerated stress condition plays a role in the vulnerability to stress-induced behavioral dysfunction [43]. Underlying neuroepigenetic mechanism during a vulnerable time window may result in neuroendocrine dysfunction that reflects abnormal behavior in adolescent rodents, which may include HPA axis reactivity, as well as elevated glucocorticoid stress responses [44].

Our laboratory has investigated the effects of social isolation in mice on several neuroendocrine and behavioral parameters as a preclinical model of stress-induced human psychopathologies, including depression, anxiety spectrum disorders, and PTSD [45–47]. Other groups have shown that post-weaning social isolation induces robust aggressive behavior in rodents, which was associated with increased glucocorticoid responses and hyperarousal [44]. In our social isolation model, we unveiled a number of stress-induced emotional behavioral dysfunctions, including exacerbated aggressiveness, increased anxiety- and depressive-like behaviors, and exaggerated contextual fear responses, impaired fear extinction, and fear recall deficits [46, 48–51]. These phenotypes are evidenced in socially isolated rodents, but more specifically, the extent of aggressive behavior and impulsivity expressed toward a same-sex intruder entails translational relevance to traits of suicidal behavior and PTSD symptoms observed in humans.

Indeed, we showed that during 3–6 weeks of social isolation, male but not female mice develop a time-related increase of aggressive behavior in the resident-intruder test [52–54]. We have also observed that the effects of social isolation in young adolescence on postnatal day 21 (PND 21) in mice are more pronounced and result in a more severe behavioral dysfunction, including severe aggressive behavior than that of this early stress, administered in mice that were isolated during late adolescence on PND 45 [54].

Several studies have demonstrated that aggressive behavior in socially isolated mice is inversely related to a significant downregulation of GABAergic neurosteroid and progesterone derivative, allopregnanolone concentrations [46, 51, 55]. The downregulation of allopregnanolone biosynthesis, which includes both the levels of allopregnanolone and the expression of several neurosteroidogenic enzymes and proteins [51], occurs after weeks of isolation in several cortico-limbic areas that participate in the neurocircuitry underlying the regulation of emotionality and affective behavior, such as the prefrontal cortex (PFC), hippocampus, and basolateral amygdala, and involves the olfactory bulb [46, 51, 55]. This brain allopregnanolone decrease (~50%) reaches significance after 3 weeks of social isolation and is maintained up to 8 weeks of social isolation. During this time, we also measured the highest levels of aggression and impulsivity in mice [51, 55].

Not only aggressive behavior is enhanced during social isolation in rodents, but also contextual fear conditioning responses are increased as well as the expression of anxiety-like behavior [46, 56]. These forms of behavior have been successfully correlated with the downregulation of allopregnanolone in neurons that participate in cortico-limbic circuitry, which encompass the pyramidal neurons of the CA3 and the dentate gyrus glutamatergic granular cells of the hippocampus, as well as the pyramidal-like neurons of the basolateral amygdala and the pyramidal neurons of layers V–VI of the medial prefrontal cortex [56]. Allopregnanolone level decrease appeared to result from the reduction of 5α-reductase type I expression and of the P450scc enzyme [51]. Collectively, these results are in support of social isolation targeting specific cortico-limbic neurons and neurosteroidogenic enzymes that mediate allopregnanolone biosynthesis. These findings provide support for social isolation as an environmental factor important to model behavioral deficits reminiscent of behavioral predictors of suicidal behaviors in humans.

4   Neurocircuitry of Suicide-Like Behavior

Current neurobiological risk factors for suicide have not been identified; hence, there is an urgent need to understand both the neural mechanisms that increase suicide risk and biological markers of suicide risk that can help in the formulation of preventive strategies to target subjects at risk. Animal models of suicide-like behavior and state-of-the-art technologies have recently facilitated the understanding of the neurobiology of suicide, and the dysfunction in brain circuitry that in pathophysiological conditions mediates behavioral traits reminiscent of suicidal behavior in humans [57] (Fig. 2). In the past years, several excellent neuroimaging investigations have helped to elucidate structural and functional brain circuitry correlates of suicidal thoughts and behaviors.

Converging evidence of several studies in the field supports a role for several brain circuitries in suicidal behavior. For instance, abnormalities in the ventral prefrontal cortex (VPFC) system, which includes affective and reward networks (anterior cingulate cortex (ACC), insula, medial and lateral orbitofrontal cortex (OFC), rostral prefrontal cortex (RPFC), ventral striatum), but also alterations in their connections may result in excessive negative states that eventually may precipitate suicidal ideations. Research on the extended VPFC system was previously implicated in suicidal

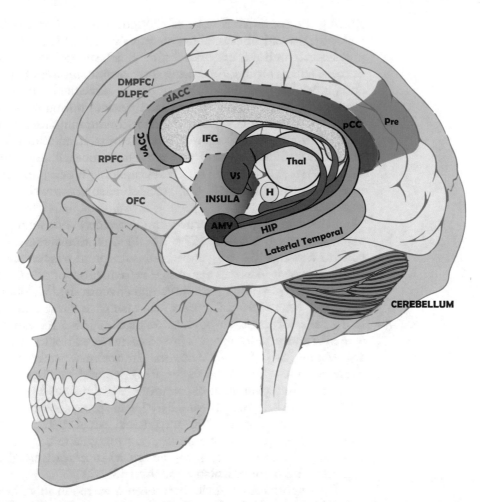

Fig. 2 Suggested circuitry driving suicidal ideation and attempts. Suicidal thoughts may be triggered by an impaired regulation of the activity of brain areas involved in emotional behavior and those involved in cognitive regulation. In the proposed circuits, rostral prefrontal cortex, dorsomedial and dorsolateral prefrontal cortex, and inferior frontal gyrus present a disrupted connectivity. The impaired connectivity is manifested by a dysregulation of the ventral anterior cingulate cortex, orbitofrontal cortex, ventral striatum, posterior cingulate cortex, precuneus, cerebellum, thalamus and limbic areas, amygdala, hippocampus, and lateral temporal area. This generates an exacerbation of negative thoughts, rumination, negative self-referencing, and impoverishment of future thinking. The interconnectivity among these areas might be mediated by dorsal anterior cingulate cortex, insula, and rostral frontal cortex. AMY, amygdala; dACC, dorsal anterior cingulate cortex; DMPFC/DLPFC, dorsomedial and dorsolateral prefrontal cortex; H, hypothalamus; HIP, hippocampus; IFG, inferior frontal gyrus; OFC, orbitofrontal cortex; pCC, posterior cingulate cortex; PRE, precuneus; RPFC, rostral prefrontal cortex; THAL, thalamus; VS, ventral striatum

behaviors including imagining future positive/negative events [58, 59]. A laterodorsal system, including dorsomedial prefrontal cortex (DMPFC), dorsolateral prefrontal cortex (DLPFC), and dorsal anterior cingulate cortex (dACC), in conjunction with the inferior frontal gyrus (IFG), IF, and rostral prefrontal cortex (RPFC), may be responsible for facilitating suicide behaviors by altering the cognitive control of thought and emotional behavior [59, 60]. Instead, abnormalities among VPFC/DPFC/IFG system may lead to high-risk episodes during which suicidal ideations may become lethal through decreasing top-down inhibition of behavior and inflexible decision-making. Thus, abnormal connections among this brain circuitry may lead suicidal thoughts to actual behaviors. The insula is another brain structure implicated in suicidal ideations and their transition to attempts. In addition, the role of dACC and insula circuitry in switching between the extended VMPFC systems and the DPFC/IFG system [59, 61] may constitute hubs that mediate the suicidal ideation transition into attempts.

In preclinical studies, cortico-limbic circuitry consistently modulates aggressive behavior [57]. The PFC represses aggression in rodents and cats [62, 63], and social isolation in rats reduces PFC neurons, which increases aggressive behavior [64]. Positron emission tomography (PET) studies suggest ventromedial, dorsomedial, and orbital frontal cortex decrement in male and female individuals subjected to physical aggression [65]. Impaired dorsomedial and anterior insula is linked to increased risk for impulsive aggression [66]. We showed that in rodents, social isolation depletes allopregnanolone in the PFC pyramidal neurons and in the amygdaloid pyramidal-like neurons, which results in decreased inhibitory $GABA_A$ receptor-mediated neurotransmission and, ultimately, a disruption of PFC-amygdaloid circuitry [56]. Hence, allopregnanolone biosynthesis downregulation in the PFC-amygdaloid circuitries may be crucial to unleash heightened aggressive behavior in isolated rodents [53, 67].

The hypothalamic nuclei have been involved in manifestation of suicide-like behavior by means of enhanced aggressive behavior in mammals [44]. Overlapping between hypothalamic neurons that mediate aggressive, sexual, and fear behaviors has been observed. Stimulation of estrogen receptor (ER)-1 neurons of the ventromedial hypothalamus-ventrolateral area (VMHvl) induces both aggression and sexual behaviors in rodents [68, 69].

Consistently, hypothalamic nuclei involved with aggressive behaviors include the medial preoptic area (MPOA) and the anteroventral periventricular nucleus (AVPV). MPOA neurons that express ER are required for aggression development in mice [70]. Likewise, neurons of the anteroventral paraventricular nucleus (APVP) that express tyrosine-hydroxylase also mediate aggressive behaviors. This is sustained by evidence showing that their stimulation inhibits aggression; on the contrary, when these neurons are removed, aggression is increased [71]. Of note, the direct connection between hypothalamus and periaqueductal gray is important for aggression as was also evidenced by clinical investigations showing that aggressive behavior correlates with the level of activity of these connections [72, 73].

The role of the mesolimbic reward circuitry is mediated by the nucleus accumbens (NAc) on aggressive behavior [74]. When the ventral tegmental area (VTA) is stimulated, dopaminergic neurons enhanced aggression in male mice [75]. Optogenetic approaches show that stimulation of GABAergic neurons that project from NAc and lateral septum (LS) to lateral habenula (LHb) enhances aggression, while their inhibition blunts it [76].

Stimulation of the LS induced impulsivity and aggression suppression, while selective lesions of LS increased aggression [77]. Medial amygdala (MeA) constitutes a crucial cortico-limbic area with a key role in modulating aggression neurocircuitry. Indeed, MeA can stimulate or block aggression and impulsivity in rodents. Stimulation of MeA GABAergic neurons induces aggression, whereas activation of glutamatergic neurons inhibits it [78]. In humans, brain imaging showed amygdalar bidirectional effects in aggressive behavior and impulsivity regulation [79, 80].

5   The Implication of Neurosteroid Biosynthesis with Suicidal Behavior

Neuroactive steroids and neurosteroids are cholesterol-derived molecules that can be either synthesized peripherally by the adrenal glands and ovaries or can be directly synthesized de novo by brain glutamatergic neurons and long-projecting GABAergic neurons [56, 81–83]. The most well-known and characterized neurosteroids include allopregnanolone and its equipotent GABAergic isomer, pregnanolone. Indeed, allopregnanolone and pregnanolone are potent, endogenous, positive, allosteric modulators of synaptic and extrasynaptic $GABA_A$ receptors [84, 85]. This action contrasts with their respective epimers that, rather than stimulating, antagonize the positive modulation of allopregnanolone and pregnanolone at $GABA_A$ receptors [86].

Neurosteroids fluctuate widely during physiological and pathophysiological conditions and during pharmacological intervention [50, 55, 84]. Several neurosteroidogenic molecules have recently been discovered that act either by stimulating downstream enzymes in the allopregnanolone biosynthetic cascade or by targeting neurosteroidogenic receptors. In 1996, Uzunov and colleagues, for the first time, showed that the SSRI antidepressants, such as fluoxetine, fluvoxamine, sertraline, and paroxetine, have potent neurosteroidogenic pharmacological actions. Several preclinical studies support rapid anxiolytic, antidepressant, analgesic, and anticonvulsant effects after allopregnanolone administration [87]. This nicely aligns with finding showing a single administration of the SSRIs, S-norfluoxetine (S-NFLX), or S-fluoxetine (S-FLX) enhances cortico-limbic allopregnanolone biosynthesis at the times that behavioral improvement of aggressive behavior is also observed [55, 88, 89].

Dysregulation in neurosteroid biosynthesis has been associated by several clinical studies with neuropsychiatric disorders [50, 90]. These include major unipolar depression and PTSD patients who show cerebrospinal fluid (CSF), plasma, serum, and postmortem brain downregulation of allopregnanolone and pregnanolone concentrations and decreased expression of steroidogenic enzymes [67, 90–96]. This allopregnanolone level decrease can be reversed by steroidogenic antidepressants, which show anxiolytic and antidepressant effects in patients who responded to treatment [88, 97]. Conversely, treatment with the 5α-reductase inhibitor, finasteride, prescribed for prostatism and hair loss induced depression and anxiety symptoms and even enhanced suicide behavior in male individuals [98–100]. This observation has been reproduced in animal models where suicide-related behavior determined by enhanced aggressiveness in mice was induced by treatment with finasteride or SKF105,111 [53, 101].

Although there is lack in the field showing a direct evidence of neurosteroid involvement in the pathophysiology of suicidal behavior, several lines of evidence support a role for neuroactive steroids that inhibit GABAergic neurotransmission in the complex mechanisms leading to suicide. For instance, the concentrations of dehydroepiandrosterone (DHEA) sulfate, which is a product of pregnenolone metabolism after the catalytic action by the P450c17 enzyme into DHEA followed by the action of sulfotransferase, were reduced in repeated suicide attempters. The levels of DHEAS significantly correlated between the neuroactive steroid dysfunction and suicidal behavior [102]. In accord with this former study, a more recent investigation also supports that DHEA and DHEA sulfate are downregulated in combat veterans with history of attempted suicide [103]. In women across the menstrual cycle, clinical studies show increased suicidal ideations and attempts at the time when levels of estradiol and progesterone, and expectedly allopregnanolone, decrease immediately preceding the menstruation [104]. Hormonal contraceptives have also been associated with suicide attempts and suicide [105, 106] and are associated with a downregulation of allopregnanolone as well as its precursors, pregnenolone and progesterone in the blood [107, 108]. These data is in accord with preclinical studies that have demonstrated a downregulation of this neurosteroid in the cerebral cortex and hippocampus of adult female rodents [109, 110].

Studies also suggest that hypothalamic-pituitary-adrenal (HPA) axis stress response in adults is a stable risk factor for suicidal behavior. Logistic multilevel models revealed altered cortisol response and elevated peer stress predicted suicidal behavior, but not ideation. Peer stress triggers suicidal ideation among female youth but only triggers suicidal behavior in young females with blunted cortisol reactivity [111].

In summary, dysregulation of neurosteroid biosynthesis appears to be associated with suicidal behavior and suicidal risk. The role of GABAergic neurosteroid in suicidal individual, especially in women during the menstrual cycle or during other conditions associated with a drop in allopregnanolone and other reproductive steroids, for instance, during the peripartum period or during menopause, deserves further investigation.

6   Neurosteroid Promise as Novel Treatments and Biomarkers for Suicide-Like Behaviors

Our studies on animal models of stress-induced aggressive behavior and impulsiveness have showed that protracted social isolation in rodents results in profound biochemical alterations. For instance, the most remarkable neurobiological change includes the altered neurosteroid biosynthesis in several cortico-limbic areas of male mice that undergo isolation [46, 55]. These alterations are also associated with changes in $GABA_A$ receptor subunit expression in prefrontal cortex and hippocampus of isolated rodents [51, 88]. These changes result in altered response to commonly prescribed anxiolytic drugs, including benzodiazepines [112]. Together, these neurobiological and pharmacological aspects resemble deficits observed in the pathophysiology of mood disorders, including depression, PTSD, and impulsivity. Indeed, decreased neurosteroid biosynthesis was observed in major unipolar depression [90, 91] and in PTSD patients [93, 94, 96, 113]. Several classes of drugs with neurosteroidogenic effects showed improvement of symptoms, which supports the hypothesis that development of novel treatments for depression, PTSD, and probably suicide may rely on neurosteroid-like compounds. We recently observed that treatment with allopregnanolone or allopregnanolone analogs (e.g., ganaxolone, BR297, BR351) improves several behavioral alterations reminiscent of PTSD/suicide-like behavior induced by social isolation in male mice [46, 49, 54]. Specifically, a single dose of the allopregnanolone analog, ganaxolone administered immediately after reactivation of fear memories, by a reconsolidation blockade, robustly facilitated extinction of fear memories. This pharmacological action of ganaxolone on contextual fear extinction appeared to be long-lasting and expanded till the recall trial when reemergence of fear was prevented by ganaxolone treatment given to block reconsolidation [49]. Ganaxolone also induced anxiolytic-like effect in socially isolated mice that were exposed to the elevated plus maze. Importantly, a single dose of ganaxolone, or the allopregnanolone analogs BR351 and BR297, strongly ameliorated suicide-like behavior, determined by means of aggressiveness of mice tested during a resident-intruder test [54]. Collectively, these and other findings in the field of neurosteroids and mood disorders have paved the way to the development of allopregnanolone in an intravenous formulation as the first specific treatment for postpartum depression [112, 114, 115]. Phase 2 and 3 clinical trials involving women affected by postpartum depression showed that intravenous infusion of allopregnanolone resulted in a remission of symptoms in 70% of patients; the pharmacological action was fast and long-lasting after only a short-course administration of 60 h [116].

Following allopregnanolone approval for the treatment of postpartum depression, the field has intensified studying new steroidogenic molecules that resemble neurosteroid mechanism of action. This includes discovering new neurosteroidogenic targets [51]. One remarkable target to induce neurosteroidogenesis constitutes the stimulation of endocannabinoid receptors, such as peroxisome proliferator-activated receptor (PPAR)-α. Recent studies conducted in our laboratory demonstrated that stimulation of PPAR-α by its endogenous modulator, palmitoylethanolamide (PEA), or by its selective synthetic agonists GW7647 and fenofibrate, stimulates allopregnanolone biosynthesis in several cortico-limbic areas, such as the hippocampus and the amygdala. This neurosteroidogenic effect results in a fast improvement of depressive and anxiety-like behavior, as well as contextual fear responses in socially isolated mice [50, 54, 117]. Importantly, PEA or fenofibrate administration dose-dependently ameliorated aggressive behavior in socially isolated mice [54]. Other studies in the field also observed that PEA reduces aggressive behavior in a traumatic brain injury rodent model [118]. Given the elevated comorbidity of PTSD with chronic pain and suicide risk in active army personnel and US veterans [119], treatment with PEA or other synthetic PPAR-α agonists could offer promising novel agents to treat PTSD and suicide. Thus, the endocannabinoid system at the interface with neurosteroid biosynthesis could offer novel pharmacological targets for mood disorders.

The question arises whether neurosteroidogenesis could provide predictive biomarkers for suicide to help diagnosis of these debilitating psychiatric disorders that shows large comorbidity and symptom overlap with MDD and PTSD. Indeed, approximately half of PTSD patients show strong comorbidity with MDD and suicide [120]. This scenario makes it more difficult to develop selective markers for suicide risk behavior without partial overlap of cellular and molecular mechanisms. The finding that allopregnanolone is downregulated in both depression and PTSD [90, 91, 93, 94] makes it challenging to point to simply allopregnanolone levels in the attempt of identifying a specific predictive marker for each of these disorders. Stimulation of allopregnanolone levels after allopregnanolone treatment itself or other drug treatments that increase its levels could be a predictor of pharmacological effectiveness for suicide and its comorbidity with other mood disorders.

We suggest that future studies should aim at identifying neurobiological alterations in suicide animal models by adopting sophisticated experimental approaches. Novel individualized treatment based on biomarker discovery maybe challenging as discussed previously [121]; however, approval of new and successful treatments may become soon available for mood disorders, such as PTSD and depression that will also impact suicidal behaviors. In biomarker discovery, investigating the interrelation among several biomarkers rather than just focusing on few isolated biomarkers may become a better approach to reveal accurate predictive markers and facilitate diagnosis and develop individualized treatments. Our preclinical studies support that the neurosteroid biosynthesis interfacing with the endocannabinoid system and their interaction may provide a path in this goal.

References

1. Olliac B, Ouss L, Charrier A (2016 Nov) Suicide attempts in children and adolescents: the place of clock genes and early rhythm dysfunction. Journal of Physiology Paris 110(4 Pt B):461–466. https://doi.org/10.1016/j.jphysparis.2017.11.001

2. Bae SM, Kang JM, Chang HY, Han W, Lee SH (2018) PTSD correlates with somatization in sexually abused children: type of abuse moderates the effect of PTSD on somatization. PLoS One 13(6):e0199138. https://doi.org/10.1371/journal.pone.0199138. eCollection 2018

3. Dardis CM, Dichter ME, Iverson KM (2018 Aug) Empowerment, PTSD and revictimization among women who have experienced intimate partner violence. Psychiatry Res 266:103–110. https://doi.org/10.1016/j.psychres.2018.05.034

4. Stone DM, Simon TR, Fowler KA, Kegler SR, Yuan K, Holland KM, Ivey-Stephenson AZ, Crosby AE. Vital signs: trends in state suicide rates - United States, 1999-2016 and circumstances contributing to suicide - 27 states,. MMWR Morb Mortal Wkly Rep. 2018 Jun 8;67(22):617-624. doi: https://doi.org/10.15585/mmwr.mm6722a1.

5. Zamora-Kapoor A, Nelson LA, Barbosa-Leiker C, Comtois KA, Walker LR, Buchwald DS (2016) Suicidal ideation in American Indian/Alaska native and White adolescents: the role of social isolation, exposure to suicide, and overweight. American Indian Alaska Native Mental Health Research 23(4):86–100. https://doi.org/10.5820/aian.2304.2016.86

6. Zalsman G (2020 Nov) Neurobiology of suicide in times of social isolation and loneliness. Eur Neuropsychopharmacol 40:1–3. https://doi.org/10.1016/j.euroneuro.2020.10.009

7. Loeffler G, Coller R, Tracy L, Derderian BR (2018 Jun 26) Prescribing trends in US active duty member with posttraumatic stress disorder: a population-based study 2007-2013. J Clin Psychiatry 79(4):pii: 17m11667. https://doi.org/10.4088/JCP.17m11667

8. Zhang L, Li H, Benedek D, Li X, Ursano R (2009 Sep) A strategy for the development of biomarker tests for PTSD. Med Hypotheses 73(3):404–409. https://doi.org/10.1016/j.mehy.2009.02.038

9. Lanzas C, Ayscue P, Ivanek R, Gröhn YT (2010 Feb) Model or meal? Farm animal populations as models for infectious diseases of humans. Nat Rev Microbiol 8(2):139–148. https://doi.org/10.1038/nrmicro2268

10. Preti A (2011 Jun 1) Animal model and neurobiology of suicide. Prog Neuropsychopharmacol Biol Psychiatry 35(4):818–830. https://doi.org/10.1016/j.pnpbp.2010.10.027

11. Malkesman O, Pine DS, Tragon T, Austin DR, Henter ID, Chen G, Manji HK (2009 Apr) Animal models of suicide-trait-related behaviors. Trends Pharmacological Sciences 30(4):165–173. https://doi.org/10.1016/j.tips.2009.01.004

12. Brezo J, Paris J, Tremblay R, Vitaro F, Zoccolillo M, Hebert M, Turecki G (2006) Personality traits as correlates of suicide attempts and suicidal ideation in young adults. Psychol Med 36:191–202

13. Coccaro EF, Fanning JR, Phan KL, Lee R (2015 Jun) Serotonin and impulsive aggression. CNS Spectr 20(3):295–302. https://doi.org/10.1017/S1092852915000310

14. de Beurs D, Have MT, Cuijpers P, de Graaf R (2019 Nov 6) The longitudinal association between lifetime mental disorders and first onset or recurrent suicide ideation. BMC Psychiatry 19(1):345. https://doi.org/10.1186/s12888-019-2328-8

15. Koolhas JM, Coopens CM, de Boer SF, Buwalda B, Meerlo P, Timmermans PJA (2013 Jul 4) The resident-intruder paradigm: a standardized test for aggression, violence and social stress. JoVE 77:e4367. https://doi.org/10.3791/4367

16. Mitchell PJ (2005 Dec 5) Antidepressant treatment and rodent aggressive behaviour. Eur J Pharmacol 526(1-3):147–162. https://doi.org/10.1016/j.ejphar.2005.09.029

17. Olfson M, Shaffer D, Marcus SC, Greenberg T (2003 Oct) Relationship between antidepressant medication treatment and suicide in adolescents. Arch Gen Psychiatry 60(10):978–982. https://doi.org/10.1001/archpsyc.60.9.978

18. Rahn KA, Cao YJ, Hendrix CW, Kaplin AI (2007) The role of 5-HT1A receptors in mediating acute negative effects of antidepressants: implications in pediatric depression. Transl Psychiatry 5:e563

19. Pattij T, Vanderschuren LJMJ (2008 Apr) The neuropharmacology of impulsive behaviour. Trends Pharmacol Sci 29(4):192–199. https://doi.org/10.1016/j.tips.2008.01.002

20. Evenden JL (1999 Oct) Varieties of impulsivity. Psychopharmacology 146(4):348–361. https://doi.org/10.1007/pl00005481

21. Winstanley CA, Eagle DM, Robbins TW (2006 Aug) Behavioral models of impulsivity in relation to ADHD: translation between clinical and preclinical studies. Clin Psychol Rev 26(4):379–395. https://doi.org/10.1016/j.cpr.2006.01.001

22. Isles A, Humby T, Wilkinson LS (2003 Dec) Measuring impulsivity in mice using a novel operant delayed reinforcement task: effects of behavioural manipulation and d-amphetamine. Psychopharmacology 170(4):376–382. https://doi.org/10.1007/s00213-003-1551-6

23. Feola TW, de Wit H, Richards JB (2000 Aug) Effects of d-amphetamine and alcohol on a measure of behavioral inhibition in rats. Behav Neurosci 114(4):838–848. https://doi.org/10.1037/0735-7044.114.4.838

24. Toohey M, DiGiuseppe R (2017 Apr) Defining and measuring irritability: construct clarification and differentiation. Clin Psychol Rev 53:93–108. https://doi.org/10.1016/j.cpr.2017.01.009

25. Orri M, Galera C, Turecki G, Forte A, Renaud J, Boivin M, Tremblay RE, Côté SM, Geoffroy MC (2018 May) Association of Childhood Irritability and Depressive/anxious mood profiles with adolescent suicidal ideation and attempts. JAMA Psychiat 75(5):465–473. https://doi.org/10.1001/jamapsychiatry.2018.0174

26. Einat H (2006 Sep) Modelling facets of mania—new directions related to the notion of endophenotypes. J Psychopharmacol 20(5):714–722. https://doi.org/10.1177/0269881106060241

27. Riitinen ML, Lindroos F, Kimanen A, Pieninkeroinen E, Pieninkeroinen I, Sippola J, Veilahti J (1986 Mar) Bergoström, Joahnson G. impoverished rearing conditions increase stress-induced irritability in mice. Dev Psychobiol 19(2):105–111. https://doi.org/10.1002/dev.420190203

28. Kimbrough A, de Guglielmo G, Kononoff J, Kallupi M, Zorrilla EP, George O (2017 Nov) CRF1 receptor-dependent increases in irritability-like behavior during abstinence from chronic intermittent ethanol vapor exposure. Alcohol Clin Exp Res 41(11):1886–1895. https://doi.org/10.1111/acer.13484

29. Vollmayr B, Gass P (2013 Oct) Learned helplessness: unique features and translational value of a cognitive depression model. Cell Tissue Res 354(1):171–178. https://doi.org/10.1007/s00441-013-1654-2

30. Landgraf D, Long J, Der-Avakian A, Streets M, Welsh DK (2015 Apr 30) Dissociation of learned helplessness and fear conditioning in mice: a mouse model of depression. PLoS One 10(4):e0125892. https://doi.org/10.1371/journal.pone.0125892

31. Greenwood BJ, Strong PV, Fleshner M (2010 Jul 29) Lesions of the basolateral amygdala reverse the long-lasting interference with shuttle box escape produced by uncontrollable stress. Behav Brain Res 211(1):71–76. https://doi.org/10.1016/j.bbr.2010.03.012

32. Shaffer D, Gould MS, Fisher P, Trautman P, Moreau D, Kleinman M, Flory M (1996 Apr) Psychiatric diagnosis in child and adolescent suicide. Archive General Psychiatry 53(4):339–348. https://doi.org/10.1001/archpsyc.1996.01830040075012

33. Renaud J, Berlim MT, McGirr A, Tousignant M, Turecki G (2008 Jan) Current psychiatric morbidity, aggression/impulsivity, and personality dimensions in child and adolescent suicide: a case-control study. J Affect Disord 105(1-3):221–228. https://doi.org/10.1016/j.jad.2007.05.013

34. Durkheim E (1967) Le suicide: etude de sociologie, 2nd edn. Presses Universitaires de France, Paris, France

35. Hämmig O (2019) Health risks associated with social isolation in general and in young, middle and old age. PLoS One 14(7): e0219663. https://doi.org/10.1371/journal.pone.0219663. eCollection 2019

36. Isometsä E (2014 Mar) Suicidal behaviour in mood disorders—who, when and why? Can J Psychiatr 59(3):120–130. https://doi.org/10.1177/070674371405900303

37. Tsai AC, Lucas M, Kawachi I (2015 Oct) Association between social integration and suicide among women in the United States. JAMA Psychiat 72(10):987–993. https://doi.org/10.1001/jamapsychiatry.2015.1002

38. Eck M, Scouflaire T, Debien C, Amad A, Sannier O, Chee CC, Thomas P, Vaiva G, Fovet T (2019 Jan) Suicide in prison: epidemiology and prevention. Presse Med 48(1 Pt 1):46–54. https://doi.org/10.1016/j.lpm.2018.11.009

39. Cacioppo JT, Cacioppo S, Capitanio JP, Cole SW (2015 Jan 3) The neuroendocrinology of social isolation. Annu Rev Psychol 66:733–767. https://doi.org/10.1146/annurev-psych-010814-015240

40. Cacioppo JT, Ernst JM, Burleson MH, McClintock MK, Malarkey WB, Hawkley LC, Kowalewski RB, Paulsen A, Hobson JA, Hugdahl K, Spiegel D, Berntson GG (2000 Mar) Lonely traits and concomitant physiological processes: the MacArthur social neuroscience studies. Int J Psychophysiol 35(2-3):143–154. https://doi.org/10.1016/s0167-8760(99)00049-5

41. Pressman SD, Cohen S, Miller GE, Barkin A, Rabin BS, Treanor JJ (2005 May) Loneliness, social network size, and immune response to influenza vaccination in college freshmen. Health Psychol 24(3):297–306. https://doi.org/10.1037/0278-6133.24.3.297

42. Cole SW (2008 Oct) Social regulation of leukocyte homeostasis: the role of glucocorticoid sensitivity. Brain Behav Immun 22(7):1049–1055. https://doi.org/10.1016/j.bbi.2008.02.006

43. Daskalakis NP, Bagot RC, Parker KJ, Vinkers CH, de Kloet ER (2013 Sep) The three-hit concept of vulnerability and resilience: toward understanding adaptation to early-life adversity outcome. Psychoneuroendocrinology 38(9):1858–1873. https://doi.org/10.1016/j.psyneuen.2013.06.008

44. Toth M, Fuzesi T, Halasz J, Tulogdi A, Haller J (2010 Dec 20) Neural inputs of the hypothalamic "aggression area" in the rat. Behav Brain Res 215(1):7–20. https://doi.org/10.1016/j.bbr.2010.05.050

45. Pinna G, Costa E, Guidotti A (2005 Feb 8) Changes in brain testosterone and allopregnanolone biosynthesis elicit aggressive behavior. Proc Nat Acad Sci U S A 102(6):2135–2140. https://doi.org/10.1073/pnas.0409643102

46. Pibiri F, Nelson M, Guidotti A, Costa E, Pinna G (2008 Apr 8) Decreased corticolimbic allopregnanolone expression during social isolation enhances contextual fear: a model relevant for posttraumatic stress disorder. Proc Nat Acad Sci U S A 105(14):5567–5572. https://doi.org/10.1073/pnas.0801853105

47. Pinna G, Rasmusson AM (2012) Up-regulation of Neurosteroid biosynthesis as a pharmacological strategy to improve behavioural deficits in a putative mouse model of post-traumatic stress disorder. J Neuroendocrinol 24(1):102–116. https://doi.org/10.1111/j.1365-2826.2011.02234.x

48. Nin MS, Martinez LA, Pibiri F, Nelson M, Pinna G (2011 Nov 21) Neurosteroids reduce social isolation-induced behavioral deficits: a proposed link with neurosteroid-mediated upregulation of BDNF expression. Front Endocrinol 2:73. https://doi.org/10.3389/fendo.2011.00073

49. Pinna G, Rasmusson AM (2014 Sep 11) Ganaxolone improves behavioral deficits in a mouse model of post-traumatic stress disorder. Front Cell Neurosci 8:256. https://doi.org/10.3389/fncel.2014.00256

50. Locci A, Pinna G (2017 Oct) Neurosteroid biosynthesis down-regulation and changes in GABAA receptor subunit composition: a biomarker axis in stress-induced cognitive and emotional impairment. Br J Pharmacol 174(19):3226–3241. https://doi.org/10.1111/bph.13843

51. Locci A, Pinna G (2019 Jun 8) Social isolation as a promising animal model of PTSD comorbid suicide: neurosteroids and cannabinoids as possible treatment options. Prog

Neuropsychopharmacol Biol Psychiat 92: 243–259. https://doi.org/10.1016/j. pnpbp.2018.12.014

52. Guidotti A, Dong E, Matsumoto K, Pinna G, Rasmusson AM, Costa E (2001 Nov) The socially-isolated mouse: a model to study the putative role of allopregnanolone and 5alpha-dihydroprogesterone in psychiatric disorders. Brain Res Brain Res Rev 37(1-3):110–115. https://doi.org/10.1016/s0165-0173(01)00129-1

53. Pinna G, Agis-Balboa RC, Pibiri F, Nelson M, Guidotti A, Costa E (2008 Oct) Neurosteroid biosynthesis regulates sexually dimorphic fear and aggressive behavior in mice. Neurochem Res 33(10):1990–2007. https://doi.org/10.1007/s11064-008-9718-5

54. Locci A, Geoffroy P, Miesch M, Mensah-Nyagan AG, Pinna G (2017 Aug 29) Social isolation in early versus late adolescent mice is associated with persistent Behavioral deficits that can be improved by Neurosteroid-based treatment. Front Cell Neurosci 11:208. https://doi.org/10.3389/fncel.2017.00208

55. Pinna G, Dong E, Matsumoto K, Costa E, Guidotti A (2003 Feb 18) In socially isolated mice, the reversal of brain allopregnanolone down-regulation mediates the anti-aggressive action of fluoxetine. Proc Nat Acad Sci U S A 100(4):2035–2040

56. Agís-Balboa RC, Pinna G, Pibiri F, Kadriu B, Costa E, Guidotti A (2007 Nov 20) Down-regulation of neurosteroid biosynthesis in corticolimbic circuits mediates social isolation-induced behavior in mice. Proc Nat Acad Sci U S A 104(47):18736–18741

57. Aleyasin H, Flanigan ME, Russo SJ (2018 Apr) Neurocircuitry of aggression and aggression seeking behavior: nose poking into brain circuitry controlling aggression. Curr Opin Neurobiol 49:184–191. https://doi.org/10.1016/j.conb.2018.02.013

58. Mann JJ, Brent DA, Arango V (2001 May) The neurobiology and genetics of suicide and attempted suicide: a focus on the serotonergic system. Neuropsychopharmacology 24(5):467–477. https://doi.org/10.1016/S0893-133X(00)00228-1

59. Schmaal L, van Harmelen AL, Chatzi V, Lippard ETC, Toenders YJ, Averill LA, Mazure CM, Blumberg HP (2020 Feb) Imaging suicidal thoughts and behaviors: a comprehensive review of 2 decades of neuroimaging studies. Mol Psychiatry 25(2):408–427. https://doi.org/10.1038/s41380-019-0587-x

60. Turecki G (2005 Nov) Dissecting the suicide phenotype: the role of impulsive-aggressive behaviours. J Psychiatry Neurosci 30(6):398–408

61. Vinod KY, Arango V, Xie S, Kassir SA, Mann JJ, Cooper TB, Hungund BL (2005 Mar 1) Elevated levels of endocannabinoids and CB1 receptor-mediated G-protein signaling in the prefrontal cortex of alcoholic suicide victims. Biol Psychiatry 57(5):480–486. https://doi.org/10.1016/j.biopsych.2004.11.033

62. Kolb B, Nonneman AJ (1974 Nov) Fronto-limbic lesions and social behavior in the rat. Physiol Behav 13(5):637–643. https://doi.org/10.1016/0031-9384(74)90234-0

63. Siegel A, Edinger H, Koo A (1977 May 20) Suppression of attack behavior in the cat by the prefrontal cortex: role of the mediodorsal thalamic nucleus. Brain Res 127(1):185–190. https://doi.org/10.1016/0006-8993(77)90392-4

64. Biro L, Toth M, Sipos E, Bruzsik B, Tulogdi A, Bendahan S, Sandi C, Haller J (2017 May) Structural and functional alterations in the prefrontal cortex after post-weaning social isolation: relationship with species-typical and deviant aggression. Brain Struct Funct 222(4):1861–1875. https://doi.org/10.1007/s00429-016-1312-z

65. New AS, Hazlett EA, Buchsbaum MS, Goodman M, Mitelman SA, Newmark R, Trisdorfer R, Haznedar MM, Koenigsberg HW, Flory J, Siever LJ (2007 Jul) Amygdala-prefrontal disconnection in borderline personality disorder. Neuropsychopharmacology 32(7):1629–1640. https://doi.org/10.1038/sj.npp.1301283

66. Young SE, Friedman NP, Miyake A, Willcutt EG, Corley RP, Haberstick BC, Hewitt JK (2009 Feb) Behavioral disinhibition: liability for externalizing spectrum disorders and its genetic and environmental relation to response inhibition across adolescence. J Abnorm Psychol 118(1):117–130. https://doi.org/10.1037/a0014657

67. Agis-Balboa RC, Guidotti A, Pinna G (2014 Sep) 5α-reductase type I expression is down-regulated in the prefrontal cortex/Brodmann's area 9 (BA9) of depressed patients. Psychopharmacology 231(17):3569–3580. https://doi.org/10.1007/s00213-014-3567-

68. Lin D, Boyle MP, Dollar P, Lee H, Lein ES, Perona P, Anderson DJ (2011 Feb 10) Functional identification of an aggression locus in the mouse hypothalamus. Nature 470(7333):221–226. https://doi.org/10.1038/nature09736

69. Lee H, Kim DW, Remedios R, Anthony TE, Chang A, Madisen L, Zeng H, Anderson DJ (2014 May 29) Scalable control of mounting

and attack by Esr1+ neurons in the ventromedial hypothalamus. Nature
509(7502):627–632. https://doi.org/10.
1038/nature13169

70. Nakata M, Sano K, Musatov S, Yamaguchi N,
Sakamoto T, Ogawa S (2016 Mar 31) Effects
of prepubertal or adult site-specific knockdown of Estrogen receptor β in the medial
preoptic area and medial amygdala on social
Behaviors in male mice. eNeuro 3(2):pii:
ENEURO.0155-15.2016. https://doi.org/
10.1523/ENEURO.0155-15.2016

71. Scott N, Prigge M, Yizhar O, Kimchi T (2015
Sep 24) A sexually dimorphic hypothalamic
circuit controls maternal care and oxytocin
secretion. Nature 525(7570):519–522.
https://doi.org/10.1038/nature15378

72. Strobel A, Zimmermann J, Schmitz A,
Reuter M, Lis S, Windmann S, Kirsch P
(2011 Jan 1) Beyond revenge: neural and
genetic bases of altruistic punishment. Neuro-
Image 54(1):671–680. https://doi.org/10.
1016/j.neuroimage.2010.07.051

73. White SF, Frick PJ, Lawing K, Bauer D (2013
Mar-Apr) Callous-unemotional traits and
response to functional family therapy in adolescent offenders. Behav Sci Law
31(2):271–285. https://doi.org/10.1002/
bsl.2041

74. Nehrenberg DL, Sheikh A, Ghashghaei HT
(2013 Jul) Identification of neuronal loci
involved with displays of affective aggression
in NC900 mice. Brain Struct Funct
218(4):1033–1049. https://doi.org/10.
1007/s00429-012-0445

75. Yu Q, Teixeira CM, Mahadevia D, Huang Y,
Balsam D, Mann JJ, Gingrich JA, Ansorge MS
(2014 Jun) Dopamine and serotonin signaling during two sensitive developmental periods differentially impact adult aggressive and
affective behaviors in mice. Mol Psychiatry
19(6):688–698. https://doi.org/10.1038/
mp.2014.10

76. Golden SA, Heshmati M, Flanigan M,
Christoffel D, Guise K, Pfau ML,
Aleyasin H, Menard C, Zhang H, Hodes
GE, Bregman D, Khibnik L, Tai J, Rebusi N,
Krawitz B, Chaudhury D, Walsh JJ, Han MH,
Shapiro ML, Russo SJ (2016 Jun 30) Basal
forebrain projections to the lateral habenula
modulate aggression reward. Nature
534(7609):688–692. https://doi.org/10.
1038/nature18601

77. Wong LC, Wang L, D'Amour JA, Yumita T,
Chen G, Yamaguchi T, Chang BC,
Bernstein H, You X, Feng JE, Froemke RC,
Lin D (2016 Mar 7) Effective modulation of
male aggression through lateral septum to

medial hypothalamus projection. Curr Biol
26(5):593–604. https://doi.org/10.1016/j.
cub.2015.12.065

78. Hong W, Kim DW, Anderson DJ (2014 Sep
11) Antagonistic control of social versus
repetitive self-grooming behaviors by separable amygdala neuronal subsets. Cell
158(6):1348–1361. https://doi.org/10.
1016/j.cell.2014.07.049

79. Kiehl KA, Smith AM, Hare RD, Mendrek A,
Forster BB, Brink J, Liddle PF (2001 Nov 1)
Limbic abnormalities in affective processing
by criminal psychopaths as revealed by functional magnetic resonance imaging. Biol Psychiatry 50(9):677–684. https://doi.org/10.
1016/s0006-3223(01)01222-7

80. Siever LJ (2008 Apr) Neurobiology of aggression and violence. Am J Psychiatr
165(4):429–442. https://doi.org/10.1176/
appi.ajp.2008.07111774

81. Baulieu EE, Robel P (1990 Nov 20) Neurosteroids: a new brain function? J Ster Biochem
Mol Biol 37(3):395–403

82. Agís-Balboa RC, Pinna G, Zhubi A,
Maloku E, Veldic M, Costa E, Guidotti A
(2006 Sep 26) Characterization of brain neurons that express enzymes mediating neurosteroid biosynthesis. Proc Nat Acad Sci U S A
103(39):14602–14607. https://doi.org/10.
1073/pnas.0606544103

83. Paul SM, Pinna G, Guidotti A (2020) Allopregnanolone: from molecular pathophysiology to therapeutics. A historical perspective.
Neurobiol Stress 12:100215. Published
online 2020 Mar 14. https://doi.org/10.
1016/j.ynstr.2020.100215

84. Pinna G, Uzunova V, Matsumoto K, Puia G,
Mienville JM, Costa E, Guidotti A (2000 Jan
28) Brain allopregnanolone regulates the
potency of the GABA(a) receptor agonist
muscimol. Neuropharmacology
39(3):440–448. https://doi.org/10.1016/
s0028-3908(99)00149-5

85. Belelli D, Lambert JJ (2005 Jul) Neurosteroids: endogenous regulators of the GABA
(a) receptor. Nat Rev Neurosci
6(7):565–575. https://doi.org/10.1038/
nrn1703

86. Matthew CC, Samba RD (2013 Nov) Neurosteroid interactions with synaptic and extrasynaptic GABAa receptors: regulation of
subunit plasticity, phasic and tonic inhibition,
and neuronal network excitability. Psychopharmacology 230(2). https://doi.org/10.
1007/s00213-013-3276-5

87. Belelli D, Harrison NL, Maguire J, Macdonald RL, Walker MC, Cope DW (2009 Oct

14) Extrasynaptic GABAA receptors: form, pharmacology, and function. J Neurosci 29(41):12757–12763. https://doi.org/10.1523/JNEUROSCI

88. Pinna G, Costa E, Guidotti A (2006 Jun) Fluoxetine and norfluoxetine stereospecifically and selectively increase brain neurosteroid content at doses that are inactive on 5-HT reuptake. Psychopharmacology 186(3). https://doi.org/10.1007/s00213-005-0213-2

89. Pinna G, Costa E, Guidotti A (2009 Feb) SSRIs act as selective brain steroidogenic stimulants (SBSSs) at low doses that are inactive on 5-HT reuptake. Curr Opin Pharmacol 9(1):24–30. https://doi.org/10.1016/j.coph.2008.12.006

90. Romeo E, Ströhle A, Spalletta G, di Michele F, Hermann B, Holsboer F, Pasini A, Rupprecht R (1998 Jul) Effects of antidepressant treatment on neuroactive steroids in major depression. Am J Psychiatr 155(7):910–913. https://doi.org/10.1176/ajp.155.7.910

91. Uzunova V, Sheline Y, Davis JM, Rasmusson A, Uzunov DP, Costa E, Guidotti A (1998 Mar 17) Increase in the cerebrospinal fluid content of neurosteroids in patients with unipolar major depression who are receiving fluoxetine or fluvoxamine. Proc Nat Acad Sci U S A 95(6):3239–3244. https://doi.org/10.1073/pnas.95.6.3239

92. van Broekhoven F, Verkes RJ (2003 Jan) Neurosteroids in depression: a review. Psychopharmacology 165(2):97–110. https://doi.org/10.1007/s00213-002-1257-1

93. Rasmusson AM, Pinna G, Paliwal P, Weisman D, Gottschalk C, Charney D, Krystal J, Guidotti A (2006 Oct 1) Decreased cerebrospinal fluid allopregnanolone levels in women with posttraumatic stress disorder. Biol Psychiatry 60(7):704–713. https://doi.org/10.1016/j.biopsych.2006.03.026

94. Rasmusson A, King A, Pineles S, Valovski I, Gregor K, Scioli-Salter E, Hamouda M, Nillni Y, Pinna G (2019) Relationships between cerebrospinal fluid GABAergic neurosteroid levels and symptom severity in men with PTSD. Psychoneuroendocrinology 102:95–104. https://doi.org/10.1016/j.psyneuen.2018.11.027

95. Pineles SL, Nillni YI, Pinna G, Irvine J, Webb A, Arditte Hall KA, Hauger R, Miller MW, Resick PA, Orr SP, Rasmusson AM (2018 Jul) PTSD in women is associated with a block in conversion of progesterone to the GABAergic neurosteroids allopregnanolone and pregnanolone measured in plasma. Psychoneuroendocrinology 93:

133–141. https://doi.org/10.1016/j.psyneuen.2018.04.024

96. Sl P, Nillni Y, Pinna G, Webb A, KAA H, Fonda JR, Irvine J, King MW, Hauger RL, Resick PA, Orr SP, Rasmusson AM (2020 May 15) Associations between PTSD-related extinction retention deficits in women and plasma steroids that modulate brain GABAA and NMDA receptor activity. Neurobiol Stress 13:100225. https://doi.org/10.1016/j.ynstr.2020.100225

97. Pinna G. Fluoxetine: pharmacology, mechanisms of action and potential side effects. New York Nova Science Publishers, Inc. 2015 Jan 1; ISBN (Print): 9781634820776, 9781634820769.

98. Altomare G, Capella GL (2002 Oct) Depression circumstantially related to the administration of finasteride for androgenetic alopecia. J Dermatol 29(10):665–669. https://doi.org/10.1111/j.1346-8138.2002.tb00200.x

99. Irwig MS (2012 Sep) Depressive symptoms and suicidal thoughts among former users of finasteride with persistent sexual side effects. J Clin Psychiatry 73(9):1220–1223. https://doi.org/10.4088/JCP.12m07887

100. Diviccaro S, Melcangi RC, Giatti S (2019 Dec 26) Post-finasteride syndrome: an emerging clinical problem. Neurobiol Stress. 12:100209. https://doi.org/10.1016/j.ynstr.2019.100209

101. Maurice-Gélinas C, Deslauriers J, Monpays C, Sarret P, Grignon S (2018 Jul 1) The 5α-reductase inhibitor finasteride increases suicide-related aggressive behaviors and blocks clozapine-induced beneficial effects in an animal model of schizophrenia. Physiol Behav 191:65–72. https://doi.org/10.1016/j.physbeh.2018.03.036

102. Bergman B, Brismar B (1994 Jul) Characteristics of violent alcoholics. Alcohol Alcoholism 29(4):451–457

103. Sher L, Flory J, Bierer L, Makotkine I, Yehuda R (2018 Jul) Dehydroepiandrosterone and dehydroepiandrosterone sulfate levels in combat veterans with or without a history of suicide attempt. Acta Psychiatr Scand 138(1):55–61. https://doi.org/10.1111/acps.12897

104. Baca-Garcia E, Diaz-Sastre C, Ceverino A, Perez-Rodriguez MM, Navarro-Jimenez R, Lopez-Castroman J, Saiz-Ruiz J, de Leon J, Oquendo MA (2010 Mar) Suicide attempts among women during low estradiol/low progesterone states. J Psychiatr Res 44(4):209–214. https://doi.org/10.1016/j.jpsychires.2009.08.004

105. Brent D (2018 Apr 1) Contraceptive conundrum: use of hormonal contraceptives is associated with an increased risk of suicide attempt and suicide. Am J Psychiatr 175(4):300–302. https://doi.org/10.1176/appi.ajp.2018.18010039

106. Skovlund CW, Mørch LS, Kessing LV, Lange T, Lidegaard Ø (2018 Apr 1) Association of Hormonal Contraception with Suicide Attempts and Suicides. Am J Psychiatry 175(4):336–342. https://doi.org/10.1176/appi.ajp.2017.17060616

107. Rapkin AJ, Biggio G, Concas A (2006 Aug) Oral contraceptives and neuroactive steroids. Pharmacol Biochem Behav 84(4):628–634. https://doi.org/10.1016/j.pbb.2006.06.008

108. Rapkin AJ, Morgan M, Sogliano C, Biggio G, Concas A (2006a May) Decreased neuroactive steroids induced by combined oral contraceptive pills are not associated with mood changes. Fertily Steriliy 85(5):1371–1378. https://doi.org/10.1016/j.fertnstert.2005.10.031

109. Santoru F, Berretti R, Locci A, Porcu P, Concas A (2014 Sep) Decreased allopregnanolone induced by hormonal contraceptives is associated with a reduction in social behavior and sexual motivation in female rats. Psychopharmacology 231(17):3351–3364. https://doi.org/10.1007/s00213-014-3539-9

110. Porcu P, Mostallino MC, Sogliano C, Santoru F, Berretti R, Concas A (2012 Aug) Long-term administration with levonorgestrel decreases allopregnanolone levels and alters GABA(a) receptor subunit expression and anxiety-like behavior. Pharmacol Biochem Behav 102(2):366–372. https://doi.org/10.1016/j.pbb.2012.05

111. Owens SA, Eisenlohr-Moul T (2018 Oct 6) Suicide risk and the menstrual cycle: a review of candidate RDoC mechanisms. Curr Psychiatry Rep 20(11):106. https://doi.org/10.1007/s11920-018-0962-3

112. Pinna G. Allopregnanolone (1938–2019): A trajectory of 80 years of outstanding scientific achievements. Neurobiol Stress. 2020 Nov; 13: 100246. doi: https://doi.org/10.1016/j.ynstr.2020.100246.

113. Kim BK, Fonda JR, Hauger RL, Pinna G, Anderson GM, Valovski IT, Ransmusson AM (2020 Apr 18) Composite contributions of cerebrospinal fluid GABAergic neurosteroids, neuropeptide Y and interleukin-6 to PTSD symptom severity in men with PTSD. Neurobiol Stress 12:100220. https://doi.org/10.1016/j.ynstr.2020.100220

114. Kanes S, Colquhoun H, Gunduz-Bruce H, Raines S, Arnold R, Schacterle A, Doherty J, Epperson CN, Deligiannidis KM, Riesenberg R, Hoffmann E, Rubinow D, Jonas J, Paul S, Meltzer-Brody S (2017a Jul 29) Brexanolone (SAGE-547 injection) in post-partum depression: a randomised controlled trial. Lancet 390(10093):480–489. https://doi.org/10.1016/S0140-6736(17)31264-3

115. Meltzer-Brody S, Colquhoun H, Riesenberg R, Epperson CN, Deligiannidis KM, Rubinow DR, Li H, Sankoh A, Clemson C, Schacterle A, Jonas J, Kanes S (2018 Sep 22) Brexanolone injection in post-partum depression: two multicentre, double-blind, randomised, placebo-controlled, phase 3 trials. Lancet 392(10152):1058–1070. https://doi.org/10.1016/S0140-6736(18)31551-4

116. Meltzer-Brody S, Kanes S (2020 Feb 3) Allopregnanolone in postpartum depression: role in pathophysiology and treatment. Neurobiology of Stress. 12:100212. https://doi.org/10.1016/j.ynstr.2020.100212

117. Pinna G (2018 Aug 6) Biomarkers for PTSD at the interface of the endocannabinoid and neurosteroid axis. Front Neurosci 12:482. https://doi.org/10.3389/fnins.2018.00482

118. Guida F, Boccella S, Iannotta M, De Gregorio D, Giordano C, Belardo C, Romano R, Palazzo E, Scafuro MA, Serra N, de Novellis V, Rossi F, Maione S, Luongo L (2017 Mar 6) Palmitoylethanolamide reduces neuropsychiatric Behaviors by restoring cortical electrophysiological activity in a mouse model of mild traumatic brain injury. Front Pharmacol 8:95. https://doi.org/10.3389/fphar.2017.00095

119. Blakey SM, Wagner HR, Naylor J, Brancu M, Lane I, Sallee M (2018 Jul) Kimbrel NA; VA mid-Atlantic MIRECC workgroup, Elbogen EB. Chronic pain, TBI, and PTSD in military veterans: a link to suicidal ideation and violent impulses? J Pain 19(7):797–806. https://doi.org/10.1016/j.jpain.2018.02.012

120. Flory JD, Yehuda R (2015 Jun) Comorbidity between post-traumatic stress disorder and major depressive disorder: alternative explanations and treatment considerations. Dialogues Clin Neurosci 17(2):141–150. https://doi.org/10.31887/DCNS.2015.17.2/jflory

121. Aspesi D, Pinna G (2018 Dec) Could a blood test for PTSD and depression be on the horizon? Expert Rev Proteomics 15(12):983–1006. https://doi.org/10.1080/14789450.2018.1544894

Chapter 7

Behavioral Paradigms for Assessing Cognitive Functions in the Chronic Social Defeat Stress Model of Depression

Marc Fakhoury, Michael Fritz, and Sama F. Sleiman

Abstract

Depression is a debilitating mental disorder that affects hundreds of millions of individuals worldwide. Also referred to as major depressive disorder or clinical depression, this disorder is mainly characterized by a persistent feeling of sadness and a loss of interest. There is also substantial evidence showing that depression is frequently accompanied by deficits in cognitive functions, including working memory, learning, and executive functions. Given the high prevalence of poor cognitive functioning in depressed patients, several studies have investigated the link between the two using animal models of depression. Despite significant progress in research, more work is still needed to better understand the association between cognitive impairment and depression. Results of such studies could lead to a better understanding of the pathophysiology of depression and could ultimately pave the way toward new and more efficient therapeutic approaches. In this chapter, we describe some of the behavioral techniques used to assess cognitive function in the chronic social defeat stress mice model, a well-established model of depression. Focus is given on the spontaneous alternation T-maze test and the Morris water maze test.

Key words Animal model of depression, Chronic social defeat stress, Cognitive test, Depression, Morris water maze, Spontaneous alternation T-maze

1 Introduction

1.1 Depression: Overview and Pathophysiology

Depression, also referred to as major depressive disorder, is a potentially life-threatening disorder resulting in enormous personal suffering. It affects hundreds of millions of individuals worldwide and can occur at any age from childhood to late life [1]. According to the World Health Organization (WHO), depression is a leading cause of disability, leading to poor quality of life and reduced productivity at work [2]. In its worst-case scenario, depression can lead to suicide; approximately 800,000 individuals die of depression-related suicide every year [2]. Other symptoms, such as sleep and psychomotor disturbances, low mood, anhedonia, and fatigue, are also frequently observed [1]. According to the *Diagnostic and Statistical Manual of Mental Disorders* (*DSM-5*), at least

Yong-Ku Kim and Meysam Amidfar (eds.), *Translational Research Methods for Major Depressive Disorder*, Neuromethods, vol. 179, https://doi.org/10.1007/978-1-0716-2083-0_7,

five of the following symptoms must be present nearly every day during the same 2-week period for a diagnosis of depression to occur: (1) depressed mood, (2) loss of interest or pleasure, (3) changes in weight or appetite, (4) insomnia or hypersomnia, (5) psychomotor agitation or retardation, (6) fatigue or loss of energy, (7) feeling of worthlessness or guilt, (8) impaired concentration or indecisiveness, and (9) recurrent thoughts of death or suicide [3].

Although depression can affect individuals of all ages, it is more prevalent in women than in men [4] and can be precipitated by several factors including genetics and environment [5]. For instance, risk factors such as poor social economic status, unemployment, alcohol and drug use, and a major life event such as the death of a relative or a loved one have all been shown to increase the likelihood of developing depression [6]. Several genes have also been associated with the likelihood of developing depression, including genes involved in serotonin regulation such as the serotonin transporter protein (5-HTT) gene, the serotonin receptor 2A (5-HTR2A) gene, and the tryptophan hydroxylase 2 (TPH2), as well as genes involved in neurotrophic factor regulation, including the brain-derived neurotrophic factor (BDNF), the insulin-like growth factor (IGF), the fibroblast growth factor (FGF), and the vascular endothelial growth factor (VEGF) [5]. However, it is important to note that depression is typically caused by a multitude of factors rather than a single gene mutation or environmental factor [5]. Moreover, accumulating evidence suggests that epigenetic factors that influence interactions between the environment and genes appear to play a primordial role in the etiology of depression [7, 8]. However, among all of the proposed hypotheses of depression, to date, the most influential one is the serotonin hypothesis, which postulates a deficit in serotonin pathways as a major cause of depression [9, 10]. This long-standing hypothesis forms the cornerstone of current treatment modalities for depression, which primarily include the use of selective serotonin reuptake inhibitors [11]. Today, it is generally accepted that depression can result from deficits in other neurotransmitter systems as well, including the dopamine, norepinephrine, glutamatergic, and GABAergic systems [12].

1.2 Cognitive Functions in Depression

In addition to the well-defined symptoms of depression that typically include low mood, anhedonia, and fatigue, depressed patients are often plagued with cognitive dysfunctions [13]. There is indeed substantial evidence from both preclinical and clinical studies showing that depression can be accompanied by deficits in working memory [14–16], learning [17, 18], executive functions [19, 20] and attention [21, 22]. These cognitive changes, along with the set of emotional and behavioral deficits already observed in depression, further contribute to the deterioration of the patients' quality of life if left untreated. Cognitive dysfunctions also constitute a major burden for depressed individuals that are in partial or full remission, with approximately 30% reporting residual cognitive symptoms that interfere with daily functioning [23]. At the neurobiological level, these cognitive symptoms may be associated with morphological and/or functional changes in certain brain structures known to play a key role in cognitive processes. Indeed, findings from brain imaging and postmortem studies have implicated numerous structures in the etiology of depression, including the prefrontal cortex, the hippocampus, the amygdala, the nucleus accumbens, and the basal ganglia [24, 25]. In particular, depression has been associated with reduced volume of the nucleus accumbens, prefrontal cortex, and the hippocampus [26, 27], decreased volumetric asymmetries in the basal ganglia [28, 29], and smaller volume of the amygdala [30, 31]. These structures have been shown to play a major role in several cognitive processes including learning, memory, and decision-making [32–35]. In addition to the structural changes observed in these brain regions, depression has been characterized by functional alterations in cortico-limbic and diencephalic structures, including decreased activation of the prefrontal cortex [36], increased activation of the amygdala [36, 37], and increased activation of the lateral habenula [38], a structure known to play a crucial role in reward learning [39].

Notwithstanding the significant progress in research, there are still many missing gaps that need to be addressed to better understand the association between cognitive dysfunctions and depression. In particular, it is still not clear whether cognitive impairments precede the onset of depressive symptoms or appear concomitantly with them. Clearly more studies are needed to address this missing gap. Results of such work could yield interesting insights into the potential use of cognitive impairments as a predictive marker for depression. In addition, more work is needed to understand the presence of residual cognitive dysfunctions in remitted depressed patients and whether these symptoms could be managed with conventional pharmacological treatments. The use of appropriate animal models of depression appears to be crucial for the better understanding of cognitive processes in depression. In this chapter, a thorough description of behavioral tests used to assess cognitive functions in animal models of depression is provided. Although there is a wide array of available tests to assess cognitive functions in rodents [40, 41], the present chapter will focus on the spontaneous alternation T-maze and the Morris water maze tests.

1.3 Animal Models of Depression

A wide array of animal models have been developed and studied in order to better understand the neurobiological mechanisms underlying depression [42]. Appropriate animal models of depression need to reliably mimic the behavioral, neurochemical, and neuroimaging phenotype of depression in humans. In addition, the animal model should recapitulate the processes that result in depression in humans and should be effective in screening therapeutic interventions that could be used for treating the disease in humans [42]. In rodents, many models, including the olfactory bulbectomy (OBX) model [43], the reserpine-induced model [44], and the unpredictable chronic mild stress (UCMS) model [45], have been validated for the study of depression. In the OBX rodent model of depression, the olfactory bulbs are surgically removed so as to mimic the behavioral and neurochemical changes observed in depression [43]. In particular, this model has been shown to induce behavioral despair [46], anhedonia [47], neuroinflammation [48], and dysfunction of various hippocampal cellular processes [49, 50]. This model is also effectively employed in research to screen for potential antidepressant drugs and therapeutic targets [51, 52]. On the other hand, the reserpine-induced model of depression is based on depletion in the levels of monoamines through repeated administration of reserpine, which has been shown to reproduce depression-like symptoms such as altered locomotor activity, behavioral despair, and cognitive impairments [44, 53, 54]. Another model of depression frequently used in research is the UCMS model, which is based on the chronic exposure of rodents to unpredictable stressors, such as food and water deprivation, overnight illumination, and cage tilt [55]. Some of the most important characteristics of this model include anhedonia, behavioral despair, reduced reward sensitivity, loss of appetite, and alteration in the activity of the hypothalamic-pituitary-adrenal axis (HPA) axis [56, 57]. Other animal models used to study depression include the learned helplessness, the early life stress model, the chronic restraint stress model, the psychological stress model, the glucocorticoid/corticosterone model, and the genetic and transgenic models (for review, see refs. [57, 58]). In this chapter, however, the focus will be on the chronic social defeat stress (CSDS) model of depression.

The CSDS model uses social conflict as a stressor to cause depression-like symptoms including anhedonia, helplessness, hyperactivity, and social avoidance [59]. In this model, a naïve rodent is repeatedly subject to bouts of social defeat by an aggressive and more dominant rodent for several days, resulting in the development of depression-like symptoms [60]. Not only does it recapitulate some of the most important features of depression, this model also enables the assessment of antidepressant activity [59, 61]. The advantage of the CSDS model over other animal models and environmental stressors is its ability to continuously activate the pituitary-adrenal axis over repeated social confrontations [62, 63] and its ability to identify neurobiological mechanisms that can induce stable changes in phenotype [64]. More importantly, defeat-induced social avoidance in rodents can persist for weeks or even months and may be more relevant to human depression because it involves a social form of stress [59].

2 Materials

2.1 Animals

Animals subjected to CSDS have been extensively used as a model of depression because of their ability to recapitulate the behavioral and neurochemical changes of the disease [59]. Models of CSDS have been reported in several animal species, including rats [65], mice [66], hamsters [67], zebrafish [68], and crickets [69]. However, this chapter will focus on the CSDS mice model, which is the most frequently used to study depression-like behavior. For this model, adult male C57BL/6 J mice (8–20 weeks old) (*see* **Note 1**) and retired breeder CD1 mice (4–6 months old) are typically used as experimental subjects, though other strains may prove suitable as well [60, 66]. Although this model has proven more feasible to implement in males, induction of the CSDS model has also been described in female mice [70]. Upon arrival to the animal facility, C57BL/6 J and CD1 mice are typically housed individually (or in group of no more than five mice per cage for C57BL/6 J mice), under a 12-h light/dark cycle and with food and water available ad libitum. All animal procedures should be approved by the institutional Animal Care and Use Committee (ACUC), and at least 1 week of acclimatization should be allowed before beginning the experiments. In addition, large group sizes of at least eigh to ten animals are generally recommended for the behavioral experiments in order to achieve statistical significance between the control and experimental groups.

2.2 Apparatus for the CSDS Mice Model

The apparatus for the CSDS mice model, as described in ref. [60], consists of the following:

- Clear rectangular cages (26.7 cm width; 48.3 cm depth; 15.2 cm height).
- Paired steel wire tops.
- Hard woodchip bedding to provide appropriate traction for the mice.
- Clear perforated Plexiglas divider (0.6 cm width; 45.7 cm depth; 15.2 cm height) to physically separate the aggressor and the experimental mice following the defeat session.
- Stopwatch for timing defeat sessions.
- Cleaning solution and wipes for disinfecting cages between each round of social defeat.

2.3 Apparatus for the Spontaneous Alternation T-Maze Test

The apparatus for the spontaneous alternation T-maze test in mice, as described in ref. [71], consists of a T-maze with the following characteristics:

- One start arm (30 cm length; 10 cm width). A start arm of a lower width would make it more difficult to place the mice into the maze.

- Two identical goal arms (30 cm length; 10 cm width) with a central partition that extend 7 cm from the center of the back of the T-maze into the start arm. Without the central partition, mice would partially sample the unchosen arm at the first sample phase of behavioral testing, which would cause interference at the choice phase.

- Three guillotine doors cut to fit maze (one for each arm). These can be cut and attached to the upper two-thirds of the wall only to facilitate cleaning and cause less distraction to the animals.

Although automated T-mazes are available (e.g., Med Associates, St. Albans, Vermont, USA), manually run mazes can be easily constructed. The T-maze is typically made from gray- or black-painted wood, medium-density fiberboard, or a plastic such as PVC [71]. It is not recommended to paint the maze in white since this would provoke anxiety in mice, and habituation would be much slower [71].

2.4 Apparatus for the Morris Water Maze Test

The apparatus for the Morris water maze test in mice, as described in ref. [72], consists of the following:

- 1× circular tank (diameter about 120 cm, height about 50 cm) (e.g., Panlab (Harvard Apparatus), Ugo Basile, TSE Systems) (*see* **Note 2**).

- 1× acrylic or PVC goal platform (square or round; 10 cm² or 10 cm in diameter), submerged ~1 cm below the surface and stably linked to a larger base at the bottom of the tank.

- Water: In general, most laboratories report a temperature that centers around 19–22 °C.

- Room configurations: It is important to consider the cues surrounding the maze, as they are crucial for the rodent to orient and navigate (*see* **Note 3**).

- Light conditions: As it is with most behavioral assays, light conditions should be indirect and dimmed. Such a light setting aids two potential problems: first, rodents perceive bright light and large open spaces as frightening and stressful, which may influence their performance (for further discussion, see ref. [73]); second, too bright light may cause reflections from the water surface and thereby negatively affect the video tracking.

- Video tracking: There exist numerous commercial providers for automated video tracking systems that can track rodent movements (e.g., EthoVisionXT from Noldus and SMART VIDEO TRACKING from Panlab). For a detailed list, please see ref. [72].

3 Methods

3.1 Chronic Social Defeat Stress Model

The CSDS mice model is an ethologically well-established model of depression that has frequently been used to study depression-like behaviors and to investigate the antidepressant effects of treatments. In this model, C57BL/6 J mice are subjected to chronic social defeat stress by a CD1 aggressor mice for 10 consecutive days [60, 66]. Discussed below is a step-by-step protocol, as described in ref. [60], illustrating the screening process of the CD1 aggressor mice and the social confrontations between the intruder and aggressor mice.

3.1.1 Screening Process for Aggressive CD1 Mice

1. Retired breeder CD1 mice are used as the resident aggressors. Upon arrival to the animal facility, male CD1 retired breeder mice are singly housed with access to food and water ad libitum and allowed 7 days of habituation prior to screening.

2. During screening, a screener C57BL/6 J mouse (8–20 weeks old) is placed directly into the home cage of the aggressor for 180 s for social confrontation (*see* **Note 4**).

3. After the 180 s of social confrontation, the latency to aggression is noted, and the screener is removed from the aggressor cage. Three screening sessions are typically performed for each aggressor, once daily, and on 3 subsequent days using different screeners.

4. To be selected as an aggressor in social defeat experiments, the CD1 mice must attack the intruder in at least two consecutive sessions, and its attack latencies must be less than 60 s.

5. Following screening, experimental aggressors are housed individually in the defeat cages overnight prior to the start of the social defeat experiments.

3.1.2 Social Defeat Experiments (Fig. 1)

- On the first day, a C57BL/6 J mouse is placed directly into the home cage compartment of the aggressor for 5–10 min.

- After 5–10 min of physical interaction, the intruder mouse is transferred in a compartment adjacent to that of the aggressor and separated with a perforated Plexiglas divider, allowing for visual, olfactory, and auditory interaction with the aggressor for 24 h (*see* **Note 5**).

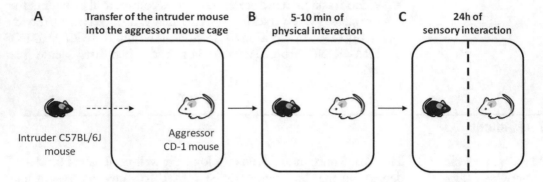

A Transfer of the intruder mouse into the aggressor mouse cage **B** 5-10 min of physical interaction **C** 24h of sensory interaction

Intruder C57BL/6J mouse

Aggressor CD-1 mouse

Fig. 1 Experimental illustration of a daily session of chronic social defeat. (**a–b**) The intruder mouse is placed directly into the home cage compartment of the aggressor mouse for 5–10 min. (**c**) During the next 24 h, the intruder is separated from the aggressor with a perforated Plexiglas divider, allowing for visual, olfactory, and auditory interaction. At the end of the experiment, the intruder mouse is exposed to a new resident mouse cage and subjected to social defeat every day for a total of 10 consecutive days

- On each of the subsequent days, the intruder C57BL/6 J mouse is exposed to a new resident CD1 mouse cage and subjected again to social defeat for 5–10 min each day followed by 24 h of sensory interaction (*see* **Note 6**).

- Non-defeated control animals are housed in pair for the entire duration of the defeat session within cages identical to those used for the socially defeated mice.

- After completion of the 10 days of social defeat experiments, all intruder mice are housed individually in standard mouse cages with ad libitum access to food and water.

3.2 Spontaneous Alternation T-Maze Test

The spontaneous alternation T-maze test is a behavioral test for measuring spatial working memory in rodents. It is based on the fact that rodents are more willing to visit the new arm of the maze, instead of the familiar and previously visited arm. Failure to visit the new arm is reflective of impaired spatial working memory. Discussed below is a step-by-step protocol, as described in ref. [71], detailing the method for using the spontaneous alternation T-maze test in mice (*see also* Fig. 2).

1. Before the beginning of the experiment, mice should be placed in the experimental room for 15–30 min for accommodation.

2. The maze should be set so that all guillotine doors are raised. No habituation to the maze is needed, since it is the novelty of the maze that drives the spontaneous exploration behavior.

3. The experimental animal is first placed in the start arm and is allowed to choose a goal arm (this is called the *sample phase*).

4. Once the animal completely enters a goal arm, the guillotine door is used to entrap it in the chosen arm for 30 s.

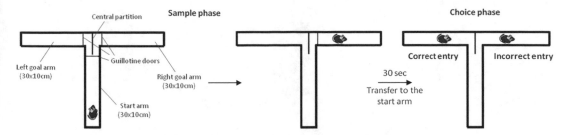

Fig. 2 The spontaneous alternation T-maze test in mice. In the sample phase, the experimental mouse is first placed in the start arm and is allowed to choose a goal arm. Once the animal completely enters a goal arm, the guillotine door is put in place for 30 s. In the choice phase, the animal is placed back into the start arm and is allowed to choose again freely between the two goal arms. A mouse with intact spatial working memory will remember the arm that was previously visited and will instinctively choose the opposite arm in an attempt to explore the new environment

5. After 30 s, the barrier is removed, and the animal is placed back into the start arm facing away from the goal arms and then allowed to choose freely between the two goal arms (this is called the *choice phase*).

6. Each trial typically lasts around 1–2 min and can be repeated several times. Between each trial, the T-maze should be thoroughly cleaned and disinfected with alcohol (*see also* **Note** 7).

Data analysis and interpretation: At the end of the experiment, the average percentage of alternation per animal is calculated and compared among the different experimental groups. With an intact spatial working memory performance, the animal will alternate between the two goal arms, reflecting exploration of a new environment and memory of the first choice. On the other hand, mice with deficits in spatial working memory will tend to choose the previously visited arm. Using this technique, Yu and colleagues showed that CSDS mice exhibit a significant decrease in the alternation rate, suggesting impairments in spatial working memory associated with depression [16].

3.3 Morris Water Maze Test

In 1981, Richard Morris developed a behavioral neuroscientific tool, what should later become known as "the water maze" or "the Morris Water Maze" [74, 75]. The Morris water maze is a widely used instrument with high validity to investigate hippocampus-dependent cognitive functions such as memory, spatial, and discrimination learning [76]. It is an easy-to-use behavioral assay with excellent interspecies translatability reaching from rodents to human beings (virtual reality mazes; [77]). The behavioral apparatus for the Morris water maze test is illustrated in Fig. 3.

As it is with all behavioral experiments, it is crucial that the experimenter is, prior to the start of the experiment, trained in handling animals properly. A mistreated animal may display stress and anxiety and hence not perform well. The circular tank of the

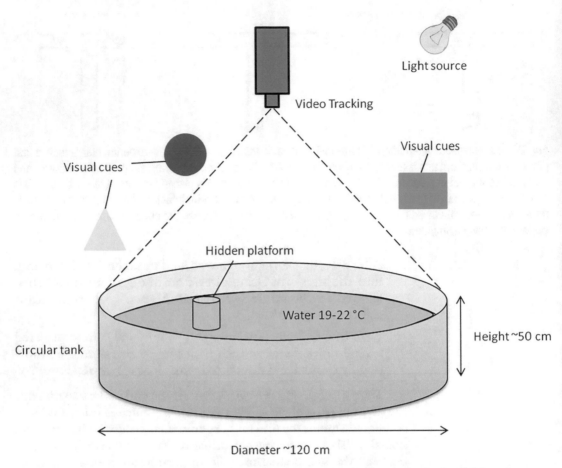

Fig. 3 Behavioral apparatus of the Morris water maze test in mice. The apparatus for the Morris water maze test in mice consists of a circular tank of approximately 120 cm in diameter and 50 cm in height, a hidden platform, water at a temperature of 19–22 °C, visual cues surrounding the maze, a video tracking system, and an appropriate light setting

Morris water maze is divided into four quadrants. This division plays an important role for all phases of training. Most commonly, each training day consists of four independent trials (60 s each), as there are the four quadrants. However, a multitude of studies used more trials per day, without demonstrating any learning improvement (for a detailed discussion, please see ref. [72]). Each trial is separated by intertrial intervals (ITIs) reaching from as little as 15–60 s or up to 30 min, during which the rodent will either be left on the goal platform or be brought back to a cage. Since mice are prone to hypothermia-dependent learning difficulties [78], it may be favorable to use longer ITIs and give the test animals the opportunity to recover under a heat lamp. To prevent drowning, each animal has to be guided to the goal platform after 60 s or be removed from the tank.

1. Place the mouse at the beginning of each trial in the desired starting quadrant facing the tank wall. It is important not to drop the mouse into the water (*see also* **Note 8**).

2. As soon as the mouse reaches the water surface, either start the timer or stop it after 60 s or as soon as the mouse touches the goal platform. The timer can, however, introduce a source of variability and investigator bias. For more accurate time measurement, a video tracking software is typically used.

3. Repeat the procedure after the initially defined ITI.

4. One variant of the Morris water maze protocol [16], which is described in this book chapter, consists of 2 training days, where the goal platform is marked by a visual attachment, allowing the animal to quickly orient itself toward it. During these two training days, the platform is regularly moved to different positions, and the mouse is always released at the opposite wall. Such an approach allows the animal to quickly understand the way to escape the water.

5. Once the animal acquired the overall task, the protocol moves to the hidden platform phase. This stage lasts 10 consecutive days. During this phase, the escape platform (no longer visually marked) remains in the same position at all times, and the animal is released from each quadrant once per training day.

6. At training day 12, a retention trial is performed. Here, the platform is removed from the tank, and the time the mouse spends searching for the platform in the right quadrant is recorded for 60 s. Such a trial should reflect the spatial memory acquired by the mouse.

7. After the retention trial, the spatial reversal phase starts. Here, the escape platform is moved to the opposite quadrant, and the initial trainings procedure is repeated for 5 consecutive days.

8. Finally, on day 17, another retention trial is performed (*see* **Note 9**).

Recent work demonstrated its usability assessing cognitive deficits in a chronic social defeat-induced model of depression [16, 79–82]. Investigating the molecular basis of cognitive impairments in a depressive-like phenotype is of clinical importance, as depression in human beings has been linked to several functionally attenuated memory dimensions [83–87]. The preclinical model of chronic social defeat is a validated model for depression [61, 88, 89], but provided so far mixed results concerning a depressive phenotype in the Morris water maze. Whereas Yu and colleagues [16] and Jin and colleagues [80] failed to detect group differences, Jianhua and colleagues [79], Jung and colleagues [81], and Monleón and colleagues [82] described a significant group effect in favor of the chronic social defeat model. These discrepancies may be explained by the degree (i.e., the intensity) of the social defeat [82] or the genetic footprint of each individual's hippocampus and the differentiation between resilient and susceptible animals [81, 90].

4   Conclusion/Notes

It is well accepted that individuals with depression most often exhibit cognitive deficits in various domains, including attention, executive functions, memory, and processing speed [91]. These cognitive symptoms often tend to persist even after remission of depression, causing severe complications to social functioning, work-related activities, and daily living [23]. However, whether these symptoms manifest themselves as a result of antidepressant medications or residual symptoms of depression remains to be further elucidated. It is also not very clear whether these symptoms are related to a specific subtype of depression and whether they precede or follow the onset of depressive symptoms. The use of appropriate animal models of depression is primordial for the better understanding of the link between cognitive dysfunctions and depressive symptoms. This chapter discusses the use of the CSDS mice model of depression for assessing cognitive functions. The advantage of using the CSDS model over other animal models is that it involves a social form of stress, which may be more relevant to human depression, and it induces depression-like behaviors that can persist for weeks and even months [59]. This chapter also provides a step-by-step description of the spontaneous alternation T-maze and the Morris water maze test for the assessment of learning and spatial working memory in CSDS mice. These additional notes are, however, worthwhile mentioning to ensure successful behavioral testing.

5   Notes

1. Social defeat can be successfully performed on adult mice of up to 20 weeks of age; however, for optimal results, it is recommended to use mice that are 7–8 weeks of age; the use of older mice will result in decreased aggression from the CD1 mice [60].

2. Commercially available swimming pools may also be used for the Morris water maze test, but caution should be taken to make sure that there are no welded seams, corrugated surfaces, or other features that could provide proximal cues to the mice [72]. A smooth interior surface is also important to reduce the risk of the animals to climb the walls [72]. Also, for optimal

video tracking, the color of the walls should be considered dependent on the color of the animals tested to ensure enough contrast for the video software to detect the animal. For example, black C57Bl/6 J mice should be tested in a white-colored maze.

3. For the Morris water maze test, it is pivotal that the cues are not moved between the trials. One basic rule is that there cannot be too many cues nor too little. Hence, if the room does not provide sufficient features, it is important for the success of the experiment to provide additional ones.

4. In the CSDS model, C57BL/6 J mice that are used for screening CD1 aggressive mice should not be used for the social defeat experiments [60]. They may, however, be used repeatedly for subsequent screenings.

5. If a serious attack occurs in the social defeat experiments, the C57BL/6 J mouse could be separated from the CD1 mice for 1 min and then returned back to the cage [16]. If physical aggression results in open wounds exceeding 1 cm, the defeated C57BL/6 J mouse should be removed from the experiment and euthanized [60].

6. The defeat cages should be thoroughly disinfected and sterilized between all rounds of social defeats.

7. A change in floor odor (e.g., by putting a fresh piece of paper under the maze) is recommended after each trial of spontaneous alternation T-maze so that the mice don't lose the motivation to explore [71].

8. Body weight, physical development, and age can influence the swimming speed and must therefore be taken into account [92]. Gender has to be taken into account as well, as male mice tend to perform better than female mice [93, 94]. Also, mice have a habit of floating and swimming along the wall of the pool (i.e., thigmotaxis), which may complicate testing.

9. It is of upmost importance to keep an eye on both the immune and stress status of experimental animals, especially if one is planning to study chronic social defeat-induced depression. If stress is considered an experimental variable, the experimenter has to ensure that supplementary stress (i.e., through inferior housing conditions or maltreatment) is minimized, as it can have a profound influence on the animals' performance and may annihilate differences between groups. For example, multiple studies were able to link adrenal stress hormones or the corticotrophin-releasing factor to impairment of spatial memory retrieval [95, 96], and Sandi hypothesized that stress directly influences memory formation [97]. Alongside stress, undiscovered viral and bacterial infections may also influence

the rodents' performance in the Morris water maze in an Interleukin-1ß-dependent manner [98–100].

Last but not least, although this chapter focuses on the CSDS model of depression for assessing cognitive functions, it is important to note that several other models of depression may be used as well. However, an important task is to ensure that the animal model used has a high predictive validity of pharmacological effects and shows neurobiological and behavioral characteristics that are very similar in humans. This will help produce results that could be reliably translated into the clinics. Animal models may probably not mirror the full spectrum of mechanistic aspects and clinical symptoms observed in humans but will definitively allow the possibility of experimentally assessing a wide array of behavioral and neurobiological subdomains and pave the way toward the development of new treatment modalities.

References

1. Brigitta B (2002) Pathophysiology of depression and mechanisms of treatment. Dialogues Clin Neurosci 4:7–20

2. WHO. (2000). Depression. Retrieved from https://www.who.int/news-room/fact-sheets/detail/depression. (Last accessed November 19, 2020)

3. American Psychiatric Association (2013) Diagnostic and statistical manual of mental disorders, 5th edn. American Psychiatric Association, Arlington

4. Albert PR (2015) Why is depression more prevalent in women? J Psychiatry Neurosci 40:219–221

5. Fakhoury M (2015) New insights into the neurobiological mechanisms of major depressive disorders. Gen Hosp Psychiatry 37:172–177

6. WHO (2017) Depression and other common mental disorders: Global Health estimates. World Health Organization, Geneva. Retrieved from https://apps.who.int/iris/bitstream/handle/10665/254610/WHO-MSD-MER-2017.2-eng.pdf (Last accessed November 19, 2020)

7. Kendler KS, Kuhn JW, Vittum J, Prescott CA, Riley B (2005) The interaction of stressful life events and a serotonin transporter polymorphism in the prediction of episodes of major depression: a replication. Arch Gen Psychiatry 62:529–535

8. Lok A, Bockting CL, Koeter MW, Snieder H, Assies J, Mocking RJ, Vinkers CH, Kahn RS, Boks MP, Schene AH (2013) Interaction between the MTHFR C677T polymorphism and traumatic childhood events predicts depression. Transl Psychiatry 3:e288

9. Fakhoury M (2016) Revisiting the serotonin hypothesis: implications for major depressive disorders. Mol Neurobiol 53:2778–2786

10. Albert PR, Benkelfat C, Descarries L (2012) The neurobiology of depression--revisiting the serotonin hypothesis. I. Cellular and molecular mechanisms. Philosophical transactions of the Royal Society of London. Series B, Biological sciences 367:2378–2381

11. Clevenger SS, Malhotra D, Dang J, Vanle B, IsHak WW (2018) The role of selective serotonin reuptake inhibitors in preventing relapse of major depressive disorder. Therapeut Adv Psychopharmacol 8:49–58

12. Fakhoury M (2018) Diagnosis of major depressive disorders: clinical and biological perspectives. In: Kim YK (ed) Understanding depression. Springer, Singapore

13. Lam RW, Kennedy SH, McIntyre RS, Khullar A (2014) Cognitive dysfunction in major depressive disorder: effects on psychosocial functioning and implications for treatment. Can J Psychiatry 59:649–654

14. Gohier B, Ferracci L, Surguladze SA, Lawrence E, El Hage W, Kefi MZ, Allain P, Garre JB, Le Gall D (2009) Cognitive inhibition and working memory in unipolar depression. J Affect Disord 116:100–105

15. Gruber O, Zilles D, Kennel J, Gruber E, Falkai P (2011) A systematic experimental neuropsychological investigation of the

functional integrity of working memory circuits in major depression. Eur Arch Psychiatry Clin Neurosci 261:179–184

16. Yu T, Guo M, Garza J, Rendon S, Sun XL, Zhang W, Lu XY (2011) Cognitive and neural correlates of depression-like behaviour in socially defeated mice: an animal model of depression with cognitive dysfunction. Int J Neuropsychopharmacol 14:303–317

17. Naismith SL, Hickie IB, Ward PB, Scott E, Little C (2006) Impaired implicit sequence learning in depression: a probe for frontostriatal dysfunction? Psychol Med 36:313–323

18. Darcet F, Mendez-David I, Tritschler L, Gardier AM, Guilloux JP, David DJ (2014) Learning and memory impairments in a neuroendocrine mouse model of anxiety/depression. Front Behav Neurosci 8:136

19. Fossati P, Ergis AM, Allilaire JF (2002) Executive functioning in unipolar depression: a review. L'Encephale 28:97–107

20. Bredemeier K, Warren SL, Berenbaum H, Miller GA, Heller W (2016) Executive function deficits associated with current and past major depressive symptoms. J Affect Disord 204:226–233

21. Pardo JV, Pardo PJ, Humes SW, Posner M (2006) Neurocognitive dysfunction in antidepressant-free, non-elderly patients with unipolar depression: alerting and covert orienting of visuospatial attention. J Affect Disord 92:71–78

22. Sommerfeldt SL, Cullen KR, Han G, Fryza BJ, Houri AK, Klimes-Dougan B (2016) Executive attention impairment in adolescents with major depressive disorder. J Clin Child Adolescent Psychol 53(45):69–83

23. Fava M, Graves LM, Benazzi F, Scalia MJ, Iosifescu DV, Alpert JE, Papakostas GI (2006) A cross-sectional study of the prevalence of cognitive and physical symptoms during long-term antidepressant treatment. J Clin Psychiatry 67:1754–1759

24. Savitz J, Drevets WC (2009) Bipolar and major depressive disorder: neuroimaging the developmental-degenerative divide. Neurosci Biobehav Rev 33:699–771

25. Duman RS (2014) Pathophysiology of depression and innovative treatments: remodeling glutamatergic synaptic connections. Dialogues Clin Neurosci 16:11–27

26. Drevets WC, Price JL, Furey ML (2008) Brain structural and functional abnormalities in mood disorders: implications for neurocircuitry models of depression. Brain Struct Funct 213:93–118

27. MacQueen G, Frodl T (2011) The hippocampus in major depression: evidence for the convergence of the bench and bedside in psychiatric research? Mol Psychiatry 16:252–264

28. Shah PJ, Glabus MF, Goodwin GM, Ebmeier KP (2002) Chronic, treatment-resistant depression and right fronto-striatal atrophy. J Mental Sci 180:434–440

29. Lacerda AL, Nicoletti MA, Brambilla P, Sassi RB, Mallinger AG, Frank E, Kupfer DJ, Keshavan MS, Soares JC (2003) Anatomical MRI study of basal ganglia in major depressive disorder. Psychiatry Res 124:129–140

30. von Gunten A, Fox NC, Cipolotti L, Ron MA (2000) A volumetric study of hippocampus and amygdala in depressed patients with subjective memory problems. J Neuropsychiatry Clin Neurosci 12:493–498

31. Hastings RS, Parsey RV, Oquendo MA, Arango V, Mann JJ (2004) Volumetric analysis of the prefrontal cortex, amygdala, and hippocampus in major depression. Neuropsychopharmacology 29:952–959

32. Funahashi S (2017) Prefrontal contribution to decision-making under free-choice conditions. Front Neurosci 11:431

33. Voss JL, Bridge DJ, Cohen NJ, Walker JA (2017) A closer look at the hippocampus and memory. Trends Cogn Sci 21:577–588

34. Yavas E., Gonzalez S., Fanselow M.S. (2019). Interactions between the hippocampus, prefrontal cortex, and amygdala support complex learning and memory. F1000Research, 8, F1000 faculty Rev-1292

35. Grahn JA, Parkinson JA, Owen AM (2009) The role of the basal ganglia in learning and memory: neuropsychological studies. Behav Brain Res 199:53–60

36. Ruhé HG, Booij J, Veltman DJ, Michel MC, Schene AH (2012) Successful pharmacologic treatment of major depressive disorder attenuates amygdala activation to negative facial expressions: a functional magnetic resonance imaging study. J Clin Psychiatry 73:451–459

37. Mingtian Z, Shuqiao Y, Xiongzhao Z, Jinyao Y, Xueling Z, Xiang W, Yingzi L, Jian L, Wei W (2012) Elevated amygdala activity to negative faces in young adults with early onset major depressive disorder. Psychiatry Res 201:107–112

38. Proulx CD, Hikosaka O, Malinow R (2014) Reward processing by the lateral habenula in normal and depressive behaviors. Nat Neurosci 17:1146–1152

39. Graziane NM, Neumann PA, Dong Y (2018) A focus on reward prediction and the lateral habenula: functional alterations and the Behavioral outcomes induced by drugs of abuse. Front Synapt Neurosci 10:12

40. Sharma S, Rakoczy S, Brown-Borg H (2010) Assessment of spatial memory in mice. Life Sci 87:521–536

41. Tanila H (2018) Testing cognitive functions in rodent disease models: present pitfalls and future perspectives. Behav Brain Res 352:23–27

42. Nestler EJ, Hyman SE (2010) Animal models of neuropsychiatric disorders. Nat Neurosci 13:1161–1169

43. Morales-Medina JC, Iannitti T, Freeman A, Caldwell HK (2017) The olfactory bulbectomized rat as a model of depression: the hippocampal pathway. Behav Brain Res 317:562–575

44. Ikram H, Haleem DJ (2017) Repeated treatment with reserpine as a progressive animal model of depression. Pak J Pharm Sci 30:897–902

45. Frisbee JC, Brooks SD, Stanley SC, d'Audiffret AC (2015) An unpredictable chronic mild stress protocol for instigating depressive symptoms, Behavioral changes and negative health outcomes in rodents. JoVE 106:53109

46. Han F, Nakano T, Yamamoto Y, Shioda N, Lu YM, Fukunaga K (2009) Improvement of depressive behaviors by nefiracetam is associated with activation of CaM kinases in olfactory bulbectomized mice. Brain Res 1265:205–214

47. Amchova P, Kucerova J, Giugliano V, Babinska Z, Zanda MT, Scherma M, Dusek L, Fadda P, Micale V, Sulcova A, Fratta W, Fattore L (2014) Enhanced self-administration of the CB1 receptor agonist WIN55,212-2 in olfactory bulbectomized rats: evaluation of possible serotonergic and dopaminergic underlying mechanisms. Front Pharmacol 5:44

48. Rinwa P, Kumar A (2013) Quercetin suppresses microglial neuroinflammatory response and induce antidepressant-like effect in olfactory bulbectomized rats. Neuroscience 255:86–98

49. Morales-Medina JC, Juarez I, Venancio-García E, Cabrera SN, Menard C, Yu W, Flores G, Mechawar N, Quirion R (2013) Impaired structural hippocampal plasticity is associated with emotional and memory deficits in the olfactory bulbectomized rat. Neuroscience 236:233–243

50. Moriguchi S, Han F, Nakagawasai O, Tadano T, Fukunaga K (2006) Decreased calcium/calmodulin-dependent protein kinase II and protein kinase C activities mediate impairment of hippocampal long-term potentiation in the olfactory bulbectomized mice. J Neurochem 97:22–29

51. Kelly JP, Wrynn AS, Leonard BE (1997) The olfactory bulbectomized rat as a model of depression: an update. Pharmacol Ther 74:299–316

52. Song C, Leonard BE (2005) The olfactory bulbectomised rat as a model of depression. Neurosci Biobehav Rev 29:627–647

53. Zhang S, Liu X, Sun M, Zhang Q, Li T, Li X, Xu J, Zhao X, Chen D, Feng X (2018) Reversal of reserpine-induced depression and cognitive disorder in zebrafish by sertraline and traditional Chinese medicine (TCM). BBF 14:13

54. Antkiewicz-Michaluk L, Wąsik A, Możdżeń E, Romańska I, Michaluk J (2014) Antidepressant-like effect of tetrahydroisoquinoline amines in the animal model of depressive disorder induced by repeated administration of a low dose of reserpine: behavioral and neurochemical studies in the rat. Neurotox Res 26:85–98

55. Willner P (2016) The chronic mild stress (CMS) model of depression: history, evaluation and usage. Neurobiol Stress 6:78–93

56. Mineur YS, Belzung C, Crusio WE (2006) Effects of unpredictable chronic mild stress on anxiety and depression-like behavior in mice. Behav Brain Res 175:43–50

57. Wang Q, Timberlake MA 2nd, Prall K, Dwivedi Y (2017) The recent progress in animal models of depression. Prog Neuro-Psychopharmacol Biol Psychiatry 77:99–109

58. Krishnan V, Nestler EJ (2011) Animal models of depression: molecular perspectives. Curr Top Behav Neurosci 7:121–147

59. Venzala E, García-García AL, Elizalde N, Delagrange P, Tordera RM (2012) Chronic social defeat stress model: behavioral features, antidepressant action, and interaction with biological risk factors. Psychopharmacology 224:313–325

60. Golden SA, Covington HE 3rd, Berton O, Russo SJ (2011) A standardized protocol for repeated social defeat stress in mice. Nat Protoc 6:1183–1191

61. Tsankova NM, Berton O, Renthal W, Kumar A, Neve RL, Nestler EJ (2006) Sustained hippocampal chromatin regulation in a mouse model of depression and antidepressant action. Nat Neurosci 9:519–525

62. Tornatzky W, Miczek KA (1993) Long-term impairment of autonomic circadian rhythms after brief intermittent social stress. Physiol Behav 53:983–993

63. Covington HE 3rd, Miczek KA (2005) Intense cocaine self-administration after episodic social defeat stress, but not after aggressive behavior: dissociation from corticosterone activation. Psychopharmacology 183:331–340

64. Meerlo P., Overkamp G.J., Daan S., Van Den Hoofdakker R.H., Koolhaas J.M. (1996). Changes in behaviour and body weight following a single or double social defeat in rats. Stress (Amsterdam, Netherlands), 1, 21–32

65. Covington HE, Miczek KA (2001) Repeated social-defeat stress, cocaine or morphine. Psychopharmacology 158:388–398

66. Nasrallah P, Haidar EA, Stephan JS, El Hayek L, Karnib N, Khalifeh M, Barmo N, Jabre V, Houbeika R, Ghanem A, Nasser J, Zeeni N, Bassil M, Sleiman SF (2019) Branched-chain amino acids mediate resilience to chronic social defeat stress by activating BDNF/TRKB signaling. Neurobiol Stress 11:100170

67. Jasnow AM, Drazen DL, Huhman KL, Nelson RJ, Demas GE (2001) Acute and chronic social defeat suppresses humoral immunity of male Syrian hamsters (Mesocricetus auratus). Horm Behav 40:428–433

68. Nakajo H, Tsuboi T, Okamoto H (2019) The behavioral paradigm to induce repeated social defeats in zebrafish. Neurosci Res S0168-0102(19):30590–30595

69. Rose J, Rillich J, Stevenson PA (2017) Chronic social defeat induces long-term behavioral depression of aggressive motivation in an invertebrate model system. PLoS One 12:e0184121

70. Harris AZ, Atsak P, Bretton ZH, Holt ES, Alam R, Morton MP, Abbas AI, Leonardo ED, Bolkan SS, Hen R, Gordon JA (2018) A novel method for chronic social defeat stress in female mice. Neuropsychopharmacology 43:1276–1283

71. Deacon RM, Rawlins JN (2006) T-maze alternation in the rodent. Nat Protoc 1:7–12

72. Vorhees CV, Williams MT (2006) Morris water maze: procedures for assessing spatial and related forms of learning and memory. Nat Protoc 1:848–858

73. Klawonn AM, Fritz M (2021) Immune-to-brain signaling effects on the neural substrate for reward: behavioral models of aversion, anhedonia, and despair. In: Fakhoury M (ed) The brain reward system. Springer, New York. https://doi.org/10.1007/978-1-0716-1146-3

74. Morris RGM (1981) Spatial localization does not require the presence of local cues. Learn Motiv 12:239–260

75. Morris RGM (1984) Developments of a water-maze procedure for studying spatial learning in the rat. J Neurosci Methods 11:47–60

76. Morris RGM (1993) An attempt to dissociate 'spatial-mapping' and 'working-memory' theories of hippocampal function. In: Seifert W (ed) Neurobiology of the hippocampus. Academic Press, New York, pp 405–432

77. Kallai J, Makany T, Karadi K, Jacobs WJ (2005) Spatial orientation strategies in Morris-type virtual water task for humans. Behav Brain Res 159:187–196

78. Iivonen H, Nurminen L, Harri M, Tanila H, Puoliväli J (2003) Hypothermia in mice tested in Morris water maze. Behav Brain Res 141:207–213

79. Jianhua F, Wei W, Xiaomei L, Shao-Hui W (2017) Chronic social defeat stress leads to changes of behaviour and memory-associated proteins of young mice. Behav Brain Res 316:136–144

80. Jin HM, Shrestha MS, Bagalkot TR, Cui Y, Yadav BK, Chung YC (2015) The effects of social defeat on behavior and dopaminergic markers in mice. Neuroscience 288:167–177

81. Jung SH, Brownlow ML, Pellegrini M, Jankord R (2017) Divergence in Morris water maze-based cognitive performance under chronic stress is associated with the hippocampal whole transcriptomic modification in mice. Front Mol Neurosci 10:275

82. Monleón S, Duque A, Vinader-Caerols C (2016) Effects of several degrees of chronic social defeat stress on emotional and spatial memory in CD1 mice. Behav Process 124:23–31

83. Cataldo MG, Nobile M, Lorusso ML, Battaglia M, Molteni M (2005) Impulsivity in depressed children and adolescents: a comparison between behavioral and neuropsychological data. Psychiatry Res 136:123–133

84. Rose EJ, Ebmeier KP (2006) Pattern of impaired working memory during major depression. J Affect Disord 90:149–161

85. Christopher G, MacDonald J (2005) The impact of clinical depression on working memory. Cogn Neuropsychiatry 10:379–399

86. Butters MA, Whyte EM, Nebes RD, Begley AE, Dew MA, Mulsant BH, Zmuda MD, Bhalla R, Meltzer CC, Pollock BG, Reynolds CF 3rd, Becker JT (2004) The nature and

determinants of neuropsychological functioning in late-life depression. Arch Gen Psychiatry 61:587–595

87. Porter RJ, Gallagher P, Thompson JM, Young AH (2003) Neurocognitive impairment in drug-free patients with major depressive disorder. J Mental Sci 182:214–220

88. Martinez M, Calvo-Torrent A, Pico-Alfonso MA (1998) Social defeat and subordination as models of social stress in laboratory rodents. Aggress Behav 24:241–256

89. Malatynska E, Knapp RJ (2005) Dominant-submissive behavior as models of mania and depression. Neurosci Biobehav Rev 29:715–737

90. Zhang TR, Larosa A, Di Raddo ME, Wong V, Wong AS, Wong TP (2019) Negative memory engrams in the hippocampus enhance the susceptibility to chronic social defeat stress. J Neurosci 39:7576–7590

91. Perini G, Cotta RM, Sinforiani E, Bernini S, Petrachi R, Costa A (2019) Cognitive impairment in depression: recent advances and novel treatments. Neuropsychiatr Dis Treat 15:1249–1258

92. Wenk GL (1998) Assessment of spatial memory using radial arm and Morris water mazes. In: Crawley JN, Gerfen C, McKay R, Rogawski M, Sibley D, Skolnick P (eds) Current protocols in neuroscience. Wiley, New York

93. Brandeis R, Brandys Y, Yehuda S (1989) The use of the Morris water maze in the study of memory and learning. Int J Neurosci 48:29–69

94. D'Hooge R, De Deyn PP (2001) Applications of the Morris water maze in the study of learning and memory. Brain Res Brain Res Rev 36:60–90

95. Walther T, Voigt JP, Fukamizu A, Fink H, Bader M (1999) Learning and anxiety in angiotensin-deficient mice. Behav Brain Res 100:1–4

96. Heinrichs SC, Stenzel-Poore MP, Gold LH, Battenberg E, Bloom FE, Koob GF, Vale WW, Pich EM (1996) Learning impairment in transgenic mice with central overexpression of corticotropin-releasing factor. Neuroscience 74:303–311

97. Sandi C (1998) The role and mechanisms of action of glucocorticoid involvement in memory storage. Neural Plast 6:41–52

98. Yayou K, Takeda M, Tsubone H, Sugano S, Doi K (1993) The disturbance of water-maze task performance in mice with EMC-D virus infection. J Vet Med Sci 55:341–342

99. McLean JH, Shipley MT, Bernstein DI, Corbett D (1993) Selective lesions of neural pathways following viral inoculation of the olfactory bulb. Exp Neurol 122:209–222

100. Gibertini M, Newton C, Friedman H, Klein TW (1995) Spatial learning impairment in mice infected with legionella pneumophila or administered exogenous interleukin-1-beta. Brain Behav Immun 9:113–128

Part III

Cellular and Molecular Research Methods for Major Depression

Chapter 8

Optogenetic Animal Models of Depression: From Mice to Men

Ayla Arslan, Pinar Unal-Aydin, Taner Dogan, and Orkun Aydin

Abstract

Optogenetics, the light-induced reversible control of specific neuronal ensembles, has revolutionized the circuit level analysis of depression, leading to the identification of relevant circuitries in several brain regions including—but not limited to—medial prefrontal cortex, ventral tegmental area, and nucleus accumbens in rodents. While it is still early to observe a direct translational utility, the continuous progress in optogenetic interrogation of specific neural populations has great potential for untangling the complex pathophysiology of depression.

Key words Optogenetics, Opsin, Virus-mediated gene expression, Depression, Mouse model, Genetic engineering, Research methods, Animal models of depression

Abbreviations

5-HT	5-Hydroxytryptamine
AAVs	Adeno-associated viruses
ACC	Anterior cingulate cortex
AD	Antidepressant
AMY	Amygdala
avBNST	Anteroventral bed nuclei of the stria terminalis
BDNF	Brain-derived neurotrophic factor
BLA	Basolateral amygdala
BMA	Basomedial amygdala
BNST	Bed nucleus of the stria terminalis
CaMKIIa	Ca^{2+}/calmodulin-dependent protein kinase II
CCK	Cholecystokinin
CCK-B	Cholecystokinin-B receptor
CeA	Central amygdala
ChR2	Channelrhodopsin-2
CMS	Chronic mild stress

Yong-Ku Kim and Meysam Amidfar (eds.), *Translational Research Methods for Major Depressive Disorder*, Neuromethods, vol. 179, https://doi.org/10.1007/978-1-0716-2083-0_8,

CRF	Corticotropin-releasing factor
CSDS	Chronic social defeat stress
D1	Dopamine 1
D2	Dopamine 2
DA	Dopamine
DG	Dentate gyrus
Drd1	Dopamine receptor 1
Drd2	Dopamine receptor 2
DRN	Dorsal raphe nucleus
EPM	Elevated plus maze
FST	Forced swim test
GABA	Gamma-aminobutyric acid
GABA(A)Rs	Gamma-aminobutyric acid A receptors
GluClR	Glutamate-gated chloride channel receptor
HPA system	Hypothalamus-pituitary-adrenal system
IL-PFC	Infralimbic prefrontal cortex
ILT	Intralaminar thalamus
LHb	Lateral habenula
MDT	Medial dorsal thalamus
mHb	Medial habenula
mPFC	Medial prefrontal cortex
MSNs	Medium spiny neurons
NAc	Nucleus accumbens
NMDAR	N-Methyl-D-aspartate receptor
NSF	Novelty-suppressed feeding test
PIT	Pavlovian-to-instrumental transfer
PR	Progressive ratio
PrL	Prelimbic area
PVH	Paraventricular hypothalamus
RMTg	Rostromedial tegmental nucleus
RN	Raphe nucleus
SDS	Social defeat stress
SPT	Sucrose preference test
SSDS	Subthreshold social defeat stress
TST	Tail suspension test
vGlut	Vesicular glutamate transporter 2
vHipp	Ventral hippocampus
vlPAG	Ventrolateral periaqueductal gray
vmPFC	Ventral medial prefrontal cortex
VP	Ventral pallidum
vSTR	Ventral striatum
VTA	Ventral tegmental area
ΔFosB	DeltaFosB

1   Introduction

Depression is estimated by the World Health Organization (WHO) to affect more than 264 million people of all ages globally.[1] This is well reflected with the results of Global Burden of Disease Study, according to which depression is the second leading cause of disability worldwide among the 354 diseases and injuries in 195 countries and territories [1, 2]. Thus, depression is a devastating mental health problem with serious burden which is on the rise globally: it is expected to be the first leading cause of disability by 2030 [3].

Despite the progress in omics technologies, neuroimaging, genome engineering, and artificial intelligence besides the thousands of genetically modified mice as neuropsychiatric disease models, the genes and associated circuitries linked to major psychiatric disorders including depression are not known [4, 5]. For example, the biological basis of depression is not well understood. This is well reflected with the existence of several theories that explain the molecular and biochemical components of depression addressing diverse mechanisms and signaling pathways. These include, but are not limited to, the monoamine hypothesis, the hypothalamus-pituitary-adrenal (HPA) system hypothesis [6, 7], the neurogenesis hypothesis [8, 9], the GABAergic deficits hypothesis [10], and the microbiota-inflammasome hypothesis [11]. This then causes the lack of validated drug targets, leading to the lack of effective and consistent treatment agents. This is well manifested by the problems for varying antidepressant (AD) response that cut across the borders of the clinical disease classifications as well as the treatment-resistant depression [12]. In addition, it takes weeks to months for patients to respond to AD pharmacotherapy, such as when patients use fluoxetine, a selective serotonin reuptake inhibitor (SSRI), one of the most prescribed AD drugs. Thus, this late onset of AD response [13] leads to the idea that current pharmacological agents do not hit to the problem at its source.

Many factors including the heterogeneity of the disease state, environmental factors, lack of biomarkers, and density of genetic architecture play a role in depression pathophysiology, making its analysis extremely difficult [4, 14–17]. There has been huge effort to search for innovative methodologies to untangle this huge complexity of depression pathophysiology. For example, endophenotypes [4] and their utilization in imaging genetics [5, 18] have been formulated in order to unveil the phenotypic complexity of neuropsychiatric disorders including depression. In addition, hypothesis-free, genome-wide association studies (GWAS) have

---

[1] https://www.who.int/news-room/fact-sheets/detail/depression retrieved in 03.04.2021

been designed to capture the most significant genetic variants among the complex architecture of genetic effects [16]. Besides, since there is about 90% homology between the genomes of mice and humans [19], genetic engineering of rodents has been important to pin down the pathophysiology of neuropsychiatric disorders, including depression [20].

Among the genetic approaches [21], several methods such as selective breeding [20], random mutation screening [20], virus-mediated gene expression [20, 22], transgenesis by injection to pronucleus [23], and gene replacement (knock-in) or silencing (knockout) by several methods such as homologous recombination and ENU-driven target-selected mutagenesis [24–28], besides Cre/Lox P systems [29] and clustered regularly interspaced short palindromic repeats (CRISPR)-based systems [30], have led to the generation of thousands of genetically engineered rodent models for the neuropsychiatric diseases. Moreover, combination of genetic modification with tools such as pharmacological intervention allows reversible and site-specific control of selected neuronal circuitry in a specific brain area [31]. Especially, progress in generation of genetically modified organism by CRISPR/CAS9 (CRISPR-associated protein 9) systems and stimulation of genetically altered neuronal ensembles by optics (optogenetics) will likely increase these genetic modification trends for the utilization of mice as neuropsychiatric disease models.

As there is an effect of genetic variation in the etiology of major neuropsychiatric disorders [4], the goal of genetic modification in a model organism is obvious. On the other hand, this genetic effect is relatively low for depression: the heritability of depression is estimated as 37% [32]. Combined with the heterogeneity of the disease states and other complexities [4, 33], such as gene environment interactions, it was not easy to identify any significant genetic risk associated with depression. This has led to the problem of "missing heritability" since both candidate gene studies [34] and initial genome-wide association studies (GWAS) [35, 36] of depression failed to identify any significant genetic factor associated with depression.

Fortunately, the groundbreaking data published by the CONVERGE Consortium [37] have led to the identification of reproducible association of two loci with major depression in humans. Located in chromosome 10, one of these loci corresponds to a locus near the SIRT1 gene, a member of the sirtuin family and characterized as a class III histone deacetylase (HDAC), which regulates the acetylation state of histones and nonhistone proteins [38]. The other one is in the intron of the LHPP gene (rs35936514). LHPP encodes an enzyme known as phospholysine phosphohistidine inorganic pyrophosphate phosphatase (LHPP) and is expressed moderately in the brain [39]. Following this, other studies further identified new genetic variants associated with depression [40–42]. Thus, recent progress in depression genetics has shed light on the issue of "missing heritability" at least to some extent.

2 Ketamine Is on the Scene: The Hope for "Rapid-Acting Antidepressants"

In the meanwhile, the ongoing research for the next-generation rapid-acting antidepressants (RAADs) open new horizons on the investigation of depression and its treatment. The idea of RAADs has gained momentum when the rapid mood lifting effect of a low dose of (R,S)-ketamine (ketamine) on depression patients was discovered [43]. This is followed by the discovery of its effect on the treatment-resistant depression [44–46].

Acting on the glutamatergic system, ketamine is the N-methyl-D-aspartate receptor (NMDAR) antagonist and first introduced as anesthetics in the 1960s [47]. Ketamine immediately improves the mood in a sub-anesthetic dose, but its abuse potential and side effects such as loss of consciousness are limitations for its consideration as a safe AD. But this does not prevent ketamine from being a valuable tool for dissecting the circuitry involved in its rapid onset AD effects. Thus, ketamine has already led to the research for RAADs, along with the other NMDAR modulators [48]. This is especially critical for animal models of depression: utilizing the pharmacological manipulation coupled with optogenetic stimulation, ketamine effects on the inducible circuitry of genetically modified mice may help in the development of safer and efficacious ADs, especially RAADs. As a result, in vivo optogenetic control of the neuronal activity may allow the identification of neurons involved in the AD actions of ketamine. Indeed, increasing evidence suggests that optogenetics emerge as a research theme focusing on the identification of neural circuits underlying depression and AD response.

3 Controlling Neuronal Operations In Vivo

The basic operation of neurons relies on the depolarization of neuronal membrane potential, which is driven by the electrochemical gradient of ions such as potassium, sodium, chloride, and calcium [49]. As the opening state of ion channels embedded in the cell membrane determines the ions' membrane permeability, the electrochemical gradient is dependent on the activity state of ion channels which are gated by chemical or physical agents. For example, in the rodent brain, GABAergic parvalbumin-positive interneurons (PPIs) in the CA3 region of the hippocampus express specific GABA(A) receptors, the heteropentameric chloride channels gated by γ-aminobutyric acid, (GABA) [33, 50]. Upon GABA binding, the receptors' channels will get open, and chloride ions will move down to their electrochemical gradient [51, 52]. As the concentration of chloride is higher in the extracellular matrix compared to cytosol in the mature brain, the movement of chloride ions will be from outside to the inside of the PPIs [51]. This will cause the neuron to hyperpolarize, leading to a decreased probability of action potential generation or silencing.

Artificial silencing of neurons in specific experimental conditions in vivo, that is, silencing targeted neuronal ensembles in a reversible manner with high spatiotemporal resolution, is an invaluable tool to dissect the function of specific neuronal circuitry. This has been done by different approaches which use genetic manipulation in combination with pharmacological intervention, for example. Such approaches involve diverse genetic manipulation strategies such as triple crossing of genetically engineered mice to achieve a Cre recombinase-induced swapping of a native ion channel (GABA$_A$ receptor) to restore a cell-specific (Purkinje cells in the cerebellum) benzodiazepine modulation in the background of a mutated mouse which was initially made benzodiazepine insensitive in the brain globally [31].

Fortunately, the unpractical strategies are replaced by more practical ones. For instance, by the virus-mediated transgene expression which makes selected neurons sensitive to a pharmacological agent, it is possible to control the activity of neuronal ensembles specifically and reversibly. Regarding this, the invertebrate glutamate-gated chloride channel receptor (GluClR), which is ivermectin-activated chloride channel, is expressed in targeted neurons by cell-specific promoters, and then the GluClR is modulated by ivermectin administration which will control the activity of targeted neurons in vivo [53, 54]. Here, the GluClR expression is achieved by viral vector, a more practical method than crossing and analyzing the mutant mice for years. In fact, virus-mediated gene expression is employed in diverse experimental paradigms to manipulate the genetic systems for reversible and selective neural control including optogenetics. Optogenetics involves a diverse set of tools and methods to manipulate and control cells of living tissue rapidly, reversibly, and specifically. Within the CNS, this manipulation allows a targeted control of neurons in defined brain regions with unprecedented spatial and temporal resolution.

4   The Principle of Optogenetics

A central theme in the methodology of optogenetics is the light-sensitive receptors or opsins found in diverse organisms. As rod cells in the rodent retina express rhodopsin [55], which is activated by light and leading to the conversion of photons to electrochemical

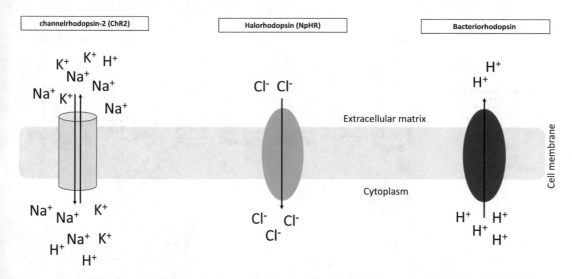

Fig. 1 Selected examples of algal (ChR2), archaeon (NpHR), and bacterial (bacteriorhodopsin) opsins and ion selectivity. In adult neurons, light activation of ChR2 will cause the nonselective cation channel to allow the cation passage down to their concentration gradient, and thus, a depolarization and excitation of the neuron are expected, while the activation of NpHR chloride pump or bacteriorhodopsin proton pump will cause the hyperpolarization and thus will likely silence the associated neuron. Figure not to scale

signals, many other eukaryotic and prokaryotic cells express their own light-sensitive rhodopsin, a G-protein-coupled receptor (GPCR) made up of seven transmembrane domains [56]. Rhodopsins are found in archaea, prokaryotes, and eukaryotes [56]. For instance, channelrhodopsin-2 (ChR2) was discovered in the green algae *Chlamydomonas reinhardtii* [57, 58]. It is permeable to cations across the membrane in both reversed and forward directions depending on the electrochemical gradient of the relevant ions. Halorhodopsin (NpHR), the chloride pump from the extracellular matrix to cytosol, was discovered in *Natronomonas pharaonis* [59, 60]. Bacteriorhodopsin, discovered in *Halobacterium halobium* [61–63], pumps protons from the cytoplasm to the extracellular matrix [64]. Figure 1 shows the selected examples of opsins and their ion selectivity.

In a typical workflow of optogenetic experiment (Fig. 2), selected opsin such as NpHR or ChR2 is cloned into viral vector as a fluorescent fusion protein. This requires designing or selecting suitable viral vector/promoter system. There are different viral systems used for targeted gene delivery such as adeno-associated viruses (AAVs) and lentiviruses [65–68]. One critical aspect of this process is the cell type-specific promoters and their compatibility with the viral vector. For example, promoters such synapsin I, CaMKIIa, the vesicular glutamate transporter 1 (VGLUT1), Dock10, and Prox1 are specific for glutamatergic neurons, but they have different expression level in the target tissue level

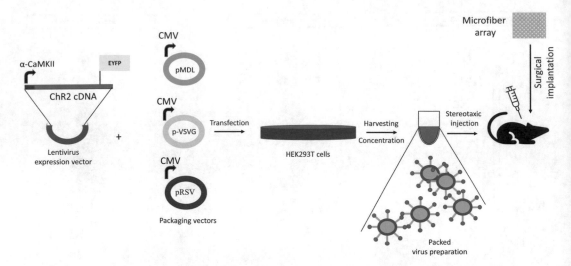

Fig. 2 Basic workflow of optogenetic experimental setup. As an example, in the backbone of the lentivirus vector, the coding DNA of the C terminally enhanced yellow fluorescent protein (EYFP)-tagged channelrhodopsin-2 (ChR2) in the presence of glutamatergic neuron-specific promoter CaMKIIa is shown as a viral construct. This construct is transfected together with other vectors (packaging vectors) into the host cells (HEK293T cells) to induce virus packaging, which is followed by the harvesting of the viruses containing EYFP-tagged channelrhodopsin-2 (ChR2). Viruses are then introduced to selected brain region by stereotaxic injection, which is followed by a surgical implantation of a microfiber device into the brain under anesthesia. Figure not to scale

depending on the viral system [68]. Thus, the optogenetics set up requires the proper selection and/or design of these systems, prior to virus packaging, which is a straightforward laboratory procedure. Then the viral preparation is administered into the desired brain region by stereotaxic injection. Also, following this, the animal is surgically implanted with a device such as wireless optrode array which allows both light delivery and measurement of neural activity, for instance [69]. Figure 2 shows such a procedure as an example for ChR2 and its delivery via lentivirus vector and its expression under the control of CaMKIIa promoter and glutamatergic neuron-specific promoter.

In addition to virus-mediated gene delivery, non-viral vectors are also exploited for optogenetic interrogation [70]. Regarding this, CRISPR-Cas9 is one example. Originally an adaptive immune system, CRISPR-Cas9 is used by many bacteria as a defense mechanism against invading agents [71]. It has been recently exploited as an effective tool for genome editing both in vivo and in vitro for a wide range of applications in molecular biology, genetics, medicine, neuroscience, and so forth [72–74]. CRISPR-Cas9 is composed of a single-guide RNA (sgRNA) for the identification of DNA targets and an endonuclease Cas9 that can bind the targeted sequence for further processing [75]. CRISPR/Cas9 system offers a powerful strategy for site-specific modifications in genome and thus assisting the precise engineering of virtually any genomic sequence in living cells [76]. Since it has genome-wide specificity and multiplexing capability, Cas9 and its variants have shown great potential in the generation of knockout animals, the functional genome screening, the correction of genetic diseases, and the treatment of infectious diseases. In addition, the current CRISPR-Cas9 systems are being explored for light-induced control [77–79].

5   Neuronal Circuitries and Depression: The Utility of Optogenetics

Increasing literatures have shed light on the neuronal circuitries involved in the depressive-like behaviors and AD effect. There is especially specific focus on glutamatergic system as rapid antidepressant effect of ketamine is mediated by NMDARs (*see* Subheading 2). In the sections below, we will examine optogenetic interrogation of mPFC, followed by the studies focusing on the ventral tegmental area and nucleus accumbens (VTA and NAc), lateral habenula (LHb) and raphe nuclei (RN), and amygdala (AMY), bed nucleus of the stria terminalis (BNST), and hippocampus.

6   Medial Prefrontal Cortex

In general, it is well known that areas of prefrontal cortex (PFC), the brain region involved in the executive functions, are altered in depression. These include alterations of structure as well as alterations in the connectivity and in the glutamatergic and GABAergic neurotransmissions [80]. Among the areas of PFC associated with depression is the medial prefrontal cortex (mPFC), which is critical for cognitive process, regulation of emotion, motivation, and sociability [81]. Conversely, there are accumulating data in the literature studying the AD effect with optogenetic stimulation of the mPFC. By coupling pharmacological manipulation with optogenetic stimulation, the infralimbic prefrontal cortex (IL-PFC) of mPFC has been studied to test if specific neuronal ensembles are involved in the AD effect of ketamine. Results confirm that CaMKIIa(Ca^{2+}/calmodulin-dependent protein kinase kinase-alpha, an enzyme that can alter the protein function by phosphorylation) expressing pyramidal neurons in IL-PFC may produce AD effect like the actions of systemic ketamine administration in rodents [82]. As expected, several other studies were conducted to prove the AD effect with optogenetic stimulation of mPFC. Mice that are challenged with social defeat stress showed enhanced social interaction and improved sucrose preference which are acknowledged to be AD-like effects when their mPFC was received optogenetic stimulation. Taken together, the findings indicate that activation of mPFC with optogenetics might aid mice in alleviating depression-like behaviors [83].

The same results were replicated in mice with forced swim test (FST) [84] and tail suspension test (TST) [85] during the optogenetic stimulation of mPFC. Son et al. [85] has shown reduced immobility in TST after the activation of vesicular glutamate transporter 2 (vGlut2) neurons. Kumar et al. [84] revealed reduced immobility in FST following optogenetic activation of Thy-1-expressing neurons. A study by Warden et al. [86] could not show the desired antidepressant-like effects in FST with stimulating the mPFC by optogenetics. However, their study revealed the importance of different pathways within mPFC; since the general activation of CaMKIIa-expressing neurons in mPFC did not have a clear effect on depression-like behavior in rats, selective activation in dorsal raphe nucleus (DRN)-mPFC neurons showed behavioral activation during FST, while optogenetic activation of LHb-mPFC axons generated a decrease in mobility [86]. The potential cause for the inconsistencies is not well-defined, but apparently, they might be attributable to targeting diverse groups of excitatory neurons with the distinct promoters implemented in these studies [87].

In one recent study, Hare et al. [88] examined the effect of optogenetic stimulation of more specific mPFC neurons such as Drd1 and Drd2 dopamine (DA) receptor-expressing pyramidal neurons. The findings demonstrated that after optogenetic stimulation of mPFC in Drd1-Cre, the cFos significantly increased in the basolateral amygdala (BLA) of the mice where photostimulated mPFC Drd1 terminals within the BLA had an AD effect in FST; however, the results were not replicated in Drd2. The previous research exhibited the BLA's vital role in stress reactions and fear learning circuitry which coordinates behavioral reactions to positive or negative stimuli owing to the reciprocal connections with mPFC. Moreover, the same study attempted to test whether antidepressant actions of ketamine necessitate Drd1-expressing pyramidal cell activation via photoinhibition method. When the photoinhibition was applied on Drd1 cells of mPFC along with the ketamine administration, it developed a non-significant decrease in immobility time and a significant decrease in latency of feed during Novelty Supressed Feeding (NSF) test which proved the essential requirement of Drd1 pyramidal cell activity for observing the antidepressant activity of ketamine [88]. mPFC circuitry is considered to have an active role in alterations of social interaction. The previous research showed the bidirectional effects of optogenetics on BLA neurons projecting to the mPFC. For instance, photoinhibition provoked the juvenile interaction, whereas photostimulation inhibited it among unstressed animals [89]. Several studies argued that stress exposure may strengthen the responses in BLA-mPFC circuitry. mPFC apical dendritic complexity along with total spine count and function was found to be decreased under chronic stress exposure, although basal dendrites targeted by BLA remained the same [90]. Therefore, BLA control may increase after stress.

Additionally, Carreno et al. [91] conducted a study to investigate the role of the vHipp-mPFC pathway in ketamine's antidepressant response. Optogenetic activation of vHipp-mPFC pathway showed the AD effects similar to ketamine. On the other hand, inhibition of the same pathway triggered amplified immobility in ketamine-treated rats, competently annulling the ketamine response. However, inhibition of the MDT-mPFC pathway did not cause any alterations in ketamine response which indicates circuit specificity [91]. In a similar vein when optogenetic stimulation is present in IL-PFC, it was found to show prompt and enduring AD-like effects which are related to improved function and increased counts of spine synapses located in V pyramidal neurons. On the contrary, the cessation of antidepressant effects was observable in rodents during the inactivation of IL-PFC, and microinfusion of ketamine into IL-PFC imitates the behavioral actions of systemic ketamine [82]. Increasing cellular activity using photostimulation with gamma bursts in IL-PFC and medial dorsal thalamus (MDT) decreased immobility in TST [92].

A subthreshold social defeat stress (SSDS) in animals revealed that social avoidant behaviors are noticeable during the inhibition of VTA projections to mPFC; thus, dopaminergic afferents to mPFC were acknowledged to be a salient factor in social avoidance [93]. There is research which supported the view of reduced D1 receptor signaling and dopamine levels in mPFC after social defeat stress [94–96]. Consequently, mPFC afferents evidently state an important position in social avoidance behaviors, and this stress reactive circuitry may be a potential target in treatment of stress-related disorders including depression.

7 Ventral Tegmental Area and Nucleus Accumbens

Mesolimbic DA pathway is consisted of ventral tegmental area (VTA) and nucleus accumbens (NAc) and has a role in reward, motivation, and learning. Over the past few years, there is an extensive research in the optogenetic stimulation of VTA exercising depression-like behaviors. For instance, in one study in which social defeat stress (SDS) model of depression was employed, it was shown that when DA neurons of VTA are activated via using optogenetics, reduced sucrose consumption and social avoidance are observed in mice [93]. On the contrary, a study by Tye et al. [97] revealed that optogenetic inhibition of DA neurons of VTA caused depression as well as optogenetic activation of them showed improvement in stress-induced depression. These two contradictory findings may be attributed to different methodology and diverse depression models [98]. In a previous study, optogenetic hyperactivation of VTA DA neurons has been shown to diminish depression-related behaviors. It was acknowledged that projection-specific and resilience-like homeostatic plasticity might be attributed to this achievement [99]. Another study that repeatedly employs optogenetic stimulation on medium spiny neurons (MSNs) in the NAc has shown enhanced social interaction in mice which were modeled to be under chronic social defeat stress (CSDS) depression. This finding supports that augmentation of the MSN activity in NAc might be responsible for antidepressant outcome since the aforementioned region is placed in the forebrain which collects inputs from VTA [100]. Another study investigating the MSNs in ventral striatum (vSTR) which receive projections from intralaminar thalamus (ILT) revealed that stress-induced postsynaptic plasticity of vSTR is mediated by elevated glutamatergic transmission in ILT-vSTR circuit, and it leads to vulnerability to social stress. Stimulation of ILT-vSTR neurons induced susceptibility after CSDS; however, when these neurons were inhibited throughout CSDS, it promoted resilience in depression-like behaviors [101].

A study by Soares-Cunha et al. [102] examined the role of activation of D1 and D2 neurons in NAc on motivation using the Pavlovian-to-instrumental transfer (PIT) and the progressive ratio (PR) tasks. Their results revealed that simultaneous stimulation of D1 and D2 neurons in NAc enhances motivational drive in mice [102].

There are also optogenetic studies that investigate the neuronal circuits of VTA and NAc on depression-like behaviors. For example, during the social interaction test, optogenetic stimulation of the neuronal pathways from ventral hippocampus (vHipp) to NAc has been found to cause depression-like behaviors in mice [103]. Another study has found out depression-like behaviors in SDS model under optogenetic stimulation of VTA afferents to NAc and when VTA neurons that project to mPFC were inhibited by optogenetics [93].

Walsh et al. [104] suggested that to mediate social avoidance behavior, it might be fundamental to increase firing pattern-dependent regulation of BDNF in VTA-NAc circuit. When CRF receptor antagonist was infused intra-NAc, the blockade of phasic firing caused social avoidance. Although BDNF levels in NAc did not change after acute or repeated phasic activation of VTA-NAc circuit in stress-naïve mice, the results showed the blocked BDNF upregulation in socially stressed mice when they were phasically activated after receiving intra-NAc CRF receptor antagonist infusion. On the contrary, the subthreshold-stressed mice exhibited an increase in upregulation of BDNF in NAc by the similar phasic activation, whereas no effects were seen among unstimulated mice. The absence of the stimulation in BDNF signaling was observed particularly in mice which lack BDNF in VTA-NAc neurons.

According to these findings, it was acknowledged that phasic firing solely is insufficient to prompt BDNF upregulation in VTA-NAc circuit and DA neurons in mesolimbic area function in a stress context-detecting approach mediated by CRF activity in NAc. Therefore, the findings shed light on mesolimbic reward circuitry's gating function in generation of specific responses to environmental stimuli [105]. Vialou et al. [106] examined the role of ΔFosB in the prelimbic (PrL) area using CSDS. ΔFosB has been found to initiate the effects partially via cholecystokinin (CCK)-B receptor. The findings showed that CCK administration into mPFC imitates the depressant-like effects of social stress, whereas the blockade of CCK-B in the same region provokes a resilient phenotype.

Among stress-exposed vulnerable mice, the optogenetic stimulation of locus coeruleus neurons projecting to VTA [107] and PPIs in ventral pallidum (VP) to VTA pathway [108] yielded a resilience-like phenotype with normalized social behavior which reflects antidepressant-like effect. Furthermore, optogenetic stimulation of VTA to NAc pathway [105, 109] and serotonergic terminals in VTA [110] also resulted in antidepressant-like affects.

8 Lateral Habenula and Raphe Nuclei

The lateral habenula (LHb) nucleus which is located in diencephalon was found to be overactive in depressive disorders [111]. This region is acknowledged to be related to the negative reward processing such as aversion [112]. Therefore, there is an accumulating interest in applying optogenetics for this nucleus and its projections in animal experiments. According to the findings when LHb neurons projecting to the rostromedial tegmental nucleus (RMTg) (the area rich in GABAergic neurons) were stimulated, it generated aversive effect in real-time place preference test. The optogenetic stimulation of the same circuitry also increased the immobility duration in FST; however, the inhibition of it decreased the duration [113]. In addition, when optogenetic inhibition was applied to PPIs in the VP-to-LHb pathway, a recovery indicating antidepressant-like effect was observed in FST [108]. Despite these promising findings, there is a need for future studies to accurately predict medial habenula (mHb)'s reactions to optogenetic stimulation since it activates during depression [98].

Raphe nucleus (RN) which has abundant serotonin neurons is placed in the brain stem [98]. RN and serotonin circuitry along with the reduced activity in serotonin pathway were suggested to be responsible for depression pathophysiology [114–116]. In one study, employing optogenetic stimulation to excitatory vmPFC input to the DRN produced an elevated social avoidant behavior. On the other hand, optogenetic inhibition of the same circuitry yielded a reduced social avoidance [117]. The researchers recommended that to filter top-down vmPFC effects on affect-regulating serotonin output, it is essential to identify GABAergic neurons and the functional organization of vmPFC-DRN pathways [117]. In a similar vein, another study practiced optogenetic inhibition on GABAergic interneurons in mPFC-DRN circuitry; as a result, antidepressant-like effect was seen in social defeat stress depression model among vulnerable mice [117]. However, they were not able to show the identical effects among mice with resilience to social defeat stress.

A more recent study demonstrated the AD effects in FST and TST with acute optogenetic stimulation of serotonergic neurons in DRN. Additionally, specific serotonergic terminals like VTA/substantia nigra are involved in antidepressant-like responses. Therefore, the findings implied that there might be functional differences between subregions of RN in the modification of depression-like behaviors via optogenetics [110].

Another potential region which was acknowledged to be related to depression pathophysiology is dorsal raphe nucleus-lateral habenula (DRN-LHb circuit). Optogenetic stimulation of this circuitry reduced the depression-like behaviors, whereas the inhibition caused a negative effect like depression in animal studies [118]. Therefore, DRN to LHb circuit was recommended to be a promising target for depression. Despite these improvements made so far, there is still a need for experimental optogenetic studies to discover various therapeutic targets in the clinical practice [98].

9   Amygdala, Bed Nucleus of the Stria Terminalis, and Hippocampus

Previous studies also focused on AMY, BNST, and hippocampus to examine their effects on depression-like behaviors via optogenetic approaches [119–121]. AMY consists of nuclei located in the medial side of the temporal lobe, and it moderates emotions and social behaviors [122]. In a recent study, when optogenetic stimulation activated the parabrachial neurons projecting to the central amygdala (CeA), it increased the depression-like behaviors in FST. If the basolateral amygdala (BLA) neurons projecting to the CeA were activated, it emerges antidepressant-like response in the same test [121]. This novel finding demonstrated that the different regions of AMY may exhibit divergent influences on moderating depression-like behaviors [121]. Further optogenetic studies are warranted to reveal the explicit role of the AMY and its pathways in depression [98].

BNST consists of a bunch of axons that travel through the ventricular side of the thalamus. It gives the main output to AMY and is acknowledged to be related to depression [123]. In a previous study, optogenetic inhibition of anteroventral BNST increased the immobility in TST as it is considered to be a behavioral marker of depression in mice experiment [120]. Furthermore, the hippocampus was also studied in optogenetic study to identify its effects in depression-like behaviors. For instance, Ramirez et al. [119] found that after activating dentate gyrus cells of the hippocampus via optogenetic stimulation, recovery was observed in depression-like behaviors in TST [119]. Despite this positive outcome, the optogenetic studies are scarce, and future studies focused on diverse anatomical subdivisions and different input/output trajectories along with specific neuronal subtypes of the hippocampus may be practical to comprehend its effect on depression [98]. The featured optogenetic experiments are summarized regarding the brain regions, depression models, stimulation protocols, and primary findings in Table 1.

Table 1
Overview of optogenetic studies in animal models of depression-like behavior

Region	Model	Stimulation protocol	Primary findings	References
VTA	CSDS	ChR2, 10 min, 20 Hz, 40 ms pulse width	Stimulation induces susceptibility in resilient mice	Chaudhury et al. [93]
VTA	CSDS	ChR2, 20 min, 20 Hz, 40 ms pulse width	Stimulation induces resilience in susceptible mice by inducing an increase in potassium channel current and normalizing hyperactivity	Friedman et al. [99]
VTA	CMS	ChR2, 20 Hz eNpHR3.0, 3 or 30 min light on	Stimulation induces resilience; inhibition induces susceptibility	Tye et al. [97]
VTA-NAc	CSDS	ChR2, 10 min, 20 Hz, 40 ms pulse width NpHR, 8 s light on, 2 s light off	Stimulation induces susceptibility following subthreshold CSDS (SCDS); inhibition induces resilience in susceptible mice	Chaudhury et al. [93]
VTA-NAc		ChR2, 20 min, 20 Hz, 40 ms pulse width	Stimulation induces resilience in susceptible mice by inducing an increase in potassium channel current and normalizing hyperactivity	Friedman et al. [99]
VTA-NAc		ChR2, 20 Hz	Stimulation increases BDNF in NAc, leading to increased susceptibility Blocking BDNF signaling blocks optically induced susceptibility. Effects require CRF release	Walsh et al. [105]
VTA-NAc		ChR2, 5 min, 20 Hz	Blocking D1 receptors blocks optically induced susceptibility following SSDS	Wook Koo et al. [109]

(continued)

Table 1
(continued)

Region	Model	Stimulation protocol	Primary findings	References
VTA-mPFC	CSDS	NpHR, 8 s light on, 2 s light off	Inhibition induces susceptibility following SSDS	Chaudhury et al. [93]
VTA-mPFC		ChR2, 20 min, 20 Hz, 40 ms pulse width	Stimulation induces resilience in susceptible mice via a different mechanism than VTA-NAc	Friedman et al. [99]
NAc	CSDS	ChETA, 15 min, 50 Hz	D1-MSN stimulation induces resilience in susceptible mice D2-MSN stimulation induces susceptibility following SSDS	Francis et al. [100]
NAc	PIT/PR	ChR2, 1 s, 40 Hz, 12.5 ms eNpHR3.0, 10 s constant light	Stimulation of both D1- and D2-MSNs increases motivation. Inhibition of D2-MSNs decreases motivation	Soares-Cunha et al. [102]
vHipp-NAc	CSDS	ChR2, 4 Hz, 5 ms pulse width (acute) ChR2, 10 min, 1 Hz, 4 ms pulse width (LTD induction)	Enhancement induces susceptibility following CSDS; attenuation induces resilience	Bagot et al. [103]
vHipp-mPFC	FST	eNpHR3.0, 20 min, 10 Hz	Inhibition blocks the antidepressant effects of ketamine administration	Carreno et al. [91]
DG-BLA-NAc	CIS	ChR2, 20 Hz, 15 ms pulse width ArchT, constant stimulation	Activating positive engrams in DG reversed depressive-like behaviors Inhibiting BLA-NAc-positive engrams blocked this rescue	Ramirez et al. [119]
AMY-NAc	CSDS	ChR2, 4 Hz, 5 ms pulse width	Stimulation induces resilience following CSDS	Bagot et al. [103]
mPFC	CSDS	ChR2, 40 ms, 100 Hz, 9.9 ms pulse width	Stimulation increases resilience following CSDS	Covington et al. [83]
mPFC-NAc	CSDS	ChR2, 4 Hz, 5 ms pulse width	Stimulation induces resilience following CSDS	Bagot et al. [103]
PrL-NAc	CSDS	ChR2, 40 ms, 100 Hz, 9.9 ms pulse width	Stimulation induces resilience and protects against the effects of CCK-B overexpression	Vialou et al. [106]
mPFC-LHb	FST	15 min; 2 min light on-light off epochs	Stimulation increases immobility	Warden et al. [86]
mPFC-DRN	FST	15 min; 2 min light on-light off epochs	Stimulation increases kicking	Warden et al. [86]
	CSDS	ChR2, 20 min, 25 Hz, 10 ms pulse width	Stimulation excites GABAergic cells to inhibit 5-HT cells in DRN and increase social avoidance	Challis et al. (2013, [117])

(continued)

Table 1
(continued)

Region	Model	Stimulation protocol	Primary findings	References
IL mPFC glutamatergic	FST/ SPT/ NSF	Chr2, 10 Hz, 15 ms, 5 mW, 1 min on–1 min off for 60 min	Antidepressant effects on depression-like behaviors	Fuchikami et al. [82]
PrL mPFC glutamatergic	FST/ SPT/ NSF	Chr2, 10 Hz 15 ms, 5 mW, 1 min on–1 min off for 60 min	No antidepressant effects on depression-like behaviors	Fuchikami et al. [82]
vmPFC Drd1 neurons	FST, EPM, NSF SPT	Chr2, 10 Hz 15 ms, 5 mW, 1 min on–1 min off for 60 min	No antidepressant effects on depression-like behaviors	Hare et al. [88]
vmPFC Drd2 neurons	FST, EPM, NSF	Chr2, 10 Hz 15 ms, 5 mW, 1 min on–1 min off for 60 min	No antidepressant effects on depression-like behaviors	Hare et al. [88]
vmPFC Drd1-BLA	FST, NSF	Chr2, 10 Hz, 15 ms, 5 mW, 1 min on–1 min off for 60 min	Antidepressant effects on depression-like behaviors	Hare et al. [88]
ILT-vSTR	CSDS	ChR2, 20 Hz, 20 ms pulse width	Stimulation induces susceptibility following SCDS. Inhibition induces resilience following CSDS	Christoffel et al. [101]
ACC	Chronic pain	ChR2, 30 min, 20 Hz, 40 ms pulse width	Stimulation increases anxio-depressive behaviors	Barthas et al. (2015)
VP-LHb	CSDS	ChR2, 20 Hz, 5 ms pulse width	Stimulation increases measures of despair	Knowland et al. [108]
VP-VTA	CSDS	NpHR, 0.1 Hz, 9 s on, 1 s off	Inhibition increases social interaction	Knowland et al. [108]
avBNST	TST/ FST	ChR2, 20 Hz, 5 ms pulse width Arch, constant illumination	Inhibition upregulates HPA axis and increases measures of despair	Johnson et al. [120]
avBNST-PVH	TST/ FST	ChR2, 20 Hz, 5 ms pulse width Arch, constant illumination	Inhibition upregulates HPA axis	Johnson et al. [120]
avBNST-vlPAG	TST/ FST	ChR2, 20 Hz, 5 ms pulse width Arch, constant illumination	Inhibition increases measures of despair	Johnson et al. [120]

10   Conclusion

The technique of optogenetics emerges as an invaluable tool to study neural circuitry more precisely than ever before, and thus, its scope and applications are rapidly growing due to the exciting results generated in the last decade. Especially progress in understanding the function of specific subtypes of neurons involved in the depression-like behaviors and AD response is significant. In essence, the earlier results manifest the multifaceted neural mechanisms in depression including diverse cells, neurotransmitters, neural circuits, and brain areas.

On the other hand, there are still several limitations to overcome such as the fundamental differences between cell-based models, animal models, and humans, safety, and off-target/side effects as summarized in Box 1. For example, due to the greater safety risks associated with gene delivery methods in humans than animal models and cell culture, many gene delivery approaches used in optogenetics might not be translatable to humans. Given that viral vector-based gene therapies are dominating the clinical studies and are already approved for some indications in current medicine including neuro-visual disorders, it is the most promising gene delivery method in optogenetic applications in human [124–126]. However, other gene delivery techniques such plasmid DNA-wrapped gold nanoparticles and cationic carbon quantum dots emerged as alternates to virus-based methods, but their efficiency and safety levels are still warranted.

Box 1 Advantages and Disadvantages of Optogenetics

Advantages	Disadvantages
Manipulation of specific cell types or neural circuits	High cost, low-throughput
Higher speed and temporal accuracy (spatiotemporal precision)	Largely unconscious and immobilized animal models
Applicable on several organs	Possible side effects such as toxicity and immunogenicity due to the differentially long-term expression of light-sensitive proteins
The types of vectors used in optogenetics are generally safe and well tolerated by patients	Physiological differences between cell culture and intact brain

(continued)

Box 1 (continued)

Advantages	Disadvantages
Enabling use of other light-sensitive proteins, e.g. G-protein modulating opsins	Possible irreversible damage to the brain areas of interest due to surgical implantations
Reducing tissue damages and its side effects due to development of recent alternative methods	Complexity in finding the optimal energy requirements for *in vivo* applications

Up to date, there is still debate in clinical utilization of optogenetics due to lack of systematical analyses of generated findings. Additionally, there is still gap in detailed identification of neural pathways thought to be related with depression-like behaviors in optogenetic studies. Further studies that collect broad preclinical data and offer generalized methodological approach in optogenetics may contribute more to its implementation into the clinical practice. As in silico modeling systems provide fast and useful information in many areas of science, their predictions should also be taken into consideration in optogenetics such as diversity in opsin expression and comparison of behavioral paradigms with different neuronal network readouts [127].

References

1. James SL, Abate D, Abate KH, Abay SM, Abbafati C, Abbasi N et al (2018) Global, regional, and national incidence, prevalence, and years lived with disability for 354 diseases and injuries for 195 countries and territories, 1990–2017: a systematic analysis for the global burden of disease study 2017. Lancet 392:1789–1858

2. Vos T, Barber RM, Bell B, Bertozzi-Villa A, Biryukov S, Bolliger I et al (2015) Global, regional, and national incidence, prevalence, and years lived with disability for 301 acute and chronic diseases and injuries in 188 countries, 1990–2013: a systematic analysis for the global burden of disease study 2013. Lancet 386:743–800

3. Malhi GS, Mann JJ (2018) Depression. Lancet 392:2299–2312. https://doi.org/10.1016/S0140-6736(18)31948-2

4. Arslan A (2015) Genes, brains, and behavior: imaging genetics for neuropsychiatric disorders. J Neuropsychiatry Clin Neurosci 27:81–92

5. Arslan A (2018) Imaging genetics of schizophrenia in the post-GWAS era. Prog Neuro-Psychopharmacology Biol Psychiatry 80:155–165

6. Coplan JD, Andrews MW, Rosenblum LA, Owens MJ, Friedman S, Gorman JM et al (1996) Persistent elevations of cerebrospinal fluid concentrations of corticotropin-releasing factor in adult nonhuman primates exposed to early-life stressors: implications for the pathophysiology of mood and anxiety disorders. Proc Natl Acad Sci U S A 93:1619–1623

7. Murphy BEP, Wolkowitz OM (1993) The pathophysiologic significance of hyperadrenocorticism: antiglucocorticoid strategies. Psychiatr Ann 23:682–690

8. Saxe MD, Battaglia F, Wang JW, Malleret G, David DJ, Monckton JE et al (2006) Ablation of hippocampal neurogenesis impairs contextual fear conditioning and synaptic plasticity in the dentate gyrus. Proc Natl Acad Sci U S A 103:17501–17506. https://doi.org/10.1073/pnas.0607207103

9. Santarelli L, Saxe M, Gross C, Surget A, Battaglia F, Dulawa S et al (2003) Requirement of hippocampal neurogenesis for the behavioral effects of antidepressants. Science 301:805–809. https://doi.org/10.1126/science.1083328

10. Luscher B, Shen Q, Sahir N (2011) The GABAergic deficit hypothesis of major depressive disorder. Mol Psychiatry 16:383–406

11. Inserra A, Rogers GB, Licinio J, Wong M (2018) The microbiota-inflammasome hypothesis of major depression. Bioessays 40:1800027

12. Fava M, Davidson KG (1996) Definition and epidemiology of treatment-resistant depression. Psychiatr Clin North Am 19:179–200

13. Machado-Vieira R, Baumann J, Wheeler-Castillo C, Latov D, Henter ID, Salvadore G et al (2010) The timing of antidepressant effects: a comparison of diverse pharmacological and somatic treatments. Pharmaceuticals (Basel) 3:19–41

14. Aydin O, Aydin PU, Arslan A (2019) Development of neuroimaging-based biomarkers in psychiatry. Adv Exp Med Biol 1192:159–195

15. Arslan A (2018) Application of neuroimaging in the diagnosis and treatment of depression. In: Understanding depression. Springer, pp 69–81

16. Unal Aydin P, Aydin O, Arslan A (2021) Genetic architecture of depression: Where do we stand now? Adv Exp Med Biol 1305:203–230

17. Kim Y-K, Park S-C (2021) An alternative approach to future diagnostic standards for major depressive disorder. Prog Neuro-Psychopharmacol Biol Psychiatry 105:110133

18. Arslan A (2018) Mapping the schizophrenia genes by neuroimaging: the opportunities and the challenges. Int J Mol Sci 19:219

19. Waterston RH, Lindblad-Toh K, Birney E, Rogers J, Abril JF, Agarwal P et al (2002) Initial sequencing and comparative analysis of the mouse genome. Nature 420:520–562

20. Nestler EJ, Hyman SE (2010) Animal models of neuropsychiatric disorders. Nat Neurosci 13:1161–1169. https://doi.org/10.1038/nn.2647

21. Barkus C (2013) Genetic mouse models of depression. Curr Top Behav Neurosci 14:55–78

22. Alexander B, Warner-Schmidt J, Eriksson TM, Tamminga C, Arango-Lievano M, Ghose S et al (2010) Reversal of depressed behaviors in mice by p11 gene therapy in the nucleus accumbens. Sci Transl Med 2:54ra76

23. Chourbaji S, Gass P (2008) Glucocorticoid receptor transgenic mice as models for depression. Brain Res Rev 57:554–560

24. Wang YM, Xu F, Gainetdinov RR, Caron MG (1999) Genetic approaches to studying norepinephrine function: Knockout of the mouse norepinephrine transporter gene. Biol Psychiatry 46:1124–1130. https://doi.org/10.1016/s0006-3223(99)00245-0

25. Holmes A, Yang RJ, Murphy DL, Crawley JN (2002) Evaluation of antidepressant-related behavioral responses in mice lacking the serotonin transporter. Neuropsychopharmacology 27:914–923

26. Urani A, Chourbaji S, Gass P (2005) Mutant mouse models of depression: candidate genes and current mouse lines. Neurosci Biobehav Rev 29:805–828. https://doi.org/10.1016/j.neubiorev.2005.03.020

27. Perona MTG, Waters S, Hall FS, Sora I, Lesch KP, Murphy DL et al (2008) Animal models of depression in dopamine, serotonin, and norepinephrine transporter knockout mice: prominent effects of dopamine transporter deletions. Behav Pharmacol 19:566–574. https://doi.org/10.1097/FBP.0b013e32830cd80f

28. Gallagher JJ, Zhang X, Hall FS, Uhl GR, Bearer EL, Jacobs RE (2013) Altered reward circuitry in the norepinephrine transporter knockout mouse. PLoS One 8:e57597

29. Stuber GD, Stamatakis AM, Kantak PA (2015) Considerations when using cre-driver rodent lines for studying ventral tegmental area circuitry. Neuron 85:439–445. https://doi.org/10.1016/j.neuron.2014.12.034

30. Liu G, Wang Y, Zheng W, Cheng H, Zhou R (2019) P11 loss-of-function is associated with decreased cell proliferation and neurobehavioral disorders in mice. Int J Biol Sci 15:1383

31. Wulff P, Goetz T, Leppä E, Linden AM, Renzi M, Swinny JD et al (2007) From synapse to behavior: rapid modulation of defined neuronal types with engineered GABAA receptors. Nat Neurosci 10:923–929. https://doi.org/10.1038/nn1927

32. Sullivan PF, Neale MC, Kendler KS (2000) Genetic epidemiology of major depression: review and meta-analysis. Am J Psychiatry 157:1552–1562

33. Arslan A (2015) The complexity of mental disorders. Period Eng Nat Sci 3

34. Bosker FJ, Hartman CA, Nolte IM, Prins BP, Terpstra P, Posthuma D et al (2011) Poor replication of candidate genes for major depressive disorder using genome-wide association data. Mol Psychiatry 16:516–532

35. Wray NR, Pergadia ML, Blackwood DHR, Penninx B, Gordon SD, Nyholt DR et al (2012) Genome-wide association study of major depressive disorder: new results, meta-

analysis, and lessons learned. Mol Psychiatry 17:36–48

36. Sullivan P, Andreassen OA, Anney RJL, Asherson P, Ashley-Koch A, Blackwood D et al (2012) Don't give up on GWAS. Mol Psychiatry 17:2–3. https://doi.org/10.1038/mp.2011.94

37. Cai N, Bigdeli TB, Kretzschmar W, Li Y, Liang J, Song L et al (2015) Sparse whole-genome sequencing identifies two loci for major depressive disorder. Nature 523:588–591

38. Vaquero A, Scher M, Erdjument-Bromage H, Tempst P, Serrano L, Reinberg D (2007) SIRT1 regulates the histone methyltransferase SUV39H1 during heterochromatin formation. Nature 450:440–444

39. Yokoi F, Hiraishi H, Izuhara K (2003) Molecular cloning of a cDNA for the human phospholysine phosphohistidine inorganic pyrophosphate phosphatase. J Biochem 133:607–614

40. Okbay A, Baselmans BML, De Neve J-E, Turley P, Nivard MG, Fontana MA et al (2016) Genetic variants associated with subjective well-being, depressive symptoms, and neuroticism identified through genome-wide analyses. Nat Genet 48:624–633

41. Direk N, Williams S, Smith JA, Ripke S, Air T, Amare AT et al (2017) An analysis of two genome-wide association meta-analyses identifies a new locus for broad depression phenotype. Biol Psychiatry 82:322–329

42. Hyde CL, Nagle MW, Tian C, Chen X, Paciga SA, Wendland JR et al (2016) Identification of 15 genetic loci associated with risk of major depression in individuals of European descent. Nat Genet 48:1031

43. Berman RM, Cappiello A, Anand A, Oren DA, Heninger GR, Charney DS et al (2000) Antidepressant effects of ketamine in depressed patients. Biol Psychiatry 47:351–354

44. Murrough JW, Iosifescu DV, Chang LC, Al Jurdi RK, Green CE, Perez AM et al (2013) Antidepressant efficacy of ketamine in treatment-resistant major depression: a two-site randomized controlled trial. Am J Psychiatry 170:1134–1142. https://doi.org/10.1176/appi.ajp.2013.13030392

45. Fava M, Freeman MP, Flynn M, Judge H, Hoeppner BB, Cusin C et al (2020) Double-blind, placebo-controlled, dose-ranging trial of intravenous ketamine as adjunctive therapy in treatment-resistant depression (TRD). Mol Psychiatry 25:1592–1603

46. Zarate CA, Singh JB, Carlson PJ, Brutsche NE, Ameli R, Luckenbaugh DA et al (2006) A randomized trial of an N-methyl-D-aspartate antagonist in treatment-resistant major depression. Arch Gen Psychiatry 63:856–864. https://doi.org/10.1001/archpsyc.63.8.856

47. Domino EF, Chodoff P, Corssen G (1965) Pharmacologic effects of CI-581, a new dissociative anesthetic, in man. Clin Pharmacol Ther 6:279–291

48. Na KS (2021) Kim YK, vol 104. Prog Neuro-Psychopharmacol Biol Psychiatry, Increased use of ketamine for the treatment of depression: benefits and concerns, p 110060

49. Hodgkin AL (1958) The Croonian lecture-ionic movements and electrical activity in giant nerve fibres. Proc R Soc London Ser B Biol Sci 148:1–37

50. Arslan A (2021) Extrasynaptic δ-subunit containing GABAA receptors. J Integr Neurosci 20:173–184

51. Goetz T, Arslan A, Wisden W, Wulff P (2007) GABAA receptors: structure and function in the basal ganglia. Prog Brain Res 160:21–41

52. Arslan A (2015) Clustering of gamma-aminobutyric acid type A receptors. Period Eng Nat Sci 3

53. Lynagh T, Lynch JW (2010) An improved ivermectin-activated chloride channel receptor for inhibiting electrical activity in defined neuronal populations. J Biol Chem 285:14890–14897

54. Islam R, Keramidas A, Xu L, Durisic N, Sah P, Lynch JW (2016) Ivermectin-activated, cation-permeable glycine receptors for the chemogenetic control of neuronal excitation. ACS Chem Neurosci 7:1647–1657

55. Molday RS, Molday LL (1998) Molecular properties of the cGMP-gated channel of rod photoreceptors. Vision Res 38:1315–1323. https://doi.org/10.1016/s0042-6989(97)00409-4

56. Terakita A (2005) The opsins. Genome Biol 6:213. https://doi.org/10.1186/gb-2005-6-3-213

57. Takahashi T, Yoshihara K, Watanabe M, Kubota M, Johnson R, Derguini F et al (1991) Photoisomerization of retinal at 13-ene is important for phototaxis of Chlamydomonas reinhardtii: simultaneous measurements of phototactic and photophobic responses. Biochem Biophys Res Commun 178:1273–1279

58. Hegemann P, Gärtner W, Uhl R (1991) All-trans retinal constitutes the functional

chromophore in Chlamydomonas rhodopsin. Biophys J 60:1477–1489

59. Hegemann P, Oesterhelt D, Steiner M (1985) The photocycle of the chloride pump halorhodopsin. I: Azidecatalyzed deprotonation of the chromophore is a side reaction of photocycle intermediates inactivating the pump. EMBO J 4:2347–2350

60. Kalaidzidis IV, Kalaidzidis YL, Kaulen AD (1998) Flash-induced voltage changes in halorhodopsin from Natronobacterium pharaonis. FEBS Lett 427:59–63

61. Oesterhelt D, Stoeckenius W (1971) Rhodopsin-like protein from the purple membrane of Halobacterium halobium. Nat New Biol 233:149–152. https://doi.org/10.1038/newbio233149a0

62. Eisenbach M, Bakker EP, Korenstein R, Caplan SR (1976) Bacteriorhodopsin: biphasic kinetics of phototransients and of light-induced proton transfer by sub-bacterial Halobacterium halobium particles and by reconstituted liposomes. FEBS Lett 71:228–232

63. Matsuno-Yagi A, Mukohata Y (1977) Two possible roles of bacteriorhodopsin; a comparative study of strains of Halobacterium halobium differing in pigmentation. Biochem Biophys Res Commun 78:237–243. https://doi.org/10.1016/0006-291x(77)91245-1

64. Lanyi JK (1986) Halorhodopsin: a light-driven chloride ion pump. Annu Rev Biophys Biophys Chem 15:11–28

65. Naldini L, Blömer U, Gallay P, Ory D, Mulligan R, Gage FH et al (1996) In vivo gene delivery and stable transduction of non-dividing cells by a lentiviral vector. Science 272:263–267. https://doi.org/10.1126/science.272.5259.263

66. Dull T, Zufferey R, Kelly M, Mandel RJ, Nguyen M, Trono D et al (1998) A third-generation lentivirus vector with a conditional packaging system. J Virol 72:8463–8471

67. Wang X, McManus M (2009) Lentivirus production. J Vis Exp 32:1499. https://doi.org/10.3791/1499

68. Watakabe A, Sadakane O, Hata K, Ohtsuka M, Takaji M, Yamamori T (2017) Application of viral vectors to the study of neural connectivities and neural circuits in the marmoset brain. Dev Neurobiol 77:354–372. https://doi.org/10.1002/dneu.22459

69. Kwon KY, Lee H-M, Ghovanloo M, Weber A, Li W (2015) Design, fabrication, and packaging of an integrated, wirelessly-powered optrode array for optogenetics application. Front Syst Neurosci 9:69

70. Wang HX, Li M, Lee CM, Chakraborty S, Kim HW, Bao G et al (2017) CRISPR/Cas9-based genome editing for disease modeling and therapy: challenges and opportunities for nonviral delivery. Chem Rev 117:9874–9906. https://doi.org/10.1021/acs.chemrev.6b00799

71. Jinek M, Chylinski K, Fonfara I, Hauer M, Doudna JA, Charpentier E (2012) A programmable dual-RNA-guided DNA endonuclease in adaptive bacterial immunity. Science 337:816–821. https://doi.org/10.1126/science.1225829

72. Niu Y, Shen B, Cui Y, Chen Y, Wang J, Wang L et al (2014) Generation of gene-modified cynomolgus monkey via Cas9/RNA-mediated gene targeting in one-cell embryos. Cell 156:836–843. https://doi.org/10.1016/j.cell.2014.01.027

73. Cox DBT, Platt RJ, Zhang F (2015) Therapeutic genome editing: prospects and challenges. Nat Med 21:121–131

74. Haas SA, Dettmer V, Cathomen T (2017) Therapeutic genome editing with engineered nucleases. Hamostaseologie 37:45–52

75. Nishimasu H, Ran FA, Hsu PD, Konermann S, Shehata SI, Dohmae N et al (2014) Crystal structure of Cas9 in complex with guide RNA and target DNA. Cell 156:935–949. https://doi.org/10.1016/j.cell.2014.02.001

76. Hsu PD, Lander ES, Zhang F (2014) Development and applications of CRISPR-Cas9 for genome engineering. Cell 157:1262–1278

77. Polstein LR, Gersbach CA (2015) A light-inducible CRISPR-Cas9 system for control of endogenous gene activation. Nat Chem Biol 11:198–200. https://doi.org/10.1038/nchembio.1753

78. Bubeck F, Hoffmann MD, Harteveld Z, Aschenbrenner S, Bietz A, Waldhauer MC et al (2018) Engineered anti-CRISPR proteins for optogenetic control of CRISPR–Cas9. Nat Methods 15:924–927

79. Hoffmann MD, Mathony J, Zu Belzen JU, Harteveld Z, Stengl C, Correia BE et al (2021) Optogenetic control of Neisseria meningitidis Cas9 genome editing using an engineered, light-switchable anti-CRISPR protein. Nucleic Acids Res 49(5):e29

80. Duman RS, Sanacora G, Krystal JH (2019) Altered connectivity in depression: GABA and glutamate neurotransmitter deficits and reversal by novel treatments. Neuron Elsevier 102:75–90

81. Xu P, Chen A, Li Y, Xing X, Lu H (2019) Medial prefrontal cortex in neurological diseases. Physiol Genomics 51:432–442

82. Fuchikami M, Thomas A, Liu R, Wohleb ES, Land BB, DiLeone RJ et al (2015) Optogenetic stimulation of infralimbic PFC reproduces ketamine's rapid and sustained antidepressant actions. Proc Natl Acad Sci U S A 112:8106–8111

83. Covington HE, Lobo MK, Maze I, Vialou V, Hyman JM, Zaman S et al (2010) Antidepressant effect of optogenetic stimulation of the medial prefrontal cortex. J Neurosci 30:16082–16090

84. Kumar S, Black SJ, Hultman R, Szabo ST, Demaio KD, Du J et al (2013) Cortical control of affective networks. J Neurosci 33:1116–1129

85. Son H, Baek JH, Go BS, Jung D-H, Sontakke SB, Chung HJ et al (2018) Glutamine has antidepressive effects through increments of glutamate and glutamine levels and glutamatergic activity in the medial prefrontal cortex. Neuropharmacology 143:143–152

86. Warden MR, Selimbeyoglu A, Mirzabekov JJ, Lo M, Thompson KR, Kim SY et al (2012) A prefrontal cortex-brainstem neuronal projection that controls response to behavioural challenge. Nature 492:428–432

87. Hare BD, Duman RS (2020) Prefrontal cortex circuits in depression and anxiety: contribution of discrete neuronal populations and target regions. Mol Psychiatry 25:2742–2758

88. Hare BD, Shinohara R, Liu RJ, Pothula S, DiLeone RJ, Duman RS (2019) Optogenetic stimulation of medial prefrontal cortex Drd1 neurons produces rapid and long-lasting antidepressant effects. Nat Commun 10:1–12. https://doi.org/10.1038/s41467-018-08168-9

89. Felix-Ortiz AC, Burgos-Robles A, Bhagat ND, Leppla CA, Tye KM (2016) Bidirectional modulation of anxiety-related and social behaviors by amygdala projections to the medial prefrontal cortex. Neuroscience 321:197–209

90. Liu RJ, Ota KT, Dutheil S, Duman RS, Aghajanian GK (2015) Ketamine strengthens CRF-activated amygdala inputs to basal dendrites in mPFC layer v pyramidal cells in the prelimbic but not infralimbic subregion, a key suppressor of stress responses. Neuropsychopharmacology 40:2066–2075

91. Carreno FR, Donegan JJ, Boley AM, Shah A, DeGuzman M, Frazer A et al (2016) Activation of a ventral hippocampus-medial prefrontal cortex pathway is both necessary and sufficient for an antidepressant response to ketamine. Mol Psychiatry 21:1298–1308

92. Carlson D, David LK, Gallagher NM, Vu MAT, Shirley M, Hultman R et al (2017) Dynamically timed stimulation of corticolimbic circuitry activates a stress-compensatory pathway. Biol Psychiatry 82:904–913

93. Chaudhury D, Walsh JJ, Friedman AK, Juarez B, Ku SM, Koo JW et al (2013) Rapid regulation of depression-related behaviours by control of midbrain dopamine neurons. Nature 493:532–536

94. Venzala E, García-García AL, Elizalde N, Tordera RM (2013) Social vs. environmental stress models of depression from a behavioural and neurochemical approach. Eur Neuropsychopharmacol 23:697–708

95. Tanaka K, Furuyashiki T, Kitaoka S, Senzai Y, Imoto Y, Segi-Nishida E et al (2012) Prostaglandin E 2-mediated attenuation of mesocortical dopaminergic pathway is critical for susceptibility to repeated social defeat stress in mice. J Neurosci 32:4319–4329

96. Shinohara R, Taniguchi M, Ehrlich AT, Yokogawa K, Deguchi Y, Cherasse Y et al (2018) Dopamine D1 receptor subtype mediates acute stress-induced dendritic growth in excitatory neurons of the medial prefrontal cortex and contributes to suppression of stress susceptibility in mice. Mol Psychiatry 23:1717–1730

97. Tye KM, Mirzabekov JJ, Warden MR, Ferenczi EA, Tsai HC, Finkelstein J et al (2013) Dopamine neurons modulate neural encoding and expression of depression-related behaviour. Nature 493:537–541

98. Fakhoury M (2021) Optogenetics: a revolutionary approach for the study of depression. Prog Neuro Psychopharmacol Biol Psychiatry 106:110094

99. Friedman AK, Walsh JJ, Juarez B, Ku SM, Chaudhury D, Wang J et al (2014) Enhancing depression mechanisms in midbrain dopamine neurons achieves homeostatic resilience. Science 344:313–319. https://doi.org/10.1126/science.1249240

100. Francis TC, Chandra R, Friend DM, Finkel E, Dayrit G, Miranda J et al (2015) Nucleus accumbens medium spiny neuron subtypes mediate depression-related outcomes to social defeat stress. Biol Psychiatry 77:212–222

101. Christoffel DJ, Golden SA, Walsh JJ, Guise KG, Heshmati M, Friedman AK et al (2015) Excitatory transmission at thalamo-striatal synapses mediates susceptibility to social stress. Nat Neurosci 18:962–964

102. Soares-Cunha C, Coimbra B, David-Pereira-A, Borges S, Pinto L, Costa P et al (2016) Activation of D2 dopamine receptor-expressing neurons in the nucleus accumbens increases motivation. Nat Commun 7:11829

103. Bagot RC, Parise EM, Peña CJ, Zhang HX, Maze I, Chaudhury D et al (2015) Ventral hippocampal afferents to the nucleus accumbens regulate susceptibility to depression. Nat Commun 6:7062

104. Walsh EC, Eisenlohr-Moul TA, Minkel J, Bizzell J, Petty C, Crowther A et al (2019) Pretreatment brain connectivity during positive emotion upregulation predicts decreased anhedonia following behavioral activation therapy for depression. J Affect Disord 243: 188–192

105. Walsh JJ, Friedman AK, Sun H, Heller EA, Ku SM, Juarez B et al (2014) Stress and CRF gate neural activation of BDNF in the mesolimbic reward pathway. Nat Neurosci 17:27–29

106. Vialou V, Bagot RC, Cahill ME, Ferguson D, Robison AJ, Dietz DM et al (2014) Prefrontal cortical circuit for depression- and anxiety-related behaviors mediated by cholecystokinin: role of ΔFosB. J Neurosci 34:3878–3887

107. Zhang X, Abdellaoui A, Rucker J, de Jong S, Potash JB, Weissman MM et al (2019) Genome-wide burden of rare short deletions is enriched in major depressive disorder in four cohorts. Biol Psychiatry 85:1065–1073

108. Knowland D, Lilascharoen V, Pacia CP, Shin S, Wang EHJ, Lim BK (2017) Distinct ventral pallidal neural populations mediate separate symptoms of depression. Cell 170: 284–297.e18

109. Wook Koo J, Labonté B, Engmann O, Calipari ES, Juarez B, Lorsch Z et al (2016) Essential role of mesolimbic brain-derived neurotrophic factor in chronic social stress–induced depressive behaviors. Biol Psychiatry 80:469–478

110. Ohmura Y, Tsutsui-Kimura I, Sasamori H, Nebuka M, Nishitani N, Tanaka KF et al (2020) Different roles of distinct serotonergic pathways in anxiety-like behavior, antidepressant-like, and anti-impulsive effects. Neuropharmacology 167:107703

111. Fakhoury M (2017) The habenula in psychiatric disorders: More than three decades of translational investigation. Neurosci Biobehav Rev 83:721–735

112. Matsumoto M, Hikosaka O (2009) Representation of negative motivational value in the primate lateral habenula. Nat Neurosci 12:77–84

113. Proulx CD, Aronson S, Milivojevic D, Molina C, Loi A, Monk B et al (2018) A neural pathway controlling motivation to exert effort. Proc Natl Acad Sci U S A 115: 5792–5797. https://doi.org/10.1073/pnas.1801837115

114. Albert PR, Benkelfat C, Descarries L (2012) The neurobiology of depression-revisiting the serotonin hypothesis. I. Cellular and molecular mechanisms. Philos Trans R Soc Lond B Biol Sci 367:2378–2381

115. Adell A (2015) Revisiting the role of raphe and serotonin in neuropsychiatric disorders. J Gen Physiol 145:257–259

116. Fakhoury M (2016) Revisiting the serotonin hypothesis: implications for major depressive disorders. Mol Neurobiol 53:2778–2786

117. Challis C, Beck SG, Berton O (2014) Optogenetic modulation of descending prefronto-cortical inputs to the dorsal raphe bidirectionally bias socioaffective choices after social defeat. Front Behav Neurosci 8: 43. https://doi.org/10.3389/fnbeh.2014.00043

118. Zhang H, Li K, Chen HS, Gao SQ, Xia ZX, Zhang JT et al (2018) Dorsal raphe projection inhibits the excitatory inputs on lateral habenula and alleviates depressive behaviors in rats. Brain Struct Funct 223:2243–2258

119. Ramirez S, Liu X, MacDonald CJ, Moffa A, Zhou J, Redondo RL et al (2015) Activating positive memory engrams suppresses depression-like behaviour. Nature 522:335–339

120. Johnson SB, Emmons EB, Anderson RM, Glanz RM, Romig-Martin SA, Narayanan NS et al (2016) A basal forebrain site coordinates the modulation of endocrine and behavioral stress responses via divergent neural pathways. J Neurosci 36:8687–8699

121. Cai YQ, Wang W, Paulucci-Holthauzen A, Pan ZZ (2018) Brain circuits mediating opposing effects on emotion and pain. J Neurosci 38:6340–6349

122. Phelps EA, LeDoux JE (2005) Contributions of the amygdala to emotion processing: from animal models to human behavior. Neuron 48:175–187

123. Crestani C, Alves F, Gomes F, Resstel L, Correa F, Herman J (2013) Mechanisms in the bed nucleus of the Stria terminalis involved in control of autonomic and neuroendocrine functions: a review. Curr Neuropharmacol 11:141–159

124. Chapin JC, Monahan PH (2018) Gene Therapy for Hemophilia: Progress to Date

125. Jean, Bennett Jennifer, Wellman Kathleen A, Marshall Sarah, McCague Manzar, Ashtari Julie, DiStefano-Pappas Okan U, Elci Daniel C, Chung Junwei, Sun J Fraser, Wright Dominique R, Cross Puya, Aravand Laura L, Cyckowski Jeannette L, Bennicelli Federico, Mingozzi Alberto, Auricchio Eric A, Pierce Jason, Ruggiero Bart P, Leroy Francesca, Simonelli Katherine A, High Albert M, Maguire (2016) Safety and durability of effect of contralateral-eye administration of AAV2 gene therapy in patients with childhood-onset blindness caused by RPE65 mutations: a follow-on phase 1 trial. The Lancet 388(10045):661–672. https://doi.org/10.1016/S0140-6736(16)30371-3

126. Naso MF, Tomkowicz B, Perry WL, Strohl WR (2017) Adeno-Associated Virus (AAV) as a vector for gene therapy. BioDrugs 31 (4):317–334. https://doi.org/10.1007/s40259-017-0234-5

127. Joshi J, Rubart M, Zhu W (2020) Optogenetics: background, methodological advances and potential applications for cardiovascular research and medicine. Front Bioeng Biotechnol 7:466

Chapter 9

The Effects of Antidepressants on Neurotransmission: Translational Insights from In Vivo Electrophysiological Studies

Meysam Amidfar and Yong-Ku Kim

Abstract

Applying electrophysiological recording techniques as considerable technological progress in neurophysiological methods has provided the major advance in the neuroscience research and the study of brain–behavior relationships. This chapter presents the main electrophysiological methods currently used to study the effects of antidepressant treatments on neurotransmission. Electrophysiological studies combined with local or systemic administration of pharmacological agents will provide worthwhile strategy elucidating the pharmacological bases for reciprocal interactions of neurotransmission systems in vivo. The chapter provides authoritative review of commonly used methodologies in the field related to electrophysiological effects of antidepressant treatments on neurotransmission in the basic research level that has translational value for clinical settings.

Key words Antidepressant, Electrophysiology, Extracellular recording, SSRIs, SNRIs, TCAs, MAOIs, NDRIs, TRIs

1 Introduction

Antidepressant drugs as a standard treatment for major depression have widespread use and clinical importance in a range of neuropsychiatric disorders [1–3]; however, their in vivo electrophysiological effects on the brain have not been completely understood. Electrophysiological studies investigating on the mechanisms of action of antidepressants have provided basic information on the reciprocal interactions of brain monoaminergic systems, including the ventral tegmental area (VTA) and substantia nigra [dopamine (DA) neurons], dorsal raphe [which contains serotonin (5-hydroxytryptamine (5-HT)) neurons], and locus coeruleus [norepinephrine (NE) neurons], that can help to develop strategies to improve the effectiveness of antidepressant drugs [4, 5]. Indeed, understanding of interconnectivity within neurotransmission

Yong-Ku Kim and Meysam Amidfar (eds.), *Translational Research Methods for Major Depressive Disorder*, Neuromethods, vol. 179, https://doi.org/10.1007/978-1-0716-2083-0_9,

networks provides important information for establishment of pharmacotherapies targeting simultaneously neurotransmitter systems and will improve the treatment of depression [4, 5]. It is worth noting that the existing studies on monoaminergic drugs have tended to focus on single-unit firing patterns and include recording data mostly from the serotonergic, noradrenergic, and dopaminergic brainstem nuclei and partly from subregions of the hippocampus [6–9]. It is well established that serotonin (5-HT), norepinephrine, and dopamine systems play a key role in the treatment of MDD; however, the etiology of depression is not yet completely understood [10]. Serotonin 5-HT neurotransmission plays a key role in pharmacotherapy of depression, and selective serotonin reuptake inhibitors (SSRIs) have been recommended as first-choice drugs for the treatment of depression. However, the SSRIs displayed limited clinical efficacy because they could lead to remission of the symptoms only in 30–40% of the patients [11]. Adaptive changes in the serotonergic system are generally believed to underlie the therapeutic effectiveness of a variety of antidepressant drugs [12]. The dual 5-HT and norepinephrine reuptake inhibitors (SNRIs) have shown lower relapse rates than SSRIs, suggesting the higher efficacy of simultaneous stimulation of 5-HT and norepinephrine neurotransmission than the solo stimulation of 5-HT tone in the treatment of depression [11]. It has proposed that triple 5-HT, norepinephrine, and dopamine reuptake inhibitors might contribute to higher clinical efficacy than SSRIs and SNRIs [13, 14]. Triple reuptake inhibitors represent a third generation of antidepressants that simultaneously inhibit the reuptake of the three monoamines norepinephrine (NE), dopamine (DA), and serotonin [5-hydroxytryptamine (5-HT)] transporters [15].

In this chapter, we review the electrophysiological recording literature related to various groups of antidepressant drugs including tricyclic antidepressants (TCAs), selective serotonin reuptake inhibitors (SSRIs), monoamine oxidase inhibitors (MAOIs), serotonin–norepinephrine reuptake inhibitors (SNRIs), norepinephrine–dopamine reuptake inhibitors (NDRIs), and triple reuptake inhibitors (TRIs). This chapter also presents the currently used electrophysiological recording techniques to study the effects of antidepressants on the firing and burst activity of LC NE, VTA DA, and DRN 5-HT neurons.

1.1 In Vivo Electrophysiological Recording Techniques

Most of our current knowledge about the neural control of behavior is based on electrophysiology [16]. Because of the various ranges of potential secondary neuronal and non-neuronal targets including receptors, ion channels, enzymes, and transporters, the CNS is one of the most difficult locations to assess drug effects [17, 18]. Neurons through integration of electrical signals and generation of an output of electrical pulses play a vital role in neural control of behavior [16]. Electrophysiology by registering such signals contributes to capture the natural language of the brain [16]. The ability of electrophysiology to capture an extensive array of neural phenomena, from the spiking activity of individual neurons to the slower network oscillations of small populations, has been favored as a means of analyzing brain activity [19, 20]. Since interspecies differences (i.e., mammalian vs. nonmammalian) may increase incorrect translation of animal data to the human condition, the selection of an animal model is an important consideration [21]. The anatomical and morphological differences are important factors observed between the rodent and human brain that might limit applicability of rodent models to humans [22]. Extracellular action potentials (spikes) on the basis of their shape and their auto- and cross-correlograms are classified into clusters of "single-spiking" and "complex-spiking" neurons [23]. One short and sharp spike at a time is a firing characteristic of single-spiking neurons, while two to seven wider spikes at a time in a rapid train (discharge event) with diminishing amplitude called "burst-like activity" are firing characteristic of complex-spiking neurons [23]. Good recordings are needed to separate neuronal groups like single-spiking neurons and complex-spiking neurons [23]. For example, in the pyramidal layer of the hippocampus CA regions, in vivo single-unit recording of well-isolated neurons is difficult because there is the dense placement of principal cells [24]. Therefore, particularly in hippocampal recordings, precise computational methods are needed to reliably isolate single units [23].

The difficulty of precise identification of the recorded neurons probably is one disadvantage of in vivo electrophysiology against in vitro techniques in most cases [23]. The technique of intracellular recording in vivo is well established for various types of neurons in the central nervous system [25]. The intracellular recording technique provides biophysical measurements such as transmembrane potential and input resistance during periods of spontaneous activity and in response to intracellular stimulation and drug administration [25]. In vivo recordings of the electrical activity of neurons in intact tissue using whole-cell patch-clamp technique utilize glass micropipettes to establish electrical and molecular access to the insides of neurons [26–28]. Sufficient signal quality and temporal fidelity provided by this methodology facilitate measuring the synaptic and ion channel-mediated subthreshold membrane potential changes enabling neurons to compute information that might be affected in brain disorders or by drug treatment [27]. Fast in vivo intracellular recording is performed by automated patch clamp that it is possible to extend this method to measure several neurons simultaneously [27]. In vivo patch-clamp and juxtacellular recording methods made it possible that specific labeling molecules are injected into the recorded neuron [29, 30]. In vivo extracellular single-unit recording combined with microiontophoresis has been suggested as an excellent method for investigating local neuropharmacological effects under in vivo conditions in different hippocampal areas [23]. Microiontophoresis is a classic method that by delivering small amounts of neuroactive compounds in the close vicinity of the recording electrode provides unique possibility for investigating local pharmacological effects in a specific brain area or on a single neuron [23, 31, 32]. Combined in vivo microiontophoresis and extracellular recordings are generally performed through the central recording channel of multibarrel carbon-fiber microelectrodes in the target brain regions of anesthetized rats such as CA1 pyramidal layer of the hippocampus of anesthetized rats, and the electrode position is labeled by means of ejecting certain dyes such as pontamine sky blue through one of the microiontophoresis pipettes [33, 34]. In addition, various drugs or experimental compounds locally are administrated by means of microiontophoresis through the surrounding micropipettes of the microelectrode [23, 32].

2　Materials

2.1　Animals and Surgery

In vivo extracellular recordings are carried out in chloral hydrate-anesthetized rodents that are placed in a stereotaxic apparatus with the skull positioned horizontally. The skull is exposed, and holes are drilled to accommodate placement of recording electrodes. The dura is opened over recording sites to prevent breakage of micropipettes. Supplemental doses of the anesthetic are given to maintain constant anesthesia and prevent nociceptive reaction to a pinching of the hind paws [35]. A lateral tail vein is cannulated for administration of drugs and additional doses of anesthetic. Prior to the electrophysiological recordings, it would be better that a catheter is inserted in a lateral tail vein for systemic intravenous injection of pharmacologic agents [35]. The extracellular recordings are performed using single- or five-barreled glass micropipettes for recordings in the discrete brain regions [36]. Micropipettes are preloaded with fiberglass strands to promote capillary filling with a 2 M NaCl solution [36].

2.2　In Vivo Electrophysiological Recordings

Extracellular recordings are carried out using conventional extracellular recording methods described previously by Hajos et al. [37, 38]. Single-unit, extracellular recordings of monoaminergic neurons are accomplished using single-barreled glass electrodes filled with 2 M NaCl solution (impedance ranging from 2.5 to 5 MΩ) [39]. In accordance with the suture lines of the skull, specific predefined coordinates are used to correctly localize the monoaminergic regions of interest. When recognized, these regions are accessed via a burr hole drilled into the skull. Neurons are individually identified based on firing characteristics such as spike shape, duration, and frequency and recorded in real time using Spike2 software (Cambridge Electronic Design, Cambridge, UK). Drugs are administered systemically to the animal intravenously through a catheter inserted into the lateral tail vein prior to the start of recording.

The extracellular recordings of norepinephrine (NE), dopamine (DA), and serotonin [5-hydroxytryptamine (5-HT)] neurons in the locus coeruleus (LC), the ventral tegmental area (VTA), and the dorsal raphe (DR), respectively, are performed using single-barreled glass micropipettes preloaded with a 2 M NaCl solution [40]. The range of their impedance usually is between 4 and 7 MV [40]. A five-barreled glass micropipette is used for the extracellular recordings of pyramidal neurons in the CA3 region of the hippocampus [39]. The central barrel contained 2 M NaCl solution, and an impedance measuring between 2 and 4 MΩ is utilized for unitary recordings. The four remaining barrels are filled with quisqualic acid (1.5 mM in 200 mM NaCl, pH 8) and 2 M NaCl solution (used for automatic current balancing), NE bitartrate (10 mM in 200 mM NaCl, pH 4), and 5-HT creatinine sulfate (15 mM in 200 mM NaCl, pH 4) [39]. Single-barreled electrodes (tip diameter of 1–3 lm and in vitro impedances of 3 $\pm$ 10 MΩ) are filled with 2 M NaCl saturated with pontamine sky blue dye. The central barrels of prefabricated iontophoretic electrodes are used for extracellular recordings, and the side barrels are filled with saline or drug solutions (impedances of 30 $\pm$ 50 MO) [38]. Ejection of NE and 5-HT as cations and Quis as an anion is performed and retained with a current of -8 to -10 nA and $+ 5$ nA, respectively [40]. The impedances of the central barrel, the balance barrel, and the side barrels are 2–5 MV, 20–30 MV, and 50–100 MV, respectively [40]. Through a high input-impedance amplifier and filters (300 Hz to 5 kHz or 10 kHz band-pass), single-unit potentials are passed [38]. At the end of each experiment, iontophoretic ejection of pontamine sky blue dye (20 μA negative current, 1 Hz pulses of 300 μs duration for 10 min) is used for marking the position of the electrode tip, and the brains were subjected to routine histological examination [38]. An audio monitor, an oscilloscope, and a chart recorder distinguish the resulting signal from background noise by a window discriminator (Digitimer) [38]. A DAT 1401plus interface system and Spike2 software (Cambridge Electronic Design) are used for computing online the discriminator output pulses [38]. Spike interval burst analysis is used for analysis of the firing patterns of DA and NE neurons that both display a bursting activity, following the criteria established by Grace and Bunney (1984). The occurrence of two spikes with an interspike interval shorter than 0.08 s is considered as the onset of a burst. An interspike interval (ISI) of ≥ 0.16 s is considered as the termination of bursts [41].

3 Methods

3.1 Recording of DRN 5-HT Neurons

Electrodes are positioned 0.9 mm anterior to lambda (λ) on the midline and lowered into the DRN, usually attained at a depth of 4.5–5.5 mm from the brain surface [42]. The positioning of the single-barreled glass micropipette is performed using the coordinates (in mm from lambda) including AP, +1.0 to 1.2; L, 0 ± 0.1; and V, 5 to 7 [40]. Stereotaxic coordinates for the DRN including 7.8 mm posterior from the bregma, lateral 0 mm, and 5 ± 6 mm below the dura in the rat brain [43]. The criteria that are used for identifying the assumed 5-HT neurons consist of a slow (0.5–2.5 Hz), regular firing rate, a positive action potential, and a long duration which ranges from 0.8 to 1.2 ms [25, 37, 39, 42]. However, 5-HT neurons may exhibit stereotyped burst firing activity [44]. Presumed 5-HT neurons are classified into two groups, depending on whether they discharged only single spikes or discharged single spikes and spikes in brief (stereotyped) bursts [37, 44, 45]. These burst firing presumed 5-HT neurons fired spike doublets or triplets with a prominent decrease in amplitude of higher-order spikes as well as with an intraburst time interval less than 20 ms [44]. Presumed 5-HT neurons in the mesencephalic raphe nuclei discharge single action potentials in a slow, regular firing pattern, but in addition, these neurons exhibit a burst firing pattern including firing of doublets or triplets of action potentials within a short time (<20 ms) interval that is also known as stereotyped burst firing [37, 44].

3.2 Recording of LC NE Neurons

LC NE neurons were recorded with single-barreled glass micropipette positioned at 1.1–1.2 mm posterior to λ and 0.9–1.3 to the midline suture. These neurons were encountered at a depth of 4.5–6.0 mm from the surface of the brain [42]. The LC is localized by positioning single-barreled glass micropipettes at the coordinates (in millimeters from lambda) including AP, −1.0 to −1.2; ML, 1.0 to 1.3; and DV, 5 to 7 [35, 40]. The following criteria are used for identifying NE neurons in this region: a biphasic action potential of long duration (2–3 milliseconds), regular firing rates (0.5–5.0 Hz), positive action potential of long duration (0.8–1.2 ms), and a characteristic burst discharge following a nociceptive pinch of the contralateral hind paw [46]. This burst discharge is typically followed by a period of quiescence, making this an easily identifiable pattern [39, 40, 47].

3.3 Recording of VTA DA Neurons

Putative DA neurons are recorded by positioning single-barreled glass micropipettes at the coordinates (in millimeters from bregma) including AP, −6 to −5.4; L, 1 to 0.6; and V, 7 to 9 [40]. DA neurons in this area are characterized by their biphasic spikes, which normally exhibit an inflection point in the rising phase, known as a "notch" [39]. The well-established in vivo electrophysiological properties of the presumed DA neurons include a characteristic long duration (>2.5 ms) often with an inflection or "notch" on the rising phase; a typical triphasic action potential with a marked negative deflection; and a slow spontaneous firing rate (0.5–5 Hz) with an irregular single-spiking pattern with slow bursting activity (characterized by spike amplitude decrement) [42, 48, 49]. In addition to displaying tonic firing, DA neurons also undergo phasic firing or bursting [39, 48]. Typically, these bursts contain three to ten spikes and demonstrate decreasing amplitude [39, 48]. An additional measure of DA neurons is the duration from the start of the spike to the center of the negative trough, which in DA neurons is typically greater than 1.1 ms [49]. The onset of a burst was defined as the occurrence of two spikes with an interspike interval shorter than 0.08 s. The termination of a burst was defined as an interspike interval of 0.16 s or longer [42, 48, 49].

3.4 Recording of Pyramidal Neurons in the CA3 Region of the Hippocampus

CA3 pyramidal neurons were recorded by positioning the five-barreled glass micropipettes at the following stereotaxic coordinates (in mm from bregma): AP, 3.8 to 4.5; L, 4 to 4.2; and V, 3 to 4.5 [40, 43]. Pyramidal neurons are found at a depth of 4.0 ± 0.5 mm below the surface of the brain. A small current of quisqualate (+1 to −6 nA) is used to activate the pyramidal neurons to fire at their physiological rate (10–15 Hz) [50], because they do not discharge spontaneously in chloral hydrate-anesthetized rats [51]. Electrophysiological characteristics of pyramidal neurons have been identified including high amplitude (0.5–1.2 mV), high frequency (8–12 Hz), and long-duration (0.8–1.2 ms) simple action potentials, alternating with complex spike discharges [52]. A + 1 to −4 nA current of quisqualate is used for activation of pyramidal neurons that make possible recording of these neurons because pyramidal neurons in chloral hydrate-anesthetized rats have a quiescent nature [39]. The current used is changed to produce activations in the physiological firing range (10–15 Hz) for these neurons [50]. Given that pyramidal neurons do not discharge spontaneously in chloral hydrate-anesthetized rats, a small current of AMPA (−2 to −5 nA) could be also constantly applied to activate CA3 pyramidal neurons within their physiological firing range (10–15 Hz) [42, 53]. Low firing rate of the recorded neuron contributes to more accurate assessment of the tonic activation of postsynaptic receptors [54]. Therefore, lowering the ejection current of quisqualate can decrease the firing rate of pyramidal neurons [51].

4   Effects of Antidepressants on Neurotransmission: Electrophysiological Evidence

In vivo recording studies showed that SSRIs could acutely diminish firing of neurons in the dorsal raphe nucleus [9, 37, 55]. Chronic and acute administration of SSRIs often have different effects on dorsal raphe firing [56] that may account for the acute (side effect-laden) versus chronic (therapeutic) effects of SSRIs in patients [57]. Electrophysiological studies found that the resistance to SSRIs could be explained at least in part by the SSRI-induced decrease of norepinephrine neuronal firing activity, while atypical antipsychotics increase norepinephrine neuronal firing that this reversal of suppression of firing might cause beneficial effect of atypical antipsychotic drugs in depressed patients not responding adequately to SSRIs [58]. These findings show that administration alone of SSRIs can suppress locus coeruleus firing, while when used in combination, 5-HT2A blocking agents paradoxically elevate firing in this nucleus [57]. This reversal may explain the beneficial effect of atypical antipsychotics in treatment-resistant depression [58]. Chronic administration of SSRIs (fluoxetine, citalopram, paroxetine) also could enhance the firing rates of spontaneously active neurons in the ventral tegmental area (VTA), a major dopaminergic nucleus [59]. In contrast, chronic administration of the SSRI citalopram did not affect the overall rate of firing, but did inhibit burst activity in VTA neurons, whereas escitalopram decreased firing rate and bursting [60]. This evidence suggests that boost dopaminergic firing or burst activity of VTA neurons might be involved in the therapeutic effects of antidepressants; however, the data on SSRIs are ambiguous [57]. It has suggested that when SSRIs probably boost VTA signaling in subjects with depression, this at least in part distinguishes responders from non-responders and mediates their antidepressant response [57].

Three electrophysiological studies that investigated on the ability of acute versus chronic SNRI, venlafaxine, at different doses, to suppress dorsal hippocampus CA3 pyramidal cells and dorsal raphe or locus coeruleus firing, compared to the SSRI paroxetine or the selective NE boosting agent desipramine, concluded that venlafaxine is more potent at boosting 5-HT than NE and is indeed more potent at blocking 5-HT than NE synaptic uptake [6, 61, 62]. Suppression of rat locus coeruleus firing by treatment with imipramine or desipramine [63, 64] and suppression of dorsal raphe firing by clomipramine administration [65], suggesting that TCAs, like the SSRIs and SNRIs, can suppress firing in either the dorsal raphe or locus coeruleus [57]. It has suggested that individual's responses to chronic stress and antidepressant action mediated by the in vivo firing patterns of mesolimbic dopamine neurons [66]. The NE/DA reuptake blocker nomifensine increased the tonic activation of α_2-adrenoceptors in the hippocampus despite decreasing NE neuronal firing activity after 2 and 14 days of administration, and prolonged nomifensine increased the firing activity of 5-HT neurons [67]. It has shown that prolonged administration of the antidepressant bupropion, as an NE and DA reuptake inhibitor (NDRI), normalized suppressed dorsal raphe firing and increased tonic activation of 5-HT1A receptors in the hippocampus [68]. Investigation on three MAOIs (including clorgyline, selective MAOI-A; deprenyl, selective MAOI-B; and phenelzine, non-selective MAOI) on the firing activity of DA neurons revealed that chronic administration of the clorgyline and phenelzine decreased firing rates and bursting activity in the dopaminergic VTA [69]. In vivo electrophysiological investigation on the effects of the triple reuptake inhibitors (TRIs) SEP-225289 and DOV216303 on the neuronal activities of locus coeruleus (LC) NE, ventral tegmental area (VTA) DA, and dorsal raphe (DR) 5-HT neurons revealed that their acute administration could dose dependently decrease the spontaneous firing rate of LC NE, VTA DA, and DRN 5-HT neurons through the activation of a2, D2, and 5-HT1A autoreceptors, respectively [40]. In addition, TRI administration could increase NE and DA but not 5-HT transmission and increased tonic activation of α_2-adrenoceptors but not 5-HT$_{1A}$ tonic activation in the hippocampus and normalized DA neuron firing and increased extracellular DA levels in the NAc [67].

It has consistently found that acute administration of drugs that boost 5-HT, NE, or DA (i.e., SSRIs, SNRIs, TCAs) inhibits dorsal raphe, locus coeruleus, and VTA firing, respectively [7, 63, 70]. Selective enhancement of 5-HT or NE by antidepressants induced opposing effects on gamma oscillations including suppressing them by 5-HT and enhancing them by NE [71–73]. It has hypothesized that 5-HT and NE are functionally opposed in a number of brain circuits, which may be a general principle describing many of their interactions in vivo [57]. A number of studies contribute to this hypothesis. For example, chronic treatment with SSRIs such as escitalopram and fluoxetine could suppress locus coeruleus firing [74, 75]. Decreased activation of the ascending 5-HT pathway onto CA3 of the dorsal hippocampus induced by the noradrenergic TCA desipramine [76] and possibly enhancement of 5-HT more than NE induced by the SNRIs venlafaxine and duloxetine 5-HT-NE [6, 61, 62, 77, 78] provide direct evidence for the functional opposition hypothesis.

5   Conclusion

By this literature review, this important issue has been raised that in vivo electrophysiological findings in rodents can be used to improve human pharmacotherapy. The principal physiological feature of the nervous tissue is the generation of electrical activity.

Through this activity, the brain is able to perform the enormous range of its actions, from the simplest to the highly complex ones. Action potential generation as an essential concept of neuroscience is the final common language of almost all neural functioning including anxiety, motivation, mood, and mood regulation. Therefore, the methods that contribute to record and analyze the brain electrical activity can supply very valuable data on the comprehending of how it acts under physiological and pathological conditions. Based on the abovementioned studies, the therapeutic effect of antidepressant treatments might be essentially mediated by adaptive changes in the 5-HT system. Future recording studies of antidepressants can consider the weakly understood neocortical effects of these drugs and thus illuminate all aspects of the circuitry underlying antidepressant drug response with the high degree of spatiotemporal precision afforded by in vivo electrophysiology, especially when accompanied by optogenetics. In vivo electrophysiology should be utilized more widely to enrich current basic neuropharmacology knowledge, cellular physiology, and both macro- and microcircuitries in order to discover and develop antidepressant drugs with translational relevance for increasing the clinical effectiveness of antidepressant treatment in major depression and perhaps other psychiatric disorders.

References

1. Cassano GB, Rossi NB, Pini S (2002) Psychopharmacology of anxiety disorders. Dialogues Clin Neurosci 4(3):271

2. Locher C, Koechlin H, Zion SR, Werner C, Pine DS, Kirsch I et al (2017) Efficacy and safety of selective serotonin reuptake inhibitors, serotonin-norepinephrine reuptake inhibitors, and placebo for common psychiatric disorders among children and adolescents: a systematic review and meta-analysis. JAMA Psychiatry 74(10):1011–1020

3. Puetz TW, Youngstedt SD, Herring MP (2015) Effects of pharmacotherapy on combat-related PTSD, anxiety, and depression: a systematic review and meta-regression analysis. PLoS One 10(5):e0126529

4. El Mansari M, Guiard BP, Chernoloz O, Ghanbari R, Katz N, Blier P (2010) Relevance of norepinephrine–dopamine interactions in the treatment of major depressive disorder. CNS Neurosci Ther 16(3):e1–e17

5. Guiard BP, El Mansari M, Merali Z, Blier P (2008) Functional interactions between dopamine, serotonin and norepinephrine neurons: an in-vivo electrophysiological study in rats with monoaminergic lesions. Int J Neuropsychopharmacol 11(5):625–639

6. Béïque JC, de Montigny C, Blier P, Debonnel G (1998) Blockade of 5-Hydroxytryptamine and noradrenaline uptake by venlafaxine: a comparative study with paroxetine and desipramine. Br J Pharmacol 125(3):526–532

7. Crespi F (2010) SK channel blocker apamin attenuates the effect of SSRI fluoxetine upon cell firing in dorsal raphe nucleus: a concomitant electrophysiological and electrochemical in vivo study reveals implications for modulating extracellular 5-HT. Brain Res 1334:1–11

8. Marcinkiewcz CA, Mazzone CM, D'Agostino G, Halladay LR, Hardaway JA, DiBerto JF et al (2016) Serotonin engages an anxiety and fear-promoting circuit in the extended amygdala. Nature 537(7618): 97–101

9. Mnie-Filali O, El Mansari M, Espana A, Sànchez C, Haddjeri N (2006) Allosteric modulation of the effects of the 5-HT reuptake inhibitor escitalopram on the rat hippocampal synaptic plasticity. Neurosci Lett 395(1):23–27

10. Kennedy SH, Lam RW, McIntyre RS, Tourjman SV, Bhat V, Blier P et al (2016) Canadian network for mood and anxiety treatments (CANMAT) clinical guidelines for the management of adults with major depressive disorder:

section 3. Pharmacological treatments. Can J Psychiatr 61(9):540–560

11. Kennedy SH, Young AH, Blier P (2011) Strategies to achieve clinical effectiveness: refining existing therapies and pursuing emerging targets. J Affect Disord 132:S21–SS8

12. Hensler JG (2003) Regulation of 5-HT1A receptor function in brain following agonist or antidepressant administration. Life Sci 72(15):1665–1682

13. Koprdova R, Csatlosova K, Durisova B, Bogi E, Majekova M, Dremencov E et al (2019) Electrophysiology and behavioral assessment of the new molecule SMe1EC2M3 as a representative of the future class of triple reuptake inhibitors. Molecules 24(23):4218

14. Sharma H, Santra S, Dutta A (2015) Triple reuptake inhibitors as potential next-generation antidepressants: a new hope? Future Med Chem 7(17):2385–2406

15. Chen Z, Skolnick P (2007) Triple uptake inhibitors: therapeutic potential in depression and beyond. Expert Opin Investig Drugs 16(9):1365–1377

16. Chorev E, Epsztein J, Houweling AR, Lee AK, Brecht M (2009) Electrophysiological recordings from behaving animals—going beyond spikes. Curr Opin Neurobiol 19(5):513–519

17. Porsolt RD, Lemaire M, Dürmüller N, Roux S (2002) New perspectives in CNS safety pharmacology. Fundam Clin Pharmacol 16(3):197–207

18. Wakefield ID, Pollard C, Redfern WS, Hammond TG, Valentin JP (2002) The application of in vitro methods to safety pharmacology. Fundam Clin Pharmacol 16(3):209–218

19. Assad JA, Berdondini L, Cancedda L, De Angelis F, Diaspro A, Dipalo M et al (2014) Brain function: novel technologies driving novel understanding. In: Bioinspired approaches for human-centric technologies. Springer, Cham, pp 299–334

20. Llinás RR (1988) The intrinsic electrophysiological properties of mammalian neurons: insights into central nervous system function. Science 242(4886):1654–1664

21. Lynch VJ (2009) Use with caution: developmental systems divergence and potential pitfalls of animal models. Yale J Biol Med 82(2):53

22. Preuss TM (2000) Taking the measure of diversity: comparative alternatives to the model-animal paradigm in cortical neuroscience. Brain Behav Evol 55(6):287–299

23. Bali Z, Budai D, Hernádi I (2014) Separation of electrophysiologically distinct neuronal populations in the rat hippocampus for neuropharmacological testing under in vivo conditions. Acta Biol Hung 65(3):241–251

24. Ranck JB Jr (1973) Studies on single neurons in dorsal hippocampal formation and septum in unrestrained rats: part I. Behavioral correlates and firing repertoires. Exp Neurol 41(2):462–531

25. Aghajanian G, Vandermaelen C (1982) Intracellular recording in vivo from serotonergic neurons in the rat dorsal raphe nucleus: methodological considerations. J Histochem Cytochem 30(8):813–814

26. Hamill OP, Marty A, Neher E, Sakmann B, Sigworth F (1981) Improved patch-clamp techniques for high-resolution current recording from cells and cell-free membrane patches. Pflugers Arch 391(2):85–100

27. Kodandaramaiah SB, Franzesi GT, Chow BY, Boyden ES, Forest CR (2012) Automated whole-cell patch-clamp electrophysiology of neurons in vivo. Nat Methods 9(6):585–587

28. Margrie TW, Brecht M, Sakmann B (2002) In vivo, low-resistance, whole-cell recordings from neurons in the anaesthetized and awake mammalian brain. Pflugers Arch 444(4):491–498

29. Furue H, Katafuchi T, Yoshimura M (2007) In vivo patch-clamp technique. In: Patch-clamp analysis. Humana Press, Totowa, pp 229–251

30. Pinault D (1996) A novel single-cell staining procedure performed in vivo under electrophysiological control: morpho-functional features of juxtacellularly labeled thalamic cells and other central neurons with biocytin or Neurobiotin. J Neurosci Methods 65(2):113–136

31. Gerhardt GA, Palmer MR (1987) Characterization of the techniques of pressure ejection and microiontophoresis using in vivo electrochemistry. J Neurosci Methods 22(2):147–159

32. Kovács P, Hernádi I (2006) Yohimbine acts as a putative in vivo α2A/D-antagonist in the rat prefrontal cortex. Neurosci Lett 402(3):253–258

33. Boakes R, Bramwell G, Briggs I, Candy J, Tempesta E (1974) Localization with pontamine sky blue of neurones in the brainstem responding to microiontophoretically applied compounds. Neuropharmacology 13(6):475–479

34. Kovács P, Dénes V, Kellényi L, Hernádi I (2005) Microiontophoresis electrode location by neurohistological marking: comparison of four native dyes applied from current balancing electrode channels. J Pharmacol Toxicol Methods 51(2):147–151

35. Herman A (2017) Alteration of monoaminergic neuronal firing by acute administration of

cariprazine: an in vivo electrophysiological study. Doctoral dissertation, Université d'Ottawa/University of Ottawa

36. Guiard BP, Mansari ME, Murphy DL, Blier P (2012) Altered response to the selective serotonin reuptake inhibitor escitalopram in mice heterozygous for the serotonin transporter: an electrophysiological and neurochemical study. Int J Neuropsychopharmacol 15(3):349–361

37. Hajos M, Gartside S, Villa A, Sharp T (1995) Evidence for a repetitive (burst) firing pattern in a sub-population of 5-hydroxytryptamine neurons in the dorsal and median raphe nuclei of the rat. Neuroscience 69(1):189–197

38. Hajós M, Hajós-Korcsok É, Sharp T (1999) Role of the medial prefrontal cortex in 5-HT1A receptor-induced inhibition of 5-HT neuronal activity in the rat. Br J Pharmacol 126(8):1741–1750

39. Crnic A (2014) Effects of acute and sustained administration of vilazodone (EMD68843) on monoaminergic systems: an in vivo electrophysiological study. Doctoral dissertation, Université d'Ottawa/University of Ottawa

40. Guiard BP, Chenu F, Mansari ME, Blier P (2011) Characterization of the electrophysiological properties of triple reuptake inhibitors on monoaminergic neurons. Int J Neuropsychopharmacol 14(2):211–223

41. Grace AA, Bunney BS (1984) The control of firing pattern in nigral dopamine neurons: burst firing. J Neurosci 4(11):2877–2890

42. El Iskandarani KS (2014) Electrophysiological investigations of the effects of a subanesthetic dose of ketamine on monoamine systems. Doctoral dissertation, Université d'Ottawa/University of Ottawa

43. Paxinos G, Watson C (2006) The rat brain in stereotaxic coordinates: hard cover edition. Elsevier, Amsterdam

44. Hajós M, Allers KA, Jennings K, Sharp T, Charette G, Sík A et al (2007) Neurochemical identification of stereotypic burst-firing neurons in the rat dorsal raphe nucleus using juxtacellular labelling methods. Eur J Neurosci 25(1):119–126

45. Hajós M, Sharp T (1996) A 5-hydroxytryptamine lesion markedly reduces the incidence of burst-firing dorsal raphe neurones in the rat. Neurosci Lett 204(3):161–164

46. Aghajanian G, VanderMaelen C (1982) Alpha 2-adrenoceptor-mediated hyperpolarization of locus coeruleus neurons: intracellular studies in vivo. Science 215(4538):1394–1396

47. Marwaha J, Aghajanian GK (1982) Relative potencies of alpha-1 and alpha-2 antagonists in the locus coeruleus, dorsal raphe and dorsal lateral geniculate nuclei: an electrophysiological study. J Pharmacol Exp Ther 222(2):287–293

48. Grace A, Bunney B (1983) Intracellular and extracellular electrophysiology of nigral dopaminergic neurons—1. Identification and characterization. Neuroscience 10(2):301–315

49. Ungless MA, Magill PJ, Bolam JP (2004) Uniform inhibition of dopamine neurons in the ventral tegmental area by aversive stimuli. Science 303(5666):2040–2042

50. Ranck JB (1975) Behavioral correlates and firing repertoires of neurons in the dorsal hippocampal formation and septum of unrestrained rats. In: The hippocampus. Springer, Boston, MA, pp 207–244

51. Chernoloz O, El Mansari M, Blier P (2012) Effects of sustained administration of quetiapine alone and in combination with a serotonin reuptake inhibitor on norepinephrine and serotonin transmission. Neuropsychopharmacology 37(7):1717–1728

52. Kandel ER, Spencer W, Brinley F Jr (1961) Electrophysiology of hippocampal neurons: I. sequential invasion and synaptic organization. J Neurophysiol 24(3):225–242

53. Redrobe JP, Bourin M (1997) Partial role of 5-HT2 and 5-HT3 receptors in the activity of antidepressants in the mouse forced swimming test. Eur J Pharmacol 325(2–3):129–135

54. Haddjeri N, Blier P, de Montigny C (1998) Long-term antidepressant treatments result in a tonic activation of forebrain 5-HT1A receptors. J Neurosci 18(23):10150–10156

55. El Mansari M, Sánchez C, Chouvet G, Renaud B, Haddjeri N (2005) Effects of acute and long-term administration of escitalopram and citalopram on serotonin neurotransmission: an in vivo electrophysiological study in rat brain. Neuropsychopharmacology 30(7):1269–1277

56. Christensen T, Bétry C, Mnie-Filali O, Etievant A, Ebert B, Haddjeri N et al (2012) Synergistic antidepressant-like action of gaboxadol and escitalopram. Eur Neuropsychopharmacol 22(10):751–760

57. Fitzgerald PJ, Watson BO (2019) In vivo electrophysiological recordings of the effects of antidepressant drugs. Exp Brain Res 237(7):1593–1614

58. Dremencov E, El Mansari M, Blier P (2007) Noradrenergic augmentation of escitalopram response by risperidone: electrophysiologic studies in the rat brain. Biol Psychiatry 61(5):671–678

59. Sekine Y, Suzuki K, Ramachandran PV, Blackburn TP, Ashby CR Jr (2007) Acute and repeated administration of fluoxetine, citalopram, and paroxetine significantly alters the activity of midbrain dopamine neurons in rats: an in vivo electrophysiological study. Synapse 61(2):72–77

60. Dremencov E, El Mansari M, Blier P (2009) Effects of sustained serotonin reuptake inhibition on the firing of dopamine neurons in the rat ventral tegmental area. J Psychiatry Neurosci 34(3):223

61. Béïque JC, De Montigny C, Blier P, Debonnel G (1999) Venlafaxine: discrepancy between in vivo 5-HT and NE reuptake blockade and affinity for reuptake sites. Synapse 32(3):198–211

62. Béïque J-C, de Montigny C, Blier P, Debonnel G (2000) Effects of sustained administration of the serotonin and norepinephrine reuptake inhibitor venlafaxine: I. in vivo electrophysiological studies in the rat. Neuropharmacology 39(10):1800–1812

63. Linnér L, Arborelius L, Nomikos GG, Bertilsson L, Svensson TH (1999) Locus coeruleus neuronal activity and noradrenaline availability in the frontal cortex of rats chronically treated with imipramine: effect of α2-adrenoceptor blockade. Biol Psychiatry 46(6):766–774

64. McMillen BA, Warnack W, German DC, Shore PA (1980) Effects of chronic desipramine treatment on rat brain noradrenergic responses to α-adrenergic drugs. Eur J Pharmacol 61(3):239–246

65. Gallager DW, Aghajanian G (1975) Effects of chlorimipramine and lysergic acid diethylamide on efflux of precursor-formed 3-H-serotonin: correlations with serotonergic impulse flow. J Pharmacol Exp Ther 193(3):785–795

66. Cao J-L, Covington HE, Friedman AK, Wilkinson MB, Walsh JJ, Cooper DC et al (2010) Mesolimbic dopamine neurons in the brain reward circuit mediate susceptibility to social defeat and antidepressant action. J Neurosci 30(49):16453–16458

67. Jiang JL, El Mansari M, Blier P (2020) Triple reuptake inhibition of serotonin, norepinephrine, and dopamine increases the tonic activation of α2-adrenoceptors in the rat hippocampus and dopamine levels in the nucleus accumbens. Prog Neuro-Psychopharmacol Biol Psychiatry 103:109987

68. Mansari ME, Manta S, Oosterhof C, El Iskandrani KS, Chenu F, Shim S et al (2015) Restoration of serotonin neuronal firing following long-term administration of bupropion but not paroxetine in olfactory bulbectomized rats. Int J Neuropsychopharmacol 18(4):pyu050

69. Chenu F, Mansari ME, Blier P (2009) Long-term administration of monoamine oxidase inhibitors alters the firing rate and pattern of dopamine neurons in the ventral tegmental area. Int J Neuropsychopharmacol 12(4):475–485

70. Svensson TH, Usdin T (1978) Feedback inhibition of brain noradrenaline neurons by tricyclic antidepressants: alpha-receptor mediation. Science 202(4372):1089–1091

71. Akhmetshina D, Zakharov A, Vinokurova D, Nasretdinov A, Valeeva G, Khazipov R (2016) The serotonin reuptake inhibitor citalopram suppresses activity in the neonatal rat barrel cortex in vivo. Brain Res Bull 124:48–54

72. Hajós M, Hoffmann WE, Robinson DD, Jen HY, Hajós-Korcsok E (2003) Norepinephrine but not serotonin reuptake inhibitors enhance theta and gamma activity of the septo-hippocampal system. Neuropsychopharmacology 28(5):857–864

73. Méndez P, Pazienti A, Szabó G, Bacci A (2012) Direct alteration of a specific inhibitory circuit of the hippocampus by antidepressants. J Neurosci 32(47):16616–16628

74. Dremencov E, El Mansari M, Blier P (2007) Distinct electrophysiological effects of paliperidone and risperidone on the firing activity of rat serotonin and norepinephrine neurons. Psychopharmacology 194(1):63–72

75. Seager MA, Huff KD, Barth VN, Phebus LA, Rasmussen K (2004) Fluoxetine administration potentiates the effect of olanzapine on locus coeruleus neuronal activity. Biol Psychiatry 55(11):1103–1109

76. Mongeau R, Blier P, de Montigny C (1993) In vivo electrophysiological evidence for tonic activation by endogenous noradrenaline of α2-adrenoceptors on 5-hydroxytryptamine terminals in the rat hippocampus. Naunyn Schmiedeberg's Arch Pharmacol 347(3):266–272

77. Kasamo K, Blier P, De Montigny C (1996) Blockade of the serotonin and norepinephrine uptake processes by duloxetine: in vitro and in vivo studies in the rat brain. J Pharmacol Exp Ther 277(1):278–286

78. Rueter LE, De Montigny C, Blier P (1998) Electrophysiological characterization of the effect of long-term duloxetine administration on the rat serotonergic and noradrenergic systems. J Pharmacol Exp Ther 285(2):404–412

Chapter 10

The Role of Transcriptional Profiling in Neurobiology and Treatment of Major Depression: A Translational Perspective on RNA Sequencing Platform

Bhaskar Roy, Praveen Korla, and Yogesh Dwivedi

Abstract

High-throughput transcriptomic profiling is a powerful experimental strategy to understand complex neurobiological mechanisms associated with many disorders, including major depressive disorder (MDD). Arguably, quantification of individual transcripts by northern blotting, reverse transcriptase quantitative PCR (RT-qPCR), and semi-high-throughput RT-PCR profiler may be able to capture a small subset(s) of a large transcriptional landscape of the MDD brain; however, it can leave many transcriptional irregularities undetected due to their limited multiplexing capabilities. Gradual understanding over the past several years suggests that neuropsychiatric disorders, such as major depression, may primarily result from system-level dysfunctions related to many genes that happen to be a part of a disorganized transcriptional network. The specific role of each gene in exacerbating the flow of genetic information under major neuronal circuits demands a careful examination strategy which often involves large-scale genome-wide transcriptomic profiling. This can be achieved using either a hybridization-based high-throughput gene chip (expression microarray) platform or massively parallel signature sequencing (MPSS)-based deep sequencing platform. This chapter provides an overview of large-scale transcriptomic profiling on the outset of the two most robustly used high-throughput expression analysis platforms, including RNA microarray and RNA-based next-generation sequencing, and their roles in the context of MDD pathophysiology.

Key words Major depression, Transcriptome, Microarray, RNA, Next-generation sequencing

1 Introduction

1.1 Altered Transcriptional Landscape of MDD Brain

MDD is a neuropsychiatric condition primarily exacerbated by chronic stress stimulus at some point of time in life ranging from child to adulthood. As stated by US Centers for Disease Control and Prevention (CDC), it may be caused by a combination of maladaptive changes related to genetic, biological, environmental, and psychological factors [1]; however, extensive research has shown that global transcriptional reprogramming is core to many maladaptive changes in MDD brain [2]. Often, transcriptional

Yong-Ku Kim and Meysam Amidfar (eds.), *Translational Research Methods for Major Depressive Disorder*, Neuromethods, vol. 179, https://doi.org/10.1007/978-1-0716-2083-0_10,
© The Author(s), under exclusive license to Springer Science+Business Media, LLC, part of Springer Nature 2022

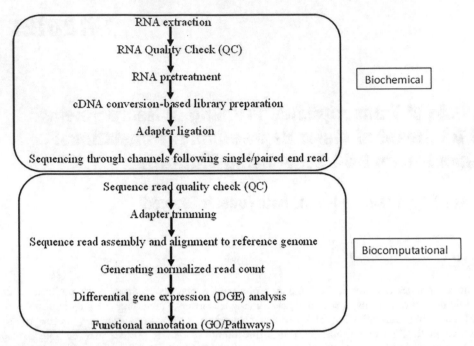

Fig. 1 Outline of RNA sequencing methodology

reprogramming is associated with molecular pathologies related to altered brain plasticity and dysregulated stress axis functioning [3]. To be more precise, many of the structural and functional abnormalities associated with plasticity and neuroendocrine system are the direct outcome of abnormal gene activities in specific regions of MDD brain that susceptible to stress responses. Studies have shown the importance of large scale expression profiling of synaptic genes in the dorsolateral prefrontal cortex (dlPFC) [4] and hippocampus [5] of MDD subjects. Over the past two decades, various technological advancements have been made to understand the gene expression profiling of the depressed brain, especially to uncover the complexities associated with the polygenic nature of this mental disorder. Recent trends in probing transcriptomic changes in specific region of MDD brain have shown promising results in identifying more therapeutic targets than investigating changes at the level of protein translation. The state-dependent changes in overall transcriptomic output of MDD brain often represent differentially expressed gene pool (compared to control) and have been considered to be the foundation for further analysis. Most of the times, a significantly altered gene pool, determined by differential expression profiling, is used to screen disease-specific pathways and to create gene ontology based on enrichment of common functions. As elaborated in Fig. 1, the overall workflow in transcriptomic profiling can be schematically visualized, starting from brain region-specific RNA isolation to gene function annotation, by following a stepwise approach involving several biochemical and biocomputational methods.

1.2 Overview of the Methods Implemented to Study the Transcriptomic Profile of MDD Brain

Various reports, based on clinical and preclinical models, have helped to identify complex gene regulatory events with the implementation of advanced higher-end transcriptional profiling methods including microarray and next-generation sequencing (NGS) platforms. Until recently, gene expression microarray offered the most powerful approach to understand the irregularities associated with gene expression changes and alternative regulatory fate which are often integral parts of complex gene expression network in the brain. Owing to the complex nature of MDD brain, microarray-based gene expression profiling has been considered a reliable and sensitive platform to discover many underlying neurochemical changes which are often treated as pleiotropic outcome of gene and environment (G x E) interactions. In this context, many groups, including our own, have successfully adapted the microarray-based multiplexing approach to show global transcriptional abnormalities in the brain of subjects with MDD or those who died by suicide and had an MDD diagnosis. Due to certain inherent technical challenges, the next-generation expression profiling platform, like massively parallel signature sequencing, has recently outpaced the need for microarray-based serial gene expression analysis in neuropsychiatric field, similar to many other domains of biological sciences.

It is important to know that microarray essentially relies on a method of indirect gene expression quantification supported by partial or complete hybridization of prefabricated probes to target genes. Inadvertently, this approach comes with certain caveats, and the most limiting issue is high background noise to signal ratio which is due to off-target cross-hybridization of probes on the microchip. The other technical issue that imposes a significant limitation on microarray platform is the lower coverage of genome due to a priori knowledge of the expressed transcripts in order to design the probes for hybridization.

On the contrary, the key advances of NGS that outweigh the need for powerful microarray-based gene expression analysis rely on its ability to discover novel gene expression de novo. The detection of low abundance transcripts is another challenge to understand the neurobiological abnormalities in MDD brain and is often limited by restrictions imposed by the biochemistry adopted for microarray platform. NGS handles these two tricky challenges very well and has helped broaden our capacity to understand the minute and sensitive changes with maximum resolution.

In this chapter, we introduce the state-of-the-art RNA sequencing technique, which offers a much more robust outcome over the microarray technique for examining transcriptomic changes. We also discuss some of the important merits and demerits of both the platforms but at the same time highlight RNA sequencing as the most preferred method to determine the transcriptomic changes in MDD brain like any other field of transcriptional biology in the current time. It is evident that this robust technique can help provide an insightful view of the expression dynamics in the brain and build an overview of the brain's regulatory code underlined by immense transcriptional complexity. We have provided detailed guidelines of the stepwise RNA sequencing methodology that have recently been adopted in our laboratory for interrogating the genome-wide alteration of transcriptional landscape using RNAs from postmortem brain samples obtained from individuals with MDD.

To address the sensitivity of gene expression profiling compounded by the complexity of the human brain, Kang and colleague [6] for the first time reported the 60-mer oligonucleotide-based microarray approach to determine MDD-associated transcriptomic changes in the dorsolateral prefrontal cortical (dlPFC) area. The study successfully reported a rich profile of dysregulated genes. Some of the genes (Strescopin and Forkhead Box D3) showed significant molecular association with MDD in a discrete brain region-specific manner. In later years, other groups included microarray platform to determine MDD-specific gene expression changes in the postmortem brain samples. For example, Tochigi et al. [7] conducted DNA microarray analysis using postmortem brain samples from MDD subjects and found 99 differentially expressed genes (DEGs). In a separate study, another group used microarray mRNA expression analysis to reveal changes in transcriptional dynamics and map genes to inflammatory, apoptotic, and oxidative stress-related pathways in MDD brain [8]. It is noted that the expression of genes involved in synaptic function, including pre- and postsynaptic genes, has also been studied in MDD postmortem brain following gene expression microarray methodology [5]. The study largely involved the identification of several synaptic gene families known for their function to regulate and support cytoskeletal structures. Mainly focusing on two hippocampal subfields, the DG (dentate gyrus) granule and CA1 (cornu Ammonis) pyramidal cell layers of subjects diagnosed with MDD, the study determined changes in gene expression that are involved in structural alterations of neuronal processes. In this process, 70-mer oligonucleotide probe-based microchip from human exonic origin was used. Collectively, all these seminal studies have suggested the underlying importance of high-throughput transcriptional profiling in MDD brain to understand the progressive pathology with a system-wide approach and more specifically the increasing use of microarray platform to generate and handle the large-scale expression data sets from human postmortem brain samples.

As suggested earlier, the microarray technique is limited in its capacity due to the fundamental nature of its detection chemistry, i.e., hybridization. Primarily, hybridization-based methods involve the binding of fluorescently labeled fragments to complementary probe sequences either in solution or on a solid surface, which requires a priori knowledge of genome sequence to design and assemble the microchip with short probe sequences. Therefore, it is heavily dependent on reference genome and cannot offer the flexibility to investigate novel transcript expression. This bottleneck has provided a fundamental thrust to invent the de novo approach, where the respective study can be designed to discover the expression status of an uncharacterized transcript due to transcriptional output of a previously unknown gene or that can arise from the result of novel transcriptional splicing event uniquely related to the disease state. Another important challenge is that microarray is generally unable to determine the alternate promoter usage, which is an important aspect of dysregulated gene transcription program in neuropsychiatric brain tissue [9], and can only be substantiated following a genome-wide deep sequencing approach. Nevertheless, the need to use minimal amount of RNA samples, which is a limiting factor for postmortem brain sample-based studies, finally halted the further usage of microarray platforms and made a gradual transition to next-generation sequencing. Until recently, an increasing number of reports on transcriptomic profiling of neuropsychiatric brain have used RNA-seq as their gene expression quantification platform. The sequencing outcomes are directly related to complex transcriptome-wide expression map and identify various drivers and regulators of transcriptomic dysregulation under diseased state [10]. Using this approach, our lab has made a significant contribution to understanding the complex synaptic neurobiology of MDD brain [11]. Primarily, our focus was to understand the activity-dependent dysregulation of synaptic transcriptome and investigate how it is different from the total transcriptomic pool under MDD pathophysiology. We used isolated mRNA pool to decipher the underlying MDD-associated changes and constructed gene-gene interaction network in dlPFC. Latest sequencing platform from Illumina (HiSeq6000) was used following a 150 bp paired-end read protocol and 40 M of sequencing depth, which is an excellent coverage for most protein-coding genes in mammals. Besides our laboratory, several other groups have taken significant leads in using RNA-based deep sequencing technology and examined MDD-associated expression changes at the exome level (genome-wide representation of transcriptomic pool exclusively represented by the exonic part of genes under a specific disease state or condition). The evidence from exome sequencing in MDD and suicide brain demonstrated marked changes in glial, endothelial, and ATPase activity [12]. In this study, authors used Illumina sequencing platform (HiSeq2500) with 100 bp paired-end reads. In another study, the authors have highlighted neuro-inflammatory changes in hippocampal region of MDD brain following Illumina paired-end read protocol and using Illumina NextSeq500 sequencing platform. The authors were able to determine differential expression of genes related to neuroplasticity and neuro-inflammation in depression [13]. Together, these studies consolidate our view of the use of next-generation sequencing platform to understand the complex nature of MDD brain, especially in their robustness to navigate data-driven exploration, which was not possible earlier with the microarray platform.

In the following section, we present the stepwise methodology that we have adapted in our laboratory, including outlines of the biochemical and the biocomputation sections (Fig. 1) that go hand in hand to determine changes in gene expression from a reliable source of extracted RNA material.

2   Methods

Six critical steps can be outlined for conducting a successful RNA sequencing experiment using postmortem brain-extracted RNA samples. These six initial steps comprise biochemical assay part of the experiment. In the biocomputational segment, the generated sequence reads are processed through various bioinformatic and biostatistical tools to determine the differential gene expression and associated gene function. This biocomputational segment primarily depends on user-defined biocomputational pipeline and requires higher-end computing power by either logging in to cloud server or using stand-alone local server. Command line interface is largely accepted by bioinformaticians to run most of the analysis packages either by using a shell script or R-statistical environment.

2.1   RNA Isolation and Quality Assessment

RNA is the critical input material which eventually determines the quality of the sequencing reads. In postmortem brain studies, shorter postmortem interval (PMI) plays a significant role in recovering good-quality RNA samples. Most importantly, the polyA tailed mRNA samples pose greater risk of losing integrity due to longer PMI. Often, the 3′-termini-based deadenylation triggers exo- and endonucleolytic degradation of mRNA molecules. Hence, the overall RNA integrity from postmortem brain samples remains a challenge for RNA sequencing-based expression studies, and great care is needed to preserve their integrity both before and after extraction.

We follow TRIzol® (Invitrogen, USA)-based liquid phase isolation methodology to extract RNA from total tissue homogenates following manufacturer's protocol. While handling the postmortem tissue samples, they are strictly maintained at −80°C temperature and accompanied with dry ice bath as needed. Post isolation,

Sample 1 with RIN=6.7

Sample 1 with RIN=7.0

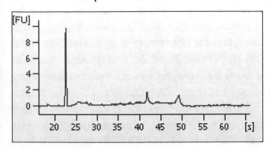
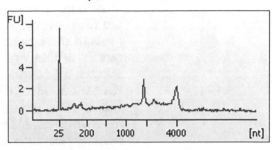

RNA electropherograms

RNA electropherograms

Fig. 2 Bioanalyzer-based electropherograms showing the quality of RNA samples with an index known as RNA Integrity Number or RIN. The RIN value ~7.0 represents good-quality RNA

we run an initial quality check to determine the purity (using NanoDrop ND-1000, ThermoScientific, Waltham, MA, USA; purity check 260/280 nm wavelength; cutoff ≥1.8) and structural integrity (following a 1% denaturing agarose gel electrophoresis) of RNA samples and take them to the next step if they pass quality control. As an alternate approach, RNA can be isolated from total tissue homogenate following a solid phase separation method. This could be done using silica-based column which are commercially available from various venders, including Qiagen (Hilden, Germany); however, this method is limited by low RNA yield but compensates for high RNA quality. Also, the overall run time is very short. In the last step, before the RNA samples are used in library preparation, a final quality check is done by following a microfluidics-based capillary electrophoresis method requiring a sensitive device called Bioanalyzer (Bioanalyzer 2100, Agilent, Santa Clara, CA). The integrity of RNA samples is determined by using a scale named RNA Integrity Number (RIN). RNAs with RIN value around 7 are considered less degraded and are useful in downstream library preparation (Fig. 2).

2.2 Library Preparation and Sequencing

In this segment, 1–2 μg total is used to quality check RNAs from each sample for RNA-seq library preparation. Briefly, mRNA enrichment is done from total RNA pool using commercially available NEBNext® Poly(A) mRNA Magnetic Isolation Module (New England Biolabs, Ipswich, MA, USA). Alternatively, rRNA can be removed from the total RNA with a Ribo-Zero Magnetic Gold Kit (Epicentre, Madison, WI, USA). Ribosomal depletion offers benefit over mRNA enrichment, i.e., it simultaneously assesses polyadenylated and non-polyadenylated RNAs from the same enriched RNA pool and allows the detection of both coding and noncoding transcripts (e.g., long non-coding RNAs). Additionally, one can also consider ribosomal depletion if the RNA samples under investigation have RIN value <6. However, a major caveat in ribosomal depletion methods is related to its prejudice to map more number of reads to the intronic regions of the transcriptome. On the contrary, polyA+ selection method enables more mappable reads to the exonic regions, which effectively helps improve accuracy in mRNA expression profiling. Next in the process, polyA+ selected mRNA pool is used for RNA-seq library preparation using KAPA Stranded RNA-Seq Library Preparation Kit (Illumina, Indianapolis, IN, USA). The library preparation procedure includes the following: (1) fragmentation of the RNA molecules, (2) reverse transcription to synthesize first strand cDNA, (3) second strand cDNA synthesis incorporating dUTP, (4) end repair and A-tailing of the double-stranded cDNA, (5) Illumina-compatible adapter ligation, and (6) PCR amplification and purification for the final RNA-seq library. During library preparation, external spike-in controls are used to ascertain the quality of the prepared library in terms of GC content. The completed libraries are quantified on Agilent 2100 Bioanalyzer for concentration, fragment size distribution between 400 and 600 bp, and adapter dimer contamination. The amount is determined by absolute quantification qPCR method. The barcoded libraries are mixed in equal amounts and used for sequencing on the instrument. At this stage, the prepared library can be sequenced in just one direction (single-end) or both directions (paired-end). Paired-end sequencing produces robust alignments and/or assemblies, which is beneficial for gene annotation and transcript isoform discovery. In our experiments, we follow 150 bp paired-end read protocol to generate 40 M reads. Practically, we denature the library with 0.1 M NaOH to generate single-stranded DNA molecules, loaded onto channels of the flow cell at 8 pM concentration, and amplify in situ using TruSeq SR Cluster Kit v3-cBot-HS (GD-401-3001, Illumina, San Diego, CA, USA). Later, the sequencing is carried out by using Illumina HiSeq 6000 (Illumina, San Diego, CA, USA) according to the manufacturer's instructions. Sequencing is carried out by running 150 cycles.

2.3 Biocomputational Pipeline in Sequence Analysis

A brief overview of the key computational tools to analyze RNA sequencing is provided in Table 1. Raw data files from the sequencer are exported in FASTQ format and set for sequence quality assessment. To provide a brief overview for the readers, we have illustrated a scheme of the pipeline that is standard in RNA-seq analysis (Fig. 3). To examine the sequencing quality, the quality score plot of each sample is plotted (Fig. 4). Sequence quality is examined using the FastQC software. The purpose of using the FastQC module is to ensure uniform quality across the sequence, especially with an emphasis on determining 5'- and 3'-termini of the reads and the GC/AT biases and finding overrepresentation of short sequence motifs or k-mers. After quality control (Fig. 5), the fragments are 5', 3'-adaptor trimmed and filtered for ≤20 bp reads

Table 1
Key computational tools used in RNA sequence analysis pipeline

FASTQ	FASTQ is the most widely used format in sequence analysis generally delivered from a sequencer. Many analysis tools require this format because it contains much more information than FastA
FastQC	FASTQ files are employed to check quality control by well-known QC software, namely, FastQC (https://www.bioinformatics.babraham.ac.uk/projects/fastqc/). FastQC aims to provide a simple way to do quality control checks on raw sequence data received from high throughput
Cutadapt	It finds and removes adapter sequences, primers, polyA tails, and other types of unwanted sequences from high-throughput sequencing reads. It does the trimming by finding the adapter or primer sequences in an error-tolerant way. Cutadapt has the ability to demultiplex the sequence reads if that has IUPAC wildcard characters. The associated web link for Cutadapt is https://cutadapt.readthedocs.io/en/stable/
HISAT2	It is a fast and sensitive spliced alignment program for mapping RNA-seq reads. HISAT2 uses a large set of small GFM indexes that collectively covers the whole genome. Each index represents a genomic region of 56 Kbp, with 55,000 indexes. The associated web link is http://daehwankimlab.github.io/hisat2/
StringTie	It is a fast and highly efficient assembler of RNA-seq alignments into potential transcripts. It uses a novel network flow algorithm as well as an optional de novo assembly step to assemble and quantitate full-length transcripts, representing multiple splice variants for each gene locus. StringTie's output can be processed by specialized software like Ballgown, Cuffdiff, or other programs (DESeq2, edgeR, etc.). The associated web link is https://ccb.jhu.edu/software/stringtie/
Ballgown	It is a program for computing differentially expressed genes in two or more RNA-seq experiments, using the output of StringTie or cufflinks. The Ballgown package provides functions to organize, visualize, and analyze expression measurements. Ballgown is written in R and is part of Bioconductor. The associated web link is https://github.com/alyssafrazee/ballgown

with cutadapt software. The trimmed reads are aligned to reference genome with HISAT2 (hierarchical indexing for spliced alignment of transcripts), which follows a two-step process. The initial alignment of the reads to the genomes starts with a short read aligner Bowtie which subsequently identifies exon-intron junctions. There are other available read aligners, including the popular one STAR (Spliced Transcripts Alignment to a Reference), which offers the mapping of longer reads. In some cases, the exact alignment of sequence to the reference can be salvaged by using other methods. Kallisto is one of the methods which can handle this issue. It does that by merging pseudoalignment and quantification into a single step, thus delivering output faster than comparable methods like TopHat/Cufflinks. In our analysis pipeline, we determined whether the results can be used for subsequent data analysis following a statistical check on mapping ratio, i.e., rRNA/mtRNA content and fragment sequence bias. The expression levels (fragments per kilobase of transcript per million [FPKM] value) of known

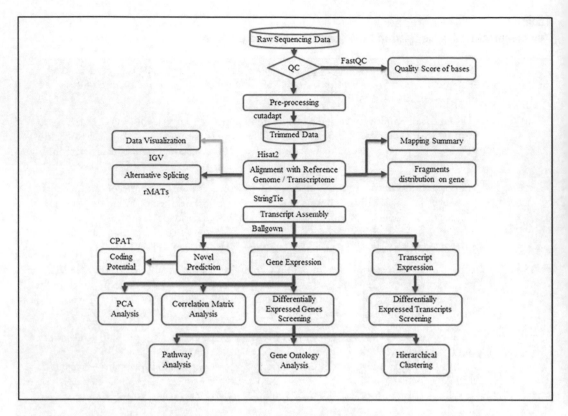

Fig. 3 RNA-seq data analysis workflow. Raw sequencing data generated from Illumina HiSeq4000 sequencer are used to generate normalized read counts following trimming and alignment to the reference genome. Differentially expressed genes were determined based on normalized data and further used in functional annotation following gene ontology (GO) and pathway analysis (KEGG). Courtesy: Arraystar Inc.

genes and transcripts are calculated using Ballgown through the transcript abundances estimated with StringTie. The number of identified genes and transcripts per group is calculated based on the mean FPKM in the group having significance ≥ 0.5. Differentially expressed gene and transcript analyses are performed with R package Ballgown sequencing pipelines. It provides a set of analyses which can be used to get a quick impression of the data. Due to the large number of genes in a typical RNA-seq data set, it is obvious to incur error at the time of differential expression analysis. Given that complexity in analysis, it is important to run correction for multiple comparisons. In this context, false discovery rate (FDR) has emerged as a popular and powerful tool for error rate control [14, 15]. The FDR as championed by Benjamini and Hochberg in 1995 provides an attractive measure of control for multiple testing in genomic settings [16]. Current days, the commonly used analysis tools for differential expression profiling offer the built-in feature of FDR correction method.

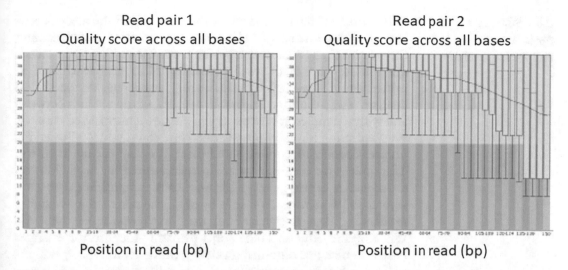

Fig. 4 RNA sequence quality score plot. The position in the sequence read is plotted on the X-axis, and the Q value is plotted on the Y-axis. The red line is the median of Q score, and the blue line is the mean of Q score. The boxplot represents the inter-quantile range from 25% to 75%, while the whiskers represent the 10% and 90%

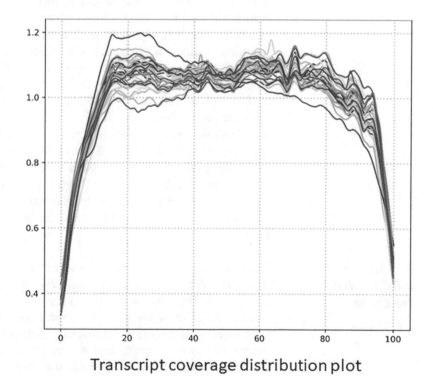

Transcript coverage distribution plot

Fig. 5 The Transcript Coverage graphs representing the fragment distribution of a transcript position. X-axis means transcript length (normalized to 100 bp), and Y-axis means mapped fragment number. A balanced coverage from 5′ to 3′ indicates a good sequencing library

2.4 Strategies for Differential Expression Analysis

Differential expression analysis is the final step in the analysis process of RNA sequencing. This segment is dedicated to determining the relative changes of transcripts between the groups based on the quantitative counts on each transcript. The relative change in expression is measured by normalizing, modeling, and statistical analysis. To get more comparable gene expression values between and within samples, it is necessary to prepare a normalized data set. Normalization is the output of scaled raw count values adjusted for unintentional artifacts generated from the sequencing platform. The three most common parameters accounted for normalization are sequence depth, gene length, and RNA composition. There are several software tools to take care of all the parameters including sequencing data normalization; the most popular of them are DESeq2 and edgeR. These tools primarily identify the changes based on negative binomial modeling of the data. DESeq is highlighted as the best performing tool in minimizing the number of false positives. However, in case of small replicate numbers (usually <12), edgeR gives true positive results compared to others. Another inherent advantage of using DESeq and edgeR method is their ability to adjust the confounding factors which are very common in human brain transcriptomic study. It is important to mention that heterogeneity in human brain tissues comes from technical, compositional, and disease-related factors. Moreover, bulk brain tissue, which is composed of many different neuronal types, brings additional complexity to the differential expression analysis. Under these circumstances, the use of regression-based analytical tools, including DESeq and edgeR, is more effective in circumventing the variabilities associated with batch effects, race, and sex, in addition to the hidden ones as described earlier. Once the expression values have been normalized for differences in sequencing depth and composition bias between the samples, they can be used to prepare differential gene counts and can be statistically clustered to determine the most significantly altered genes.

2.5 Visualization of DEGs with Plots and Graphs

Visualization of analyzed results is an integrated part of RNA-seq data representation. There are various graphical tools that are used to represent the RNA-seq data in a meaningful way. These days, the most popular static visualization tools explicitly use modern statistical indices to generate advanced plots in order to explore the differential gene expression data. Many of them use pre-scripted packages running on R platform and freely available on public domains like GitHub (https://github.com) or Bioconductor (https://www.bioconductor.org).

- *Heatmap*: Heatmaps are commonly used to visualize RNA-seq results. They are useful for visualizing the expression of genes across the samples (Fig. 6a). Additionally, it can be used to

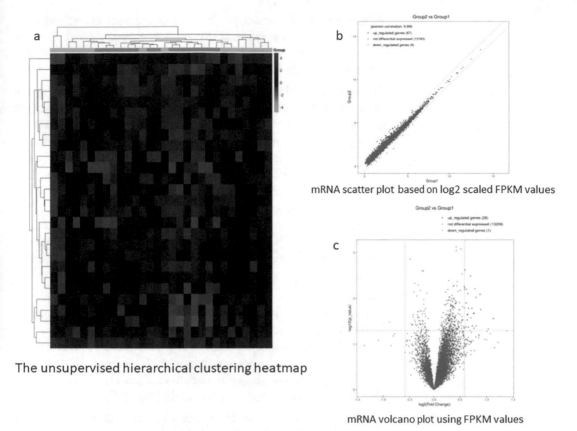

The unsupervised hierarchical clustering heatmap

mRNA scatter plot based on log2 scaled FPKM values

mRNA volcano plot using FPKM values

Fig. 6 Visualizing differential expression of genes. (**a**) The color-coded heatmap generated using significant differentially expressed genes. Each row represents a gene, and each column represents a sample. The color represents the relative expression level (log2 scaled FPKM), the red indicates higher values, and the green indicates lower values. (**b**) The scatterplot is based on group-wise comparison of expressed genes with log2 scaled FPKM values. The values of X- and Y-axes in the scatterplot are the averaged FPKM values of each group (log2 scaled). Genes above the top line (red dots, upregulation) or below the bottom line (green dots, downregulation) indicate more than 1.5 fold change between the two compared groups. Gray dots indicate non-differentially expressed genes. (**c**) The volcano is demonstrating the magnitude of expression changes which are statistically significant. The display is based on group-wise comparison of expressed genes with log2 scaled FPKM values. The values of X- and Y-axes in the volcano plot are log2 transformed fold change and -log10 transformed p-values between the two groups, respectively. Red/green circles indicate statistically significant differentially expressed genes with fold change no less than 1.5 and p-value ≤ 0.05 (red, upregulated; green, downregulated). Gray circles indicate non-differentially expressed genes, with FC and/or q-value not meeting the cutoff thresholds

visualize color-coded differences between the samples for most significantly altered genes. Currently, various heatmap tools are available to generate heatmaps with heatmap.2 function from the R gplots package. To make a heatmap of the top differentially expressed (DE) genes in an RNA-seq experiment, a file of normalized counts is needed. These values can be obtained from DESeq2 or edgeR programs. The log2 scale transformed

FPKM values of transcript abundance are also visualized with scatterplot. However, the scatterplots are mostly used to show up- and downregulated groups of genes above and below a certain threshold level (Fig. 6b).

- *Volcano plot*: Volcano plots are commonly used to display the results of RNA-seq data (Fig. 6c). It is a type of scatterplot that shows statistical significance (P-value) versus magnitude of change (fold change). It enables a quick visual identification of genes with large fold changes that are also statistically significant. These may be the most biologically significant genes. In a volcano plot, the most upregulated genes are toward the right, the most downregulated genes are toward the left, and the most statistically significant genes are toward the top. To generate a volcano plot of RNA-seq results, a file of differentially expressed results from differential expression program like DESeq2 is a prerequisite.

- *MA plot*: MA plot which is also called as ratio intensity plot is used to represent read counts between two groups of samples. In RNA-seq data analysis, the MA plot has been used as scatterplot with Y-axis displaying base-2 log fold change and X-axis to present normalized mean expression. This plot is useful in visualizing reproducibility between samples and also helps to identify any inconsistency across technical replicates.

- *Other visualization tools*: More advanced visualization tools can also be implemented to develop Manhattan, Q-Q, and circular (Circos) plots which are designed to present genome-wide differentially expressed gene (DEG) annotation with their expression level (hyper/hypo). Phenogram, Manhattan, Q-Q, and Circos plots are useful in visualizing chromosome-wise key significant (p-value <0.05) genes (hyper and hypo) from RNA-seq analysis.

2.6 Functional Annotation

The gene ontology (GO) enrichment analysis and the KEGG pathway enrichment analysis are the common downstream procedures to interpret the differential expression results in a biological context. Given a set of genes that are up- or downregulated under a certain contrast of interest, a GO (or pathway) enrichment analysis will find which GO terms (or pathways) are over- or underrepresented using annotations for the genes in that set.

- *Gene ontology (GO) Enrichment Analysis*: A common downstream method of differential gene expression analysis is gene set testing, which aims to understand the association of DEGs to specific pathways. DAVID/g:Profiler annotation tools (https://david.ncifcrf.gov/) are the most commonly used for functional annotation and pathway analysis. These tools are developed to statistically adjust the score for gene function enrichment by applying Benjamini-Hochberg multiple testing corrections within the gene set. Many other online tools (Panther GO, WebGestalt, Metascape, etc.) offer similar functional analysis based on a list of significant genes derived from differential expression analysis of an RNA-seq experiment and often cataloged as common functional descent under three broad categories, i.e., molecular function, biological process, and cellular component. However, the choice of selecting a web tool or self-scripted program completely depends on the aims that the study wants to achieve.

3   Conclusions and Future Perspectives

The human brain is a complex structure; hence, studying the transcriptional repatterning at any stage of the disease onset and progression is key to understanding the molecular pathology associated with brain disorders. MDD is not any different, given that it is associated with a significant loss of synaptic plasticity and morphometric attenuation of susceptible brain areas with systematic changes in gene expression. This necessitates an overarching need to actively investigate the gene expression characteristics across various regions of the MDD brain. Developing a high-quality brain transcriptomic atlas may help integrate interventions that an emotionally compromised brain receives during depressive disorder. A well-thought-out and carefully designed transcriptomic profiling can elucidate many aspects of transcriptional network reorganization during the course of MDD development and help decipher key driver genes involved in this process. However, as suggested earlier, the brain is an extremely complex organ with various modes of regulatory activities operating at different scales to synchronize spatiotemporal functioning. Desynchronization of this process may be the key in MDD pathogenesis, and it requires an appropriate quantitative method to explore the function of a specific set of genes or multiple clusters of different genes at the system level. Hence, the application of high-throughput methods like transcriptome-wide RNA-seq could get us better answer and abrogate any chance of speculation which are very common in candidate gene approach. So far, massively parallel sequencing platform has helped in determining many unknown genes (neuropeptides, substance P [TAC1], cholecystokinin [CCK], cocaine- and amphetamine-regulated transcript [CARTPT], galanin [GAL], and receptors for mu and kappa opioids [OPRM1, OPRK1]) and their clustered functions in depression [17]. Gene function enrichment analysis from sequenced data has also helped to identify multiple biological pathways, including neuropeptides, GPCR binding, and related intracellular MAPK, ERK, and cAMP signaling [17]. The robustness of this powerful technique has recently helped to open up new avenues in postmortem studies. The need to understand complex interactions between brain regions and distinct cell types within the same brain region has been greatly acknowledged. Given the wide variety of cell types in the brain and their complex interactions, investigative approaches using cell-type specificity are especially needed to gain insight into psychiatric phenotypes, including MDD [18]. Recently developed techniques for high-throughput single-cell RNA sequencing (scRNA-seq) have shown enormous potential and helped in determining cell type-specific gene expression changes in MDD prefrontal cortex. The results are exciting and have reported for the first time the predominant involvement of two cell types, i.e., oligodendrocyte precursor cells (OPCs) and deep layer excitatory neurons, in MDD pathology [18]. The usage of higher-order statistics, namely, unsupervised graph-based clustering on single-cell seq data, was able to identify 26 MDD-specific distinct cell-type clusters in PFC area. In fact, refined cellular subtypes as determined by this scRNA-seq method helped to configure a MDD-specific cortical cellular architecture. The technological advancement that is tied with this sequencing methodology is the adaptation of proprietary $10\times$ Genomics Chromium Controller platform that utilizes the power of single nuclei isolation and bar tagging the transcriptome before they are sequenced. Collectively, it shows the power and ability of the RNA sequencing technique with a far-reaching effect to quantify the transcriptome of a single cell, providing major opportunities to parse the complex cellular composition of the brain under a complex mental condition like MDD [19].

Acknowledgments

The research was partly supported by grants from the National Institute of Mental Health (MH082802, MH101890, MH100616, MH107183, MH112014, MH118884) and the American Foundation for Suicide Prevention (DIG-0-041-18) to Dr. Dwivedi.

References

1. CDC (2021) Mental health conditions: depression and anxiety. https://www.cdc.gov/tobacco/campaign/tips/diseases/depression-anxiety.html

2. Bagot RC, Cates HM, Purushothaman I, Lorsch ZS, Walker DM, Wang J, Huang X, Schluter OM, Maze I, Pena CJ, Heller EA, Issler O, Wang M, Song WM, Stein JL, Liu X, Doyle MA, Scobie KN, Sun HS, Neve RL, Geschwind D, Dong Y, Shen L, Zhang B, Nestler EJ (2016) Circuit-wide transcriptional profiling reveals brain region-specific gene networks regulating depression susceptibility. Neuron 90(5):969–983. https://doi.org/10.1016/j.neuron.2016.04.015

3. Girgenti MJ, Pothula S, Newton SS (2021) Stress and its impact on the transcriptome.

Biol Psychiatry 90(2):102–108. https://doi.org/10.1016/j.biopsych.2020.12.011

4. Kang HJ, Voleti B, Hajszan T, Rajkowska G, Stockmeier CA, Licznerski P, Lepack A, Majik MS, Jeong LS, Banasr M, Son H, Duman RS (2012) Decreased expression of synapse-related genes and loss of synapses in major depressive disorder. Nat Med 18(9):1413–1417. https://doi.org/10.1038/nm.2886

5. Duric V, Banasr M, Stockmeier CA, Simen AA, Newton SS, Overholser JC, Jurjus GJ, Dieter L, Duman RS (2013) Altered expression of synapse and glutamate related genes in post-mortem hippocampus of depressed subjects. Int J Neuropsychopharmacol 16(1):69–82. https://doi.org/10.1017/S1461145712000016

6. Kang HJ, Adams DH, Simen A, Simen BB, Rajkowska G, Stockmeier CA, Overholser JC, Meltzer HY, Jurjus GJ, Konick LC, Newton SS, Duman RS (2007) Gene expression profiling in postmortem prefrontal cortex of major depressive disorder. J Neurosci 27(48):13329–13340. https://doi.org/10.1523/JNEUROSCI.4083-07.2007

7. Tochigi M, Iwamoto K, Bundo M, Sasaki T, Kato N, Kato T (2008) Gene expression profiling of major depression and suicide in the prefrontal cortex of postmortem brains. Neurosci Res 60(2):184–191. https://doi.org/10.1016/j.neures.2007.10.010

8. Shelton RC, Claiborne J, Sidoryk-Wegrzynowicz M, Reddy R, Aschner M, Lewis DA, Mirnics K (2011) Altered expression of genes involved in inflammation and apoptosis in frontal cortex in major depression. Mol Psychiatry 16(7):751–762. https://doi.org/10.1038/mp.2010.52

9. Gandal MJ, Zhang P, Hadjimichael E, Walker RL, Chen C, Liu S, Won H, van Bakel H, Varghese M, Wang Y, Shieh AW, Haney J, Parhami S, Belmont J, Kim M, Moran Losada P, Khan Z, Mleczko J, Xia Y, Dai R, Wang D, Yang YT, Xu M, Fish K, Hof PR, Warrell J, Fitzgerald D, White K, Jaffe AE, Psych EC, Peters MA, Gerstein M, Liu C, Iakoucheva LM, Pinto D, Geschwind DH (2018) Transcriptome-wide isoform-level dysregulation in ASD, schizophrenia, and bipolar disorder. Science 362(6420):eaat8127. https://doi.org/10.1126/science.aat8127

10. Ramaker RC, Bowling KM, Lasseigne BN, Hagenauer MH, Hardigan AA, Davis NS, Gertz J, Cartagena PM, Walsh DM, Vawter MP, Jones EG, Schatzberg AF, Barchas JD, Watson SJ, Bunney BG, Akil H, Bunney WE, Li JZ, Cooper SJ, Myers RM (2017) Post-mortem molecular profiling of three psychiatric disorders. Genome Med 9(1):72. https://doi.org/10.1186/s13073-017-0458-5

11. Yoshino Y, Roy B, Kumar N, Shahid Mukhtar M, Dwivedi Y (2021) Molecular pathology associated with altered synaptic transcriptome in the dorsolateral prefrontal cortex of depressed subjects. Transl Psychiatry 11(1):73. https://doi.org/10.1038/s41398-020-01159-9

12. Pantazatos SP, Huang YY, Rosoklija GB, Dwork AJ, Arango V, Mann JJ (2017) Whole-transcriptome brain expression and exon-usage profiling in major depression and suicide: evidence for altered glial, endothelial and ATPase activity. Mol Psychiatry 22(5):760–773. https://doi.org/10.1038/mp.2016.130

13. Mahajan GJ, Vallender EJ, Garrett MR, Challagundla L, Overholser JC, Jurjus G, Dieter L, Syed M, Romero DG, Benghuzzi H, Stockmeier CA (2018) Altered neuro-inflammatory gene expression in hippocampus in major depressive disorder. Prog Neuro-Psychopharmacol Biol Psychiatry 82:177–186. https://doi.org/10.1016/j.pnpbp.2017.11.017

14. Korthauer K, Kimes PK, Duvallet C, Reyes A, Subramanian A, Teng M, Shukla C, Alm EJ, Hicks SC (2019) A practical guide to methods controlling false discoveries in computational biology. Genome Biol 20(1):118. https://doi.org/10.1186/s13059-019-1716-1

15. Bi R, Liu P (2016) Sample size calculation while controlling false discovery rate for differential expression analysis with RNA-sequencing experiments. BMC Bioinformatics 17(1):146. https://doi.org/10.1186/s12859-016-0994-9

16. Benjamini Y, Hochberg Y (1995) Controlling the false discovery rate: a practical and powerful approach to multiple testing. J R Stat Soc Ser B 57(1):289–300

17. Anderson KM, Collins MA, Kong R, Fang K, Li J, He T, Chekroud AM, Yeo BTT, Holmes AJ (2020) Convergent molecular, cellular, and cortical neuroimaging signatures of major depressive disorder. Proc Natl Acad Sci U S A 117(40):25138–25149. https://doi.org/10.1073/pnas.2008004117

18. Nagy C, Maitra M, Tanti A, Suderman M, Theroux JF, Davoli MA, Perlman K, Yerko V, Wang YC, Tripathy SJ, Pavlidis P, Mechawar N, Ragoussis J, Turecki G (2020) Single-nucleus transcriptomics of the prefrontal cortex in major depressive disorder implicates oligodendrocyte precursor cells and excitatory neurons. Nat Neurosci 23(6):771–781. https://doi.org/10.1038/s41593-020-0621-y

19. Reiner BC, Crist RC, Stein LM, Weller AE, Doyle GA, Arauco-Shapiro G, Ferraro TN, Hayes MR, Berrettini WH (2020) Single-nuclei transcriptomics of schizophrenia prefrontal cortex primarily implicates neuronal subtypes. bioRxiv:2020.2007.2029.227355. https://doi.org/10.1101/2020.07.29.227355

Part IV

Neuroimaging Research Methods for Major Depression

Chapter 11

Molecular Imaging of Major Depressive Disorder with Positron Emission Tomography: Translational Implications for Neurobiology and Treatment

Mikael Tiger and Johan Lundberg

Abstract

Here we discuss the positron emission tomography (PET) method applied to research on the pathophysiology of major depressive disorder (MDD). The principles for PET quantification of radioligand binding are described together with its benefits and limitations. The concept of MDD, its historical development, and the consequences for research in the biology of depression are addressed. The most commonly applied radioligands are reviewed, as are the typical findings. A particular interest is given to the serotonin system as this represents the main focus of the research up to recently. The inherent challenge of small patient samples in PET studies of MDD is addressed. Finally, future perspectives are critically discussed. Here both current trends and areas of particular interest are reviewed.

Key words PET, Positron emission tomography, Major depressive disorder, MDD, Serotonin, 5-HT_{1B}, Serotonin transporter, Synaptic density

1 Introduction

1.1 Basics of Positron Emission Tomography (PET)

The positron emission tomography (PET) method relies on a photon detector ring surrounding a positron-emitting source generating photons through the annihilation of positrons with naturally occurring electrons. The detectors consist of scintillating crystals, which emit visible light when γ particles pass through them. The crystals are coupled to a photomultiplier tube, which generates an electric response and registers a detection whenever the crystals scintillate. When two photons are detected simultaneously, this is recorded as a coincidence, and it is assumed that the positron-electron annihilation occurred at some position along the line of response between the two detectors. The detection of photons from the same annihilation (i.e., coincidences) during the time of a PET experiment (typically 60–90 min) enables reconstruction of the position and number of annihilations. These

Yong-Ku Kim and Meysam Amidfar (eds.), *Translational Research Methods for Major Depressive Disorder*, Neuromethods, vol. 179, https://doi.org/10.1007/978-1-0716-2083-0_11,

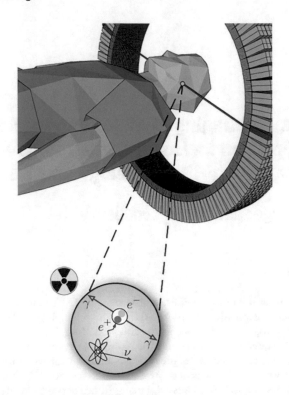

Fig. 1 Illustration of decay of positron emitting source, annihilation, generation of one pair of gamma particles that are detected by the camera ring. (Source: Wikimedia Commons)

represent the position of the positron-emitting source in the given experiment, typically a radiolabeled ligand with affinity for a target of interest, i.e., a specific protein within the human brain (Fig. 1).

Radioligands are usually administered through intravenous injection and always in tracer doses (i.e., without any pharmacological effect). PET radioisotopes undergo β+ decay, meaning that one of the protons in the nucleus of the parent nuclide becomes a neutron, and a positron and a neutrino are emitted. The kinetic energy of the emitted particles is dependent on the specific radionuclide. PET enables in vivo quantification of molecular concentrations in the picomolar range with a spatial and temporal resolution of millimeters and hours, respectively [1]. The specificity of PET is related to its use of radiolabeled ligands for target molecules of interest. These radioligands are selected for their high binding affinity for the target molecule which is ideally several orders of magnitude greater than for other related molecules. This means that the vast majority of the signal should be attributable to the availability of the target molecule.

Scalar quantification of the molecule investigated in a predefined volume of the tissue examined (i.e., target region) using PET and a given radioligand is done by kinetic modeling where the time-activity curve (TAC) of radioactivity concentration in the target region for the time of the experiment is compared to that of a reference radioactivity concentration in, e.g., arterial plasma or in a reference region of the tissue examined by PET.

The aim of kinetic modeling is to derive an analytical expression which describes the measured TAC and to use this mathematical description to estimate outcome measures which describe the concentration of the investigated molecule in the target region. Typically, applied PET research do not involve arterial blood sampling, and the kinetic models applied do therefore produce the unitless compound outcome measure non-displaceable binding potential (BP_{ND}). The outcome measures can then be statistically compared between individuals, groups, or conditions.

BP reflects target protein density and/or concentration of endogenous ligand. Molecular PET has mostly been used for quantification of target proteins [2]. However, for many radioligands, it is not possible to rule out displacement of radioligand binding by endogenous ligands [3]. Radioligand displacement has typically been demonstrated with drug challenges pharmacologically increasing concentrations of endogenous ligands [4–7]. The challenge is to determine how sensitive each radioligand is to endogenous ligand concentrations. For example, the 5-HT$_{1B}$ receptor selective radioligand [^{11}C]AZ10419369 is sensitive to a supratherapeutic dose of the selective serotonin reuptake inhibitor (SSRI) escitalopram [8]. In contrast, no displacement of [^{11}C]AZ10419369 binding was found in serotonin projection areas with a single dose of escitalopram [8]. Better knowledge on sensitivities for endogenous ligands for different radioligands would facilitate interpretation of PET results, including studies of MDD.

PET is an invaluable method for drug occupancy studies in the living brain. Here PET measurements are repeated without and with the presence of a competitor for the same target as the radioligand. The difference in specific binding in the two experimental conditions together with information on (oral or intravenous) dose and/or plasma concentration of the examined drug gives information on the relation between drug dose, drug plasma concentration, and drug occupancy of the target protein. Occupancy study designs include non-human [9] and human control subjects [10] and patients [11], as well as within- [12] and between-subject [13] comparisons. They are done both to test hypotheses on in vivo mechanism of action and for estimation of dose-response relationships before phase 2b or phase 3 trials of novel compounds.

1.2 Pathophysiology of Depression

The pathophysiology of major depressive disorder (MDD) remains elusive, despite more than half a century of intense research. The most established hypotheses for MDD pathophysiology are derived from the mechanism of action of antidepressant drugs. An early

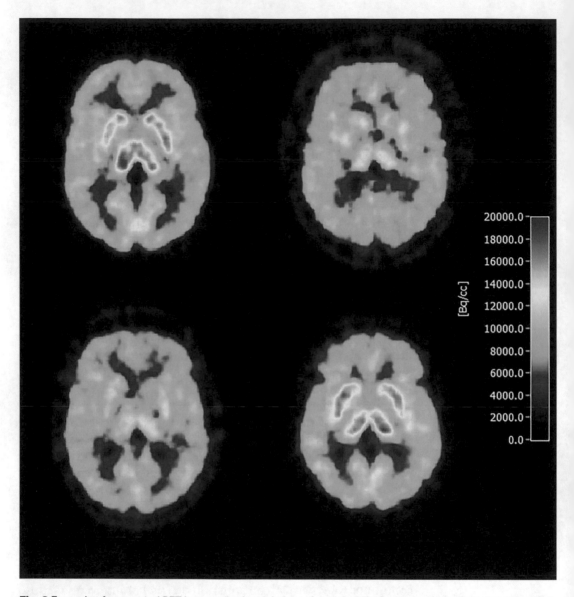

Fig. 2 Example of summated PET images (horizontal plane; frame 6–20) of an untreated reference subject (top left), of subjects treated with a selective serotonin reuptake inhibitor (citalopram, top right), a tricyclic antidepressant (clomipramine, bottom left) and mirtazapine (bottom right) [19]

observation was that antidepressants such as monoamine oxidase inhibitors and tricyclic antidepressants increased synaptic mono-amine levels either by decreasing degradation or by inhibiting presynaptic monoamine reuptake [14–16]. In his influential paper from 1967, Alec Coppen suggested that depression at least partly may be caused by monoamine deficiency [17]. Lapin and Oxenkrug argued that serotonin was the most important monoamine in the pathophysiology of depression and the main mediator of mood-elevating effect of antidepressant drugs [18]. We have corroborated the serotonin transporter as a target for a wide range of antidepressants using PET (Fig. 2) [13]. The serotonin hypothesis of MDD has not yet been tested, as it is not yet feasible to measure serotonin in the living brain in MDD patients. With the development of a radioligand sensitive to endogenous concentrations of serotonin, the serotonin hypothesis could be tested with PET.

2 Materials

One of the main challenges in MDD research is how to define the depressive state. Today criteria-based definitions are applied, to increase reliability. The MDD criteria in use today stem from the third edition of the American Psychiatric Association's *Diagnostic and Statistical Manual of Mental Disorders* (DSM-III). The DSM-III criteria for MDD are a slight revision of the Feighner criteria presented earlier as a research tool for studies of depression [20]. The present definitions of MDD in DSM-5 and in the World Health Organization's International Classification of Diseases (ICD-11) are quite similar to the landmark MDD criteria in DSM-III, except that in DSM-5, the original bereavement exclusion criterion has been omitted. The two main limitations of the current MDD definitions are the heterogeneity of the condition and the disregard of current life context. The hallmark of MDD is a disproportionally lowered mood and/or loss of interest [21]. None of the MDD symptoms are pathognomonic, and proportional sadness in reaction to a stressful life situation is not necessarily best classified as MDD, even if at least five out of nine symptom criteria are fulfilled [22]. Moreover, the combinatorics of the DSM MDD definition underscores the heterogeneity of the diagnosis that can be assigned in 1497 different ways [23]. The false MDD positives without disproportional lowering of mood and the heterogeneous MDD diagnosis have hampered attempts to pinpoint the molecular pathophysiology of MDD and may have contributed to inconsistent results in PET studies of neuroreceptors in MDD [24]. In response to shortcomings of today's mostly symptom-based diagnostic criteria, the National Institute of Mental Health in the United States in 2010 set out to develop research domain criteria (RDoC) based on neuroscientific observations and genetic data, to better match underlying pathophysiology [25]. Attempts have been made to apply RDoC in depression research, with studies of rumination as a model [26]. However, an optimized definition of MDD or relevant endophenotypes within the condition remains a research priority.

In humans, there is a range of PET studies that could provide valuable knowledge on MDD. There are PET studies of sadness, one of the cardinal symptoms of MDD. The other core MDD symptom anhedonia may reflect a reactive state, although it is not to be considered a normal emotional state suitable for studies in healthy individuals. MDD populations can be grouped according to severity as mild, moderate, or severe, with or without psychotic symptoms. MDD grouping may also be based on level of psychiatric care, ranging from general practice to psychiatric outpatients and patients hospitalized for MDD. Comorbidities are common in a psychiatric setting, so it is rare to exclude patients with anxiety disorders, as long as MDD remains the primary diagnosis. However, there are PET studies reporting on other diagnoses, with large subpopulations with MDD comorbidity [27]. Furthermore, there are studies of major depressive episodes, including patients with MDD as well as with bipolar disorder [28]. Pooling of these inherently different mood disorders is mostly limited to patient populations that are expected to respond similarly to a studied treatment, such as electroconvulsive therapy (ECT) [28]. Moreover, MDD patients can be sorted as drug naïve or not or according to family history of MDD.

Radioligands previously and currently used in PET research related to MDD and its treatment can be separated into markers for the monoamine systems, markers for the GABA and glutamate systems, markers for neuroinflammation, and markers for synaptic density (*see* Table 1 for a non-exhaustive list of examples).

3 Methods

Sadness is typically the prevailing emotion during MDD. In early PET studies, increased regional blood flow in the anterior cingulate cortex as measured with radioactive water ($H_2{}^{15}O$) has been reported in healthy women with sadness [54, 55]. Today, functional PET examinations of healthy volunteers have mostly been replaced by methods such as functional magnetic resonance imaging.

Studies on MDD patients are either cross-sectional or longitudinal. Case-control studies dominate the cross-sectional MDD literature. There are different types of MDD case populations, as outlined in "Materials." The main divide is arguably medication status. Even though patients with medications with known action on the PET target of interest are excluded, medication may still alter the results. For example, 5-HT_{1B} receptor binding has been reported low in the anterior cingulate cortex in MDD patients [27, 32]. Increased cortical 5-HT_{1B} receptor binding has been reported in healthy volunteers after administration of a selective serotonin reuptake inhibitor [8]. Ongoing antidepressant medication may thus cancel out the pathophysiological effect of MDD, as in an autoradiography study of the anterior cingulate cortex in medicated MDD patients [56]. Ongoing medication may even yield false positives, as the low 5-hydroxyindoleacetic acid

Table 1
Radioligands previously and currently used in PET research related to MDD

Target protein	Radioligand name	Chemical compound intrinsic activity	Research application in MDD populations (example). CC = case-control design; LO = longitudinal design; OC = drug occupancy study
Serotonin 1A receptor	[Carbonyl-^{11}C] WAY-100635	Antagonist	DOI: https://doi.org/10.1016/j.biopsych.2012.11.012 (LO) [29]
	[^{11}C]CUMI-101	Agonist	DOI: https://doi.org/10.1038/s41380-020-0733-5 (CC) [30]
Serotonin 1B receptor	[^{11}C] AZ10419369	Antagonist	DOI: https://doi.org/10.1038/s41398-020-0844-4 (LO) [31] DOI: https://doi.org/10.1016/j.pscychresns.2016.04.016 (CC) [32] DOI: https://doi.org/10.1016/j.pscychresns.2016.04.016 (LO) [33]
	[^{11}C]P943	Antagonist	DOI: https://doi.org/10.1007/s00213-010-1881-0 (CC) [34]
Serotonin 2A receptor	[^{11}C] MDL100,907	Antagonist	DOI: https://doi.org/10.1176/ajp.2006.163.9.1580 (CC) [35]
	[^{18}F]altanserin	Antagonist	DOI: https://doi.org/10.1016/j.biopsych.2003.08.015 (CC) [36]
Serotonin transporter	[^{11}C]MADAM	Inhibitor	DOI: https://doi.org/10.1017/S1461145711001945 (CC, OC) [13]
	[^{11}C]DASB	Inhibitor	DOI: https://doi.org/10.2967/jnumed.117.189654 (CC) [37] DOI: https://doi.org/10.1007/s00213-009-1660-y (CC) [38]
	[^{11}C](+) McN5652	Inhibitor	DOI: https://doi.org/10.1002/syn.20637 (CC) [39]
Monoamine oxidase A	[^{11}C]Harmine	Inhibitor	DOI: https://doi.org/10.1016/j.brs.2018.12.976 (CC, LO) [40]
Monoamine oxidase B	[^{11}C] SL25.1188	Inhibitor	DOI: https://doi.org/10.1001/jamapsychiatry.2019.0044 (CC) [41]
Dopamine D1 receptor	[^{11}C]SCH 23,390	Antagonist	DOI: https://doi.org/10.1002/da.20168 (CC) [42]
Dopamine D2/D3 receptor	[^{11}C]Raclopride	Antagonist	DOI: https://doi.org/10.1111/pcn.12980 (CC, LO) [28]
	[^{11}C]FLB-457	Antagonist	DOI: https://doi.org/10.4088/JCP.08m04746blu (CC, LO) [43]
Dopamine transporter	[^{18}F]FE-PE2I	Inhibitor	DOI: https://doi.org/10.1016/j.pscychresns.2020.111086 (LO) [44]
	[^{11}C]RTI-32	Inhibitor	DOI: https://doi.org/10.1097/00001756-200112210-00052 (CC) [45]
Norepinephrine transporter	[^{18}F]FMeNER-D2	Inhibitor	DOI: https://doi.org/10.1176/appi.ajp.2016.15101334 (CC) [46] DOI: https://doi.org/10.1093/ijnp/pyz003 (CC, OC) [11]

(continued)

Table 1
(continued)

Target protein	Radioligand name	Chemical compound intrinsic activity	Research application in MDD populations (example). CC = case-control design; LO = longitudinal design; OC = drug occupancy study
Muscarinic 2 receptor	$[^{18}F]$FP-TZTP	Antagonist	DOI: https://doi.org/10.1001/archpsyc.63.7.741 (CC) [47]
GABA-A receptor	$[^{11}C]$flumazenil	Antagonist	DOI: https://doi.org/10.1007/s00259-009-1292-9 (CC, LO) [48]
Metabotropic glutamate receptor 5	$[^{18}F]$FPEB	Antagonist	DOI: https://doi.org/10.1016/j.bpsc.2017.03.019 (CC) [49]
18 kDa translocator protein	$[^{11}C]$PBR-28	Na	DOI: https://doi.org/10.1186/s13550-018-0401-9 (CC) [50]
	$[^{18}F]$FEPPA	Na	DOI: https://doi.org/10.1016/j.pnpbp.2017.12.011 (CC, LO) [51]
	$[^{11}C]$-(R)-PK11195	Na	DOI: https://doi.org/10.1016/j.biopsych.2017.08.005 (CC) [52]
Synaptic vesicle protein 2A	$[^{11}C]$UCB-J	Na	DOI: https://doi.org/10.1038/s41467-019-09562-7 (CC) [53]

concentrations reported in the cerebrospinal fluid in pharmacologically treated MDD patients [57, 58]. Ideally, patients should be unmedicated at the time of PET, to avoid confounding by medication, in case-control studies of MDD. Studies on monoaminergic targets dominate the MDD case-control PET literature [24]. With many radioligands, results are inconsistent, with either high, low, or non-significantly different BP in the MDD group [24]. For the 5-HT$_{1A}$ receptor radioligand $[^{11}C]$WAY-100635, the discrepant results may in part be related to methodological issues. In $[^{11}C]$ WAY-100635 PET studies applying an arterial input function, BP tends to be high in MDD patients, while it is mostly reported as low in limbic and cortical regions in studies using reference tissue models [24]. It has been suggested that this discrepancy may be due to specific radioligand binding in the cerebellum, the standard reference region, confounding the results [59]. The heterogeneity of the current MDD definition may also yield substantial noise to case-control studies. Furthermore, due to high cost, PET studies tend to be small, reporting data on 10–30 patients, with inherent risk of type II errors. The field would thus benefit from meta-analyses, pooling compatible MDD case-control PET studies, to get a sufficiently large study sample to be able to conclude something about the results, as has been done with studies of serotonin transporter binding in MDD [60]. In the study by Gryglewski et al., the low or non-significantly different serotonin transporter binding in the included studies is transformed into a low serotonin transporter binding in the midbrain, in MDD patients, in the pooled analysis [60]. In PET studies of the 5-HT$_{1B}$ receptor, low binding in limbic brain regions in MDD has consistently been reported, although the number of studies is small [61].

The vast majority of longitudinal PET studies of MDD contains examinations before and after a given treatment. Hence, these longitudinal studies are better described as mechanistic, addressing mode of action of MDD treatments. Most mechanistic PET studies examine the treatment effect of different antidepressant drugs, with the serotonin transporter as the most common target [13, 62–66]. However, there are also PET studies of the mechanism of action of non-pharmacological MDD treatments such as electroconvulsive therapy (ECT) [43, 67–70] and psychotherapy [33, 71]. To our knowledge, there are no longitudinal PET studies of the natural course of MDD, and likely there will be no such studies in the future, for ethical reasons.

4 Conclusions

PET remains the most accurate tool for in vivo quantification of radioligand binding in man. The development of radioligands for quantification of novel targets continues and is expected to follow the path of the research fields where it is applied. In the area of mood disorders, preclinical research on pathophysiology and drug development has hitherto paved the way for radioligand development. We predict that this will remain the most important pathway in the foreseeable future to test preclinical hypotheses on pathophysiology and drug-target interaction in the living human brain. One example is the preclinical and postmortem research on synaptic density in relation to MDD that via the development of the SV2A marker [^{11}C]UCB-J now can be tested in control subjects and MDD patients in relation to disease severity, behavioral changes related to disease, and the association between symptom relief and SV2A availability during and following treatment interventions. The report on a correlation between [^{11}C]UCB-J binding in the hippocampus and depressive symptoms in 24 patients with MDD and PTSD is an example of the potential of this tool [32].

Despite promising results with new radioligands, the inherent challenges of defining depression most likely have hampered research on the biology of the disorder. The heterogeneous and unspecific current diagnostic criteria may have contributed to inconsistent results in previous PET studies of MDD. In order to optimize the use of data on the pathophysiology of depression, the condition should be better defined. Studies of subtypes of depression or domains within the condition, such as RDoC, might progress the field, with increased likelihood of identifying common biological underpinnings of depression. Characterizing patients according to response to MDD treatments such as ketamine [31] or SSRI [72] is clinically meaningful and may contribute to knowledge about antidepressant mechanisms of action. Treatment that points in the direction of normalizing radioligand binding in the MDD group may provide indirect support of the validity of the finding. For example, in patients with SSRI-resistant depression, subanesthetic doses of ketamine have been shown to increase $5\text{-}HT_{1B}$ receptor binding in the hippocampus [31], a region with reported low $5\text{-}HT_{1B}$ receptor binding in MDD [27, 32].

A common limitation in many PET studies is low statistical power. Although sample sizes are typically larger in more recent studies than what was common previously, the often somewhat inconsistent findings remain a problem in the field. Recently, examples of meta-analyses including individual participant data have been published [73]. This approach is promising and should be encouraged.

In essence, the main contribution of in vivo molecular imaging in MDD research hitherto has been to contribute to a better understanding of the mechanism of action of MDD treatments in man in vivo and to identify group-level differences between cases and controls with regard to markers for the monoamine systems and more recently also markers for the 18 kDa translocator protein (TSPO). The development of radioligands labeling potential markers crucial for disease pathology such as TSPO and synaptic vesicle protein 2A may pave the way for future biomarkers in the field of psychiatry.

References

1. McArthur RA (2012) Translational neuroimaging: tools for CNS drug discovery, development and treatment. Academic Press, London

2. Farde L (1996) The advantage of using positron emission tomography in drug research. Trends Neurosci 19(6):211–214

3. Laruelle M (2000) Imaging synaptic neurotransmission with in vivo binding competition techniques: a critical review. J Cereb Blood Flow Metab 20(3):423–451

4. Finnema SJ, Varrone A, Hwang TJ, Gulyas B, Pierson ME, Halldin C et al (2010) Fenfluramine-induced serotonin release decreases [11C]AZ10419369 binding to 5-HT1B-receptors in the primate brain. Synapse 64(7):573–577

5. Finnema SJ, Varrone A, Hwang TJ, Halldin C, Farde L (2012) Confirmation of fenfluramine effect on 5-HT(1B) receptor binding of [(11)C]AZ10419369 using an equilibrium approach. J Cereb Blood Flow Metab 32(4):685–695

6. Tedroff J, Pedersen M, Aquilonius SM, Hartvig P, Jacobsson G, Langstrom B (1996) Levodopa-induced changes in synaptic dopamine in patients with Parkinson's disease as measured by [11C]raclopride displacement and PET. Neurology 46(5):1430–1436

7. Hartvig P, Torstenson R, Tedroff J, Watanabe Y, Fasth KJ, Bjurling P et al (1997) Amphetamine effects on dopamine release and synthesis rate studied in the Rhesus monkey brain by positron emission tomography. J Neural Transm (Vienna) 104(4–5):329–339

8. Nord M, Finnema SJ, Halldin C, Farde L (2013) Effect of a single dose of escitalopram on serotonin concentration in the non-human

and human primate brain. Int J Neuropsychopharmacol 16:1577–1586

9. Yang KC, Stepanov V, Amini N, Martinsson S, Takano A, Bundgaard C et al (2019) Effect of clinically relevant doses of vortioxetine and citalopram on serotonergic PET markers in the nonhuman primate brain. Neuropsychopharmacology 44(10):1706–1713

10. Lundberg J, Christophersen JS, Buchberg Petersen K, Loft H et al (2007) PET measurement of serotonin transporter occupancy: a comparison of escitalopram and citalopram. Int J Neuropsychopharmacol 10:777–785

11. Arakawa R, Stenkrona P, Takano A, Svensson J, Andersson M, Nag S et al (2019) Venlafaxine ER blocks the norepinephrine transporter in the brain of patients with major depressive disorder: a PET study using [18F]FMeNER-D2. Int J Neuropsychopharmacol 22(4):278–285

12. Stenkrona P, Halldin C, Lundberg J (2013) 5-HTT and 5-HT(1A) receptor occupancy of the novel substance vortioxetine (Lu AA21004). A PET study in control subjects. Eur Neuropsychopharmacol 23:1190

13. Lundberg J, Tiger M, Landen M, Halldin C, Farde L (2012) Serotonin transporter occupancy with TCAs and SSRIs: a PET study in patients with major depressive disorder. Int J Neuropsychopharmacol 15(8):1167–1172

14. Schildkraut JJ (1965) The catecholamine hypothesis of affective disorders: a review of supporting evidence. Am J Psychiatry 122(5): 509–522

15. Carlsson A, Corrodi H, Fuxe K, Hokfelt T (1969) Effect of antidepressant drugs on the depletion of intraneuronal brain 5-hydroxytryptamine stores caused by 4-methyl-alpha-ethyl-meta-tyramine. Eur J Pharmacol 5(4):357–366

16. Carlsson A, Corrodi H, Fuxe K, Hokfelt T (1969) Effects of some antidepressant drugs on the depletion of intraneuronal brain catecholamine stores caused by 4,alpha-dimethyl-meta-tyramine. Eur J Pharmacol 5(4):367–373

17. Coppen A (1967) The biochemistry of affective disorders. Br J Psychiatry 113(504): 1237–1264

18. Lapin IP, Oxenkrug GF (1969) Intensification of the central serotoninergic processes as a possible determinant of the thymoleptic effect. Lancet 1(7586):132–136

19. Tiger M (2014) PET studies of the serotonin system in depression and its treatment. PhD thesis, Karolinska Institutet

20. Feighner JP, Robins E, Guze SB, Woodruff RA Jr, Winokur G, Munoz R (1972) Diagnostic criteria for use in psychiatric research. Arch Gen Psychiatry 26(1):57–63

21. Belmaker RH, Agam G (2008) Major depressive disorder. N Engl J Med 358(1):55–68

22. Horwitz AV, Wakefield JC (2007) The loss of sadness - how psychiatry transformed normal sorrow into depressive disorder. Oxford University Press, Oxford

23. Ostergaard SD, Jensen SO, Bech P (2011) The heterogeneity of the depressive syndrome: when numbers get serious. Acta Psychiatr Scand 124(6):495–496

24. Savitz JB, Drevets WC (2013) Neuroreceptor imaging in depression. Neurobiol Dis 52:49–65

25. Insel T, Cuthbert B, Garvey M, Heinssen R, Pine DS, Quinn K et al (2010) Research domain criteria (RDoC): toward a new classification framework for research on mental disorders. Am J Psychiatry 167(7):748–751

26. Woody ML, Gibb BE (2015) Integrating NIMH research domain criteria (RDoC) into depression research. Curr Opin Psychol 4:6–12

27. Murrough JW, Czermak C, Henry S, Nabulsi N, Gallezot JD, Gueorguieva R et al (2011) The effect of early trauma exposure on serotonin type 1B receptor expression revealed by reduced selective radioligand binding. Arch Gen Psychiatry 68(9):892–900

28. Tiger M, Svensson J, Liberg B, Saijo T, Schain M, Halldin C et al (2020) [(11) C] raclopride positron emission tomography study of dopamine-D2/3 receptor binding in patients with severe major depressive episodes before and after electroconvulsive therapy and compared to control subjects. Psychiatry Clin Neurosci 74(4):263–269

29. Gray NA, Milak MS, DeLorenzo C, Ogden RT, Huang YY, Mann JJ et al (2013) Antidepressant treatment reduces serotonin-1A autoreceptor binding in major depressive disorder. Biol Psychiatry 74(1):26–31

30. Schneck N, Tu T, Falcone HR, Miller JM, Zanderigo F, Sublette ME et al (2020) Large-scale network dynamics in neural response to emotionally negative stimuli linked to serotonin 1A binding in major depressive disorder. Mol Psychiatry 26:2393

31. Tiger M, Veldman ER, Ekman CJ, Halldin C, Svenningsson P, Lundberg J (2020) A randomized placebo-controlled PET study of ketamine s effect on serotonin 1B receptor binding in patients with SSRI-resistant depression. Transl Psychiatry 10(1):159

32. Tiger M, Farde L, Ruck C, Varrone A, Forsberg A, Lindefors N et al (2016) Low serotonin 1B receptor binding potential in the

anterior cingulate cortex in drug-free patients with recurrent major depressive disorder. Psychiatry Res 253:36–42

33. Tiger M, Ruck C, Forsberg A, Varrone A, Lindefors N, Halldin C et al (2014) Reduced 5-HT1B receptor binding in the dorsal brain stem after cognitive behavioural therapy of major depressive disorder. Psychiatry Res 223(2):164–170

34. Murrough JW, Henry S, Hu J, Gallezot JD, Planeta-Wilson B, Neumaier JF et al (2011) Reduced ventral striatal/ventral pallidal serotonin 1B receptor binding potential in major depressive disorder. Psychopharmacology 213(2–3):547–553

35. Bhagwagar Z, Hinz R, Taylor M, Fancy S, Cowen P, Grasby P (2006) Increased 5-HT (2A) receptor binding in euthymic, medication-free patients recovered from depression: a positron emission study with [(11)C]MDL 100,907. Am J Psychiatry 163(9):1580–1587

36. Mintun MA, Sheline YI, Moerlein SM, Vlassenko AG, Huang Y, Snyder AZ (2004) Decreased hippocampal 5-HT2A receptor binding in major depressive disorder: in vivo measurement with [18F]altanserin positron emission tomography. Biol Psychiatry 55(3):217–224

37. Ananth MR, DeLorenzo C, Yang J, Mann JJ, Parsey RV (2018) Decreased pretreatment amygdalae serotonin transporter binding in unipolar depression remitters: a prospective PET study. J Nucl Med 59(4):665–670

38. Selvaraj S, Murthy NV, Bhagwagar Z, Bose SK, Hinz R, Grasby PM et al (2011) Diminished brain 5-HT transporter binding in major depression: a positron emission tomography study with [11C]DASB. Psychopharmacology 213(2–3):555–562

39. Miller JM, Kinnally EL, Ogden RT, Oquendo MA, Mann JJ, Parsey RV (2009) Reported childhood abuse is associated with low serotonin transporter binding in vivo in major depressive disorder. Synapse 63(7):565–573

40. Baldinger-Melich P, Gryglewski G, Philippe C, James GM, Vraka C, Silberbauer L et al (2019) The effect of electroconvulsive therapy on cerebral monoamine oxidase A expression in treatment-resistant depression investigated using positron emission tomography. Brain Stimul 12(3):714–723

41. Moriguchi S, Wilson AA, Miler L, Rusjan PM, Vasdev N, Kish SJ et al (2019) Monoamine oxidase B total distribution volume in the prefrontal cortex of major depressive disorder: an [11C]SL25.1188 positron emission tomography study. JAMA Psychiatry 76(6):634–641

42. Dougherty DD, Bonab AA, Ottowitz WE, Livni E, Alpert NM, Rauch SL et al (2006) Decreased striatal D1 binding as measured using PET and [11C]SCH 23,390 in patients with major depression with anger attacks. Depress Anxiety 23(3):175–177

43. Saijo T, Takano A, Suhara T, Arakawa R, Okumura M, Ichimiya T et al (2010) Electroconvulsive therapy decreases dopamine D(2) receptor binding in the anterior cingulate in patients with depression: a controlled study using positron emission tomography with radioligand [(1)(1)C]FLB 457. J Clin Psychiatry 71(6):793–799

44. Moriya H, Tiger M, Tateno A, Sakayori T, Masuoka T, Kim W et al (2020) Low dopamine transporter binding in the nucleus accumbens in geriatric patients with severe depression. Psychiatry Clin Neurosci 74(8):424–430

45. Meyer JH, Kruger S, Wilson AA, Christensen BK, Goulding VS, Schaffer A et al (2001) Lower dopamine transporter binding potential in striatum during depression. NeuroReport 12(18):4121–4125

46. Moriguchi S, Yamada M, Takano H, Nagashima T, Takahata K, Yokokawa K et al (2017) Norepinephrine transporter in major depressive disorder: a PET study. Am J Psychiatry 174(1):36–41

47. Cannon DM, Carson RE, Nugent AC, Eckelman WC, Kiesewetter DO, Williams J et al (2006) Reduced muscarinic type 2 receptor binding in subjects with bipolar disorder. Arch Gen Psychiatry 63(7):741–747

48. Klumpers UM, Veltman DJ, Drent ML, Boellaard R, Comans EF, Meynen G et al (2010) Reduced parahippocampal and lateral temporal GABAA-[11C]flumazenil binding in major depression: preliminary results. Eur J Nucl Med Mol Imaging 37(3):565–574

49. Abdallah CG, Hannestad J, Mason GF, Holmes SE, DellaGioia N, Sanacora G et al (2017) Metabotropic glutamate receptor 5 and glutamate involvement in major depressive disorder: a multimodal imaging study. Biol Psychiatry Cogn Neurosci Neuroimaging 2(5):449–456

50. Richards EM, Zanotti-Fregonara P, Fujita M, Newman L, Farmer C, Ballard ED et al (2018) PET radioligand binding to translocator protein (TSPO) is increased in unmedicated depressed subjects. EJNMMI Res 8(1):57

51. Li H, Sagar AP, Keri S (2018) Translocator protein (18kDa TSPO) binding, a marker of microglia, is reduced in major depression during cognitive-behavioral therapy. Prog Neuro-Psychopharmacol Biol Psychiatry 83:1–7

52. Holmes SE, Hinz R, Conen S, Gregory CJ, Matthews JC, Anton-Rodriguez JM et al (2018) Elevated translocator protein in anterior cingulate in major depression and a role for inflammation in suicidal thinking: a positron emission tomography study. Biol Psychiatry 83(1):61–69

53. Holmes SE, Scheinost D, Finnema SJ, Naganawa M, Davis MT, DellaGioia N et al (2019) Lower synaptic density is associated with depression severity and network alterations. Nat Commun 10(1):1529

54. Mayberg HS, Liotti M, Brannan SK, McGinnis S, Mahurin RK, Jerabek PA et al (1999) Reciprocal limbic-cortical function and negative mood: converging PET findings in depression and normal sadness. Am J Psychiatry 156(5):675–682

55. Liotti M, Mayberg HS, Brannan SK, McGinnis S, Jerabek P, Fox PT (2000) Differential limbic--cortical correlates of sadness and anxiety in healthy subjects: implications for affective disorders. Biol Psychiatry 48(1): 30–42

56. Veldman ER, Svedberg MM, Svenningsson P, Lundberg J (2017) Distribution and levels of 5-HT1B receptors in anterior cingulate cortex of patients with bipolar disorder, major depressive disorder and schizophrenia - an autoradiography study. Eur Neuropsychopharmacol 27(5):504–514

57. Asberg M, Thoren P, Traskman L, Bertilsson L, Ringberger V (1976) "Serotonin depression"--a biochemical subgroup within the affective disorders? Science 191(4226):478–480

58. Asberg M (1997) Neurotransmitters and suicidal behavior. The evidence from cerebrospinal fluid studies. Ann N Y Acad Sci 836:158–181

59. Shrestha S, Hirvonen J, Hines CS, Henter ID, Svenningsson P, Pike VW et al (2012) Serotonin-1A receptors in major depression quantified using PET: controversies, confounds, and recommendations. NeuroImage 59(4):3243–3251

60. Gryglewski G, Lanzenberger R, Kranz GS, Cumming P (2014) Meta-analysis of molecular imaging of serotonin transporters in major depression. J Cereb Blood Flow Metab 34(7): 1096–1103

61. Tiger M, Varnas K, Okubo Y, Lundberg J (2018) The 5-HT1B receptor - a potential target for antidepressant treatment. Psychopharmacology 235(5):1317–1334

62. Voineskos AN, Wilson AA, Boovariwala A, Sagrati S, Houle S, Rusjan P et al (2007) Serotonin transporter occupancy of high-dose selective serotonin reuptake inhibitors during major depressive disorder measured with [11C]DASB positron emission tomography. Psychopharmacology 193(4):539–545

63. Meyer JH, Wilson AA, Ginovart N, Goulding V, Hussey D, Hood K et al (2001) Occupancy of serotonin transporters by paroxetine and citalopram during treatment of depression: a [(11)C]DASB PET imaging study. Am J Psychiatry 158(11):1843–1849

64. Meyer JH, Kapur S, Eisfeld B, Brown GM, Houle S, DaSilva J et al (2001) The effect of paroxetine on 5-HT(2A) receptors in depression: an [(18)F]setoperone PET imaging study. Am J Psychiatry 158(1):78–85

65. Meyer. (2007) Imaging the serotonin transporter during major depressive disorder and antidepressant treatment. Rev Psychiatr Neurosci 32(2):86–102

66. Meyer JH, Wilson AA, Sagrati S, Hussey D, Carella A, Potter WZ et al (2004) Serotonin transporter occupancy of five selective serotonin reuptake inhibitors at different doses: an [11C]DASB positron emission tomography study. Am J Psychiatry 161(5):826–835

67. Masuoka T, Tateno A, Sakayori T, Tiger M, Kim W, Moriya H et al (2020) Electroconvulsive therapy decreases striatal dopamine transporter binding in patients with depression: a positron emission tomography study with [(18)F]FE-PE2I. Psychiatry Res Neuroimaging 301:111086

68. Saijo T, Takano A, Suhara T, Arakawa R, Okumura M, Ichimiya T et al (2010) Effect of electroconvulsive therapy on 5-HT1A receptor binding in patients with depression: a PET study with [11C]WAY 100635. Int J Neuropsychopharmacol 13(6):785–791

69. Lanzenberger R, Baldinger P, Hahn A, Ungersboeck J, Mitterhauser M, Winkler D et al (2013) Global decrease of serotonin-1A receptor binding after electroconvulsive therapy in major depression measured by PET. Mol Psychiatry 18(1):93–100

70. Yatham LN, Liddle PF, Lam RW, Zis AP, Stoessl AJ, Sossi V et al (2010) Effect of electroconvulsive therapy on brain 5-HT(2) receptors in major depression. Br J Psychiatry 196(6):474–479

71. Karlsson H, Hirvonen J, Kajander J, Markkula J, Rasi-Hakala H, Salminen JK et al (2010) Research letter: psychotherapy increases brain serotonin 5-HT1A receptors in patients with major depressive disorder. Psychol Med 40(3):523–528

72. Svenningsson P, Berg L, Matthews D, Ionescu DF, Richards EM, Niciu MJ et al (2014) Preliminary evidence that early reduction in p11 levels in natural killer cells and monocytes predicts the likelihood of antidepressant response to chronic citalopram. Mol Psychiatry 19(9): 962–964

73. Plaven-Sigray P, Matheson GJ, Collste K, Ashok AH, Coughlin JM, Howes OD et al (2018) Positron emission tomography studies of the glial cell marker translocator protein in patients with psychosis: a meta-analysis using individual participant data. Biol Psychiatry 84(6):433–442

Chapter 12

Magnetic Resonance Imaging as a Translational Research Tool for Major Depression

Chien-Han Lai

Abstract

Major depressive disorder (MDD, also known as major depression) is a major health issue in the modern society. The translational research in MDD can help us understand the pathophysiology of MDD. In the neuroimaging field, magnetic resonance imaging (MRI) will be a major tool for the translational research. In this chapter, several methods of MRI category will be addressed, such as the task functional MRI (T-FMRI), resting-state functional MRI (Rs-FMRI), diffusion tensor imaging (DTI), diffusion spectrum imaging (DSI), voxel-based morphometry (VBM), and magnetic resonance spectroscopy (MRS). The theory, preparation, and details of these MRI-related methods will be addressed in this review article. Basically, these methods can compensate each other in the representations of biological meaning. The functional characteristics of MRS, T-FMRI, and Rs-FMRI can enrich the functional fundamentals of structural characteristics of DTI and VBM. Therefore, theoretically, the "multimodal MRI" methods will be a future trend of neuroimaging research to help us make a sophisticated differentiation of pathophysiology subtype of MDD.

Key words Magnetic resonance imaging, Major depressive disorder, Fronto-limbic, Task functional MRI, Resting-state functional MRI, Diffusion tensor imaging, Voxel-based morphometry, Magnetic resonance spectroscopy

1 Introduction: Brain Pathophysiology of MDD

The burden of major depressive disorder (MDD, also known as major depression) is more significant in modern society. The burden includes the influences of life quality and mental and physical health [1, 2]. The significant social and occupational function impairments of MDD also contribute to a significant impact on the economics [3]. Since the impacts of MDD are so significant, therefore, the task to understand the pathophysiology model of MDD will be necessary. Among the potential targets of pathophysiology, the brain will be the first choice. For example, MDD patients usually have the impairments of cognition and emotion

Yong-Ku Kim and Meysam Amidfar (eds.), *Translational Research Methods for Major Depressive Disorder*, Neuromethods, vol. 179, https://doi.org/10.1007/978-1-0716-2083-0_12,
© The Author(s), under exclusive license to Springer Science+Business Media, LLC, part of Springer Nature 2022

regulation, which are related to the structure and function of the brain. The structural and functional pathophysiology of MDD might be related to the alterations of fronto-limbic network, which includes frontal regions and limbic regions, such as the anterior cingulate cortex (ACC), dorsolateral prefrontal cortex (DLPFC), amygdala, and hippocampus [4–12].

In the fronto-limbic model, ACC is a significant component for the pathophysiology of MDD. The ACC is responsible for many important functions, such as the attention, problem-solving, motivation, and decision-making, which are significantly altered in MDD [13–15]. The ACC has two components, which are the affective and cognitive subdivisions [14, 16]. The affective subdivision can modulate emotions via the connection with the limbic regions, such as the amygdala and brainstem [17]. The cognitive subdivision is responsible for the cognitive processing in MDD [16] and might have important roles in the pathophysiology of MDD [18, 19]. Therefore, the MDD pathophysiology biomarkers might include the ACC, and recent ketamine treatment study in MDD also focused on this fingerprint region [20]. Other regions of the frontal lobe, such as DLPFC [21–26], would be responsible for working memory and form the cognitive control network with the ACC [27] to inhibit the excessive negative emotions from the limbic regions, such as the hippocampus and amygdala [22, 28–33]. The hippocampus network might predict the antidepressant response to electroconvulsive therapy for MDD [34]. A recent meta-analysis of structural and functional MRI studies also showed the convergent findings in the ACC, amygdala, and hippocampus for the pathophysiology study in MDD [35]. From the viewpoint of cognitive control of emotional responses, the fronto-limbic network model might be most suitable to explain the symptoms of cognition impairments and the exaggerated negative emotion responses from the limbic regions in MDD.

1.1 MRI Methods to Study the Brain Pathophysiology of MDD

The neuroimaging method will be a crucial tool for the researchers to understand the brain. There are several pros and cons in the dimensions of spatial and temporal resolutions for the different kinds of neuroimaging methods. For example, the positron emission tomography has a great temporal resolution but with a limited spatial resolution. For the compensation of the limitation, magnetic resonance imaging (MRI) is usually applied simultaneously to localize the significant clusters revealed by the positron emission tomography. The MRI has a relative balanced profile of spatial resolution and temporal resolution, which makes this method more applicable and popular in the translational medicine study to understand the structure and function of the brain. The MRI is an important tool to investigate the structure and function of the brain [36]. The MRI contrast comes from the hydrogen ion fluctuations derived from the directional magnetic field or moment. The images result from the MRI contrast. MRI can survey the brain structure, such as the total brain, cortical thickness, gray matter (GM), white matter (WM), cerebrospinal fluid volume, and other structural targets. In addition, MRI can explore the brain function in many kinds of perspectives, such as magnetic resonance spectroscopy (MRS), functional MRI (FMRI), arterial spin labeling, and other methods. In this chapter, the focus will be put on the MRI methods in the translational medicine study usually applied in the pathophysiology of MDD. Basically it is divided into two major categories, which are functional MRI (FMRI) and structural MRI. For the FMRI category, it includes the task FMRI (T-FMRI) and resting-state FMRI (Rs-FMRI). The T-FMRI depends on the task-on-hand characteristics and accompanying changes in brain activities in the specific regions. The Rs-FMRI depends on the wandering mind, which means without any task on hand. This method will focus on the default brain activities without any stimulation by the task. Therefore, the major distinction point between T-FMRI and Rs-FMRI will be the existence and absence of task on hand. In addition, the pulse sequence and parameters of T-FMRI and Rs-FMRI will be different, which will be addressed in the sections below. The results of T-FMRI and Rs-FMRI can also be compared between two groups, such as healthy controls and MDD patients, which can detect the endophenotype of MDD.

For the structural MRI, GM and WM will be the major structures to study in MDD. For WM, the diffusion tensor imaging (DTI) and diffusion spectrum imaging (DSI) will be the major methods to understand the micro-integrity of WM tracts, such as the fasciculus in the brain [37]. The theory base of this method category is to suggest that the water diffusion property and water molecule can move freely within the fasciculus, which is known as Brownian motion. Therefore, the diffusion MRI-based techniques can obtain the neuroanatomical structures of white matter tract by analyzing the fluency of water diffusion within the tract. If the diffusion is fluent within the tract, it suggests that the WM tract will be with better micro-integrity [38]. In addition, the directions and pathways of WM fasciculus will be demonstrated by this category of method. Diffusion MRI and tractography techniques can be used to probe the architecture of WM tracts, which shows their significant influences in the investigation of the macro-, meso-, and microscopic organization of anatomical brain connectivity. The diffusion MRI methods also can compare the WM of two groups, such as the healthy controls and MDD patients. For the GM, the voxel-based morphometry will be the popular method to investigate the volume and density of GM. This method can compare the GM density and volume of two groups using a nonparametric t-test to obtain the significant differences of GM between two groups, such as the healthy controls and MDD subjects.

2   Materials

2.1   T-FMRI Theory and Preparation

The increased neural activity can produce two primary consequences, which both can be detected by MRI. They are the increases of local cerebral blood flow and changes in oxygenation concentration (blood oxygen level-dependent (BOLD) contrast). The changes in CBF can be observed noninvasively using the perfusion-weighted MRI by a method called arterial spin labeling (ASL) with an injected contrast agent. However, the disadvantages of increased acquisition time, reduced sensitivity, and increased sensitivity to motion will make ASL inferior to BOLD contrast method. Therefore, the application of ASL will be focused on obtaining quantitative measurements of baseline cerebral blood flow for studies modeling the neurobiological mechanisms of activation or calibration of vaso-reactivity, rather than in routine mapping of brain function.

The BOLD contrast was first demonstrated in rats and later in humans. It is the contrast that is used in nearly all T-FMRI experiments. BOLD contrast is derived from the change in magnetic field around the red blood cells, which will depend on the oxygenated state of the hemoglobin. The changes in the oxygenation of blood hemoglobin will produce the fluctuations of magnetic effects, which can be used for the production of BOLD signals to detect the brain activity [39]. The neuronal activation will be associated with reduced oxygen extraction, which leads to an increase in the ratio of oxy- to deoxyhemoglobin. The origin of signal changes in the BOLD FMRI depends on different magnetic properties of hemoglobin-carrying oxygen and deoxygenated hemoglobin. The BOLD signals can indirectly measure the parameters of the language, cognition, and memory using the above characteristics [36, 40].

The oxygenated hemoglobin is diamagnetic, which makes it magnetically indistinguishable from brain tissue. On the contrary, the deoxygenated hemoglobin has four unpaired electrons and is highly paramagnetic, which results in local gradients in magnetic field whose strength depends on the concentration of the deoxygenated hemoglobin. From the above theory, the deoxygenated hemoglobin is paramagnetic, and oxygenated hemoglobin is diamagnetic. Therefore, the vessels containing oxygenated arterial blood used to produce limited distortion to the magnetic field in the surrounding tissue. However, the capillaries and veins containing partially deoxygenated hemoglobin will distort the magnetic field. These endogenous gradients of hemoglobin then modulate the T2 and T2* relaxation times of intra- and extravascular blood through diffusion and intra-voxel dephasing, respectively. When the gradient refocused echo MRI pulse sequence is applied, the acquisition is sensitive to the T2* and T2 relaxation time. The inhomogeneities of microscopic field will be associated with the presence of deoxygenated hemoglobin, which causes destructive interference from signal within the tissue voxel and shortens the T2* relaxation time. Therefore, greater neuronal activity occurs in a region, and enhanced local blood flow will increase oxygenated hemoglobin. Then the T2* relaxation time becomes longer, and the MRI signal intensity increases relative to the baseline state [41].

At the 1.5 and 3 T magnetic field strength, the T2* contrast is predominant and is largest in venules. Because signals are generated preferentially in capillaries, therefore, the diffusion-weighted contrast of T2 relaxation becomes more important at higher field strength. The contrast of T2* will reveal the tissue with spin-echo acquisitions and provide greater spatial specificity. Since most FMRI is currently performed at 3 T or below, most BOLD T-FMRI utilizes primarily gradient refocused echo MRI methods because of the increased T2* contrast. Therefore, for a T-FMRI study, the preparation of a stable MRI machine with magnetic field strength at least 1.5 T (3 T will be better now) will be the basic requirement.

Another important preparation of T-FMRI will be the task on hand. The type of task will determine the study results. The tasks of T-FMRI in MDD will be divided into several categories, such as emotion-related, cognition-related, reward-related, and other categories according to the methods of tasks used in the study. Since different kinds of tasks will activate different kinds of regions, therefore, the T-FMRI would depend on the methods of tasks. These characteristics are the major distinction between T-FMRI and Rs-FMRI. The T-FMRI will be significantly influenced by task categories and underlying mechanisms of different tasks. BOLD contrast FMRI has a good spatial resolution for the localization of activated brain areas and neighboring regions, which are usually modulated by a few million neurons. In addition, the time lags of 1–2 s behind the stimulus for BOLD response will be needed for the vascular system to respond and reach peaks at 5 s after the stimulus. The continued same stimulus will downregulate the BOLD response, and a refractory period of just a few seconds is frequently inadequate for BOLD imaging after activation to fade, which depends on the characteristics of tasks. Therefore, if we want to eliminate noise in the recording, the stimulus will be repeated several times, which often takes a few minutes to complete the tasks [39]. In this way, the task performance time will be relatively long. In addition, the implementation of tasks will be tested in different intervals. The pulse sequence of 1.5 or 3 T machines for T-FMRI is used to be the fast method of echo-planar imaging, which will collect whole brain data in a brief duration, such as a few seconds. However, the spatial resolution is considerably lower (typically $4 \times 4 \times 4$ mm^3). In addition, the image intensity is also reduced in frontal and temporal regions with substantial distortions of brain shape, which is associated with the sensitivity of the echo-planar imaging to magnetic susceptibility differences. This distortion will be significant in the air sinus and tissue interfaces, which will be more severe with increasing field strength [42]. The typical examples of echo-planar imaging for T-FMRI is as follows: 2D echo-planar imaging (slice thickness 3 mm, matrix 64 × 64, TR (repetition time) 3000 ms, TE (echo time) 30 ms, flip angle 90°) and T2*-weighted echo-planar imaging sequence (TR = 2300 ms, TE = 45 ms, flip angle = 90°, 30 oblique transverse slices in ascending order, and matrix size = 3 × 3 × 3.5 mm) or gradient-echo echo-planar imaging (TR 2.6 s, TE 29 ms, FOV 256 mm × 256 mm, acquisition matrix 128 × 128, 36 axial slices, voxel dimensions 2 mm × 2 mm × 3 mm). The typical scanning time of T-FMRI will be longer than Rs-FMRI, which will depend on the task existence and characteristics and details of pulse sequence parameters.

2.2 Rs-FMRI Theory and Preparation

The FMRI techniques of the brain have been evolved to survey the Rs-FMRI signal changes in recent years, which can detect the resting-state brain activities [43–46]. The Rs-FMRI signals can reveal default brain activities, which will be related to the mechanism of astrocyte-neuron coupling. In addition, the recent advance of analysis method can formulate the representations of brain network, which will be similar with the complex social system networks. Therefore, the term "small world" has been added on the brain network. This network with a high efficiency (with a very low wiring and energy cost) is believed to have a high level of adaptation. In addition, the whole-network functioning will be related to the degree distribution of brain networks. The highly connected hubs will have better brain function [44].

One of the major differences between T-FMRI and Rs-FMRI will be that subjects receiving Rs-FMRI will not need to perform any explicit task [47]. The participating subjects are usually requested to close their eyes with relaxing manner and not sleep while scanning. In addition, the participating subjects will be instructed to move as little as possible and stay fully awake while scanning.

The acquisition time of Rs-FMRI data is typically around 5–15 min, which has the image intensity reflecting the variations in local neural activity. The variations will lead to the local blood flow and changes in oxygenation. To achieve this sensitivity and acquire the data rapidly, it is also common to utilize echo-planar imaging. A temporal resolution of 2–3 s with a spatial resolution of 3–5 mm will be achieved by standard acquisitions working in 3 T MRI machines. Apart from the typical multi-sectional echo-planar imaging (TE 30–40 ms, TR 2000–3000 ms, voxel size 3–4 mm, flip angle 80–90°), a faster acquisition method, "multiband accelerated echo-planar imaging," can acquire multiple slices simultaneously with a faster speed and shorter duration of scanning (TE 30–40 ms, TR 500–1500 ms, voxel size 2–3 mm, flip angle 50–70°). It will enable major improvements in spatial and temporal resolution with acquiring data with 2 mm spatial resolution in less than a second. In addition, the higher temporal resolution of the FMRI data can improve overall statistical sensitivity and increase the information content of the data, such as the richness of the neural dynamics. However, the sluggish response of the brain's hemodynamics to neural activity will ultimately place a plateau on the improvements in temporal resolution [48].

Apart from the above stationary Rs-FMRI, another kind of Rs-FMRI needing shorter acquisition time can also detect the non-stationary state of the brain, which can provide more data of the brain and can establish the diseased brain-specific pattern of alterations for specific neuropsychiatric illnesses. This kind of Rs-FMRI is known as "dynamic Rs-FMRI" [47]. It can detect the neural communication and the subsequent local metabolism from the microscopic viewpoint, which can provide the data of global metabolic fluctuations. Therefore, its more sophisticated characteristics and shorter acquisition time will provide the advantages for researchers to obtain more data of neuro-metabolic changes and cerebrovascular activities in the brain. The biomarker and physiology of specific neuropsychiatric illness, such as MDD, can be more precisely localized. In addition, the diagnosis and drug discovery will also be beneficial from this new technique. However, if using correlation alone, the dynamic Rs-FMRI is still difficult to delineate the cause and effect. The hypothesis-driven study is still needed to explain the findings of dynamic Rs-FMRI.

The commonly used pulse sequences of Rs-FMRI will be as follows: echo-planar imaging sequence (with greater slice thickness) acquired in 20 axial slices (TR = 2000 ms; TE = 40 ms; flip angle = 90°; field of view = 24 cm; 5 mm thickness and 1 mm gap; the sequence duration was 300 s for each subject; 150 time points were acquired; voxel dimension $64 \times 64 \times 20$); a T2*-weighted gradient-echo echo-planar imaging sequence (TR = 1700 ms; TE = 33 ms; matrix size = 64×64; FOV = 230 mm; flip angle = 90°; slice thickness = 4 mm, no gap; in-plane pixel size = 3.59 mm $\times$ 3.59 mm; and axial slices = 32 or repeat time = 2000 ms, echo time = 30 ms, field of view (FOV) = 240 mm $\times$ 240 mm, flip angle = 90°, matrix = 64×64, voxel size = 3 mm $\times$ 3 mm $\times$ 3 mm, axial slice = 36, slice thickness = 4 mm, scanning interval = 0 mm, and 250 time points); or a T2*-weighted echo-planar imaging sequence [repetition time (TR) = 2000 ms; time echo (TE) = 30 ms; flip angle = 90°; field of view (FOV) = 220 mm; matrix = 64×64, 32 slices; 3 mm slice thickness; voxel size = 3.4 mm $\times$ 3.4 mm $\times$ 3 mm]. The scanning time depends on the slice thickness, TR, and TE. If the researchers want to detect the more sophisticated region of default mode activity of the brain, the thinner slice thickness will be more appropriate. However, the disadvantage will be a longer acquisition time.

2.3 DTI Theory and Preparation

The diffusion of water, which is caused by random thermal fluctuations, occurs in biological tissues inside, outside, around, and through cellular structures. The interactions between cellular membranes and subcelluar organelles will also determine the diffusion efficiency. For the cellular membranes blocking the diffusion of water, the water diffusion will need to take more tortuous paths, which will lead to the decreases in the mean squared displacement. Therefore, the hindering condition, such as cellular swelling or increased cellular density, will increase the diffusion tortuosity and corresponding apparent diffusivity. Apart from the hindered status, the restricted diffusion will also decrease the apparent diffusivity and increase the diffusion time. The diffusion tensor can be applied for anisotropic diffusion behavior inside the brain, especially WM tracts. In this model, the diffusion is described by a multivariate normal distribution, and the diffusion tensor can be placed in a covariance matrix. The diagonalization of the diffusion tensor can produce eigenvalues and corresponding eigenvectors of the diffusion tensor, which describe the directions and apparent diffusivities. The diffusion tensor is anisotropic when the eigenvalues are significantly different in magnitude, which will be altered by the local tissue microstructure and related micro-integrity within WM tracts. Therefore, the diffusion tensor is a sensitive probe for characterizing both normal and abnormal tissue microstructures. This directional relationship of major diffusion eigenvector is the basis for estimating the trajectories of WM pathways with tractography algorithms [49]. The DTI technique of MRI can derive the white matter tract [37]. Diffusion MRI and tractography techniques can be used to probe the architecture of both white and gray matter. These techniques showed their significant influences in the investigation of the macro-, meso-, and microscopic organization of brain connectivity and anatomy. Diffusion MRI-based techniques can therefore obtain the neuroanatomical structures of white matter tract by analyzing the water diffusion property and water molecule free movement (Brownian motion) within the fasciculus [38].

The most common pulse sequence of DTI (one kind of diffusion-weighted image) is the pulsed-gradient spin-echo pulse sequence with a single-shot, echo-planar imaging readout. A pair of large-gradient pulses placed on both sides of the 180° refocusing pulse can perform the simplest configuration of this pulse sequence. The first gradient pulse can dephase the magnetization across the voxel, and the second pulse can rephase the magnetization. There are a lot of choices of diffusion tensor encoding strategies, with six or more encoding directions. The optimum range the b-value (diffusion weighting) for the brain is between 700 and 1300 s/mm^2. The scanning duration, the availability of encoding direction sets, and the maximum number of images that can be obtained in a series can determine the selection of the number of encoding directions. The sum of the diagonal elements can measure the magnitude of diffusion. The mean diffusivity (apparent diffusion coefficient) is the sum of diagonal elements divided by 3, which represents the average of the eigenvalues. From the eigenvalues and mean diffusivity, the fractional anisotropy can be calculated. In addition, the radial diffusivity and axial diffusivity can support the researchers to determine the WM micro-integrity environment of fasciculus [49]. In advance, the DT images can be used to establish, configure, and visualize the WM fasciculus pathway in a method called fiber tractography, which can demonstrate the micro-integrity and orientation of WM fibers in vivo. It can also be applied in many kinds of neuropsychiatric illnesses [50] and in the research of development of the brain [51] and the supplementation of structure data for the FMRI [52]. The application of DTI in MDD study is also common [53–57]. The typical example of pulse sequence parameters of DTI will be as follows:

- A single-shot, twice-refocused, spin-echo echo-planar imaging pulse sequence DTI with 30 diffusion-sensitized gradient directions with the following parameters was performed at first visit: TR, 7900 ms; TE, 79 ms; number of excitations = 3; directions = 30; number of acquisitions in axial orientation, 70; FOV, 256 mm × 256 mm; slice thickness = 2 mm; matrix = 128 × 128; and b-value = 0 and 900 s/mm^2. One non-diffusion-weighted (b0) image was also acquired.

- A spin-echo single-shot EPI sequence with the following parameters: TR = 7646 ms; TE = 60 ms; 90° flip angle; FOV 224 × 224 mm^2, yielding 60 slices, no gap; and voxel size 2 × 2 × 2 mm^3 (one b0 image was acquired, and diffusion gradients were applied in 21 non-collinear directions (b = 1000 s/mm^2)) or TE = 88 ms; TR = 9600 ms; acquisition matrix = 128 × 128; FOV, 256 × 256; and slice thickness = 2 mm with no gap.

2.4 VBM Theory and Preparation

The VBM method can help the researcher investigate the status of GM and WM. For most applications, VBM can detect the differences of GM volume or density between two groups. The subjects need to receive the structural image scanning, such as the pulse sequence of three-dimensional fast spoiled gradient-echo recovery. Basically the researcher will prepare two groups of subjects, such as the patients versus controls. After brain skull extraction in the image and image co-registration steps, the VBM can obtain the GM and WM volume or density from the segmentation step [58]. The voxels of GM volume of two groups will be input as the subsequent calculation parameters. Then the nonparametric independent two-sample test was performed for the group comparison to identify the significant differences in the GM or WM volume or density between two groups, which will be translated as the significant alterations of structural endophenotype for specific kind of illness [59]. The basic group will be the comparison between two groups. However, if the researchers want to perform the comparisons between three groups, the ANOVA test can be applied to complete this kind of comparison.

The common pulse sequence parameters for the VBM methods will be as follows (at least two groups will receive the same scanning parameters): the structural MR imaging brain scans were obtained using the 1.5, 3, or 7 T MRI machines with three-dimensional fast spoiled gradient-echo recovery (3D-FSPGR) T1W1 (TR, 25.30 ms; TE, 3.03 ms; slice thickness = 1 mm (no gap); 192 slices; matrix = 224 × 256; field of view, 256 mm; number of excitation = 1) or the 3D T1-weighted magnetization-prepared rapid acquisition gradient echo (MPRAGE) structural image (repetition time, 8.70 ms; echo time, 3.1 ms; 8° flip angle; the field of view, 256 × 256 × 180 mm; and voxel size, 0.7 × 0.7 × 0.7 mm).

2.5 MRS Theory and Preparation

The MRS method can noninvasively detect and quantify the neural metabolites, such as the N-acetyl aspartate, choline creatine, choline, glutamate, myoinositol, lactate, and GABA, in a specific region of the brain. The key output of MRS is a magnetic resonance spectrum which arises from nuclei in the atoms, such as the proton, of the individual molecules of the tissue sample. An intrinsic magnetic moment of the nucleus will generate an MRS signal. It will be influenced by the external magnetic fields. The voxel size required to obtain reasonable signal will be negatively correlated with the concentration of the measured metabolites. A MRS typical figure will have two axes, which are the amplitude on the vertical axis and the temporal frequency on the horizontal axis. The MRS can detect several types of interactions, such as chemical shift, J-coupling, and the parameters of spectral peaks. For the chemical shift, it represents the difference between the Larmor or gyromagnetic frequency of the nuclei and the reference molecule. In addition, the chemical shift will be influenced by also the dynamical properties of the molecules and bonds. The typical chemical shifts of MRS will show all the observable nuclei in the molecules of the sample. The J-coupling means that the electron orbitals of the molecule generate the linking or coupling between adjacent nuclei. The presentations of J-coupling are complex and will produce complicated patterns of spectral peaks. The J-coupling effects of clinical MRS are observed in the coupling between protons on adjacent carbon atoms via the carbon-carbon bond. The glutamate, lactate, glutamine, myoinositol, GABA, and glutathione usually have J-coupling effects, especially for the GABA and glutathione in the application of MRS to detect the reliable signals. The spectral peak has different patterns in MRS. For the single peak (singlet), it represents the peak signal of nuclei that are not coupled with adjacent nuclei, such as the *N*-acetyl-aspartate. For the double peak (doublet), the nuclei with J-coupling effects with other nuclei will lead to the splitting of peak into two peaks, such as the lactate. The relaxation time of T1 and T2 in each metabolite of MRS will be different, which can help researchers determine which signal belongs to which metabolite or nuclei [60].

The typical examples of imaging methods of MRS are as follows: single-voxel spectroscopy and chemical shift imaging [61]. The single-voxel spectroscopy is the popular method to get the metabolite concentration in a single voxel. The single-voxel spectroscopy pulse sequence can help detect the brain volume with signals during data acquisition. In addition, the radiofrequency and field gradient pulses (saturation pulses) are applied in the pulse sequence to ensure that magnetization outside of this defined voxel is not disturbing the targeted signals of targeted metabolites. For this type of pulse sequence, the point-resolved spectroscopic sequence is the most commonly used method in MRS. In this kind of MRS images, it includes the NAA peak at 2.01 ppm, creatine with singlet peaks at 3.92 and 3.03 ppm individually, and choline singlet peak at 3.21 ppm. The typical measured voxel will be $2 \times 2 \times 2$ in size (around 8 cm^3). However, for the metabolites with multiple peaks and J-coupling effects, such as lactate and GABA, the localized voxel will need a larger size. The relatively poor spatial resolution of measured voxel size will be the limitation of MRS [60, 62].

Apart from the single-voxel spectroscopy, another kind of MRS method, the chemical shift imaging, can cover larger brain regions and provide the spectra throughout the selected plane or volume. It is not like the signle-voxel spectroscopy used the radiofrequency to produce the "excitation voxel." The chemical shift imaging is subdivided into multiple "resolution voxels" by the phase and spatial encoding process. This method can use the pulse sequence to acquire a spectrum at each voxel within a two- or three-dimensional grid. The primary advantage of chemical shift imaging over single-voxel spectroscopy is spatial resolution. The chemical shift imaging acquisition can measure metabolite concentrations across a range of spatial locations. Since the scanning time of chemical shift imaging will be too long, the modified pulse sequence is developed to decrease the scanning duration. The modified sequence is the proton echo-planar spectroscopic imaging pulse sequence, which interleaves the acquisition of the echo for spectral estimation with spatial encoding. This technique replaces one direction of phase encoding with frequency encoding. This pulse sequence was made possible by the recent development of very fast gradients in the new MRI machine which can perform frequency encoding in that same time frame [63, 64].

The parameters of typical pulse sequences in MRS methods are as follows: MEGA-PRESS (repetition time (TR) = 2 s; echo time (TE) = 68 ms; 320 averages; acquisition bandwidth = 2000 Hz; total acquisition time 10 min and 56 s); MOIST water suppression and Philips pencil-beam (PB-auto) shimming; stimulated echo acquisition mode (TE = 7.8 ms, 128 averages, TR = 2 s); and semi-localized by adiabatic selective refocusing (TE = 28 ms, 16 averages, TR = 5 s). Voxels (2 cm × 2 cm × 2 cm) were located in the left frontal and left occipital lobe. Non-water-suppressed spectra were obtained for quantification (carrier frequency was set to the chemical shift of H_2O, acquisition time = 10 s). Prior to the MRS exams, second order B_0 shimming was applied using the FASTERMAP algorithm at the voxel of interest.

3 Methods

3.1 T-FMRI Methods

The most popular tool for T-FMRI will be the statistical parametric mapping (SPM, https://www.fil.ion.ucl.ac.uk/spm/). When the T-FMR images have been acquired, the time series data will be processed to obtain maps of brain activation. Since the BOLD contrast is small, therefore, the noise will influence the analyzed results, which will produce the false-positive and false-negative results. The noise results from many situations, such as the motion of the head, the heat sources in the subject, and the cardiac beat and respiratory sound. Since the noise can sometimes be larger than the signal of interest, T-FMRI analyses compare the signal difference between the states using a statistical test. The statistical tests for activation include several theories, such as a general linear model, cross-correlation with a modeled regressor, or data-driven approaches (independent component analysis). The experimental design of interest and "nuisance regressors" of no interest such as signal drift, motion, and noise reflected in global or white matter signals will be analyzed to minimize the impacts from these factors. The activation testing is preceded by the following preprocessing steps:

First, time-slice correction step will be performed to eliminate differences between the time of acquisition of each slice in the volume. Second, the motion co-registration will be performed to detect the affine head motion. Then the time series of volumes will be resampled to register each time frame to a reference frame. Third, a low-pass and/or high-pass temporal filtering will be used to remove spectral components of no interest, such as the physiological noise from breathing and cardiovascular function. Fourth, a spatial smoothing step will be applied to improve the signal to noise ratio. In addition, the normality of the noise distribution will be improved after smoothing steps. Fifth, a pre-whitening step will be used to correct for autocorrelation in the time series. The T-FMRI analysis will also depend on the task characteristics. For example, a kind of task will activate specific brain regions in all subjects, such as the visual task. The activation of occipital lobe should be modified to decrease the amplification of activation in this region for the explanation of pathophysiology in MDD. Therefore, in addition to the above five major steps of T-FMRI analysis, the characteristics of tasks should be taken into consideration for the interpretations of the results.

3.2 Rs-FMRI Methods

There are many kinds of measurements derived from Rs-FMRI. These parameters include the fractional amplitude of regional homogeneity [65], low-frequency oscillations [66], functional connectivity, and voxel-mirrored homotopic connectivity [67].

1. Regional homogeneity is a measure of similarities of several time series from signals of Rs-FMRI. It can measure the synchronizing abilities of neuronal activations in a specific region to understand the complexity of the regional stability of a specific region [65].

   The analysis of regional homogeneity is calculated as Kendall's coefficient to measure the similarity of ranked time series of a given voxel with its nearest 26 neighboring voxels in a voxel-wise way. Kendall's coefficient value can be calculated to this voxel. Then the individual Kendall's coefficient map can be obtained for each subject. Then the individual ReHo maps are generated by assigning each voxel a value corresponding to Kendall's coefficient value of its time series around the nearest voxels. So the ReHo analysis is a voxel-wise analysis for Kendall's coefficient of the time series of a given voxel with 26 nearest neighboring voxels. Kendall's coefficient formula (where W is the KCC among given voxels, ranging from 0 to 1) is shown as follows:

$$W = \frac{\sum (R_i)^2 - n(\overline{R})^2}{\frac{1}{12} K^2 (n^3 - n)}$$

$$\overline{R} = \frac{(n+1) \times K}{2}$$

   R_i is the sum rank of the ith time point, is the mean of the R_i, K is the number of time series within a measured cluster ($K = 27$, one given voxel plus the number of its neighbors), and n is the number of ranks. Then the intracranial voxels are extracted to make a mask, which can assure the matching of normalization steps and removed the noise and non-brain tissue on the regional homogeneity map. Then each regional homogeneity map is divided by its own mean regional homogeneity within the mask for the standardization purpose. The modulated regional homogeneity map image files are then smoothed by FWHM $4 \times 4 \times 4$ Gaussian kernel to reduce the noise and residual differences in the gyral anatomy. Then the outputted regional homogeneity between two groups or three groups can be compared using the t test or ANOVA test to find the differences of regional homogeneity between groups of subjects.

2. The low-frequency oscillations of the human brain can reveal the baseline brain activities [68]. Rs-FMRI studies on low-frequency oscillations will detect the neuronal fluctuations [68, 69]. The low-frequency oscillations can weaken the biases of physiological noise and provide more accurate measures of impacts of amplitude of low-frequency fluctuations within a specific band of frequency [66]. The calculation will be as follows: for a given voxel, the filtered task-residual and resting-state time series are transformed into the frequency domain using the fast Fourier transform. Since the power is proportional to [amplitude] square at a given frequency, the power spectrum obtained by fast Fourier transform is square rooted to obtain the amplitude. The average squared root was termed ALFF at a given voxel. A ratio of the amplitude averaged across 0.01–0.08 Hz to that of the entire frequency range (0–0.25 Hz) is computed at each voxel to obtain the fALFF, creating an amplitude map for the whole brain, which is then normalized by the following formula:

$$\text{Normalized fALFF} = \frac{(\text{fALFF} - \text{global mean fALFF})}{\text{Standard deviation of global mean power spectrum density}}$$

The sum of amplitudes across 0.01–0.08 Hz is then divided by the amplitude across the entire frequency range. The amplitude of low-frequency fluctuations of each voxel is divided by the individual global mean of amplitude of low-frequency fluctuations within a brain mask. The individual data will be converted into Z-scores by subtracting the global mean and dividing by the global standard deviation. The outputted fractional amplitude of low-frequency fluctuations between two groups or three groups can be compared using the t test or ANOVA test to find the differences of regional homogeneity between groups of subjects.

3. Functional connectivity is the most popular indicator derived from the Rs-FMRI. It can measure the connectivity between two spots or two regions using the Pearson correlation method. However, the researchers will have more interests in the whole picture of functional connectivity, such as the small world connectivity or functional connectome. Some toolboxes, such as GRETNA toolbox (https://www.nitrc.org/projects/gretna), can calculate the connectivity matrix for the preprocessed Rs-FMRI data of each subject. GRETNA toolbox has

been designed for the graph theoretical network analysis of Rs-FMRI data. The connectivity matrix can be used to calculate the functional connectome, which can identify significant differences between patients and controls for pairwise associations using whole brain approach. It can incorporate graph model to identify the pairwise association, such as the connection or link, between different pairs of nodes [70]. The processing steps were as follows: the design matrix is set at first to define the group and number for subjects of two groups. Then the connectivity matrix file is put for each subject as the input data. From the probability point, the $N(N-1)/2$ unique pairwise associations would be possible for an $N \times N$ connectivity matrix. For each pairwise association, the test statistic of interest can be calculated independently using the values stored in each subject's connectivity matrix, such as Fisher's r-to-z transform for a correlation-based measure of association to ensure normality. The AAL-90 nodes and labels are used to define and locate the significant nodes within the significant subnetwork in a graph model. The reason to choose the AAL-90 is due to its relatively accurate localization of coordinates, and similar application in network-based statistics methodology was also mentioned in previous report [71]. A nonparametric test is used for a breadth-first search. This step is repeated for multiple permutations to estimate the null distribution. The random permutations were generated independently, and the group to which each subject belonged will be randomly exchanged for each permutation. The number or size of links the potential nodes comprised according to the threshold (corrected $p < 0.05$). Permutation testing will be used to ascribe a p-value controlled for the FWE to each connected component based on its size. For each permutation, the test statistic of interest is recalculated, after which the same threshold is applied to define a set of suprathreshold links. The test detected the potential connected structures of suprathreshold links, which could demonstrate the topological extent of all significant structures. The maximal component size in the set of suprathreshold links is derived from each of multiple permutations that is then determined and stored; thereby, an empirical estimate of the null distribution of maximal component size can be obtained. Finally, the p-value (0.05) of an observed component of size k is estimated by finding the total number of permutations to identify maximal component size.

4. The voxel-mirrored homotopic connectivity method can measure the synchrony of spontaneous brain activities between geometrically corresponding regions in each hemisphere [72]. The voxel-mirrored homotopic connectivity assumes symmetric morphology between hemispheres. The

preprocessed images are transformed by applying that Rs-FMR images to anatomical data. Then an image is averaged with its left-right mirrored version to create identical mirrored hemispheres. Therefore, the Rs-FMRI data is transformed to fit the new symmetrical anatomical image. The voxel-mirrored homotopic connectivity is computed as the resting-state functional connectivity between any pair of symmetric inter-hemispheric voxels. For each subject, the Pearson correlation coefficient between the residual time series of each voxel and its symmetrical inter-hemispheric counterpart is performed. Correlation values are then Fisher z-transformed to improve normality. The resultant values constituted the voxel-mirrored homotopic connectivity and are used for the group analyses. In addition, since voxel-mirrored homotopic connectivity results were bilaterally identical or symmetric, a unilateral hemisphere mask will be used to confirm that the voxel-mirrored homotopic connectivity is indeed found in one hemisphere (one-sided results).

3.3   DTI Methods

The DTI analysis can be performed by the FDT (FMRIB's Diffusion Toolbox v2.0) function that is implemented in the FSL (FMRIB Software Library) [73, 74], which is developed by the Oxford Centre for Functional MRI of the Brain (FMRIB), London, UK. The merged DT images are preprocessed by the step of eddy current correction to reduce the stretches and shears in diffusion-weighted images, to correct the motion between images, and to adjust the gradient directions for rotations [75]. A brain extraction tool [76] can also be used to remove the non-brain tissue of the b0 image to obtain the nodif brain mask for the following DTIFIT process to fit a diffusion tensor model at each voxel. Fractional anisotropy, eigenvector, and eigenvalue maps were computed by the above procedure with the b vector and b-value of gradient directions.

Fractional anisotropy images are then visually inspected for the orientation and image quality. Then, the fractional anisotropy volumes are skeletonized and transformed into common space [77], and all of the fractional anisotropy volumes are warped to the template by FMRIB's nonlinear image registration. The mean fractional anisotropy volume of all of the individuals is thinned to create a mean fractional anisotropy skeleton that represented the centers of all of the WM tracts. This step can provide the background WM tract map for the presentations of the TBSS (tract-based spatial statistics) results. The mean fractional anisotropy skeleton is thresholded and binarized at 0.2. Individual fractional anisotropy values are warped onto the mean fractional anisotropy skeleton.

After the above steps, a permutation-based nonparametric inference (randomise function of FSL, http://www.fmrib.ox.ac.uk/fsl/randomise, version 2.1) will be used to perform the voxel-wise analyses for the fractional anisotropy skeletons to compare two groups' fractional anisotropy, such as MDD vs. controls. Nonparametric computations will be used due to the relatively small sample size, and the method is comparable to multiple comparisons in random field theory [78]. For the main purpose of group comparisons, a factor analysis with group as the main random factor over all subjects will be used. The randomise function applies the general linear model for permutations, and global brain volume, age, gender, and duration of illness will be included as covariates to control possible confounding factors. The family-wise error threshold will be used to obtain results for continuous random processes to find p-values. Statistical image after multiple comparisons is explored to find regions of FA deficits. Statistical threshold is set as family-wise error p-value <0.05 (corrected for multiple comparisons).

A correlation between the scores of clinical rating scales (such as Hamilton Rating Scales for Depression) and fractional anisotropy in the general linear model considering the voxel-wise matrix with age and gender as covariates in design matrix of FSL correlation analysis would be performed (threshold, corrected $p < 0.05$, multiple comparisons). This step could help us confirm which white matter tract correlates with depressive symptoms and which regions may be important in the physiopathology of MDD.

3.4 VBM Methods

VBM methods have a lot of choices, such as DARTEL (Diffeomorphic Anatomical Registration Through Exponentiated Lie Algebra) [79] and FSLVBM. The following methods will be based on the FSLVBM steps. Structural MR images are preprocessed with FSLVBM (http://www.fmrib.ox.ac.uk/fsl/fslvbm/, version 1.1) function of FSL (FMRIB Software Library, version 4.1.1) to compare the differences of GMV between patients and healthy controls.

First, brain skull or other non-brain tissue is removed to discard the confounding factors of non-brain tissues. Second, FSL Automated Segmentation Tool v4 performs the tissue-specific segmentation to produce partial volume images of gray matter [80]. The affine registered images are averaged and concatenated to establish a 4D self-template of gray matter from all the subjects in this study. Third, the brain would be nonlinearly registered to the self-template, and the quality of registration will be visually inspected. All the Jacobian modulated and segmented gray matter images are concatenated into a 4D multi-subject concatenated image. The modulated 4D image will be smoothed by Gaussian kernels (sigma 3 mm in FSLVBM protocol, which approximately equals to full width at half maximum 7.5 mm) [81]. In addition, a gray matter mask is created by unsmoothed segmentations and unmodulated normalized segmentations. Smoothing 4D modulated image and gray matter mask will be necessary for the following step of permutations.

The randomise function of FSL (http://www.fmrib.ox.ac.uk/fsl/randomise) is performed with gray matter mask and 4D image by threshold-free cluster enhancement method to compare two groups' GMV. Nonparametric computations are used due to the relatively small sample size, and the method is comparable to multiple comparisons in random field theory [78]. For the main purpose of group comparisons, the factor analysis with group as the main random factor over all subjects will be used. The randomise function used general linear model for permutations, and we included global brain volume, age, gender, and duration of illness as covariates to control possible confounding factors. Threshold-free cluster enhancement is a new method for finding clusters in data without having to define clusters in a binary way, which can avoid the bias related to the arbitrary threshold. Cluster-like structures were enhanced, but the image remained fundamentally voxel-wise. This procedure would produce test statistic images and sets of p-value images. The neighborhood connectivity parameters have been optimized and should be left unchanged to avoid edge effects of the border between gray matter and white matter. We used family-wise error to obtain results for continuous random processes to find p-values. Statistical image after multiple comparisons was explored to find regions of gray matter volume deficits. Statistical threshold was set as FWE p-value <0.05 (corrected for multiple comparisons).

A correlation between the scores of clinical rating scales and gray matter volume in the general linear model considering the voxel-wise matrix with age and gender as covariates in design matrix of FSL correlation analysis would be performed (threshold, corrected $p < 0.05$, multiple comparisons).

3.5 MRS Methods

The MRS can use the quantification and relative quantification methods to measure the neural metabolites. The quantification method can use several software, such as Advanced Method for Accurate, Robust, and Efficient Spectral fitting program incorporated into the Magnetic Resonance User Interface software package and Linear Combination Model software and the quantitation based on Quantum ESTimation program incorporated into the Magnetic Resonance User Interface software package. The peak integration can be started when the frequency range containing the peak of interest will be defined, such as 2.01 ± 0.15 ppm for the N-acetyl-aspartate peak. Then the area under the curve can be calculated after summing the values across that frequency range and subtracting an estimate of the baseline. This simple method is more suitable for longer TE sequences, not suitable for overlapping peaks.

Another method, peak fitting method, can provide more accurate quantification of the metabolite signal intensity. Under this method, the important peak is fitted to a model peak shape defined mathematically, and an optimal combination of peak integral values is iteratively calculated for the entire set of peaks. This method is most useful when combined with prior knowledge, such as the ratios of amplitude, frequency relationships, and scalar coupling, about the metabolite signals in the spectrum. The information for each metabolite can be measured using iterative peak fitting calculations for the MRS data. Since the scaling factor of MRS is unknown and variable, the raw signal intensity values for each measurement are unable to be directly converted to absolute concentration values. Therefore, the relative quantification of MRS metabolites will be more applicable. The relative quantification in brain 1H-MRS experiments can calculate the ratio normalization using an endogenous metabolite. The signal intensity values from each metabolite of interest in a given 1H-MRS acquisition will be normalized to the signal value of a reference metabolite in the same voxel. The total creatine (creatine plus phosphocreatine) is most commonly used as the reference metabolite due to its strong signal characteristics and stable with less variability. The ratio normalization using the creatine signal value is as the reference metabolite value. It can decrease the impacts from unknown characteristics of scaling factor. In addition, since the proportion of cerebrospinal fluid within the voxel affects the measured signal similarly for each parenchymal metabolite, the creatine normalization procedure is also a useful way to correct for variation due to this partial volume effect. Since total creatine values appear to be stable over short-term repeated measurements, dynamic changes in a creatine-normalized metabolite are unlikely to be due to changes in creatine. However, the false positives and false negatives of this method still need to be considered in the analysis. For the absolute concentration of metabolites, the calculation can be based on the inner water and phantom calibration. However, the absolute concentration calculation might be altered under its basis of theory.

4   Conclusion

From the above content, we can understand there are rich analyzing methods and parameters for the MRI research in MDD. For the structural perspectives, the researchers can perform the VBM and DTI to investigate the gray matter volume or density and the micro-integrity of white matter tracts in MDD. For the functional perspectives, the T-FMRI and Rs-FMRI can help the researchers understand the task-related brain activities and resting-state brain activities in MDD. The task-related brain network, which can be derived from the T-FMRI, might be associated with cognitive and emotional functions in MDD. The resting-state brain network, which can be obtained from the Rs-FMRI, might represent the baseline characteristics of default brain function in MDD. For the neural metabolites, the MRS can provide a pathway to understand the quantification of the neural metabolites. For the complete understanding of the pathophysiology of MDD, the comprehensive viewpoints of structural and functional perspectives will be the appropriate way to explore. Therefore, the functional characteristics of T-FMRI, Rs-FMRI, and MRS can be used as the compensatory methods to combine with structural findings revealed by VBM and DTI in MDD subjects. Therefore, the "multimodal MRI" analysis will be the future trend for researcher to understand the pathophysiology of MDD in a whole-picture viewpoint.

References

1. Murray CJ, Vos T, Lozano R, Naghavi M, Flaxman AD, Michaud C, Ezzati M, Shibuya K, Salomon JA, Abdalla S, Aboyans V, Abraham J, Ackerman I, Aggarwal R, Ahn SY, Ali MK, Alvarado M, Anderson HR, Anderson LM, Andrews KG, Atkinson C, Baddour LM, Bahalim AN, Barker-Collo S, Barrero LH, Bartels DH, Basanez MG, Baxter A, Bell ML, Benjamin EJ, Bennett D, Bernabe E, Bhalla K, Bhandari B, Bikbov B, Bin Abdulhak A, Birbeck G, Black JA,https://doi.org/10.1016/S0140-6736(12)61689-4

2. Greenberg PE, Fournier AA, Sisitsky T, Pike CT, Kessler RC (2015) The economic burden of adults with major depressive disorder in the United States (2005 and 2010). J Clin Psychiatry 76(2):155–162. https://doi.org/10.4088/JCP.14m09298

3. Souery D, Oswald P, Massat I, Bailer U, Bollen J, Demyttenaere K, Kasper S, Lecrubier Y, Montgomery S, Serretti A, Zohar J, Mendlewicz J (2007) Clinical factors associated with treatment resistance in major depressive disorder: results from a European multicenter study. J Clin Psychiatry 68(7): 1062–1070

4. Alexopoulos GS, Hoptman MJ, Kanellopoulos D, Murphy CF, Lim KO, Gunning FM (2012) Functional connectivity in the cognitive control network and the default mode network in late-life depression. J Affect Disord 139(1):56–65. https://doi.org/10.1016/j.jad.2011.12.002

5. van Tol MJ, van der Wee NJ, van den Heuvel OA, Nielen MM, Demenescu LR, Aleman A, Renken R, van Buchem MA, Zitman FG, Veltman DJ (2010) Regional brain volume in depression and anxiety disorders. Arch Gen Psychiatry 67(10):1002–1011. https://doi.org/10.1001/archgenpsychiatry.2010.121

6. Lai CH, Hsu YY, Wu YT (2010) First episode drug-naive major depressive disorder with panic disorder: gray matter deficits in limbic and default network structures. Eur Neuropsychopharmacol 20(10):676–682. https://doi.org/10.1016/j.euroneuro.2010.06.002

7. de Kwaasteniet B, Ruhe E, Caan M, Rive M, Olabarriaga S, Groefsema M, Heesink L, van Wingen G, Denys D (2013) Relation between structural and functional connectivity in major depressive disorder. Biol Psychiatry 74(1): 40–47. https://doi.org/10.1016/j.biopsych.2012.12.024

8. Sheline YI, Barch DM, Price JL, Rundle MM, Vaishnavi SN, Snyder AZ, Mintun MA, Wang S, Coalson RS, Raichle ME (2009) The default mode network and self-referential processes in depression. Proc Natl Acad Sci U S A 106(6):1942–1947. https://doi.org/10.1073/pnas.0812686106

9. Gorka SM, Young CB, Klumpp H, Kennedy AE, Francis J, Ajilore O, Langenecker SA, Shankman SA, Craske MG, Stein MB, Phan KL (2019) Emotion-based brain mechanisms and predictors for SSRI and CBT treatment of anxiety and depression: a randomized trial. Neuropsychopharmacology 44(9): 1639–1648. https://doi.org/10.1038/s41386-019-0407-7

10. Connolly CG, Ho TC, Blom EH, LeWinn KZ, Sacchet MD, Tymofiyeva O, Simmons AN, Yang TT (2017) Resting-state functional connectivity of the amygdala and longitudinal changes in depression severity in adolescent depression. J Affect Disord 207:86–94. https://doi.org/10.1016/j.jad.2016.09.026

11. Smoski MJ, Keng SL, Ji JL, Moore T, Minkel J, Dichter GS (2015) Neural indicators of emotion regulation via acceptance vs reappraisal in remitted major depressive disorder. Soc Cogn Affect Neurosci 10(9):1187–1194. https://doi.org/10.1093/scan/nsv003

12. Groenewold NA, Roest AM, Renken RJ, Opmeer EM, Veltman DJ, van der Wee NJ,

de Jonge P, Aleman A, Harmer CJ (2015) Cognitive vulnerability and implicit emotional processing: imbalance in frontolimbic brain areas? Cogn Affect Behav Neurosci 15(1): 69–79. https://doi.org/10.3758/s13415-014-0316-5

13. Allman JM, Hakeem A, Erwin JM, Nimchinsky E, Hof P (2001) The anterior cingulate cortex. The evolution of an interface between emotion and cognition. Ann N Y Acad Sci 935:107–117

14. Bush G, Luu P, Posner MI (2000) Cognitive and emotional influences in anterior cingulate cortex. Trends Cogn Sci 4(6):215–222. S1364-6613(00)01483-2 [pii]

15. Rushworth MF, Behrens TE, Rudebeck PH, Walton ME (2007) Contrasting roles for cingulate and orbitofrontal cortex in decisions and social behaviour. Trends Cogn Sci 11(4): 168–176. https://doi.org/10.1016/j.tics.2007.01.004. S1364-6613(07)00053-8 [pii]

16. Yucel M, Wood SJ, Fornito A, Riffkin J, Velakoulis D, Pantelis C (2003) Anterior cingulate dysfunction: implications for psychiatric disorders? J Psychiatry Neurosci 28(5): 350–354

17. Devinsky O, Morrell MJ, Vogt BA (1995) Contributions of anterior cingulate cortex to behaviour. Brain 118(Pt 1):279–306

18. Mayberg HS (1997) Limbic-cortical dysregulation: a proposed model of depression. J Neuropsychiatr Clin Neurosci 9(3):471–481

19. Ressler KJ, Mayberg HS (2007) Targeting abnormal neural circuits in mood and anxiety disorders: from the laboratory to the clinic. Nat Neurosci 10(9):1116–1124. https://doi.org/10.1038/nn1944. nn1944 [pii]

20. Li M, Demenescu LR, Colic L, Metzger CD, Heinze HJ, Steiner J, Speck O, Fejtova A, Salvadore G, Walter M (2017) Temporal dynamics of antidepressant ketamine effects on glutamine cycling follow regional fingerprints of AMPA and NMDA receptor densities. Neuropsychopharmacology 42(6): 1201–1209. https://doi.org/10.1038/npp.2016.184

21. Bae JN, MacFall JR, Krishnan KR, Payne ME, Steffens DC, Taylor WD (2006) Dorsolateral prefrontal cortex and anterior cingulate cortex white matter alterations in late-life depression. Biol Psychiatry 60(12):1356–1363. https://doi.org/10.1016/j.biopsych.2006.03.052

22. Frodl TS, Koutsouleris N, Bottlender R, Born C, Jager M, Scupin I, Reiser M, Moller HJ, Meisenzahl EM (2008) Depression-related variation in brain morphology over 3 years: effects of stress? Arch Gen Psychiatry 65(10): 1156–1165. https://doi.org/10.1001/archpsyc.65.10.1156

23. Li CT, Lin CP, Chou KH, Chen IY, Hsieh JC, Wu CL, Lin WC, Su TP (2010) Structural and cognitive deficits in remitting and non-remitting recurrent depression: a voxel-based morphometric study. NeuroImage 50(1):347–356. https://doi.org/10.1016/j.neuroimage.2009.11.021

24. Liao C, Feng Z, Zhou D, Dai Q, Xie B, Ji B, Wang X, Wang X (2012) Dysfunction of fronto-limbic brain circuitry in depression. Neuroscience 201:231–238. https://doi.org/10.1016/j.neuroscience.2011.10.053

25. Zavorotnyy M, Zollner R, Rekate H, Dietsche P, Bopp M, Sommer J, Meller T, Krug A, Nenadic I (2020) Intermittent theta-burst stimulation moderates interaction between increment of N-Acetyl-Aspartate in anterior cingulate and improvement of unipolar depression. Brain Stimul 13(4):943–952. https://doi.org/10.1016/j.brs.2020.03.015

26. Watters AJ, Carpenter JS, Harris AWF, Korgaonkar MS, Williams LM (2019) Characterizing neurocognitive markers of familial risk for depression using multi-modal imaging, behavioral and self-report measures. J Affect Disord 253:336–342. https://doi.org/10.1016/j.jad.2019.04.078

27. Tozzi L, Goldstein-Piekarski AN, Korgaonkar MS, Williams LM (2019) Connectivity of the cognitive control network during response inhibition as a predictive and response biomarker in major depression: evidence from a randomized clinical trial. Biol Psychiatry 87: 462. https://doi.org/10.1016/j.biopsych.2019.08.005

28. Sheline YI (2000) 3D MRI studies of neuroanatomic changes in unipolar major depression: the role of stress and medical comorbidity. Biol Psychiatry 48(8):791–800

29. Egger K, Schocke M, Weiss E, Auffinger S, Esterhammer R, Goebel G, Walch T, Mechtcheriakov S, Marksteiner J (2008) Pattern of brain atrophy in elderly patients with depression revealed by voxel-based morphometry. Psychiatry Res 164(3):237–244. https://doi.org/10.1016/j.pscychresns.2007.12.018

30. Gatt JM, Nemeroff CB, Dobson-Stone C, Paul RH, Bryant RA, Schofield PR, Gordon E, Kemp AH, Williams LM (2009) Interactions between BDNF Val66Met polymorphism and early life stress predict brain and arousal pathways to syndromal depression and anxiety. Mol Psychiatry 14(7):681–695. https://doi.org/10.1038/mp.2008.143

31. van Eijndhoven P, van Wingen G, Fernandez G, Rijpkema M, Verkes RJ,

Buitelaar J, Tendolkar I (2011) Amygdala responsivity related to memory of emotionally neutral stimuli constitutes a trait factor for depression. NeuroImage 54(2):1677–1684. https://doi.org/10.1016/j.neuroimage.2010.08.040

32. van Tol MJ, Demenescu LR, van der Wee NJ, Kortekaas R, Marjan MAN, Boer JA, Renken RJ, van Buchem MA, Zitman FG, Aleman A, Veltman DJ (2012) Functional magnetic resonance imaging correlates of emotional word encoding and recognition in depression and anxiety disorders. Biol Psychiatry 71(7):593–602. https://doi.org/10.1016/j.biopsych.2011.11.016

33. Herringa RJ, Birn RM, Ruttle PL, Burghy CA, Stodola DE, Davidson RJ, Essex MJ (2013) Childhood maltreatment is associated with altered fear circuitry and increased internalizing symptoms by late adolescence. Proc Natl Acad Sci U S A 110(47):19119–19124. https://doi.org/10.1073/pnas.1310766110

34. Leaver AM, Vasavada M, Kubicki A, Wade B, Loureiro J, Hellemann G, Joshi SH, Woods RP, Espinoza R, Narr KL (2020) Hippocampal subregions and networks linked with antidepressant response to electroconvulsive therapy. Mol Psychiatry 26:4288. https://doi.org/10.1038/s41380-020-0666-z

35. Gray JP, Muller VI, Eickhoff SB, Fox PT (2020) Multimodal abnormalities of brain structure and function in major depressive disorder: a meta-analysis of neuroimaging studies. Am J Psychiatry 177(5):422–434. https://doi.org/10.1176/appi.ajp.2019.19050560

36. Le Bihan D (1996) Functional MRI of the brain principles, applications and limitations. J Neuroradiol 23(1):1–5

37. Richardson FM, Price CJ (2009) Structural MRI studies of language function in the undamaged brain. Brain Struct Funct 213(6):511–523. https://doi.org/10.1007/s00429-009-0211-y

38. Bastiani M, Roebroeck A (2015) Unraveling the multiscale structural organization and connectivity of the human brain: the role of diffusion MRI. Front Neuroanat 9:77. https://doi.org/10.3389/fnana.2015.00077

39. Chow MS, Wu SL, Webb SE, Gluskin K, Yew DT (2017) Functional magnetic resonance imaging and the brain: a brief review. World J Radiol 9(1):5–9. https://doi.org/10.4329/wjr.v9.i1.5

40. Chen W, Liu X, Zhu XH, Zhang N (2009) Functional MRI study of brain function under resting and activated states. Annu Int Conf IEEE Eng Med Biol Soc 2009:4061–4063. https://doi.org/10.1109/IEMBS.2009.5333175

41. Glover GH (2011) Overview of functional magnetic resonance imaging. Neurosurg Clin N Am 22(2):133–139., vii. https://doi.org/10.1016/j.nec.2010.11.001

42. Matthews PM, Jezzard P (2004) Functional magnetic resonance imaging. J Neurol Neurosurg Psychiatry 75(1):6–12

43. Andellini M, Cannata V, Gazzellini S, Bernardi B, Napolitano A (2015) Test-retest reliability of graph metrics of resting state MRI functional brain networks: a review. J Neurosci Methods 253:183–192. https://doi.org/10.1016/j.jneumeth.2015.05.020

44. Guye M, Bettus G, Bartolomei F, Cozzone PJ (2010) Graph theoretical analysis of structural and functional connectivity MRI in normal and pathological brain networks. MAGMA 23(5–6):409–421. https://doi.org/10.1007/s10334-010-0205-z

45. Joo SH, Lim HK, Lee CU (2016) Three large-scale functional brain networks from resting-state functional MRI in subjects with different levels of cognitive impairment. Psychiatry Investig 13(1):1–7. https://doi.org/10.4306/pi.2016.13.1.1

46. Smyser CD, Snyder AZ, Neil JJ (2011) Functional connectivity MRI in infants: exploration of the functional organization of the developing brain. NeuroImage 56(3):1437–1452. https://doi.org/10.1016/j.neuroimage.2011.02.073

47. Thompson GJ (2018) Neural and metabolic basis of dynamic resting state fMRI. NeuroImage 180(Pt B):448–462. https://doi.org/10.1016/j.neuroimage.2017.09.010

48. Smith SM, Vidaurre D, Beckmann CF, Glasser MF, Jenkinson M, Miller KL, Nichols TE, Robinson EC, Salimi-Khorshidi G, Woolrich MW, Barch DM, Ugurbil K, Van Essen DC (2013) Functional connectomics from resting-state fMRI. Trends Cogn Sci 17(12):666–682. https://doi.org/10.1016/j.tics.2013.09.016

49. Alexander AL, Lee JE, Lazar M, Field AS (2007) Diffusion tensor imaging of the brain. Neurotherapeutics 4(3):316–329. https://doi.org/10.1016/j.nurt.2007.05.011

50. Lee SK, Kim DI, Kim J, Kim DJ, Kim HD, Kim DS, Mori S (2005) Diffusion-tensor MR imaging and fiber tractography: a new method of describing aberrant fiber connections in developmental CNS anomalies. RadioGraphics 25(1):53–65.; discussion 66–58. https://doi.org/10.1148/rg.251045085

51. Cascio CJ, Gerig G, Piven J (2007) Diffusion tensor imaging: application to the study of the developing brain. J Am Acad Child Adolesc Psychiatry 46(2):213–223. https://doi.org/10.1097/01.chi.0000246064.93200.e8

52. Hennig J, Speck O, Koch MA, Weiller C (2003) Functional magnetic resonance imaging: a review of methodological aspects and clinical applications. J Magn Reson Imaging 18(1):1–15. https://doi.org/10.1002/jmri.10330

53. Meinert S, Leehr EJ, Grotegerd D, Repple J, Forster K, Winter NR, Enneking V, Fingas SM, Lemke H, Waltemate L, Stein F, Brosch K, Schmitt S, Meller T, Linge A, Krug A, Nenadic I, Jansen A, Hahn T, Redlich R, Opel N, Schubotz RI, Baune BT, Kircher T, Dannlowski U (2020) White matter fiber microstructure is associated with prior hospitalizations rather than acute symptomatology in major depressive disorder. Psychol Med:1–9. https://doi.org/10.1017/S0033291720002950

54. Liu X, He C, Fan D, Zhu Y, Zang F, Wang Q, Zhang H, Zhang Z, Zhang H, Xie C (2020) Disrupted rich-club network organization and individualized identification of patients with major depressive disorder. Prog Neuro-Psychopharmacol Biol Psychiatry 2020:110074. https://doi.org/10.1016/j.pnpbp.2020.110074

55. van Velzen LS, Kelly S, Isaev D, Aleman A, Aftanas LI, Bauer J, Baune BT, Brak IV, Carballedo A, Connolly CG, Couvy-Duchesne B, Cullen KR, Danilenko KV, Dannlowski U, Enneking V, Filimonova E, Forster K, Frodl T, Gotlib IH, Groenewold NA, Grotegerd D, Harris MA, Hatton SN, Hawkins EL, Hickie IB, Ho TC, Jansen A, Kircher T, Klimes-Dougan B, Kochunov P, Krug A, Lagopoulos J, Lee R, Lett TA, Li M, MacMaster FP, Martin NG, McIntosh AM, McLellan Q, Meinert S, Nenadic I, Osipov E, Penninx B, Portella MJ, Repple J, Roos A, Sacchet MD, Samann PG, Schnell K, Shen X, Sim K, Stein DJ, van Tol MJ, Tomyshev AS, Tozzi L, Veer IM, Vermeiren R, Vives-Gilabert Y, Walter H, Walter M, van der Wee NJA, van der Werff SJA, Schreiner MW, Whalley HC, Wright MJ, Yang TT, Zhu A, Veltman DJ, Thompson PM, Jahanshad N, Schmaal L (2020) White matter disturbances in major depressive disorder: a coordinated analysis across 20 international cohorts in the ENIGMA MDD working group. Mol Psychiatry 25(7):1511–1525. https://doi.org/10.1038/s41380-019-0477-2

56. Lai CH, Wu YT (2014) Alterations in white matter micro-integrity of the superior longitudinal fasciculus and anterior thalamic radiation of young adult patients with depression. Psychol Med 44(13):2825–2832. https://doi.org/10.1017/S0033291714000440

57. Lai CH, Wu YT (2016) The white matter microintegrity alterations of neocortical and limbic association fibers in major depressive disorder and panic disorder: the comparison. Medicine 95(9):e2982. https://doi.org/10.1097/MD.0000000000002982

58. Ashburner J, Friston KJ (2000) Voxel-based morphometry--the methods. NeuroImage 11(6 Pt 1):805–821. https://doi.org/10.1006/nimg.2000.0582

59. Bigler ED (2015) Structural image analysis of the brain in neuropsychology using magnetic resonance imaging (MRI) techniques. Neuropsychol Rev 25(3):224–249. https://doi.org/10.1007/s11065-015-9290-0

60. Buonocore MH, Maddock RJ (2015) Magnetic resonance spectroscopy of the brain: a review of physical principles and technical methods. Rev Neurosci 26(6):609–632. https://doi.org/10.1515/revneuro-2015-0010

61. Brady TJ, Wismer GL, Buxton R, Stark DD, Rosen BR (1986) Magnetic resonance chemical shift imaging. In: Magnetic resonance annual. Raven Press, New York, NY, pp 55–80

62. Klose U (2008) Measurement sequences for single voxel proton MR spectroscopy. Eur J Radiol 67(2):194–201. https://doi.org/10.1016/j.ejrad.2008.03.023

63. Dreher W, Erhard P, Leibfritz D (2011) Fast three-dimensional proton spectroscopic imaging of the human brain at 3 T by combining spectroscopic missing pulse steady-state free precession and echo planar spectroscopic imaging. Magn Reson Med 66(6):1518–1525. https://doi.org/10.1002/mrm.22963

64. Posse S, Otazo R, Tsai SY, Yoshimoto AE, Lin FH (2009) Single-shot magnetic resonance spectroscopic imaging with partial parallel imaging. Magn Reson Med 61(3):541–547. https://doi.org/10.1002/mrm.21855

65. Zang Y, Jiang T, Lu Y, He Y, Tian L (2004) Regional homogeneity approach to fMRI data analysis. NeuroImage 22(1):394–400. https://doi.org/10.1016/j.neuroimage.2003.12.030

66. Zou QH, Zhu CZ, Yang Y, Zuo XN, Long XY, Cao QJ, Wang YF, Zang YF (2008) An improved approach to detection of amplitude of low-frequency fluctuation (ALFF) for resting-state fMRI: fractional ALFF. J Neurosci Methods 172(1):137–141. https://doi.org/

10.1016/j.jneumeth.2008.04.012.   S0165-0270(08)00245-8 [pii]

67. Zuo XN, Kelly C, Di Martino A, Mennes M, Margulies DS, Bangaru S, Grzadzinski R, Evans AC, Zang YF, Castellanos FX, Milham MP (2010) Growing together and growing apart: regional and sex differences in the lifespan developmental trajectories of functional homotopy. J Neurosci 30(45):15034–15043. https://doi.org/10.1523/JNEUROSCI.2612-10.2010

68. Biswal B, Yetkin FZ, Haughton VM, Hyde JS (1995) Functional connectivity in the motor cortex of resting human brain using echoplanar MRI. Magn Reson Med 34(4):537–541

69. Fox MD, Raichle ME (2007) Spontaneous fluctuations in brain activity observed with functional magnetic resonance imaging. Nat Rev Neurosci 8(9):700–711. https://doi.org/10.1038/nrn2201. nrn2201 [pii]

70. Zalesky A, Fornito A, Bullmore ET (2010) Network-based statistic: identifying differences in brain networks. NeuroImage 53(4):1197–1207. https://doi.org/10.1016/j.neuroimage.2010.06.041

71. Hong SB, Zalesky A, Cocchi L, Fornito A, Choi EJ, Kim HH, Suh JE, Kim CD, Kim JW, Yi SH (2013) Decreased functional brain connectivity in adolescents with internet addiction. PLoS One 8(2):e57831. https://doi.org/10.1371/journal.pone.0057831

72. Salvador R, Suckling J, Coleman MR, Pickard JD, Menon D, Bullmore E (2005) Neurophysiological architecture of functional magnetic resonance images of human brain. Cereb Cortex 15(9):1332–1342. https://doi.org/10.1093/cercor/bhi016

73. Smith SM, Jenkinson M, Woolrich MW, Beckmann CF, Behrens TE, Johansen-Berg H, Bannister PR, De Luca M, Drobnjak I, Flitney DE, Niazy RK, Saunders J, Vickers J, Zhang Y, De Stefano N, Brady JM, Matthews PM (2004) Advances in functional and structural MR image analysis and implementation as FSL. NeuroImage   23(Suppl   1):S208–S219. https://doi.org/10.1016/j.neuroimage.2004.07.051

74. Woolrich MW, Jbabdi S, Patenaude B, Chappell M, Makni S, Behrens T, Beckmann C, Jenkinson M, Smith SM (2009) Bayesian analysis of neuroimaging data in FSL. NeuroImage   45(1   Suppl):S173–S186. https://doi.org/10.1016/j.neuroimage.2008.10.055. S1053-8119(08)01204-4 [pii]

75. Jenkinson M, Smith S (2001) A global optimisation method for robust affine registration of brain images. Med Image Anal 5(2):143–156. S1361841501000366 [pii]

76. Smith SM (2002) Fast robust automated brain extraction. Hum Brain Mapp 17(3):143–155. https://doi.org/10.1002/hbm.10062

77. Smith SM, Johansen-Berg H, Jenkinson M, Rueckert D, Nichols TE, Miller KL, Robson MD, Jones DK, Klein JC, Bartsch AJ, Behrens TE (2007) Acquisition and voxelwise analysis of multi-subject diffusion data with tract-based spatial statistics. Nat Protoc 2(3):499–503. https://doi.org/10.1038/nprot.2007.45

78. Nichols TE, Holmes AP (2002) Nonparametric permutation tests for functional neuroimaging: a primer with examples. Hum Brain Mapp 15(1):1–25. https://doi.org/10.1002/hbm.1058

79. Colloby SJ, Firbank MJ, Vasudev A, Parry SW, Thomas AJ, O'Brien JT (2011) Cortical thickness and VBM-DARTEL in late-life depression. J Affect Disord 133(1–2):158–164. https://doi.org/10.1016/j.jad.2011.04.010

80. Thomas AG, Marrett S, Saad ZS, Ruff DA, Martin A, Bandettini PA (2009) Functional but not structural changes associated with learning: an exploration of longitudinal Voxel-Based Morphometry (VBM). NeuroImage 48:117

81. Seidman LJ, Biederman J, Liang L, Valera EM, Monuteaux MC, Brown A, Kaiser J, Spencer T, Faraone SV, Makris N (2011) Gray matter alterations in adults with attention-deficit/hyperactivity disorder identified by voxel based morphometry. Biol Psychiatry 69(9):857–866. https://doi.org/10.1016/j.biopsych.2010.09.053.   S0006-3223(10)01054-1 [pii]

Chapter 13

Neurobiochemistry Alterations Associated with Major Depression: A Review of Translational Magnetic Resonance Spectroscopic Studies

Darren William Roddy, John R. Kelly, Thomas Drago, Kesidha Raajakesary, Madeline Haines, and Erik O'Hanlon

Abstract

Conventional magnetic resonance imaging uses water resonance to generate images of brain tissue. Here, we describe how a spectrum of potentially useful chemicals in the brain may be generated by examining the resonance from molecules other than water. This is usually achieved by analyzing proton (H^+) nuclei resonance from a limited number of small mobile molecules using a technique called proton or H^+ magnetic resonance spectroscopy (H^+ MRS). However, due to low concentrations and overlapping signals, this technique is prone to signal-to-noise ratio difficulties. Consideration is given to methods to boost this ratio and increase the likelihood of generating robust, reliable spectra useful in the study of depression. Additionally, a mini-review of H^+ MRS findings of glx (a combination of glutamate and glutamine) and N-acetylaspartate in the prefrontal, anterior cingulate, and hippocampal regions in major depressive disorder is described.

Key words MRS, Magnetic resonance spectroscopy, Depression, Glutamate, Glutamine, glx, *N*-Acetylaspartate, NAA

1 The Theory of MRS

1.1 Basic MRI Physics

Water is the most abundant molecule in the brain. Conventional magnetic resonance imaging (MRI) maps the distribution and interactions of water in the brain by analyzing the protons (hydrogen nuclei) that make up water molecules [1]. Magnetic resonance spectroscopy (MRS) analyzes nuclei other than protons around water molecules. This most commonly involves the evaluation of signals from protons in molecules other than water, the so-called ^1H, H^+, or proton MRS. However, occasionally, other nuclei such as ^{31}P and ^{13}C can be used to generate an alternative MR spectrum.

Yong-Ku Kim and Meysam Amidfar (eds.), *Translational Research Methods for Major Depressive Disorder*, Neuromethods, vol. 179, https://doi.org/10.1007/978-1-0716-2083-0_13,

When exposed to a strong magnetic field (B_0), as found in an MRI machine, nuclear spins become polarized in equilibrium along the axis of the magnetic field. Following an appropriate radiofrequency (RF) pulse, these nuclei further polarize temporarily with the RF field, tipping the nuclei out of their equilibrium position, causing them to spin along the new axis of the RF field. Termination of the RF pulse allows the spins to "relax" and precess (wobble) along the axis of the original magnet. These precessing nuclei create a rotating magnetic field around the original equilibrium direction. This rotating field fluctuates with a characteristic frequency (resonance) creating an oscillating voltage in the receiver coil.

These rotating nuclei do not exist independently of the local electrochemical environment. Electrons surrounding the nucleus and adjacent nuclei can cause small shielding effects of the main magnetic field. This can result in slightly different resonance frequencies for a given nucleus in different molecules and different positions of the nucleus in the same molecule. This change in frequency is known as chemical shift. Consequently, each nucleus in each molecule exhibits a unique chemical shift due to its unique chemical environment. The total of all chemical shifts from the different nuclei results in a spectrum of chemical shifts. MRS involves measuring the frequency and intensity of the different chemical shifts from a particular nuclear species (e.g., H^+) to generate a spectrum.

As well as chemical shifts, the spectrum can be modified by the phenomenon known as J-coupling [2]. This occurs when one nuclear spin perturbs another nuclear spin through the electron cloud of the molecule. This can change the signal intensity, "breaking up" a signal resonance peak into a more complex pattern of multiple peaks. J-coupling is dependent on the echo time (*TE*) of the acquisition sequence (see below). Conversely, the chemical shift of different nuclei may occasionally overlap, with a single peak "hiding" more than one metabolite. Together, these factors can make interpretation of the spectrum more complicated.

1.2 Single-Voxel Vs. Multivoxel MRS

Two broad methods of MRS exist. Essentially, the technique focuses on either one area or multiple areas at once (Fig. 1). These areas are termed voxels and need to be defined prior to the MRS scan. Single-voxel MRS acquires the spectrum from a single predefined region of interest, and signals outside this area are ignored or suppressed [3]. During this technique, parameters are optimized to produce the best spectrum possible from a relatively small area. A single spectrum, averaging the resonance from all the signals within the voxel, is generated. In contrast, multivoxel MRS measures the spectrum from many adjacent smaller regions at once [4]. In this technique, multiple spectra are obtained along a single column of voxels (1D), a sheet or plane of voxels (2D), or a block of

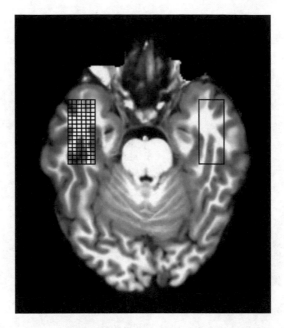

Fig. 1 Single- and multivoxel MRS. Scout axial T_1 brain MR image with representative temporal lobe multivoxel (left) and single-voxel MRS (right). Note in this example that some of the individual voxels of the multivoxel MRS are almost fully encompassing sulci rather than gray or white matter. Apart from partial volume effects and contaminated spectra in these specific voxels, contamination will "bleed" into adjacent voxels

voxels (3D). These voxels are considerably smaller than those used in single-voxel MRS, although the overall brain area covered by these many voxels may be larger. The spectral signal within each component voxel is localized using spatial encoding (application of another RF gradient pulse along a different axis to the original RF gradient). Finally, a composite image is generated with each of the small multiple voxels showing a unique spectrum. This technique is also known as MRSI (magnetic resonance spectroscopic imaging) or CSI (chemical shift imaging).

Both single- and multivoxel MRS have advantages and disadvantages. Single-voxel MRS is quicker and easier to acquire, but most importantly, because it averages the signal emitted from a larger area, the spectra produced are more robust. Although the spectrum generated is high quality, it can only provide an "overview" of the brain area studied with no subregion specificity. Multivoxel MRS allows spatial coverage over wider areas of the brain and a more granular resolution (i.e., a spectrum at each component voxel). However, it is more time-consuming and technically difficult to undertake, and the individual spectra obtained from each voxel are of a lower quality, suffering contamination from adjacent areas. This leads to issues with identification and quantification of metabolites.

Table 1
Comparison of single- and multivoxel magnetic resonance spectroscopy (MRS)

	Single-voxel MRS	Multivoxel MRS
Region characteristics		
Region size	Smaller	Larger
Tissue homogeneity	Assumed homogeneous	Can be heterogeneous
Data characteristics		
Signal-to-noise ratio	High	Low
Readily quantifiable	Yes	With difficulty
Output	Single spectrum	Multiple spectra
Scanning characteristics		
Setup	Easy, fast	Intricate, slow
Speed	3–7 min per voxel	2D up to 10 min, 3D up to 20 min
Shimming	Reliable	Less reliable

Single-voxel MRS is superior when a reliable investigation of specific brain areas assumed to be homogeneous is required [3]. It is especially useful when scanning time may be an issue (e.g., due to potential patient movement or cost). MRSI is best for larger, inhomogeneous brain areas or if multiple brain areas need to be surveyed at once [5]. It can also be useful to detect broad spatial patterns of spectral changes that may warrant more precise localized investigation with single-voxel MRS. A comparison of single-voxel versus multivoxel MRS is detailed in Table 1.

1.3 Proton MR Spectroscopy

MRS can be performed on any nucleus that produces a resonance frequency as part of a molecule in an MRI scanner. Although H^+ is the most common nucleus in the brain, being a constituent of every molecule, the resonance signal from water can overwhelm an H^+ MRS spectrum. Early hardware and technique limitations meant that early MRS studies focused on nuclear isotopes that rendered a clear useable spectrum such ^{31}P (present in membranes and neuroenergetically important molecules). However, advances in MRI hardware (field strengths, receiver coils), curve-fitting software, and particularly optimized sequences with water suppression techniques have allowed reliable and useful exploration of H^+ nuclei in a range of molecules other than water [6]. Proton (^1H or H^+) MRS is by far the most common form of MRS in use today in both clinical practice and research. H^+ MRS does not require any additional hardware or modifications to the standard MRI machines used in clinical and research facilities. It can be easily incorporated into a standard scanning protocol without any particular extra technical knowledge on behalf of the radiographer or researcher and does not require isotope contrast agents like some other forms of MRS such ^{13}C MRS [7]. The vast majority of modern MRS depression studies are performed using H^+ MRS; however, a few studies have investigated the unique spectra obtained from studying ^{31}P [8], ^{13}C [7], and ^{19}F [9] in depression. The remainder of this review will concentrate on the techniques and results from H^+ MRS unless otherwise specified.

1.4 Spectrum and Peaks: What Can be Measured?

In conventional MRI, only a single "peak" of water is mapped. However, in MRS, the output is a collection of peaks at different radiofrequencies, each representing the H^+ nucleus in different chemical environments. This collection of peaks is called the spectrum. Only small and mobile (i.e., not attached to cell membranes) molecules can be measured; anything that is too large and/or immobile is rendered invisible to MRS. Unfortunately, in practice, this leaves MRS only useful for studying a very small number of molecules. These small, mobile molecules are all involved in metabolic processes in some way in the brain and are termed "metabolites."

Each metabolite exhibits one or more unique chemical shifts on the MR spectrum (Fig. 2). A given metabolite can therefore be identified by the pattern of one or more peaks along the x-axis of the spectrum, resulting in a signature distinctive to that molecule. This is usually expressed as parts per million (ppm). Water has a large peak at 4.7 ppm along the x-axis. By measuring the intensity or amplitude of the signal at the peak along the y-axis, relative concentration of the metabolite can be estimated. Therefore, the MR spectrum can be used for both the identification and quantification of the metabolite under consideration.

The number of distinct metabolite peaks will vary according to the *TE* of the acquisition sequence. As a general rule for brain MRS, shorter *TE*s (15–20 ms) will generate more peaks of interest [10]. This, however, can lead to complex and often superimposed peaks leading to quantification and interpretation issues. The concentrations of some metabolites, such as free lipids, are only perturbed with very severe physiological changes such as in necrotic tumors [11]. Also, some require acquisition sequences with longer *TE*s [12]. As such, the practical usefulness of investigating these metabolites in depression is limited. Of the 20 or so principle metabolites easily rendered using H^+ MRS, only a few are truly useful in depression research. A list of potentially useful metabolites is given in Table 2.

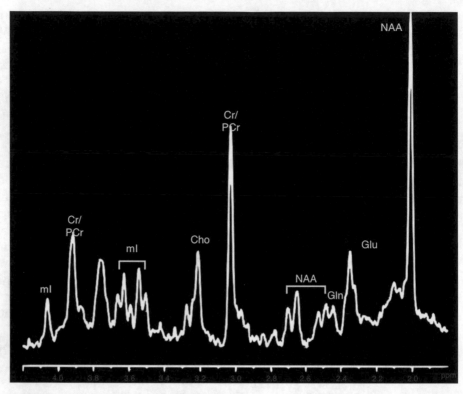

Fig. 2 Representative H$^+$ MRS spectrum. Example of a simplified representative spectrum taken at the medial prefrontal cortex (3T Philips MRI, MEGAPRESS sequence). The *x*-axis represents the chemical shift and can be used to identify the metabolites. The *y*-axis (not labeled) is the relative concentration. Only the readily identifiable peaks are labeled. The compound glx peak is resolved in this case into glutamate and glutamine. Note the multiple peaks for some metabolites. *Cho* choline, *Cr* creatine, *gln* glutamine, *glu* glutamate, *ml* myo-inositol, *NAA* N-acetylaspartate, *PCr* phosphocreatine, *ppm* parts per million

2   The Practice of MRS

As with all good research, MRS particularly benefits from well-planned groundwork prior to scanning. MRS is a complex process and requires a homogeneous magnetic field, a predefined voxel of interest, suppression of extraneous interfering resonance signals (water and surrounding fat), a sufficient number of measurements using a sequence devised to generate a useful spectrum, and post-processing software with the ability to extract meaningful data from the spectrum. Scanning is expensive and time-consuming and relies on subjects' cooperation in remaining still for long periods of time. Therefore, all efforts need to be directed toward reducing scanning time while simultaneously increasing the capture of meaningful information. Meticulous preparation and study design before entering the scanner will result in money savings, time savings, and enhanced participant goodwill and, most importantly, result

Table 2
H⁺ MRS metabolites in depression

Metabolite	Major chemical shift	Physiological function	Potential interpretation
Lactate	1.33 ppm	End of product of glycolysis [12]	Mitochondrial or metabolic dysfunction [13]
N-Acetylaspartate (NAA)	2.02 ppm Second peak at 2.6 ppm	Found in neurons, axons, and glia [14]	Neuronal and glial marker [15], synaptogenesis or dendritogenesis [16]
Glutamate	2.35 ppm, second peak at 3.65–3.8 ppm, together with glutamine forms single glx peak	Excitatory neurotransmitter GABA precursor [17]	Excitotoxicity, glutamate-glutamine cycling [18]
Glutamine	2.43 ppm, second peak at 3.65–3.8 ppm, together with glutamate forms single glx peak	Reserve and precursor for glutamate and GABA [19]	Glutamate-glutamine cycling, glial activity [20]
Glutathione	2.80 ppm	Antioxidant and redox balance [21]	Potential inflammatory marker [21]
GABA	3.0 ppm	Inhibitory neurotransmitter [17]	Cortical inhibition-excitation imbalance [22]
Creatine	3.03 ppm Second peak at 3.9 ppm	Energetics [12]	Used as a reference for metabolic ratios
Choline	3.2 ppm and 3.52 ppm	Phospholipid and acetylcholine formation [12]	Membrane turnover [23]
Myo-inositol	3.55 ppm Second peak at 4.06 ppm	Osmoregulation, abundant in astrocytes [24]	Glial marker [25]

Note many interpretations of metabolite changes are possible and those listed here are only a guide. *GABA* gamma-aminobutyric acid, *ppm* parts per million

in good-quality robust useful data. In planning any MRS study, consideration needs to be given to the signal-to-noise ratio.

2.1 Signal-to-Noise Ratio

Insufficient signal-to-noise ratio is the most significant challenge of in vivo MRS. Investigating proton resonance results in water completely dominating the spectrum. H⁺ MRS has to disregard this overpowering resonance to instead focus on the weaker and often overlapping resonances from numerous low concentration molecules. These tiny resonances (signal) can often be difficult to distinguish from background resonances (noise) in a sample. The signal-to-noise ratio is defined as the ratio between the amplitude of a resonance peak and the amplitude of random noise elsewhere in the spectrum. The ultimate goal of MRS planning is to increase this ratio and produce a visible, quantifiable peak for each metabolite along the spectrum. Approaches to amplify the MRS signal and/or decrease irrelevant noise include using scanners of higher field strength, increasing the scanning time and voxel volume, careful choice and placement of the voxel, optimizing echo times and scanning sequences, water and fat suppression techniques, targeted magnetic field homogenization, suitable RF coils, and targeted postprocessing software.

2.2 Increasing Field Strengths

Increasing the MRI scanner magnetic field strength increases the signal-to-noise ratio and spectral resolution. As field strength increases, the spectrum stretches along the x-axis (chemical shift axis). This makes the peaks more distinct with reduced overlap allowing for enhanced identification of metabolites. However, higher field strengths and the subsequent stretching of the x-axis allow previously undetectable peaks to "emerge" from the spectrum (chemical shift artifacts) [26]. This can complicate interpretation of spectra. Previously singlet peaks at lower field strengths may fragment into multiple smaller indistinct peaks at higher field strengths, leading to both identification and quantification difficulties [27]. However, these issues notwithstanding, increased signal-to-noise ratio and spectral resolution are almost always useful with increased field strength.

2.3 Acquisition Time

Meaningful signal-to-noise improvements require increases in acquisition. To double the signal-to-noise ratio, acquisition time needs to be four times longer, whereas to triple the signal-to-noise ratio, acquisition time needs to be nine times longer [28]. However, acquisition time cannot be increased indefinitely due to cost, magnetic instabilities, and subject movement.

Magnetic instabilities such as signal drift and localized eddy currents due to high shim currents or rapid gradient switching [29, 30] can reduce the signal-to-noise ratio. Such magnetic fluxes can lead to ineffective water suppression and distorted spectral shapes [31, 32]. This is especially likely following high-gradient cycles (e.g., diffusion imaging) and in busy MRI scanners. Ideally, MRS should be undertaken at the beginning of the day or at very least at the same time every day. Similarly, the procedure should be scheduled first in a sequence of scans if more than one scan modality is being performed.

Too much movement and the prescribed voxel may not correspond exactly to the expected brain area under investigation. This can lead to conflation of spectra from two or more adjacent regions, reducing the validity of the study. Similarly, if the region of interest is close to the bone, cerebrospinal fluid or air movement may nudge

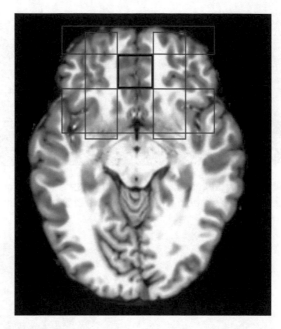

Fig. 3 Fat suppression bands. Scout axial T$_1$ brain MRI showing single-voxel MRS placed in the medial prefrontal gray matter (black lines). Surrounding the medial prefrontal voxel are four widely placed fat suppression bands (red lines). These block (or significantly restrict) signals from adjacent brain areas, for a purer spectrum from the voxel of interest

the voxel into these "non-brain tissue" areas, complicating or even collapsing the spectrum. The consequences of motion include improper spatial localization, phase fluctuations, incoherent averaging, and spectral peak broadening [33]. These may all lead to interpretation and quantification errors. Post-scanning techniques exist for correction of signal loss from motion [32, 34]; however, such techniques are subject to errors and cannot compensate for spatial localization errors. Similar methods exist where motion can be detected during the scan (using cameras) and adjustments made contemporaneously [35]. Ideally, subject movement should be limited as much as possible.

As a general rule, scans any longer than 20–25 min result in movement in all but the most cooperative of subjects. Additionally, before the actual MRS scanning process, scout images (rapid anatomical guidance scans) need to be acquired (Figs. 3, 4, and 5), and areas of interest identified on these scans and voxels are constructed accurately in 3D space (see below), a process taking some minutes. The length of these procedures needs to be taken into account when estimating subject time in the scanner.

2.4 Voxel Size

Unlike other modalities of MRI, which regularly explore resolutions of 1 mm^3 or less in research, MRS is restricted to analysis of much larger volumes of brain tissue (1–10 cm^3). This is due to low

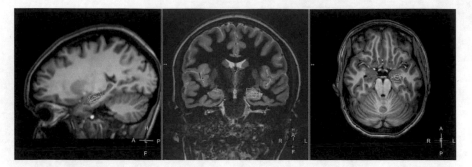

Fig. 4 Shimming. Sagittal, coronal, and axial scout T_1 brain images showing single-voxel MRS of the hippocampal region (yellow box) and simultaneous automated shimming around the voxel (white box)

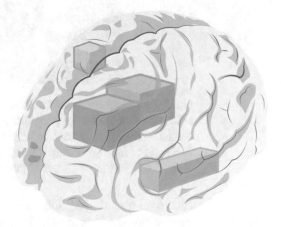

Fig. 5 Schematic image of reviewed brain areas and voxels. Cubes are representative of typical voxel placements in the medial prefrontal cortex (orange), dorsolateral prefrontal cortex (blue), anterior cingulate cortex (purple), and hippocampus (green)

absolute concentrations of MRS metabolites requiring sizeable segments of brain tissue to generate a clear useable spectrum. The signal-to-noise ratio is directly (linearly) proportional to the voxel volume [36]. The voxel size needed will directly impact the choice of brain area of interest.

2.5 Brain Area of Interest

Choosing a brain area of interest for MRS in depression can be problematic. The brain region needs to be of sufficient volume to generate a robust and useable spectrum. This and the restricted number of metabolites measured by H^+ MRS are the most limiting factors in the usefulness of the technique in the study of depression. Any brain chemistry differences need to be extensive over large enough brain areas to be detected by MRS. Yet, these same brain areas need to be small enough that meaningful functional differences in brain chemistry will be present in depression. In choosing an appropriate brain area for investigation, the following questions need to be addressed:

- Is this area potentially important for depression? Have any pre-clinical or clinical studies investigated this region? Is there a hypothesis for why MRS metabolites might be implicated in this region?

- Is the area large enough that a robust spectrum will be generated from it? The larger, the better, but anything less than 3 cm^3 can be troublesome in generating useable spectra.

- Is the region near or overlapping water, bone, or air that will interfere with the spectrum? If a voxel is placed over these non-brain tissue areas (or if movement occurs), the spectrum may be contaminated or even collapse completely.

- Can the area be reliably captured in three-dimensional space using a standard shaped volume (i.e., cubic or rectangular)?

Good spectra are obtained if the voxel of interest is of sufficient volume to overcome the signal-to-noise ratio, yet small enough to be scientifically useful. Ideally, a previous MRS study will have documented the pitfalls of investigating the brain region of interest. Generating a robust spectrum of a specific brain region is best done through trial and error. Previously published MRS anatomical definitions are a good starting point and will save effort and money. The precise anatomical characteristics of the voxel of interest on scout MR images should be clearly defined before the actual scan and ideally protocolized allowing methodical and accurate representation across subjects. Practicing voxel placement prior to actual MRS acquisition can increase reliability and decrease the length of time subjects need to remain still.

2.6 Echo Times

As a general rule, the shortest possible *TE* is always the best choice. Single-voxel MRS performs robustly with short echo times [10]. This increases the signal-to-noise ratio of spectra, minimizing signal loss of rapidly decaying metabolite peaks such as myo-inositol, glutamate, and glutamine. Short *TE*s can result in a fluctuating undulating baseline, but this can often be compensated by postprocessing software. Shorter *TR* can cause T_1 saturation, but this has little practical effect on MRS signal-to-noise ratio. As such, any practical TR should suffice.

2.7 MRS Sequences

The choice of MRS sequence is dependent on the application required and the metabolite of interest [37]. This will require careful selection and assessment of sequence parameters, the metabolites under investigation, field strengths, and brain areas of interest. There is no "one-sequence-fits-all" rule here, and careful evaluation with respect to the research question is essential. Some sequences are best suited to single-voxel applications and others to

CSI-type environments. Sequences can be very sensitive to certain physical properties, while others are more forgiving and "robust" in the application. The more widely used sequences are discussed below.

Point RESolved Spectroscopy (PRESS) is commonly used for H^+ spectroscopy using three slice selection RF pulses and features spin echo (SE) for its signal [38]. One of the disadvantages of this sequence is the limitation to utilize a *TE* less than 25 ms (25–35 ms). PRESS is useful for both single-voxel and multivoxel approaches but not suitable for metabolites that have relatively short T2s, so ^{31}P imaging is not possible. In neonatal or young child participants the additional energy contained in the 180-degree PRESS pulses can cause specific absorption rate (SAR) limit issues [39]. More recent advances have seen adaptations to the PRESS sequence in an attempt to reduce acquisition times in CSI applications.

Reduction of Acquisition time by Partition of sIgnal Decay in Spectroscopic Imaging (RAPID-SI) adds two additional gradient blips during the PRESS acquisition [40]. Signal-to-noise loss is known to occur with RAPID-SI, but this may be compensated to some extent during preprocessing. This technique has reported reduction of typical CSI times from 17 min to 2 min.

MEshcher-GArwood Point-RESolved Spectroscopy (MEGA-PRESS) has become the standard sequence for GABA investigations by adapting a PRESS sequence with two further "editing" pulses added around the second 180-degree pulse [41]. Here a *TE* is in the order of 68 ms.

STimulated Echo Acquisition Mode (STEAM) uses three slice-selective 90-degree pulses and utilizes stimulated echoes for its signal [42]. STEAM has the advantage that it can utilize shorter *TE*s (<10 ms) but has reduced signal (up to 50%) compared to other sequences such as PRESS [37, 43]. As such, awareness of other factors that decrease the signal-to-noise ratio (such as field strength) is essential.

2.8 Water Suppression

Water suppression is a vital process for successful spectral identification due to the abundance of water concentrations relative to the metabolites of interest. The proton MRS signal of water is up to five orders of magnitude greater than other detectable metabolites in the brain [44]. Numerous sequences adopt a strategy where reference scans of tissue water are acquired independently prior to the acquisition of interest. The signal from these scans is removed or modeled during data preprocessing, allowing a greater signal-to-noise ratio for the metabolites of interest. This process is referred to as water suppression.

CHEmical Shift Selective (CHESS) saturation is the most common technique to suppress the water signal and is akin to fat suppression in standard MRI (see below) [45]. Spoiler gradients are applied to saturate the water signal prior to the MRS sequence. Multiple spoiler gradients (usually three) resonating at the resonant frequency of water are needed to adequately suppress the water signal. An alternative approach includes water suppression enhanced through T1 effects (WET) [46] where the MRS acquisition alternates between positive and negative water signal residuals that cancel each other out to leave the spectra of interest remaining.

2.9 Fat Suppression

Scalp and marrow fat can interfere with spectra from MRS voxels close to the brain surface [47]. Various methods of fat suppression exist; however, the most common in brain MRS is the use of outer volume suppression bands around the MRS voxel [46]. These are introduced at the time of MRS voxel placement and suppress all signals from tissues surrounding the MRS voxel (Fig. 3). This is usually sufficient for most brain MRS; however, inversion recovery methods targeting fat-specific frequencies to nullify the effects of fat may also be used. These include sequences introduced before the MRS sequence (e.g., short tau inversion recovery [STIR] [48], spectral presaturation with inversion recovery [SPIR] [49], and spectral attenuated inversion recovery [SPAIR] [50]) or pulses inserted mid-MRS sequence (e.g., band selective inversion with gradient dephasing [BASING] [47]).

2.10 Magnetic Field Shimming

Shimming in MRS involves optimizing the homogeneity of the B_0 magnetic field specifically around the volume of interest [51] (Fig. 4). This is achieved through targeted deployment of electrical currents through specialized coils to "correct" the magnetic field where needed. Uniform magnetic fields in and around the area of interest increase the signal-to-noise ratio. In practice, the magnetic field is best optimized around voxels that are smaller and more uniform. Large, irregularly shaped voxels are less amenable to robust shimming and are susceptible to diminished signal-to-noise ratios. Most modern scanners can shim voxels automatically according to predefined criteria [52].

2.11 Radiofrequency Coils

Smaller, tighter head coils, while also potentially disrupting homogeneity, may also increase the signal-to-noise ratio [53]. Whereas the signal is directly proportional to the volume of the voxel of interest, all noise is produced from the entirety of space within the RF coil. Reducing the volume within the RF coil can reduce the overall noise.

2.12 Considerations on the Day

Assuming an optimal scanning protocol has been devised between the researcher and the radiographer, a general good practice procedure for MRS would consist of the following:

2.12.1 Scheduling

Ideally MRS should be done early in the morning, and MRS would be the first sequence if more than one scanning modality is being undertaken. This is due to the phenomenon of signal drift due to increased workload of the scanner.

2.12.2 Caffeine

Although hippocampal myo-inositol [54] and hypothalamic glutamate [55] concentration in rats has been shown to increase following acute caffeine administration, there is no evidence from human studies that normal everyday caffeine intake has any impact on MRS metabolites [56]. Patients are often advised not to drink coffee or tea before the scan if possible, but caffeine is not a contraindication to MRS.

2.12.3 Head Stabilization

Movement will interfere with spectral acquisition and interpretation. The subject's head should be as stable as possible. Techniques to aid this include mock trial scanning before the actual scan and physical impediments such as foam pads/cushions/sheets around the head, forehead taping, and bite bars during the scan.

2.12.4 Good-Quality Scout Images

Good-quality whole head multiplanar scout images allow for the accurate placement of the acquisition voxel (Figs. 2, 3, and 4). The exact details of these should be worked out in advance depending on the anatomical and tissue composition of the brain region of interest and could include T_1, T_2, or T_2-FLAIR (fluid-attenuated inversion recovery) images. The resolution of these scout images should be tweaked to the particulars of the area of interest. Resolution should be high enough to allow accurate identification of anatomical landmarks, yet low enough to be achieved in the least possible time.

2.12.5 Accurate Anatomically Placed Voxels of Interest

Using the scout images, the region of interest should be quickly and accurately identified and the voxel of interest defined (Fig. 2) according to a predefined and well-practiced protocol. The images should be quickly checked in all three dimensions to identify potential pitfalls within or close to the region that could collapse or contaminate the spectra, e.g., bone, fat, and cerebrospinal fluid.

2.12.6 Outer Volume Bands

Outer volume suppression bands (Fig. 3) should be placed in a methodical manner, especially over adjacent fat dense regions to the voxel of interest. Ideally, this should be part of the voxel of interest identification and placement protocol.

2.12.7 Shimming

Inhomogeneities in the magnetic field can be smoothed over by the use of active shimming where magnetic inconsistencies around the voxel are identified and resolved by the use of magnetic currents (Fig. 4). Most modern scanners have software to methodically shim around voxels.

2.12.8   Intrascan Motion Limitation

The subject should be reminded with regular clear instructions and prompting not to move during the scan.

2.13   Post-Scan Processing

Ideally, the area under a spectral peak should be directly proportional to the concentration of nuclei that give rise to that peak and would involve simply calculating a value from the graph. However, peaks are dependent on numerous factors including individual metabolite relaxation times, magnetic and RF inhomogeneities, eddy currents, and factors relating to the scanning protocol (e.g., pulse sequences, *TR*, and *TE*). As mentioned, many resonances overlap and/or have complex splitting patterns. An "undulating baseline" and metabolite inhomogeneities add further complexity to spectrum interpretation. However, through the use of standardized solutions and trial and error, sophisticated spectral curve-fitting software has become available. These programs allow a more quantifiable spectrum to be extracted from the raw often noisy data. The most commonly cited software in the literature is the commercially available LCModel [57] (Linear Combination of Model Spectra) [http://s-provencher.com/lcmodel.shtml]. Other freeware products are also available depending on the metabolites required, e.g., jMRUI toolbox [58] (Java-based Magnetic Resonance User Interface) [http://www.jmrui.eu] or TARQUIN [59] (Totally Automatic Robust Quantitation in NMR) [http://tarquin.sourceforge.net].

2.13.1   Relative Vs. Absolute Concentrations

However, the area under a peak can only estimate the relative concentrations of each metabolite. In order to generate more meaningful information, metabolite data is usually expressed in one of two ways: metabolic ratios or absolute concentrations. Metabolic ratios involve expressing one metabolite as a ratio of another. Creatine is often chosen as the ratio denominator as it is traditionally assumed to be constant across normal brain and most pathological states; however, evidence suggests that creatine is not as stable as previously thought [60]. Ratios rather than absolute concentrations can control for scanner differences, regional magnetic instabilities, and partial volume effects. They may also be useful in situations where metabolites are expected to change relative to each other [61]. Although ratios are useful when sizeable changes in metabolites are expected (e.g., cancer), ratio calculations can reduce both the sensitivity and specificity of measurements. In depression, less sweeping metabolite differences would be expected, and subtle changes may remain hidden under the increased type 1 and type 2 errors of ratio calculations. Recently, absolute metabolite concentrations have become a much more common method for interpreting and presenting MRS results in depression.

Absolute metabolite concentrations can be estimated in one of three ways: (1) attaching a model solution to the subject's head during the MRS acquisition, (2) exploiting the known brain water content as an internal standard, or (3) using comparisons of in vivo spectra with spectra of pure compounds measured independently. A model solution taped to the subject's head is the gold standard for spectral curve fitting but can introduce magnetic and RF inhomogeneity and other susceptibility effects. It is now mostly used for non-H^+ MRS where robust spectral fitting software is uncommon. Brain tissue is assumed to contain water at 55.5 mol/kg and is often used to estimate metabolite concentrations through acquisition of non-water-suppressed reference sequences along with MRS [62]. Software such as LCModel uses model spectra to make the measured spectrum "fit" to known metabolite spectra supplied along with the software package. However, absolute concentration software may exhibit problems with systematic bias, as features such as J-coupling or distortions are assumed identical across the sample [63].

2.13.2 *Choosing Software*

Over the last few years, many advances in MRS postprocessing software packages have been made. Both relative and absolute metabolite concentration estimation techniques have been refined with ever-tighter algorithms and decades of experience. When considering a software package for depression MRS data processing, the following needs to be taken into consideration, with each package offering different benefits: the nuclei of interest (H+ or other nuclei), the metabolites of interest, the MRS acquisition method (single vs. multivoxel, different pulse sequences, etc.), the level of software automation versus the level of customization required, the preferred user interface (graphical or text based), the availability of comprehensive documentation or user forums to answer queries, and finally the cost (free for academic or commercial).

3   Review of Proton MRS Studies in Major Depressive Disorder

A full review of all metabolites in all brain areas in depression is outside the scope of this chapter. Great difficulty also exists in comparing results across studies due to heterogeneity of MRS techniques (single-voxel vs. MRSI, assorted field strengths and sequences, varied relaxation times, different curve-fitting software, etc.), heterogeneity of brain areas (variously sized voxels, differing anatomical definitions, and partial volume effects from white matter, gray matter, and CSF all within a single voxel), as well as heterogeneity of clinical characteristics (small sample sizes, first-episode depression vs. chronic vs. recurrent, medicated vs. unmedicated, etc.).

The most common brain areas studied using MRS in depression are those regions believed to be involved in emotion and behaviors [64]. This mini-review will examine the three most common regions investigated in MRS studies of depression: the prefrontal cortex, the anterior cingulate cortex, and the hippocampus. NAA and glx are among the most explored metabolites in depression MRS studies. NAA is a marker of neuronal and glial integrity [15] and possibly synaptodendritic function [16], whereas the compound metabolite glx with its constituent glutamate and glutamine can be a marker of abnormalities of glial glutamine-glutamate cycling [65], neuroenergetics [66], and excitotoxicity [67]. This review will be limited to studies that focus on NAA and glx metabolites in depression.

3.1 Method

A PubMed and Google Scholar search was performed with the following terms:"Magnetic Resonance Imaging" *OR* "MRS" *OR* "NAA" *OR* "N-acetylaspartate" *OR* "glx" *OR* "glutamate" *OR* "glutamine" *AND* "Depression" *OR* "Major Depressive Disorder."

Only MRS studies that compared patients between the ages of 18 and 65 years with a current diagnosis of major depressive disorder to healthy controls were included. Other diagnoses such as bipolar depression and dysthymia were excluded. Studies involving remitted and recovered patients, children, adolescents, or elderly were also excluded. Treatment studies were only included if they compared controls to depressed patients at baseline as well as following treatment. Studies were only included if a clear PFC, ACC, or hippocampal voxel was described in the text or in a diagram. Voxels that covered mostly white matter were excluded. Only the NAA and glx results were retrieved from the text. The results of glx studies also included the constituent metabolites glutamate and glutamine where calculated.

3.2 Prefrontal Cortex

The prefrontal cortex (PFC) is situated at the most rostral (anterior) part of the frontal lobe. This large, highly evolved area in humans straddles many Brodmann areas (BA), including BA 8–14, 24, 25, and 44–47. The interconnected neuroanatomy of this area is highly complex, resulting in many overlapping and often contradictory definitions of the area. However, as a useful approximation, the PFC can be divided into two broad complementary regions—a medial and lateral PFC—each subdivided into two subregions.

The medial PFC in particular suffers from inconsistent border definitions as well as frequent conflation with the anterior cingulate cortex. For the purpose of a review of MRS, the medial PFC may be largely subdivided into two regions. A ventral region or orbitofrontal cortex lies over the orbital plate consisting of BA 11, 13, and 10. This region is involved in reward behavior and emotion, and its dysfunction has been implicated in depression [68]. However, robust orbitofrontal MRS is problematic. A slim elongated voxel shape and proximity over bone, air, and CSF pockets often result in inadequate spectra. A more dorsal medial area lies above the ventral region, anterior to the genu of the corpus callosum. In MRS studies, this region is often just termed the medial PFC (mPFC). In MRS literature, due to the need to have large cubic voxels, there is often spatial overlap between the mPFC and anterior cingulate cortex. This can lead to difficulty in mPFC and ACC MRS interpretation. The ventral medial prefrontal cortex will not be considered for this review.

The lateral PFC can also be broadly subdivided into ventrolateral and dorsolateral regions. The ventrolateral PFC consists of BA 44, 45, and 47 and is involved in motor functions such as motor inhibition, set shifting, and action updating [69]. The dorsolateral PFC (dlPFC) consists of BA 8, 9, 10, and 46 and is involved in executive functions such as decision-making [70], working memory [71], abstract reasoning [72], and social cognition [73]. Gross damage to this brain area can result in some symptoms of depression, such as lethargy, amotivation, memory difficulties, and psychomotor slowing [74]. The dlPFC has also been implicated in depression studies, showing reduced activity in disease and increasing following recovery [75]. Only the dlPFC will be considered for this review.

3.2.1 Medial Prefrontal glx

Nine studies have investigated glx metabolites in the mPFC (Table 3). Of the seven reporting a compound glx peak, only two showed a glx reduction in this region [25, 76], with five showing no change [77–81]. However, of the studies reporting the glutamate metabolite, three larger studies (total depressed $n = 72$) found glutamate reduction [25, 77, 78] in the mPFC in depression, compared to two smaller studies (total depressed $n = 17$) that found no change [81, 82]. Of these glutamate studies, glutamate but not glx was found to be lower in two studies [77, 78], with this finding confined to chronic patients only in one study [77]. A third study found that both glx and glutamate but not glutamine were reduced in the mPFC region [25]. No study to date has reported reduced glutamine [25, 82].

Only two studies have reported glx metabolites in the mPFC following treatment (Table 5). No difference in mPFC glutamate was seen following ECT (electroconvulsive therapy) [82] or glutamate or glutamine following antidepressants after 1 year of treatment [78].

In summary, reduced mPFC glutamate but not compound glx seems to be associated with depression, although treatment appears to have no effect on glutamate.

Table 3
Prefrontal cortex glx in depression

Author	Year	PFC region	p-value	MDD (N)	MDD age	Female/Male	MDD duration	Controls (N)	Control age	Tesla	MRS type	Software
Glx decreased												
Hasler	2007	dmPFC	[glx] 0.02	20	34.0	13/7	225.6	20	34.8	3 T	SV	n/a
Hasler	2007	vmPFC	[glx] 0.02	20	34.0	13/7	225.6	20	34.8	3 T	SV	n/a
Portella (CD)	2011	vmPFC	[glu] 0.001	19	51.0	15/4	273.6	15	40.5	3 T	SV	LCModel
Draganov	2020	vmPFC	[glu] 0.02	31	37.3	18/13	n/a	63	41.8	3 T	SV	SPM12
Shirayama	2017	mPFC	[glx] 0.042, [glu] 0.018	22	40.9	5/17	n/a	27	36.8	3 T	SV	LCModel
Michael	2003	dlPFC	[glx] 0.002 (left)	12	63.4	8/4	152.4	12	62.0	1.5 T	SV	LCModel
Şendur	2020	dlPFC	[glx] 0.044 (left)	30	38.8	23/7	118.0	30	33.4	1.5 T	SV	n/a
Glx unchanged												
Portella (FED)	2011	vmPFC	[glx] NS, [glu] NS	19	51.0	15/4	273.6	15	40.5	3 T	SV	LCModel
Portella (CD)	2011	vmPFC	[glx] NS	19	51.0	15/4	273.6	15	40.5	3 T	SV	LCModel
Draganova	2020	vmPFC	[glx] NS	31	37.3	18/13	n/a	63	41.8	3 T	SV	SPM12
Taylor	2012	mPFC	[glx] NS	39	31.5	22/17	n/a	27	31.4	3 T	SV	LCModel
Li	2016	mPFC	[glx] NS	20	28.0	13/7	56.4	20	31.7	3 T	2D MRSI	LCModel
Zhang	2016	mPFC	[glx] 0.799, [glu] 0.518	11	34.1	11/0	n/a	11	33.6	3 T	SV	LCModel
Shirayama	2017	mPFC	[gln] 0.109	22	40.9	5/17	n/a	27	36.8	3 T	SV	LCModel

(continued)

Table 3
(continued)

Author	Year	PFC region	p-value	MDD (N)	MDD age	Female/ Male	MDD duration	Controls (N)	Control age	Tesla	MRS type	Software
Knudsen	2018	mPFC	[glu] NS, [gln] NS	6	38.4	2/4	n/a	11	38.8	3 T	SV	LCModel
Nery	2009	dlPFC	[glx] 0.3, [glu] 0.3 (left)	37	36.6	24/13	189.6	40	40.0	1.5 T	SV	LCModel
Merkl	2011	dlPFC	[glu] 0.53 (left)	25	49.1	20/5	n/a	27	36.3	3 T	SV	n/a
Grimm	2012	dlPFC	[glu] 0.062 (left)	14	38.8	7/7	n/a	14	32.3	3 T	SV	n/a
Chen	2014	dlPFC	[glx] NS (left), NS (right)	15	27.9	9/6	14.8	15	27.5	1.5 T	SV	LCModel
Şendur	2020	dlPFC	[glx] 0.371 (right)	30	38.8	23/7	118.0	30	33.4	1.5 T	SV	GE workstation

The table is divided into studies that show glx decreased or glx unchanged. Data is subdivided into mPFC and dlPFC studies *CD* chronic depression, *dmPFC* dorsomedial prefrontal cortex, *FED* first-episode depression, *gln* glutamine, *glu* glutamate, *glx* composite glutamate–glutamine, *MDD* major depressive disorder, *mPFC* medial prefrontal cortex, *MRS* magnetic resonance spectroscopy, *MRSI* magnetic resonance spectroscopic imaging, *N* number, *n/a* not available, *NS* non-significant, *PFC* prefrontal cortex, *SV* single voxel, *vmPFC* ventromedial prefrontal cortex

3.2.2 Dorsolateral Prefrontal glx

Six studies have investigated glx metabolites in the dlPFC (Table 3). The left dlPFC compound glx peak was lower in two studies [83, 84] with two studies also reporting no differences in compound glx [85, 86]. Interestingly, glutamate was found to be unchanged in all studies investigating the metabolite [85, 87, 88]. Of the two studies investigating the right dlPFC, no glx metabolite differences were found [84, 86]. No study to date has investigated glutamine in this region.

Five of the six aforementioned studies investigated dlPFC glx metabolites following treatment for depression (Table 5). Only two studies showed an increase, both on the left side, with one reporting a compound glx increase following ECT [83] and another reporting glutamate increase in the left dlPFC but not the right dlPFC following 8 weeks of SSRI treatment [84]. The remainder of studies reported no changes in glutamate in the left or right dlPFC following treatments with either ECT [86, 87] or antidepressants [88].

In summary, dlPFC compound glx but not glutamate seems to be associated with depression on the left side. Most studies suggest that treatment appears to have little effect on glx metabolites. Further studies investigating glutamine may shed light on this issue. There is no evidence for glx involvement in the right dlPFC.

3.2.3 Medial PFC NAA

Eight studies investigated NAA in the mPFC (Table 4). Half demonstrated no NAA difference in depression [25, 76, 80, 82]. Two showed a NAA decrease in depression in both left and right mPFC [89, 90], while a third showed a decrease in chronic patients but not first-episode patients [77]. One study found a NAA increase in treated compared to placebo patients [79]. Only two studies have investigated mPFC NAA following treatment for depression (Table 5). Results are mixed, with one large study finding NAA increased following escitalopram treatment [79] and another smaller study finding NAA reduced following ECT [82].

In summary, mPFC NAA results in depression are somewhat mixed with both increases and decreases at baseline and following treatment found.

3.2.4 Dorsolateral PFC NAA

Seven studies investigated NAA in the dlPFC (Table 4), with no study finding any change in the metabolite in the region in depression. Five studies investigated dlPFC NAA following treatment for depression (Table 5) with one study finding a decrease after ECT [87].

3.3 Anterior Cingulate Cortex

The human cingulate cortex is cytoarchitecturally and functionally varied and is divided into anterior cingulate, posterior cingulate, and subgenual cortices. Due to its role in attention, reward, and emotion, the anterior cingulate (ACC) is the subregion most implicated in depression [91]. The ACC consists of three areas: dorsal

Table 4
Prefrontal cortex NAA in depression

Author	Year	PFC region	p-value	MDD (N)	MDD age	Female/Male	MDD duration	Controls (N)	Control age	Tesla	MRS type	Software
NAA decreased												
Portella (CD)	2011	vmPFC	0.04	19	51.0	15/4	273.6	15	40.5	3 T	SV	LCModel
Shan	2017	mPFC	0.002 (left), 0.002 (right)	31	30.3	15/16	445.8	31	26.5	3 T	2D MRSI	n/a
Liu	2020	mPFC	0.001 (left), 0.091 (right)	29	31.2	21/8	35.0	32	32.4	3 T	2D MRSI	GE workstation
NAA increased												
Taylor (treated)	2012	mPFC	$p < 0.01$	39	31.5	22/17	n/a	27	31.4	3 T	SV	LCModel
NAA unchanged												
Hasler	2007	dmPFC	0.11	20	34.0	13/7	225.6	20	34.8	3 T	SV	n/a
Hasler	2007	vmPFC	0.45	20	34.0	13/17	225.6	20	34.8	3 T	SV	n/a
Portella (FED)	2011	vmPFC	NS	19	51.0	15/4	273.0	15	40.5	3 T	SV	LCModel
Taylor (placebo)	2012	mPFC	NS	39	31.5	22/17	n/a	27	31.4	3 T	SV	LCModel
Li	2016	mPFC	NS	20	28.0	13/7	56.0	20	31.7	3 T	2D MRSI	LCModel
Shirayama	2016	mPFC	0.94	22	40.9	5/17	n/a	27	36.8	3 T	SV	LCModel
Knudsen	2018	mPFC	NS	6	38.4	2/4	n/a	11	38.8	3 T	SV	LCModel
Michael	2003	dlPFC	NS (left)	12	63.4	8/4	152.0	12	62.0	1.5 T	SV	LCModel

Brambilla	2005	dlPFC	0.35 (left)	19	37.4	13/6	13.3	19	37.0	1.5 T	SV	n/a
Kaymak	2009	dlPFC	NS (left)	17	30.9	17/0	4.4	13	29.1	3 T	SV	Siemens workstation
Nery	2009	dlPFC	0.51 (left)	37	36.6	24/13	189.0	40	40.0	1.5 T	SV	LCModel
Merkl	2011	dlPFC	0.11 (left)	25	49.1	20/5	n/a	27	36.3	3 T	SV	SPM2
Chen	2014	dlPFC	NS (left), NS (right)	15	27.9	9/6	177.0	15	27.5	1.5 T	SV	LCModel
Şendur	2020	dlPFC	0.075 (left), 0.81 (right)	30	38.8	23/7	118.0	30	33.4	1.5 T	SV	GE workstation

The table is divided into studies that show NAA decreased, increased, or unchanged. Data is subdivided into mPFC and dlPFC studies. *CD* chronic depression, *dmPFC* dorsomedial prefrontal cortex, *FED* first-episode depression, *MDD* major depressive disorder, *mPFC* medial prefrontal cortex, *MRS* magnetic resonance spectroscopy, *MRSI* magnetic resonance spectroscopic imaging, *N* number, *n/a* not available, *NAA* N-acetylaspartate, *PFC* prefrontal cortex, *SV* single voxel, *vmPFC* ventromedial prefrontal cortex

Table 5
Prefrontal cortex glx and NAA post treatment for depression

Author	Year	PFC region	Treatment	Treatment effect (p-value)	MDD (N)	Mean Age	Female/Male	MDD duration	Controls (N)	Control age	Comment
NAA											
Taylor	2012	mPFC	Antidepressants	↑ (<0.01)	39	31.5	22/17	n/a	27	31.4	7 day escitalopram treatment
Knudsen	2018	mPFC	ECT	↓ (0.0038)	6	38.4	2/4	n/a	11	38.8	NAA decreased as ECT number increased
Michael	2003	dlPFC	ECT	No change (left)	12	63.4	8/4	152.4	12	62.0	9.9 average ECT treatments
Kaymak	2009	dlPFC	Antidepressants	No change (left)	17	30.9	17/0	4.4	13	29.1	8 weeks SSRI treatment
Merkl	2011	dlPFC	ECT	↓ (0.03) (left)	25	49.0	20/5	16.1	27	36.3	NAA doesn't predict response to treatment
Chen	2014	dlPFC	Antidepressants	No change (left), no change (right)	15	26.2	9/6	14.8	15	27.6	8 weeks SSRI treament
Şendur	2020	dlPFC	Antidepressants	No change (left), no change (right)	30	38.8	23/7	118.8	30	33.4	8 weeks SSRI treament
Glx											
Draganova	2020	vmPFC	Antidepressants	[glu], [gln] no change	31	37.3	18/13	n/a	63	41.8	One year follow up
Knudsen	2018	mPFC	ECT	[glu] no change	6	38.4	2/4	n/a	11	38.8	ECT 3 times per week
Michael	2003	dlPFC	ECT	[glx] ↑(0.016) (left)	12	63.4	8/4	152.4	12	62.0	9.9 average ECT treatments
Merkl	2011	dlPFC	ECT	[glu] no change (left)	25	49.1	20/5	n/a	27	36.3	ECT 3 times per week
Grimm	2012	dlPFC	Antidepressants	[glu] no change (left)	14	38.8	7/7	n/a	14	n/a	4 weeks treatment, various antidepressants
Chen	2014	dlPFC	ECT	[glu] no change (left), no change (right)	15	27.9	9/6	14.8	15	27.6	Glu baseline predicted response
Şendur	2020	dlPFC	Antidepressants	[glu] increase (left), no change (right)	30	38.8	23/7	118.8	30	33.4	8 weeks SSRI treament

The table is divided into studies that investigate NAA or glx. Data is subdivided into mPFC and dlPFC studies. *ECT* electroconvulsive therapy, *glu* glutamate, *glx* composite glutamate-glutamine *MDD* major depressive disorder, *mPFC* medial prefrontal cortex. *N* number, *n/a* not available, *NAA* N-acetylaspartate, *PFC* prefrontal cortex, *SSRI* selective serotonin reuptake inhibitors, *vmPFC* ventromedial prefrontal cortex

(BA 32), ventral (BA 24), and pregenual (BA 33). The dorsal area (dACC) has connections with the adjacent prefrontal and parietal cortices and is involved in attentional allocation [92], whereas the ventral ACC has connections with the amygdala, hypothalamus, hippocampus, and insula suggesting an involvement with emotional salience and motivation [93]. The pregenual region (pgACC) also has involvement with emotions [94]. Although often considered part of the ACC in some literature, the subgenual cingulate cortex (sgACC) has cytoarchitectural and hodological distinctions from the ACC, and overactivity has been shown in depression [95]. Unfortunately, due to the spatial limitations of MRS, the majority of studies investigate a large cubic voxel placed somewhere in the midline anterior to the genu of the corpus callosum, overlapping some or all three ACC areas, occasionally the sgACC as well as mPFC areas. This can lead to difficulty in MRS interpretation in the region.

3.3.1 Anterior Cingulate glx

Nineteen studies have investigated glx metabolites in the ACC (Table 6). The composite glx peak shows a reduction [86, 96–99] in five studies (total depressed $n = 109$) with two studies (total depressed $n = 52$) reporting no difference [80, 100] and one study showing an increase [101]. Interestingly, one large study ($n = 43$) investigating ACC subregions found that glx was reduced in the sgACC but not the dACC in depression [99]. Overactivity of sgACC has been associated with depression in fMRI studies [95]. In studies reporting the glutamate peak, four studies (total depressed $n = 106$) revealed a decrease [87, 102–104], compared with seven studies (total depressed $n = 177$) reporting no change [94, 105–110] and one study showing an increase in glutamate [101]. Of these, six studies also reported on glutamine with most showing no change [105, 106, 108, 109], with one study each reporting a decrease [94] and increase [101].

Following treatment for depression, ACC glx metabolites increased in four studies [86, 97, 99, 102] and remained unchanged in four studies [87, 100, 107, 108] (Table 8). All increases were noted on a baseline of lower pretreatment glx. No study reported a decrease in ACC glx following treatment. ECT demonstrated a glx increase in three [97, 99, 102] out of four studies. Again, interestingly, the dACC but not the sgACC revealed glx increase following ECT [99]. The only study to report laterality showed no change in either left or right ACC glx following rTMS. The only study to report glutamine revealed no change in glutamine following 8 weeks of SSRI therapy [108].

In summary, the evidence for ACC glx metabolite disturbance in depression is mixed and may be region dependent. Large MRS voxels straddling multiple cytoarchitecturally and functionally distinct areas may not have the resolution to discover complex region-specific metabolite differences. Any changes in glx metabolites

Table 6
Anterior cingulate cortex glx in depression

Author	Year	p-value	MDD (N)	MDD age	Female/Male	MDD duration	Controls (N)	Control age	Tesla	MRS type	Software
Glx decreased											
Auer	2000	[glx] 0.043	19	50.2	13/6	n/a	18	43.2	1.5 T	SV	LCModel
Pfleiderer	2003	[glx] 0.001 (left only)	17	61.0	12/5	178.8	17	60.1	1.5 T	SV	LCModel
Price (TRD)	2009	[glx] 0.03	15	46.8	7/8	>24.0	24	37.3	3 T	SV	n/a
Walter	2009	[gln] 0.01 (right)	19	40.0	11/8	n/a	24	34.6	3 T	SV	LCModel
Merkl	2011	[glu] 0.01	25	49.0	20/5	n/a	27	36.3	3 T	SV	n/a
Zhang	2013	[glu] 0.01	10	44.0	6/4	n/a	10	39.0	3 T	SV	LCModel
Chen	2014	[glx] 0.006	15	26.2	9/6	n/a	15	27.6	1.5 T	SV	LCModel
Njau (sgACC)	2017	[glx] 0.025	43	43.8	27/23	213.4	33	39.3	3 T	SV	LCModel
Wise	2018	[glu] 0.03	20	29.6	18/2	76.2	20	30.1	3 T	SV	LCModel
Benson	2020	[glu] 0.01	51	33.2	33/18	n/a	25	33.9	4 T	SV	LCModel
Glx increased											
Abdallah	2017	0.009 [glx] 0.03 [glu] 0.03 [gln]	16	36.7	n/a	n/a	23	35.7	3 T	SV	n/a
Glx unchanged											
Price (nTRD)	2009	[glx] 0.99	18	38.3	6/12	>24.0	24	37.3	3 T	SV	n/a
Walter	2009	[glu] NS (right)	19	40.0	11/8	n/a	24	34.6	3 T	SV	LCModel
Zheng	2015	[glx] 0.46 (left), 0.43 (right)	32	26.9	12/20	56.4	28	27.6	3 T	3D MRSI	NUMARIS/4

Study	Year		N				N				
Li	2016	[glx] NS	20	28.0	13/7	56.4	20	31.7	3 T	2D MRSI	LCModel
Njau (dACC)	2017	[glx] NS	43	43.8	27/23	213.4	33	39.3	3 T	SV	LCModel
Taylor	2017	[glu] 0.354, [gln] 0.592 (left)	17	22.5	11/6	28.6	18	23.9	7 T	SV	Custom
Godlewska	2018	[glu] 0.51, [gln] 0.075	55	31.3	31/24	n/a	50	31.3	7 T	SV	LCModel
Evans	2018	[glu] 0.7	20	36.2	12/8	n/a	17	34.7	7 T	SV	Custom
Brennan	2017	[glu] NS, [gln] NS	19	38.5	8/11	116.0	10	38.4	3 T	SV	LCModel
Colic	2019	[glu] 0.5, [gln] 0.21	32	40.9	19/13	n/a	32	33.1	7 T	SV	LCModel
Nugent	2019	[glu] 0.22	15	35.0	16/14	228.0	14	34.0	3 T	SV	n/a

The table is divided into studies that show glx decreased or glx unchanged. *dACC* dorsal anterior cingulate cortex, *gln* glutamine, *glu* glutamate, *glx* composite glutamate-glutamine, *MDD* major depressive disorder, *MRS* magnetic resonance spectroscopy, *MRSI* magnetic resonance spectroscopic imaging, *N* number, *n/a* not available, *NS* non-significant, *nTRD* non-treatment-resistant depression, *gACC* subgenual anterior cingulate cortex, *SV* single voxel, *TRD* treatment-resistant depression

following treatment occur on a baseline of lower concentrations, suggesting that pretreatment glx may be a marker for potential treatment efficacy. The sgACC appears to be associated with reduced glx in depression but shows no glx response to treatment.

3.3.2 Anterior Cingulate NAA

Nineteen studies have also investigated NAA in the ACC in depression (Table 7). No studies show NAA increased in depression. A minority of studies (depressed patients $n = 118$) report NAA decreased [86, 87, 100, 111–113], with two of these actually reporting at trend level rather than significance [111, 113]. The majority of studies (depressed patients = 241) show no change in NAA in depression [24, 80, 94, 96, 97, 99, 102, 105, 107, 114, 115]. The only study investigating the sgACC region shows no NAA differences in depression.

The effects of depression treatment on ACC NAA are ambiguous with four studies showing an increase [86, 87, 100, 107] and three showing a decrease [99, 102, 116] (Table 8). ECT was more associated with NAA decrease following treatment, whereas other treatments (antidepressants, rTMS [repetitive transcranial magnetic stimulation], and ketamine) are associated with NAA increase. This may reflect a differential effect of ECT on ACC NAA concentration but similarly may also reflect the severity of patients referred for this therapy. The only study to investigate laterality reported increased left ACC NAA but not right following a course of rTMS [100]. The only study to investigate the subgenual ACC found no effect of ECT on NAA concentration in this region [99].

In summary, the majority of MRS evidence points to limited ACC NAA involvement in depression. The sgACC region shows no NAA changes. There may, however, be a differential effect of treatment on ACC NAA response, with ECT associated with lower NAA and non-ECT treatments associated with higher NAA post treatment.

3.4 Hippocampus

The hippocampus is one of the most consistently implicated studies in neuroimaging studies of depression [117, 118]. This temporal lobe structure has key roles in memory and spatial coordination. It is a fundamental hub within the limbic system, with connectivity throughout the brain, including the cingulate cortex, thalamus, prefrontal regions, and brain stem [119, 120]. As such, the hippocampus lies at the interface between memory, incoming signals, cognition, and behavior. Until recently, relatively few studies have investigated the hippocampus in depression using MRS. Many reasons exist for this. The hippocampus is a heterogeneous structure consisting of many overlapping and intercoiled neuronal "fields" [121]. Similarly, the hippocampal shape doesn't lend itself easily to MRS, being long and thin and potentially curving rather than a large homogeneous block. As such, smaller voxels focusing on just the head or body of the hippocampus are frequently used,

Table 7
Anterior cingulate cortex NAA in depression

Author	Year	p-value	MDD (N)	MDD age	Female/Male	Illness duration	Controls (N)	Control age	Tesla	MRS type	Software
NAA decreased											
Gonul	2006	0.07 (trend)	20	32.1	17/3	8.7	18	28.3	1.5 T	SV	n/a
Merkl	2011	0.005	25	49.0	20/5	n/a	27	36.3	3 T	SV	n/a
Chen	2014	0.004	15	26.2	9/6	n/a	15	27.6	1.5 T	SV	LCModel
Zheng	2015	0.02 (left), NS (right)	32	26.9	12/20	56.4	28	27.6	3 T	3D MRSI	NUMARIS/4
Li (b)	2016	0.044 (left), NS (right)	16	30.2	11/5	n/a	10	32.1	7 T	3D MRSI	LCModel
Nase	2016	0.069 (trend)	15	39.1	47/29	n/a	41	36.6	3 T	SV	LCModel
Gules	2019	0.006	10	42.3	5/5	n/a	10	42.1	3 T	SV	Philips workstation
NAA unchanged											
Auer	2000	NS	19	50.2	13/6	n/a	18	43.2	1.5 T	SV	LCModel
Pfleiderer	2003	0.28 (left)	17	61.0	12/5	178.8	17	60.1	1.5 T	SV	LCModel
Coupland	2005	0.1	13	37.0	7/6	n/a	13	32.0	3 T	SV	LCModel
Walter	2009	NS	19	40.0	11/8	n/a	24	34.6	3 T	SV	LCModel
Wang	2012	0.870 (left), 0.995 (right)	24	30.2	14/10	9.2	13	29.7	1.5 T	2D MSRI	GE workstation
Zhang	2013	NS	10	44.0	6/4	n/a	10	39.0	3 T	SV	LCModel
Jia	2015	0.767 (left), 0.908 (right)	26	32.9	15/11	30.4	13	28.1	1.5 T	2D MRSI	GE workstation

(continued)

Table 7
(continued)

Author	Year	p-value	MDD (N)	MDD age	Female/ Male	Illness duration	Controls (N)	Control age	Tesla	MRS type	Software
Li	2016	0.608	20	28.0	13/7	56.4	20	31.7	3 T	2D MRSI	LCModel
Njau (daCC)	2017	NS	43	43.8	27/23	213.4	33	39.3	3 T	SV	LCModel
Njau (sgACC)	2017	NS	43	43.8	27/23	213.4	33	39.3	3 T	SV	LCModel
Taylor	2017	0.23 (left)	17	22.5	11/6	28.6	18	23.9	7 T	SV	Custom
Evans	2018	0.9	20	36.2	12/8	n/a	17	34.7	7 T	SV	LCModel
Tosun	2020	0.876	13	45.9	7/6	n/a	14	41.4	3 T	SV	n/a

The table is divided into studies that show NAA decreased or NAA unchanged. *dACC* dorsal anterior cingulate cortex, *MDD* major depressive disorder, *MRS* magnetic resonance spectroscopy, *MRSI* magnetic resonance spectroscopic imaging, *N* number, *n/a* not available, *NAA* N-acetylaspartate, *NS* non-significant, *nTRD* non-treatment-resistant depression, *sgACC* subgenual anterior cingulate cortex, *SV* single voxel, *TRD* treatment-resistant depression

Table 8
Anterior cingulate cortex glx and NAA post treatment for depression

Author	Year	Treatment	Treatment effect (p-value)	MDD (N)	Mean age	Female/Male	MDD duration	Controls (N)	Control age	Comment
NAA										
Merkl	2011	ECT	↑ (0.03)	25	49.0	20/5	n/a	27	36.3	NAA doesn't predict response to treatment
Zhang	2013	ECT	↓ (0.048)	10	44.0	6/4	n/a	10	39.0	Assessed at: Baseline, second and sixth ECT sessions
Chen	2014	Antidepressants	↑ (0.017)	15	26.2	9/6	n/a	15	27.6	8 weeks SSRI treatment
Zheng	2015	rTMS	↑ Left (0.02), no change right	32	26.9	12/20	56.4	28	27.6	Neurocognitive improvement with increased left NAA
Njau (dACC)	2017	ECT	↓ (0.001)	43	43.8	27/23	213.4	33	39.3	Higher NAA predicted greater HAM_D reductions
Njau (sgACC)	2017	ECT	No change	43	43.8	27/23	213.4	33	39.3	ECT 3 times per week
Evans	2018	Ketamine infusion	↑ (0.05)	20	36.2	12/8	n/a	17	34.7	Infusions 2 weeks apart
Tosun	2019	ECT	↓ (0.044)	13	45.9	7/6	n/a	10	42.1	6 sessions ECT
Glx										
Chen	2014	Antidepressants	[glx] ↑ (0.017)	15	26.2	9/6	n/a	15	27.6	8 weeks SSRI treatment
Pfleiderer	2003	ECT	[glx] ↑ (0.04)	17	61.0	12/5	178.8	17	60.1	Glx increase in responders only
Evans	2018	Ketamine infusion	[glx] no change (0.1)	20	36.2	12/8	n/a	17	34.7	Response not predicted by glutamate levels
Merkl	2011	ECT	Glu no change (0.87)	25	49.0	20/5	n/a	27	36.3	Glu baseline predicted response to treatment

(continued)

Table 8
(continued)

Author	Year	Treatment	Treatment effect (p-value)	MDD (N)	Mean age	Female/ Male	MDD duration	Controls (N)	Control age	Comment
Njau (dACC)	2017	ECT	[glx] ↑ (0.025)	43	43.8	27/23	213.4	33	39.3	Glx increase correlate with mood rating improvements
Njau (sgACC)	2017	ECT	[glx] no change	43	43.8	27/23	213.4	33	39.3	ECT 3 times per week
Zhang	2013	ECT	[glu] (0.037)	10	44.0	6/4	n/a	10	39.0	Glu variations correlate with depression scores
Zheng	2015	rTMS	[glx] no change left (0.46), right (0.43)	32	26.9	12/20	56.4	28	27.6	20 sessions within 4 weeks
Brennan	2017	Antidepressants	[glu] [gln] no change	19	38.5	8/11	116.5	10	38.4	6 weeks treatment with escitalopram

The table is divided into studies that investigate NAA or glx. *dACC* dorsal anterior cingulate cortex, *ECT* electroconvulsive therapy, *gln* glutamine, *glu* glutamate, *glx* composite glutamate-glutamine, *MDD* major depressive disorder, *N* number, *n/a* not available, *NAA* N-acetylaspartate, *NS* non-significant, *rTMS* repetitive transcranial magnetic stimulation, *sgACC* subgenual anterior cingulate cortex, *SSRI* selective serotonin reuptake inhibitors

potentially leading to noisy spectra. Finally, lying at the base and medial aspect of the temporal lobe, it lies close to the bone and CSF. However, recent increases in field strengths and shimming advances are allowing capture of adequate robust spectra in the important limbic structure.

3.4.1 *Hippocampal glx*

Seven studies have investigated glx metabolites in the hippocampus (Table 9) with most showing no change (depressed $n = 145$) in glx metabolites in the hippocampus [25, 122–124]. However, the composite glx peak was smaller (depressed $n = 81$) in three studies [99, 125, 126]. One report of depression chronicity revealed composite glx reduction in both left and right hippocampus in chronic but not first-episode depression [126]. However, this was not supported in an alternate left hippocampal study of first-episode and recurrent depression [124]. Two further studies have found a decrease in compound glx in the left hippocampus [99, 125]. The one study to investigate hippocampal glx, glutamate, and glutamine in depression reported no difference in glx and glutamate but did suggest a potential hippocampal reduction in glutamine in depression [25].

Only three studies have investigated the effects of depression treatment on glx metabolites in the hippocampus (Table 11). Antidepressants were reported to have no effect on compound glx in either side [125], while a course of ECT was associated both with a glx reduction [99] and increase in the left hippocampus [127].

In summary, the evidence for hippocampal glx involvement in either depression or its treatment is very mixed with only one study of glx components. Additional studies investigating glutamate and glutamine may reveal a further pattern in depression.

3.4.2 *Hippocampal NAA*

Of the 11 hippocampal NAA studies (Table 10) examined, a predominance of no differences between controls and depression [25, 114, 115, 123–125, 128, 129] was observed. Only two studies showed a decrease in NAA in this region. One study showed reduced NAA in the left but not right hippocampus [99], while another reported a NAA reduction in both the left and right hippocampus in chronic depression but not first-episode depression [126].

Of the five treatment studies investigating hippocampal NAA (Table 11), only one found a change in the left hippocampus with treatment, with NAA decreasing following ECT [127]. However, of the two studies investigating the right hippocampus, one study did show an NAA increase following antidepressants [114], while another found an NAA decrease following ECT [99].

In summary, there is limited MRS evidence for NAA disruption in the hippocampus or resolution with treatment.

Table 9
Hippocampal glx in depression

Author	Year	p-value	MDD (N)	MDD age	Female/Male	MDD duration	Controls (N)	Control age	Tesla	MRS type	Software
Glx decreased											
Block	2009	[glx] 0.003, [gln] 0.015 (left)	18	36.0	8/10	n/a	10	36.0	3 T	SV	jMRUI
De Diego-Adeliño (CD)	2013	[glx] 0.008 (right), 0.024 (left)	20	49.6	16/4	280.0	16	43.4	3 T	SV	LCModel
Njau	2017	[glx] 0.049 (left)	43	43.8	27/23	213.0	33	33.9	3 T	SV	LCModel
Glx unchanged											
Milne (FED)	2009	[glx] NS (left)	14	32.1	5/9	121.7	13	30.0	3 T	SV	LCModel
Milne (RD)	2009	[glx] NS (left)	14	47.3	10/4	293.1	14	41.8	3 T	SV	LCModel
De Diego-Adeliño (FED)	2013	[glx] NS (left), NS (right)	14	42.5	9/5	6.8	16	43.4	3 T	SV	LCModel
Hermens	2015	[glx] NS (left)	160	18.0–30.0	105/55	n/a	40	19.0–30.0	3 T	SV	LCModel
Poletti	2016	[glx] (NS) (no laterality)	19	51.5	14/5	225.0	17	26.5	3 T	SV	LCModel
Njau	2017	[glx] NS (right)	43	43.8	27/23	213.4	33	33.9	3 T	SV	LCModel
Shirayama	2017	[glx] 0.152, [glu] 0.352 (right)	22	40.9	5/17	n/a	27	36.8	3 T	SV	LCModel

The table is divided into studies that show glx decreased or glx unchanged. *CD* chronic depression, *FED* first-episode depression, *glx* composite glutamate-glutamine, *MDD* major depressive disorder, *MRS* magnetic resonance spectroscopy, *MRSI* magnetic resonance spectroscopic imaging, *N* number, *n/a* not available, *NS* non-significant, *RD* recurrent depression, *SV* single voxel

Hippocampal NAA in depression

Author	Year	p-value	MDD (N)	MDD age	Female/Male	MDD duration	Controls (N)	Control age	Tesla	MRS type	Software
NAA decreased											
De Diego-Adeliño (CD)	2013	0.001 (right), 0.07 (left)	20	49.6	16/4	280.3	16	43.4	3 T	SV	LCModel
Njau	2017	0.001 (left)	43	43.8	27/23	213.4	33	39.3	3 T	SV	LCModel
NAA unchanged											
Ende	2000	NS (bilateral)	23	61.3	12/11	11.3	24	35.3	1.5 T	2D MRSI	n/a
Block	2009	0.405 (left)	18	36.1	8/10	n/a	10	36.1	3 T	SV	jMRUI
Milne (FED)	2009	NS (left)	14	32.1	5/9	121.7	13	30.0	3 T	SV	LCModel
Milne (RD)	2009	NS (left)	14	47.3	10/4	293.1	14	41.8	3 T	SV	LCModel
Wang	2012	0.085 (left), 0.318 (right)	26	32.1	15/11	2.5	13	29.7	1.5 T	2D MRSI	GE workstation
De Diego-Adeliño (FED)	2013	NS (left), NS (right)	14	42.5	9/5	6.8	16	43.4	3 T	SV	LCModel
Hermens	2015	NS (left)	160	18.0–30.0	105/55	n/a	40	19.0–30.0	3 T	SV	LCModel
Jia	2015	0.8 (left), 0.065 (right)	26	32.9	15/11	30.4	13	28.1	1.5 T	2D MRSI	GE workstation
Njau	2017	NS (right)	43	43.8	27/23	213.4	33	39.3	3 T	SV	LCModel
Shirayama	2017	0.984 (right)	22	40.9	5/17	n/a	27	36.8	3 T	SV	LCModel
Xu	2018	NS (bilateral)	15	34.6	15/0	n/a	12	34.1	3 T	SV	LCModel
Yingying	2018	0.78 (left), 0.58 (right)	12	35.4	12/0	26.1	12	35.1	3 T	SV	LCModel

The table is divided into studies that show NAA decreased or NAA unchanged. *CD* chronic depression, *FED* first-episode depression, *MDD* major depressive disorder, *MRS* magnetic resonance spectroscopy, *MRSI* magnetic resonance spectroscopic imaging, *N* number, *n/a* not available, *NAA* N-acetylaspartate, *NS* non-significant, *RD* recurrent depression, *SV* single voxel

Table 11
Hippocampal glx and NAA post treatment for depression

Author	Year	Treatment	Treatment effect (*p*-value)	MDD (N)	Mean age	Female/Male	MDD duration	Controls (N)	Control age	Comment
NAA										
Ende	2000	ECT	No change	23	61.3	11/12	11.3	24	35.3	5–18 ECT treatments
Block	2009	Antidepressants	No change 0.561 (left)	18	36.0	8/10	n/a	10	36.0	8 weeks of treatment;
Wang	2012	Antidepressants	No change (left), ↑ (0.003) right	26	32.1	15/11	2.5	13	29.7	12 weeks treatment (SNRI)
Cano	2017	ECT	↓ (0.015) left	12	59.2	6/6	n/a	10	54.4	Decreased after ninth ECT session
Njau	2017	ECT	No change (left), ↓ (0.001) right	43	43.8	27/23	213.4	33	39.3	ECT 3 times per week
Glx										
Block	2009	Antidepressants	[glx] 0.585, [gln] 0.743 (left)	18	36.0	8/10	n/a	10	36.0	8 weeks of treatment
Cano	2017	ECT	[glx] ↑ 0.05 left	12	59.2	6/6	n/a	10	54.4	Increased after ninth session (trend)
Njau	2017	ECT	[glx] ↓ 0.016 (left), no change right	23	61.3	12/11	213.4	24	35.3	ECT 3 times per week

The table is divided into studies that investigate NAA or glx. *ECT* electroconvulsive therapy, *glx* composite glutamate-glutamine, *MDD* major depressive disorder, *N* number, *n/a* not available, *NAA* *N*-acetylaspartate, *NS* non-significant, *SNRI* serotonin-norepinephrine reuptake inhibitors

3.5 Summary of PFC, ACC, and Hippocampal glx and NAA Findings in Depression

The compound glx peak can be considered to comprise glutamate and glutamine components. With increasing hardware advances, optimized scanning protocols, and postprocessing curve-fitting software, glutamate and glutamine signals are being separated with increasing reliability. In this review of MRS glx in depression, the compound *glx but not glutamate* appears to be reduced in the left dlPFC (not the right) in depression. Conversely, *glutamate but not compound glx* appears to be reduced in the mPFC in depression. This may reflect a differential abnormality in the glutamate-glutamine cycle in the left dlPFC and mPFC glial cells [19]. Further studies focusing on the glutamine peak may resolve these findings. The evidence for glx metabolite depression differences in the ACC and hippocampus is less robust. Treatment appears to have minimal/mixed effects on glx and its components in all areas reviewed. Further studies, focusing on separating glutamate and glutamine accurately from glx and/or investigating glutamate/glutamine ratios in all regions, may further illuminate the role of these metabolites in depression.

NAA is considered a marker of neuronal density [15]. As it has also been found in high concentrations in myelin and oligodendrocytes, this metabolite may also be used as a white matter marker [14]. There is limited MRS evidence for reduced NAA as marker of reduced neuronal or white matter viability in the PFC, ACC, or hippocampus.

However, intriguingly, NAA may show a differential response to treatment modality. When differences exist, NAA mostly increases in the mPFC and ACC but not the dlPFC and hippocampus following non-ECT treatments (antidepressants, ketamine, and rTMS). It has already been noted that considerable overlap exists between the mPFC and ACC voxels in many of the MRS studies reviewed. Both the mPFC and ACC regions have been implicated in depression and also in the response to treatment [130–133] with repeated stress causing dendritic loss in both the ACC and mPFC in animal models [134, 135]. Antidepressant treatments, irrespective of action, have been associated with dendritic remodeling, synaptogenesis [136], and white matter regeneration [137]. The NAA increases found in both the mPFC and ACC following non-ECT treatment could reflect increased neuronal, oligodendritic, and myelin viability.

Conversely, when differences exist, ECT is mostly associated with reduced NAA post treatment in all three areas. This may suggest a potential neurotoxic effect of the treatment. Epilepsy has been associated with reduced NAA following seizure activity [138], suggesting a similar mechanism with ECT. As the MRS scans were performed relatively soon after the course of ECT was completed, follow-up studies are needed to explore if NAA remains reduced or rises in the longer term post ECT.

Further studies using more defined neuroanatomical definitions of ACC and mPFC voxels following both ECT and non-ECT treatments for depression may help clarify the existence of any differential effect of treatment on ACC/mPFC NAA concentration.

4 Conclusion

Although underutilized as a neuroimaging modality, recent improvements in MRI hardware, focused sequences, and post-scanning software are expanding the role of MRS across psychiatric research. MRS requires meticulous planning with awareness of signal-to-noise considerations. Small incremental increases or decreases in the signal-to-noise ratio along each step of the process may make the difference between generating robust and reliable spectra or contaminated and uninterpretable spectra.

MRS allows the investigation of molecules other than water in the brain. Unfortunately, the number and usefulness of these molecules are limited, particularly in psychiatric research. Two metabolites with potential utility in depression research are NAA and the compound metabolite glx (with its constituent components glutamate and glutamine). NAA is a marker of general neuronal, glial, and white matter integrity. Glx and its constituents may be markers of glial-neuronal glutamate-glutamine cycling or excitotoxicity. A mini-review of MRS studies of these two metabolites suggests involvement in key limbic regions in depression. Although glx appears to have a limited role in the hippocampus and anterior cingulate cortex in depression, a differential glx-glutamine MRS expression may be present in the prefrontal cortex. Glx but not glutamate appears reduced in the left dorsolateral prefrontal region, whereas glutamate but not glx appears reduced in the medial prefrontal region. Glutamine estimation in these areas is unknown, and resolving the precise resonances of glx, glutamate, and glutamine may untangle the role of the glx in the prefrontal cortex. NAA shows nonspecific findings in the prefrontal, anterior cingulate, and hippocampal regions in depression. However, NAA may exhibit a differential effect following ECT or non-ECT treatments. ECT treatment appears more associated with NAA decrease following treatment, whereas non-ECT treatment appears more associated with NAA increase. This may reflect a generalized neuronal injury immediately after ECT (NAA decrease) compared to a general neuropil increase after antidepressant treatment (NAA increase). Longitudinal studies following ECT and non-ECT treatments may help elucidate this finding.

References

1. Plewes DB, Kucharczyk W (2012) Physics of MRI: a primer. J Magn Reson Imaging 35:1038–1054

2. Jung WI, Staubert A, Widmaier S, Hoess T, Bunse M, van Erckelens F et al (1997) Phosphorus J-coupling constants of ATP in human brain. Magn Reson Med 37:802–804

3. Öz G, Deelchand DK, Wijnen JP, Mlynárik V, Xin L, Mekle R et al (2020) Advanced single voxel 1H magnetic resonance spectroscopy techniques in humans: Experts' consensus recommendations. NMR Biomed:e4236

4. Posse S, Otazo R, Dager SR, Alger J (2013) MR spectroscopic imaging: principles and recent advances. J Magn Reson Imaging 37: 1301–1325

5. Maudsley AA, Andronesi OC, Barker PB, Bizzi A, Bogner W, Henning A et al (2020) Advanced magnetic resonance spectroscopic neuroimaging: Experts' consensus recommendations. NMR Biomed:e4309

6. Wijnen JP, Klomp DW (2013) Advances in magnetic resonance spectroscopy. PET Clin 8:237–244

7. Rothman DL, De Feyter HM, de Graaf RA, Mason GF, Behar KL (2011) 13C MRS studies of neuroenergetics and neurotransmitter cycling in humans. NMR Biomed 24: 943–957

8. Harper DG, Joe EB, Jensen JE, Ravichandran C, Forester BP (2016) Brain levels of high-energy phosphate metabolites and executive function in geriatric depression. Int J Geriatr Psychiatry 31:1241–1249

9. Strauss WL, Layton ME, Dager SR (1998) Brain elimination half-life of fluvoxamine measured by 19F magnetic resonance spectroscopy. Am J Psychiatry 155:380–384

10. Cianfoni A, Law M, Re T, Dubowitz D, Rumboldt Z, Imbesi S (2011) Clinical pitfalls related to short and long echo times in cerebral MR spectroscopy. J Neuroradiol 38: 69–75

11. Feingold KR, Elias PM (2014) Role of lipids in the formation and maintenance of the cutaneous permeability barrier. Biochim Biophys Acta Mol Cell Biol Lipids 1841:280–294

12. Verma A, Kumar I, Verma N, Aggarwal P, Ojha R (2016) Magnetic resonance spectroscopy—revisiting the biochemical and molecular milieu of brain tumors. BBA Clin 5: 170–178

13. Lin DD, Crawford TO, Barker PB (2003) Proton MR spectroscopy in the diagnostic evaluation of suspected mitochondrial disease. Am J Neuroradiol 24:33–41

14. Nordengen K, Heuser C, Rinholm JE, Matalon R, Gundersen V (2015) Localisation of N-acetylaspartate in oligodendrocytes/ myelin. Brain Struct Funct 220:899–917

15. Moffett JR, Ross B, Arun P, Madhavarao CN, Namboodiri AM (2007) N-Acetylaspartate in the CNS: from neurodiagnostics to neurobiology. Prog Neurobiol 81:89–131

16. Lentz MR, Kim JP, Westmoreland SV, Greco JB, Fuller RA, Ratai EM et al (2005) Quantitative neuropathologic correlates of changes in ratio of N-acetylaspartate to creatine in macaque brain. Radiology 235:461–468

17. Stanley JA, Raz N (2018) Functional magnetic resonance spectroscopy: the "new" MRS for cognitive neuroscience and psychiatry research. Front Psychiatry 9:76

18. Zhao J, Verwer R, Van Wamelen D, Qi X-R, Gao S-F, Lucassen P et al (2016) Prefrontal changes in the glutamate-glutamine cycle and neuronal/glial glutamate transporters in depression with and without suicide. J Psychiatr Res 82:8–15

19. Newsholme P, Procopio J, Lima MMR, Pithon-Curi TC, Curi R (2003) Glutamine and glutamate—their central role in cell metabolism and function. Cell Biochem Funct 21:1–9

20. Rodríguez A, Ortega A (2017) Glutamine/ glutamate transporters in glial cells: much more than participants of a metabolic shuttle. Adv Neurobiol 16:169–183

21. Shukla D, Mandal PK, Ersland L, Grüner ER, Tripathi M, Raghunathan P et al (2018) A multi-center study on human brain glutathione conformation using magnetic resonance spectroscopy. J Alzheimers Dis 66:517–532

22. Fogaça MV, Duman RS (2019) Cortical GABAergic dysfunction in stress and depression: new insights for therapeutic interventions. Front Cell Neurosci 13:87

23. Riley CA, Renshaw PF (2018) Brain choline in major depression: a review of the literature. Psychiatry Res Neuroimaging 271:142–153

24. Coupland NJ, Ogilvie CJ, Hegadoren KM, Seres P, Hanstock CC, Allen PS (2005) Decreased prefrontal Myo-inositol in major depressive disorder. Biol Psychiatry 57: 1526–1534

25. Shirayama Y, Takahashi M, Osone F, Hara A, Okubo T (2017) Myo-inositol, glutamate, and glutamine in the prefrontal cortex,

hippocampus, and amygdala in major depression. Biol Psychiatry Cogn Neurosci Neuroimaging 2:196–204

26. Lufkin R, Anselmo M, Crues J, Smoker W, Hanafee W (1988) Magnetic field strength dependence of chemical shift artifacts. Comput Med Imaging Graph 12:89–96

27. Ladd ME, Bachert P, Meyerspeer M, Moser E, Nagel AM, Norris DG et al (2018) Pros and cons of ultra-high-field MRI/MRS for human application. Prog Nucl Magn Reson Spectrosc 109:1–50

28. Blüml S, Panigrahy A (2012) MR spectroscopy of pediatric brain disorders. Springer Science & Business Media, Berlin

29. Foerster BU, Tomasi D, Caparelli EC (2005) Magnetic field shift due to mechanical vibration in functional magnetic resonance imaging. Magn Reson Med 54:1261–1267

30. Near J, Edden R, Evans CJ, Paquin R, Harris A, Jezzard P (2015) Frequency and phase drift correction of magnetic resonance spectroscopy data by spectral registration in the time domain. Magn Reson Med 73:44–50

31. Tal A, Gonen O (2013) Localization errors in MR spectroscopic imaging due to the drift of the main magnetic field and their correction. Magn Reson Med 70:895–904

32. Thiel T, Czisch M, Elbel GK, Hennig J (2002) Phase coherent averaging in magnetic resonance spectroscopy using interleaved navigator scans: compensation of motion artifacts and magnetic field instabilities. Magn Reson Med 47:1077–1082

33. Saleh MG, Edden RA, Chang L, Ernst T (2020) Motion correction in magnetic resonance spectroscopy. Magn Reson Med 84:2312–2326

34. Helms G, Piringer A (2001) Restoration of motion-related signal loss and line-shape deterioration of proton MR spectra using the residual water as intrinsic reference. Magn Reson Med 46:395–400

35. Keating B, Deng W, Roddey JC, White N, Dale A, Stenger VA et al (2010) Prospective motion correction for single-voxel 1H MR spectroscopy. Magn Reson Med 64:672–679

36. Li BS, Regal J, Gonen O (2001) SNR versus resolution in 3D 1H MRS of the human brain at high magnetic fields. Magn Reson Med 46:1049–1053

37. Wilson M, Andronesi O, Barker PB, Bartha R, Bizzi A, Bolan PJ et al (2019) Methodological consensus on clinical proton MRS of the brain: review and recommendations. Magn Reson Med 82:527–550

38. Bottomley PA (1987) Spatial localization in NMR spectroscopy in vivo. Ann N Y Acad Sci 508:333–348

39. Xu D, Vigneron D (2010) Magnetic resonance spectroscopy imaging of the newborn brain—a technical review. In: Seminars in perinatology. Elsevier, Amsterdam, pp 20–27

40. Bhaduri S, Clement P, Achten E, Serrai H (2018) Reduction of acquisition time using partition of the sIgnal decay in spectroscopic imaging technique (RAPID-SI). PLoS One 13:e0207015

41. Mescher M, Tannus A, Johnson MN, Garwood M (1996) Solvent suppression using selective echo dephasing. J Magn Reson Ser A 123:226–229

42. Frahm J, Merboldt K-D, Hänicke W (1987) Localized proton spectroscopy using stimulated echoes. J Magn Reson 72:502–508

43. Frahm JA, Bruhn H, Gyngell M, Merboldt K, Hänicke W, Sauter R (1989) Localized high-resolution proton NMR spectroscopy using stimulated echoes: initial applications to human brain in vivo. Magn Reson Med 9:79–93

44. Dong Z (2015) Proton MRS and MRSI of the brain without water suppression. Prog Nucl Magn Reson Spectrosc 86:65–79

45. Haase A, Frahm J, Hanicke W, Matthaei D (1985) 1H NMR chemical shift selective (CHESS) imaging. Phys Med Biol 30:341

46. Ogg RJ, Kingsley R, Taylor JS (1994) WET, a T1-and B1-insensitive water-suppression method for in vivo localized 1H NMR spectroscopy. J Magn Reson B 104:1–10

47. Star-Lack J, Nelson SJ, Kurhanewicz J, Huang LR, Vigneron DB (1997) Improved water and lipid suppression for 3D PRESS CSI using RF band selective inversion with gradient dephasing (BASING). Magn Reson Med 38:311–321

48. Krinsky G, Rofsky NM, Weinreb JC (1996) Nonspecificity of short inversion time inversion recovery (STIR) as a technique of fat suppression: pitfalls in image interpretation. AJR Am J Roentgenol 166:523–526

49. Kaldoudi E, Williams SC, Barker GJ, Tofts PS (1993) A chemical shift selective inversion recovery sequence for fat-suppressed MRI: theory and experimental validation. Magn Reson Imaging 11:341–355

50. Tannús A, Garwood M (1997) Adiabatic pulses. NMR Biomed 10:423–434

51. Juchem C, de Graaf RA (2017) B0 magnetic field homogeneity and shimming for in vivo magnetic resonance spectroscopy. Anal Biochem 529:17–29

52. Zhang Y, Li S, Shen J (2009) Automatic high-order shimming using parallel columns mapping (PACMAP). Magn Reson Med 62:1073–1079

53. Gruber B, Froeling M, Leiner T, Klomp DW (2018) RF coils: a practical guide for nonphysicists. J Magn Reson Imag 48:590–604

54. Duarte JM, Carvalho RA, Cunha RA, Gruetter R (2009) Caffeine consumption attenuates neurochemical modifications in the hippocampus of streptozotocin-induced diabetic rats. J Neurochem 111:368–379

55. John J, Kodama T, Siegel JM (2014) Caffeine promotes glutamate and histamine release in the posterior hypothalamus. Am J Physiol Regul Integr Comp Physiol 307:R704–R710

56. Oeltzschner G, Zöllner HJ, Jonuscheit M, Lanzman RS, Schnitzler A, Wittsack HJ (2018) J-difference-edited MRS measures of γ-aminobutyric acid before and after acute caffeine administration. Magn Reson Med 80:2356–2365

57. Provencher SW (2001) Automatic quantitation of localized in vivo 1H spectra with LCModel. NMR Biomed 14:260–264

58. Naressi A, Couturier C, Castang I, De Beer R, Graveron-Demilly D (2001) Java-based graphical user interface for MRUI, a software package for quantitation of in vivo/medical magnetic resonance spectroscopy signals. Comput Biol Med 31:269–286

59. Wilson M, Reynolds G, Kauppinen RA, Arvanitis TN, Peet AC (2011) A constrained least-squares approach to the automated quantitation of in vivo 1H magnetic resonance spectroscopy data. Magn Reson Med 65:1–12

60. Li BS, Wang H, Gonen O (2003) Metabolite ratios to assumed stable creatine level may confound the quantification of proton brain MR spectroscopy. Magn Reson Imaging 21:923–928

61. Hoch SE, Kirov II, Tal A (2017) When are metabolic ratios superior to absolute quantification? A statistical analysis. NMR Biomed 30:e3710

62. Gasparovic C, Song T, Devier D, Bockholt HJ, Caprihan A, Mullins PG et al (2006) Use of tissue water as a concentration reference for proton spectroscopic imaging. Magn Reson Med 55:1219–1226

63. Mandal PK (2012) In vivo proton magnetic resonance spectroscopic signal processing for the absolute quantitation of brain metabolites. Eur J Radiol 81:e653–e664

64. Drago T, O'Regan PW, Welaratne I, Rooney S, O'Callaghan A, Malkit M et al (2018) A comprehensive regional neurochemical theory in depression: a protocol for the systematic review and meta-analysis of 1H-MRS studies in major depressive disorder. Syst Rev 7:158

65. Schousboe A, Waagepetersen HS (2005) Role of astrocytes in glutamate homeostasis: implications for excitotoxicity. Neurotox Res 8:221–225

66. Rothman DL, Behar KL, Hyder F, Shulman RG (2003) In vivo NMR studies of the glutamate neurotransmitter flux and neuroenergetics: implications for brain function. Annu Rev Physiol 65:401–427

67. Shutter L, Tong KA, Holshouser BA (2004) Proton MRS in acute traumatic brain injury: role for glutamate/glutamine and choline for outcome prediction. J Neurotrauma 21:1693–1705

68. Rolls ET (2016) A non-reward attractor theory of depression. Neurosci Biobehav Rev 68:47–58

69. Levy BJ, Wagner AD (2011) Cognitive control and right ventrolateral prefrontal cortex: reflexive reorienting, motor inhibition, and action updating. Ann N Y Acad Sci 1224:40

70. Knoch D, Fehr E (2007) Resisting the power of temptations: the right prefrontal cortex and self-control. Ann N Y Acad Sci 1104:123–134

71. Barbey AK, Koenigs M, Grafman J (2013) Dorsolateral prefrontal contributions to human working memory. Cortex 49:1195–1205

72. Frangou S, Haldane M, Roddy D, Kumari V (2005) Evidence for deficit in tasks of ventral, but not dorsal, prefrontal executive function as an endophenotypic marker for bipolar disorder. Biol Psychiatry 58:838–839

73. Petrican R, Schimmack U (2008) The role of dorsolateral prefrontal function in relationship commitment. J Res Pers 42:1130–1135

74. Stuss DT, Knight RT (2013) Principles of frontal lobe function. Oxford University Press, Oxford

75. Mayberg HS, Lozano AM, Voon V, McNeely HE, Seminowicz D, Hamani C et al (2005) Deep brain stimulation for treatment-resistant depression. Neuron 45:651–660

76. Hasler G, van der Veen JW, Tumonis T, Meyers N, Shen J, Drevets WC (2007) Reduced prefrontal glutamate/glutamine and gamma-aminobutyric acid levels in major depression determined using proton magnetic resonance spectroscopy. Arch Gen Psychiatry 64:193–200

77. Portella MJ, de Diego-Adeliño J, Gómez-Ansón B, Morgan-Ferrando R, Vives Y, Puigdemont D et al (2011) Ventromedial

prefrontal spectroscopic abnormalities over the course of depression: a comparison among first episode, remitted recurrent and chronic patients. J Psychiatr Res 45:427–434

78. Draganov M, Vives-Gilabert Y, de Diego-Adeliño J, Vicent-Gil M, Puigdemont D, Portella MJ (2020) Glutamatergic and GABAergic abnormalities in first-episode depression. A 1-year follow-up 1H-MR spectroscopic study. J Affect Disord 266:572–577

79. Taylor MJ, Godlewska BR, Norbury R, Selvaraj S, Near J, Cowen PJ (2012) Early increase in marker of neuronal integrity with antidepressant treatment of major depression: 1H-magnetic resonance spectroscopy of N-acetyl-aspartate. Int J Neuropsychopharmacol 15:1541–1546

80. Li H, Xu H, Zhang Y, Guan J, Zhang J, Xu C et al (2016) Differential neurometabolite alterations in brains of medication-free individuals with bipolar disorder and those with unipolar depression: a two-dimensional proton magnetic resonance spectroscopy study. Bipolar Disord 18:583–590

81. Zhang X, Tang Y, Maletic-Savatic M, Sheng J, Zhang X, Zhu Y et al (2016) Altered neuronal spontaneous activity correlates with glutamate concentration in medial prefrontal cortex of major depressed females: an fMRI-MRS study. J Affect Disord 201:153–161

82. Knudsen MK, Near J, Blicher AB, Videbech P, Blicher JU (2018) Magnetic resonance (MR) spectroscopic measurement of gamma-aminobutyric acid (GABA) in major depression before and after electroconvulsive therapy. Acta Neuropsychiatr 31:17–26

83. Michael N, Erfurth A, Ohrmann P, Arolt V, Heindel W, Pfleiderer B (2003) Metabolic changes within the left dorsolateral prefrontal cortex occurring with electroconvulsive therapy in patients with treatment resistant unipolar depression. Psychol Med 33:1277–1284

84. Şendur I, Kalkan Oğuzhanoğlu N, Sözeri Varma G (2020) Study on dorsolateral prefrontal cortex neurochemical metabolite levels of patients with major depression using 1H-MRS technique. Turk J Psychiatry 31:75–83

85. Nery FG, Stanley JA, Chen H-H, Hatch JP, Nicoletti MA, Monkul ES et al (2009) Normal metabolite levels in the left dorsolateral prefrontal cortex of unmedicated major depressive disorder patients: a single voxel 1H spectroscopy study. Psychiatry Res Neuroimaging 174:177–183

86. Chen LP, Dai HY, Dai ZZ, Xu CT, Wu RH (2014) Anterior cingulate cortex and cerebellar hemisphere neurometabolite changes in

depression treatment: a 1H magnetic resonance spectroscopy study. Psychiatry Clin Neurosci 68:357–364

87. Merkl A, Schubert F, Quante A, Luborzewski A, Brakemeier EL, Grimm S et al (2011) Abnormal cingulate and prefrontal cortical neurochemistry in major depression after electroconvulsive therapy. Biol Psychiatry 69:772–779

88. Grimm S, Luborzewski A, Schubert F, Merkl A, Kronenberg G, Colla M et al (2012) Region-specific glutamate changes in patients with unipolar depression. J Psychiatr Res 46:1059–1065

89. Shan Y, Jia Y, Zhong S, Li X, Zhao H, Chen J et al (2017) Correlations between working memory impairment and neurometabolites of prefrontal cortex and lenticular nucleus in patients with major depressive disorder. J Affect Disord 227:236–242

90. Liu X, Zhong S, Li Z, Chen J, Wang Y, Lai S et al (2020) Serum copper and zinc levels correlate with biochemical metabolite ratios in the prefrontal cortex and lentiform nucleus of patients with major depressive disorder. Prog Neuropsychopharmacol Biol Psychiatry 99:109828

91. Rolls ET (2019) The cingulate cortex and limbic systems for emotion, action, and memory. Brain Struct Funct 224:3001–3018

92. Morgane PJ, Galler JR, Mokler DJ (2005) A review of systems and networks of the limbic forebrain/limbic midbrain. Prog Neurobiol 75:143–160

93. Baleydier C, Mauguiere F (1980) The duality of the cingulate gyrus in monkey. Neuroanatomical study and functional hypothesis. Brain 103:525–554

94. Walter M, Henning A, Grimm S, Schulte RF, Beck J, Dydak U et al (2009) The relationship between aberrant neuronal activation in the pregenual anterior cingulate, altered glutamatergic metabolism, and anhedonia in major depression. Arch Gen Psychiatry 66:478–486

95. Hamani C, Mayberg H, Stone S, Laxton A, Haber S, Lozano AM (2011) The subcallosal cingulate gyrus in the context of major depression. Biol Psychiatry 69:301–308

96. Auer DP, Putz B, Kraft E, Lipinski B, Schill J, Holsboer F (2000) Reduced glutamate in the anterior cingulate cortex in depression: an in vivo proton magnetic resonance spectroscopy study. Biol Psychiatry 47:305–313

97. Pfleiderer B, Michael N, Erfurth A, Ohrmann P, Hohmann U, Wolgast M et al (2003) Effective electroconvulsive therapy reverses glutamate/glutamine deficit in the

left anterior cingulum of unipolar depressed patients. Psychiatry Res Neuroimaging 122: 185–192

98. Price RB, Shungu DC, Mao X, Nestadt P, Kelly C, Collins KA et al (2009) Amino acid neurotransmitters assessed by proton magnetic resonance spectroscopy: relationship to treatment resistance in major depressive disorder. Biol Psychiatry 65:792–800

99. Njau S, Joshi SH, Espinoza R, Leaver AM, Vasavada M, Marquina A et al (2017) Neurochemical correlates of rapid treatment response to electroconvulsive therapy in patients with major depression. J Psychiatry Neurosci 42:6

100. Zheng H, Jia F, Guo G, Quan D, Li G, Wu H et al (2015) Abnormal anterior cingulate N-acetylaspartate and executive functioning in treatment-resistant depression after rTMS therapy. Int J Neuropsychopharmacol 18: pyv059

101. Abdallah CG, Hannestad J, Mason GF, Holmes SE, DellaGioia N, Sanacora G et al (2017) Metabotropic glutamate receptor 5 and glutamate involvement in major depressive disorder: a multimodal imaging study. Biol Psychiatry Cogn Neurosci Neuroimaging 2:449–456

102. Zhang J, Narr KL, Woods RP, Phillips OR, Alger JR, Espinoza RT (2013) Glutamate normalization with ECT treatment response in major depression. Mol Psychiatry 18:268

103. Wise T, Taylor MJ, Herane-Vives A, Gammazza AM, Cappello F, Lythgoe DJ et al (2018) Glutamatergic hypofunction in medication-free major depression: secondary effects of affective diagnosis and relationship to peripheral glutaminase. J Affect Disord 234:214–219

104. Benson KL, Bottary R, Schoerning L, Baer L, Gonenc A, Jensen JE et al (2020) 1H MRS measurement of cortical GABA and glutamate in primary insomnia and major depressive disorder: relationship to sleep quality and depression severity. J Affect Disord 274: 624–631

105. Taylor R, Osuch EA, Schaefer B, Rajakumar N, Neufeld RW, Théberge J et al (2017) Neurometabolic abnormalities in schizophrenia and depression observed with magnetic resonance spectroscopy at 7 T. BJPsych open 3:6–11

106. Godlewska BR, Masaki C, Sharpley AL, Cowen PJ, Emir UE (2018) Brain glutamate in medication-free depressed patients: a proton MRS study at 7 tesla. Psychol Med 48: 1731–1737

107. Evans JW, Lally N, An L, Li N, Nugent AC, Banerjee D et al (2018) 7T (1)H-MRS in major depressive disorder: a ketamine treatment study. Neuropsychopharmacology 43(9):1908–1914

108. Brennan BP, Admon R, Perriello C, LaFlamme EM, Athey AJ, Pizzagalli DA et al (2017) Acute change in anterior cingulate cortex GABA, but not glutamine/glutamate, mediates antidepressant response to citalopram. Psychiatry Res Neuroimaging 269: 9–16

109. Colic L, von Düring F, Denzel D, Demenescu LR, Lord AR, Martens L et al (2019) Rostral anterior cingulate glutamine/glutamate disbalance in major depressive disorder depends on symptom severity. Biol Psychiatry Cogn Neurosci Neuroimaging 4:1049–1058

110. Nugent AC, Farmer C, Evans JW, Snider SL, Banerjee D, Zarate CA Jr (2019) Multimodal imaging reveals a complex pattern of dysfunction in corticolimbic pathways in major depressive disorder. Hum Brain Mapp 40: 3940–3950

111. Gonul AS, Kitis O, Ozan E, Akdeniz F, Eker C, Eker OD et al (2006) The effect of antidepressant treatment on N-acetyl aspartate levels of medial frontal cortex in drug-free depressed patients. Prog Neuropsychopharmacol Biol Psychiatry 30:120–125

112. Li Y, Jakary A, Gillung E, Eisendrath S, Nelson SJ, Mukherjee P et al (2016) Evaluating metabolites in patients with major depressive disorder who received mindfulness-based cognitive therapy and healthy controls using short echo MRSI at 7 tesla. MAGMA 29: 523–533

113. Nase S, Köhler S, Jennebach J, Eckert A, Schweinfurth N, Gallinat J et al (2016) Role of serum brain derived neurotrophic factor and central N-acetylaspartate for clinical response under antidepressive pharmacotherapy. Neurosignals 24:1–14

114. Wang Y, Jia Y, Chen X, Ling X, Liu S, Xu G et al (2012) Hippocampal N-acetylaspartate and morning cortisol levels in drug-naive, first-episode patients with major depressive disorder: effects of treatment. J Psychopharmacol 26:1463–1470

115. Jia Y, Zhong S, Wang Y, Liu T, Liao X, Huang L (2015) The correlation between biochemical abnormalities in frontal white matter, hippocampus and serum thyroid hormone levels in first-episode patients with major depressive disorder. J Affect Disord 180:162–169

116. Tosun Ş, Tosun M, Akansel G, Gökbakan AM, Ünver H, Tural Ü (2020) Proton magnetic resonance spectroscopic analysis of

changes in brain metabolites following electroconvulsive therapy in patients with major depressive disorder. Int J Psychiatry Clin Pract 24:96–101

117. Nolan M, Roman E, Nasa A, Levins KJ, O'Hanlon E, O'Keane V et al (2020) Hippocampal and Amygdalar volume changes in major depressive disorder: a targeted review and focus on stress. Chronic Stress 4: 2470547020944553

118. Roddy DW, Farrell C, Doolin K, Roman E, Tozzi L, Frodl T et al (2018) The hippocampus in depression: more than the sum of its parts? Advanced hippocampal substructure segmentation in depression. Biol Psychiatry 85:487–497

119. Weininger JK, Roman E, Tierney P, Barry D, Gallagher H, Levins KJ et al (2019) Papez's forgotten tract: 80 years of unreconciled findings concerning the thalamocingulate tract. Front Neuroanat 13:14

120. Roman E, Weininger J, Lim B, Roman M, Barry D, Tierney P et al (2020) Untangling the dorsal diencephalic conduction system: a review of structure and function of the stria medullaris, habenula and fasciculus retroflexus. Brain Struct Funct 225:1437–1458

121. Roddy D, O'Keane V (2019) Cornu Ammonis changes are at the Core of hippocampal pathology in depression. Chronic Stress 3: 2470547019849376

122. Poletti S, Locatelli C, Falini A, Colombo C, Benedetti F (2016) Adverse childhood experiences associate to reduced glutamate levels in the hippocampus of patients affected by mood disorders. Prog Neuropsychopharmacol Biol Psychiatry 71:117–122

123. Hermens DF, Naismith SL, Chitty KM, Lee RS, Tickell A, Duffy SL et al (2015) Cluster analysis reveals abnormal hippocampal neurometabolic profiles in young people with mood disorders. Eur Neuropsychopharmacol 25: 836–845

124. Milne A, MacQueen GM, Yucel K, Soreni N, Hall GB (2009) Hippocampal metabolic abnormalities at first onset and with recurrent episodes of a major depressive disorder: a proton magnetic resonance spectroscopy study. Neuroimage 47:36–41

125. Block W, Traber F, von Widdern O, Metten M, Schild H, Maier W et al (2009) Proton MR spectroscopy of the hippocampus at 3 T in patients with unipolar major depressive disorder: correlates and predictors of treatment response. Int J Neuropsychopharmacol 12:415–422

126. de Diego-Adeliño J, Portella MJ, Gómez-Ansón B, López-Moruelo O, Serra-Blasco M, Vives Y et al (2013) Hippocampal abnormalities of glutamate/glutamine, N-acetylaspartate and choline in patients with depression are related to past illness burden. J Psychiatry Neurosci 38:107

127. Çano M, Martínez-Zalacaín I, Bernabéu-Sanz Á, Contreras-Rodríguez O, Hernández-Ribas R, Via E et al (2017) Brain volumetric and metabolic correlates of electroconvulsive therapy for treatment-resistant depression: a longitudinal neuroimaging study. Transl Psychiatry 7:e1023–e1023

128. Ende G, Braus DF, Walter S, Weber-Fahr W, Henn FA (2000) The hippocampus in patients treated with electroconvulsive therapy: a proton magnetic resonance spectroscopic imaging study. Arch Gen Psychiatry 57:937–943

129. Tang Y, Zhang X, Sheng J, Zhang X, Zhang J, Xu J et al (2018) Elevated hippocampal choline level is associated with altered functional connectivity in females with major depressive disorder: a pilot study. Psychiatry Res Neuroimaging 278:48–55

130. Kennedy SH, Konarski JZ, Segal ZV, Lau MA, Bieling PJ, McIntyre RS et al (2007) Differences in brain glucose metabolism between responders to CBT and venlafaxine in a 16-week randomized controlled trial. Am J Psychiatry 164:778–788

131. Godlewska BR, Browning M, Norbury R, Igoumenou A, Cowen PJ, Harmer CJ (2018) Predicting treatment response in depression: the role of anterior cingulate cortex. Int J Neuropsychopharmacol 21: 988–996

132. Yucel K, McKinnon MC, Chahal R, Taylor VH, Macdonald K, Joffe R et al (2008) Anterior cingulate volumes in never-treated patients with major depressive disorder. Neuropsychopharmacology 33:3157–3163

133. Belleau EL, Treadway MT, Pizzagalli DA (2019) The impact of stress and major depressive disorder on hippocampal and medial prefrontal cortex morphology. Biol Psychiatry 85:443–453

134. Radley JJ, Rocher AB, Miller M, Janssen WG, Liston C, Hof PR et al (2006) Repeated stress induces dendritic spine loss in the rat medial prefrontal cortex. Cereb Cortex 16:313–320

135. Cerqueira JJ, Pêgo JM, Taipa R, Bessa JM, Almeida OF, Sousa N (2005) Morphological correlates of corticosteroid-induced changes in prefrontal cortex-dependent behaviors. J Neurosci 25:7792–7800

136. Aston C, Jiang L, Sokolov BP (2005) Transcriptional profiling reveals evidence for signaling and oligodendroglial abnormalities in the temporal cortex from patients with major depressive disorder. Mol Psychiatry 10: 309–322

137. Boda E (2021) Myelin and oligodendrocyte lineage cell dysfunctions: new players in the etiology and treatment of depression and stress-related disorders. Eur J Neurosci 53: 281–297

138. Cendes F, Knowlton RC, Novotny E, Min LL, Antel S, Sawrie S et al (2002) Magnetic resonance spectroscopy in epilepsy: clinical issues. Epilepsia 43:32–39

Chapter 14

Deficits of Neurotransmitter Systems and Altered Brain Connectivity in Major Depression: A Translational Neuroscience Perspective

Je-Yeon Yun and Yong-Ku Kim

Abstract

Recent studies that examined brain characteristics for major depressive disorder (MDD) have reported deficient neurotransmitter system, altered brain morphology and patterns of white matter-based brain inter-regional connections, and differential brain inter-regional coherence of oscillatory patterns during resting and task performance status compared to healthy controls. For better understanding of the MDD pathophysiology that underlies clinical symptoms, the current review provides practical approaches of multimodal brain imaging encompassing the regional deficiency of neurotransmitter receptors or synaptic density to altered patterns of brain inter-regional connection or communication for MDD. First, to elucidate the deficits of neurotransmitter system in MDD, the current review illustrates how to acquire the brain molecular positron emission tomography (PET) images and estimate the synaptic density in addition to the binding potential (or receptor availabilities) of serotonergic (5-HT transporter and 5-HT_{1A} autoreceptor), glutamatergic (metabotropic glutamate receptor 5), and dopaminergic (D_2 receptor) system across the whole brain. Second, the current review demonstrates how to explore the possible associations between the regional deficits of neurotransmitter binding potential and altered resting-state functional connectivity [voxel-to-whole brain (intrinsic functional connectivity) or region-to-region (seed-based functional connectivity)], structural connectivity [brain white matter-based region-to-region structural connectivity, estimated using the probabilistic fiber tracking system], and directed functional connectivity [region-to-region] during task performance in MDD. Third, opened resources of software and pipelines that could be applied in running these analytical procedures are also provided.

Key words Major depressive disorder, Neurotransmitter, Brain, Connectivity, Magnetic resonance imaging, Positron emission tomography

---

1 Introduction: Neurotransmitter and Brain Connectivity in Major Depression

In relevance to the serotonin hypothesis of depression that posits diminished activity of serotonin pathway as a pathophysiology of depression [1], multiple positron emission tomography (PET) studies examined the receptor availability of serotonin (5-HT)

Yong-Ku Kim and Meysam Amidfar (eds.), *Translational Research Methods for Major Depressive Disorder*, Neuromethods, vol. 179, https://doi.org/10.1007/978-1-0716-2083-0_14,
© The Author(s), under exclusive license to Springer Science+Business Media, LLC, part of Springer Nature 2022

system. First, radio-ligands of [^{11}C]CUMI-101 (CUMI) and [^{11}C] WAY100635 have been used to quantify the serotonin 1A (5-HT$_{1A}$) autoreceptor binding. 5-HT$_{1A}$ autoreceptors regulate serotonin neuron firing and release at nerve terminal, moderating brain-wide response for stimuli that elicit negative emotion [2–4], and possible associations between the decreased amygdala response to negative stimuli and greater 5-HT$_{1A}$ autoreceptor binding [5, 6] were reported. Serotonin released by neurons of the raphe nucleus (RN) activates both 5-HT$_{1A}$ heteroreceptors on the dendrites of postsynaptic target neurons and presynaptic 5-HT$_{1A}$ autoreceptors on the dendrites and cell bodies of raphe neurons [7]. Somato-dendritic 5-HT$_{1A}$ autoreceptors hyperpolarize serotonin neurons in RN, providing negative feedback control of RN firing rate and thus of synaptic 5-HT release. Moreover, 5-HT$_{1A}$ autoreceptors in the median and dorsal midbrain raphe nuclei (RN) are associated with pathophysiology of major depressive disorder (MDD) [8, 9]. Second, the radiotracer [^{11}C]-3-amino-4-(2-dimethylami-nomethyl-phenylsulfanyl)-benzonitrile ([^{11}C]-DASB) is used to measure the availability of serotonin transporter (5-HTT), which regulates serotonin reuptake at the synaptic cleft [10, 11]. Third, as the number of vesicles per nerve terminal is a stable feature of neurons [12], radio-ligand binding to synaptic vesicle glycoprotein 2A (SV2A) such as [^{11}C]UCB-J [13] serves as a proxy for the quantification of synaptic density [14]. SV2A is a transmembrane protein ubiquitously and homogeneously located in presynaptic secretory vesicles across the whole brain and is vital for normal synaptic function including the release of neurotransmitter [14–16]. In addition, possible involvements of glutamate (monoamine) and dopaminergic systems for reward processing and pathophysiology of depression are also explored by way of the PET studies that apply radio-ligands of [^{11}C]ABP688 [for metabotropic receptor 5 (mGluR5)] and [^{11}C]raclopride [for dopaminergic D$_2$ receptor], respectively [17–19].

Further, multimodal imaging studies that examined associations among the neurotransmitter system (measured using PET), white matter tract-based structural connectivity [constructed from the diffusion-weighted image (DWI)], and inter-regional correlations of time-varying oscillating patterns in functional activation among the brain regions during resting status (resting-state functional connectivity) or task performance (task-related functional connectivity) [acquired as T2* magnetic resonance imaging (MRI)] enable deeper understanding of brain-based pathophysiology for depression [20, 21]. Aiming for facilitating these neurotransmitter-to-brain network multifaceted studies for depression, which could be utilized as more accurate biomarkers for precision medicine in clinical psychiatry [22], this chapter would provide detailed illustration of how to acquire (refer to Subheading

2) and analyze (refer to Subheading 3) the PET and MRI data in multimodal imaging studies for depressive patients.

2   Materials: Data Acquisition

2.1   PET Image Acquisition

Participants were scanned at the PET center using a cerium-doped lutetium oxyorthosilicate (Lu25i05[Ce], LSO) detector-based, dedicated high-resolution human brain PET scanner—second-generation High-Resolution Research Tomograph scanner (HRRT, Siemens Healthcare, Knoxville, TN) [23]—or a Biograph 6 PET scanner (Siemens Medical Imaging Systems, Knoxville, USA). Specifically for the radio-ligand of [^{11}C]ABP688, all PET scans were performed at the same time of the day (10:00 a.m.) to avoid possible diurnal variations in glutamate levels that may affect mGluR5 surface localization and ligand accessibility [24, 25]. Each subject was fitted with a thermoplastic mask modeled to their face to reduce head motion during the PET study.

Attenuation maps (for estimating the attenuation of the PET data) were generated using a transmission scan with a [^{137}Cs] point source [26], a computed tomography (CT)-based transmission scan acquired immediately prior to the tracer injection. In other case, MR-based attenuation correction was performed with the clinical atlas-based method as implemented on the SIGNA PET/MR system, where individual T1-weighted MR images were rigidly and non-rigidly registered to a CT-based head atlas [27]. After the acquisition of a transmission scan, radio-ligand prepared according to the protocol ([^{11}C]UCB-J [16], [^{11}C]CUMI-101 [28], [^{11}C]ABP688 [29], and [^{11}C]raclopride [30]) was administered intravenously (via the left antecubital vein) as a bolus using an automated infusion pump (Harvard PHD 22/2000, Harvard Apparatus) [13]. Mean injected doses of radiotracers are as follows: 6.49 ± 0.4 mCi for [^{11}C]WAY100635, 20 mCi ± 10% for [^{11}C]-DASB, 636.2 ± 71.4 MBq for [^{11}C]ABP688, and 10 mCi for [^{11}C]raclopride. After the intravenous infusion of radio-ligand, acquisition of dynamic PET data in list mode or 3D mode started immediately and lasted for 120 min using 33 frames (6 × 12 s, 3 × 1 min, 2 × 2 min, and 22 × 5 min, for [^{11}C]UCB-J), for 120 min using 21 frames (3 × 20 s, 3 × 1 min, 3 × 2 min, 2 × 5 min, and 10 × 10 min [26], for [^{11}C]CUMI-101), for 110 min using 20 frames (for [^{11}C]-WAY100635), for 90 min (for [^{11}C]-DASB [26]), for 60 min ([^{11}C]ABP688), or for 42 min ([^{11}C]raclopride) of increasing duration. The field of view (FOV) was 22 cm with a matrix size of 128 × 128.

The images were reconstructed using the MOLAR algorithm, the iterative ordered-subset expectation maximization (OSEM) algorithm (with six iterations and 16 subsets), the two-dimensional   ordered-subset   expectation   maximization

(OSEM-2D) algorithm, or a fully 3D TOF iterative ordered-subset expectation maximization algorithm (28 subsets, 3 iterations), with correction for and corrected for attenuation, scatter, point spread function, dead time, randoms, and radioactive decay [31]. To calculate the radio-ligand binding potential with respect to non-displaceable compartment (BP_{ND}), corrected images were re-binned into 30 frames (four 15 s, four 30 s, three 1 min, two 2 min, five 4 min, and 12 5 min), into 21 frames (two 15 s, three 30 s, three 1 min, two 1.5 min, three 2 min, two 3 min, four 5 min, and two 10 min, total 60 min for $[^{11}C]$ABP688), or into 30-s time frames. Event-by-event motion correction was included in the reconstruction based on motion detection with a Polaris Vicra optical tracking system (NDI Systems, Waterloo, Canada). The reconstructed image space consisted of 256 (left-to-right) by 256 (nasion-to-inion) by 207 (neck-to-cranium) cubic voxels, each 1.22 mm in dimension [26]. In other cases, the reconstructed PET images had a matrix size of $256 \times 256 \times 109$ and a voxel size of $1.33 \times 1.33 \times 1.50$ mm^3. The final spatial resolution was less than 2.5 mm full width at half maximum (FWHM) in three directions [26].

2.2 Acquisition of Brain Magnetic Resonance Imaging: T1WI

To exclude structural abnormality and for coregistration with PET images, T1-weighted MRI scans are acquired on 3-T scanner (such as the Siemens Prisma). Image acquisition parameters are able to be provided from venders or also could be customized for specific needs of researchers, for example, a high-resolution, three-dimensional magnetization-prepared rapid acquisition gradient echo (MPRAGE) T1-weighted sequence (TR = 1500 ms, TE = 2.83 ms, FOV = 256×256 mm^2, matrix = 256×256 mm^2, slice thickness = 1.0 mm without gap, 160 slices, voxel size $1.0 \times 1.0 \times 1.0$ mm^3) [13]; the magnetization-prepared rapid acquisition gradient echo (MPRAGE) pulse sequence (TE = 4, TR = 8.9, flip angle = 8 degrees, NSA = 1, 0.7 mm isotropic voxel size) using a Phillips 3.0 T Achieva MRI instrument with an 8-channel head coil (Philips Medical Systems, Best, Netherlands) [26]; a T1-weighted sequence (1 mm isotropic voxels, 200 slices, field of view = 256 mm) using a GE SIGNA 3 T scanner (GE Healthcare, Milwaukee, WI) with 32-channel head coil [32]; a high-resolution spoiled gradient echo T1-weighted anatomical images collected using the parameters of FOV = 220 mm, matrix = 256×256, through-plane resolution = 1.2 mm, in-plane resolution = 0.86×0.86 mm, slice spacing = 0, TR = 8.52 ms, TE = 3.32 ms, TI = 450 ms, flip angle = 12°, and slices = 124, acquired sequentially left to right in sagittal plane [27]; or a three-dimensional spoiled gradient (repetition time 7.3 ms, echo time 3.0 ms, slice thickness 1 mm with 1 mm gap, flip angle 9 degrees, 168 slices, field of view $260 \times 260 \times 168$ mm, 256×256 matrix, with a voxel size of $1.0 \times 1.0 \times 1.0$ mm) [33].

2.3 Acquisition of Brain Magnetic Resonance Imaging: Resting-State T2* Image

For the acquisition of T2* image to estimate the resting-state functional connectivity, all subjects were instructed to relax, lie still, and stay awake in the scanner, with their head movement comfortably restricted by sponges [34]. Functional runs included 340 whole-brain volumes acquired using a multiband echo-planar imaging sequence with the following parameters: TR = 1 s, TE = 30 ms, flip angle = 62°, matrix = 84 × 84, in-plane resolution = 2.5 mm^2, 51 axial-oblique slices parallel to the ac-pc line, slice thickness = 2.5, multiband = 3, and acceleration factor = 2 (3-T Siemens Prisma scanner) [13]. On the other hand, for rs-fMRI recordings optimized for detecting changes in blood-oxygen-level-dependent (BOLD) signal levels, the 3-T rs-fMRI images comprised of a total of 180 volumes (9 min in length) were acquired using echo-planar imaging (EPI) with the following parameters: repetition time = 3000 ms, echo time = 30 ms, flip angle = 90°, pixel size = 3.5 × 3.5 mm^2, thickness = 3.5 mm, matrix size = 72 × 72, and number of slices = 45 MRI (MAGNETOM Verio, Siemens, Erlangen, Germany).

2.4 Concurrent PET and Task-Based fMRI Scanning

Simultaneous PET-MRI examination could be conducted using devices such as the time-of-flight (TOF) PET-MRI scanner (SIGNA PET-MRI, GE Medical Systems, USA). PET scanning took place for the entire time of the scanning session. A bolus injection of radiotracer (such as [^{11}C]raclopride) was administered 1 min after the PET scan commenced, and, at this same time, brain structural and MRI-based attenuation correction data were acquired for 9 min. In the following 32-min-length fMRI scan, participants were instructed to rest with their eyes open for the first and last 6 min of this scan and to complete trials of the monetary incentive delay (MID) task for 20 min between these two rest conditions. Minutes 2–17 of PET data were used to estimate basal BP_{ND}; PET data from minute 2 onward (baseline plus task) were used to estimate endogenous ligand displacement by the task [27]. Whole-brain fMRI data were collected with the following specifications: FOV = 220 mm, matrix = 64 × 64, through-plane resolution = 3.5 mm, in-plane resolution = 3.44 × 3.44 mm, slice spacing = 0, TR = 2500 ms, TE = 30 ms, flip angle = 80°, and slices = 27 sequential ascending/axial [27].

3 Methods: Data Analyses

3.1 Arterial Sampling and PET Tracer Kinetic Modeling

For the radiochemistry and input function measurement of the radio-ligand, all subjects underwent arterial cannulation, and blood was collected after the radiotracer injection [14, 35]. For the [^{11}C]UCB-J, samples were drawn every 10 s for the first 90 s and at 1.75, 2, 2.25, 2.5, 2.75, 3, 4, 5, 6, 8, 10, 15, 20, 25, 30,

45, 60, 75, 90, 105, and 120 min. For $[^{11}C]$-WAY100635, arterial blood was sampled every 5 s automatically for the first 2 min and manually at longer intervals after (6, 12, 20, 40, and 60 min). Collected blood samples were centrifuged for 10 min, and radioactivity in the separated plasma was counted using a gamma counter (Wallac 1480 Wizard 3M Automatic Gamma Counter). Radiometabolite analyses were performed for plasma samples collected at 3, 8, 15, 30, 60, and 90 min (for $[^{11}C]$UCB-J and $[^{11}C]$CUMI-101) and at 2, 6, 12, 20, 40, and 60 min (for $[^{11}C]$-WAY100635, $[^{11}C]$-DASB, $[^{11}C]$ABP688, and $[^{11}C]$raclopride) which were processed by protein precipitation with acetonitrile and purified via automatic column-switching high-pressure liquid chromatography (HPLC) system. By centrifuging 200 L aliquots of plasma mixed with tracer, the plasma-free fraction (f_p) was measured in triplicate using an ultrafiltration method (Millipore Centrifree micropartition device 4104, Billerica, MA, USA).

Using these samples, the unmetabolized parent fraction or the plasma-free fraction (f_p) was determined as the radioactivity ratio of ultrafiltrate to total plasma activity concentrations [13] collected and fitted with an inverted integrated gamma function. Again, these unmetabolized fractions were fit to the Hill function, with the value at time zero constrained to 100%. The curve was normalized with the time-varying extraction efficiency, which was determined by corresponding reference plasma samples and fitted with an exponential function. The arterial plasma input function was calculated as the product of the total plasma activity, the (unmetabolized) radiotracer HPLC fraction curve, and the extraction efficiency curve. The calculated input function values were fit to a sum of three exponentials from the time of the peak to the final data point, whereas the early rising part of the curve was fit to a straight line. These fitted values were then used as input to the kinetic analysis.

3.2 PET Image Preprocessing for Applying the MRI-Derived ROIs

Transfer of MRI-derived ROI masks onto the PET images was conducted using PET-to-MRI coregistration parameters that were obtained by way of the coregistration module in SPM12 (Wellcome Trust Center for Neuroimaging, UK) running on MATLAB 7.10 (MathWorks, Natick, Massachusetts) or the extension to Functional Magnetic Resonance Imaging of the Brain's (FMRIB's) Linear Image Registration Tool (FLIRT v.5.2 [36]) [37]. First, after the motion correction by frame-by-frame rigid body registration to a reference frame [33], the mean PET image was coregistered to each participant's 3D T1MPRAGE MR images [26]. Second, coregistered T1-weighted MRI was skull-stripped using the Brain Extraction Tool v1.2 [38]. Third, tissue segmentation delineated the gray matter, white matter, and CSF within the skull-stripped T1-weighted image [39]. Fourth, spatial normalization with the nonlinear deformation field estimated function that matches brain regional coordinates within the segmented T1-weighted image into the Montreal Neurological Institute (MNI) template, to be applied in warping the corresponding PET images into the high-resolution template space [40]. Spatial smoothing of the normalized PET images by a Gaussian kernel of 8 mm FWHM was also performed [41].

Finally, brain ROIs to be warped onto the PET images (using the coregistration function to match the mean PET image to the T1-weighted MR image) could be derived from the (1) automated multi-label approach algorithms of the probabilistic atlas [such as the Automated Anatomical Labeling (AAL) atlas] or (2) individual-specific ROIs estimated by way of the automated parcellation of MRI scans (coregistered to the PET images) using the FreeSurfer (FS v6.0) into the several brain regions comprising the Desikan-Killiany atlas [42], the bilateral hippocampus and amygdala, and the cerebellar gray matter. Moreover, group-wise averaging of the PET images could also be used for ROI creation. For instance, the raphe nucleus (RN) ROI for the [^{11}C]CUMI-101 or [^{11}C]WAY100635 was labeled using a mask of the average location of the RN in 52 healthy subjects, which was created using [^{11}C]WAY100635 voxel-binding maps warped into standard space [43]. PET images were warped into a high-resolution MNI template space [40] and averaged. As binding of [^{11}C]WAY100635 is higher in the RN compared to surrounding areas, thresholding technique was used to extract the RN. The RN was then applied to individual participant's MRI by first warping their MRI to the MNI template space and then using those parameters to inverse-warp the RN ROI into the individual participant's space [33].

3.3 Estimation of Receptor Availability: Time Activity Curves and BP$_{ND}$

Distribution volume (V_T) is the tissue-to-plasma concentration ratio at equilibrium and reflects the total uptake (specific plus nonspecific binding) of the radio-ligand. After producing the high-quality parametric maps for the radio-ligand binding [16, 35], total volume of distribution (V_T) for radiotracer is computed parametrically using a metabolite-corrected arterial input function (refer to Subheading 3.1) and the PET time activity curves. Time activity curves could be generated from dynamic PET images by averaging the measured activity over all the voxels within each region of interest (ROI)—which were coregistered to the corresponding MR images—for the duration of the PET acquisition [26]. PET time activity curves for the radio-ligand obtained from the ROI were fit with a two-compartment model (for [^{11}C] WAY100635) or a one-tissue (1T) compartment model (for [^{11}C] UCB-J) [14], with rate constants of radiotracer between blood and tissue compartments constrained to that of the cerebellar white matter. Finally, based on the parameter estimation and PET time activity curves, voxel-wise BP$_F$ and BP$_{ND}$ were then estimated.

BP_F is defined as specific brain regional binding [regional volume of distribution (V_T) minus nonspecific volume of distribution (V_{ND})] normalized to the arterial plasma-free fraction of radio-tracer (f_P); thus, $BP_F = (V_T - V_{ND})/f_P$. To derive the nonspecific volume of distribution (V_{ND}), reference region of cerebellar gray matter with no specific binding for the radio-ligands of [^{11}C] WAY100635 [44, 45], [^{11}C]-raclopride, and [^{11}C]ABP688 [46, 47] was used. The multilinear reference tissue model implemented in PMOD software version 3.8 (PMOD Technologies Ltd., Zürich, Switzerland; http://www.pmod.com) was used with the cerebellum as the reference tissue [27].

The BP_{ND} is defined as the ratio of specifically bound to non-displaceable radio-ligand at equilibrium (= $V_T - V_{ND}$) [48] normalized to the nonspecific binding in the reference region (V_{ND}); thus, $BP_{ND} = (V_T - V_{ND})/V_{ND}$. Binding potential ($BP_{ND}$) is a measure of receptor availability (such as 5-HT$_{1A}$ and mGluR5) estimated using the simplified reference tissue model (SRTM) [49] or simplified reference tissue model 2 (SRTM2) [50]. Standard errors (SE) for both measures were calculated by bootstrapping both PET and plasma data [51]. In case of quantifying the magnitude of the [^{11}C]raclopride ligand displacement (and hence endogenous dopamine release) due to the performance of the monetary incentive delay task in comparison to the baseline PET data acquisition, the linear simplified reference tissue model (LSRTM) [52] was applied to the motion-corrected dynamic PET data. Parametric images of the main outcome measure, the amplitude (γ) of ligand displacement, were computed using the start of task onset as an input parameter and the cerebellar gray matter as the reference tissue.

3.4 Preprocessing and White Matter Fiber Tracking of the DWI

To generate the average number of white matter tracts between the ROI (such as raphe nucleus) and other brain regions, brain diffusion-weighted imaging (DWI) was gathered. Prior to the image preprocessing, DW images were assessed for quality by trained technicians to check for artifacts such as venetian blind, gradient-wise motion, ghost, ringing, and slice-wise intensity artifacts [53]. Images that passed quality control process were corrected for motion and gradient coil-induced distortions using the eddy current correction routine in FSL (FMRIB Software Library, http://www.fmrib.ox.ac.uk/fsl/), and non-brain tissue was removed through FSL's Brain Extraction Tool. To assess the white matter-based fiber tracts, probabilistic fiber tractography was computed using the FMRIB Diffusion Toolbox (FDT) [54]. In other words, the probability of connections from the seed (such as raphe nucleus ROI delineated by PET) to all target brain regions (such as the Desikan-Killiany-defined MRI labels) was estimated by repeated sampling of streamlines (for 5000 times, a maximum of 2000 steps per sample) from each principal diffusion direction through each voxel with the threshold criteria [a step length of 0.5 mm and a tract curvature threshold of 0.2 mm]. The number of the white matter-based fiber tracts was calculated by multiplying the number of the 5000 sample fibers that met the threshold criteria by the number of voxels from the seed region (fixed for all study participants) [33].

3.5 Preprocessing of the Brain Resting-State T2* Image

The first ten volumes of each resting-state functional run were discarded to allow for the magnetization to reach a steady state. Preprocessing of the resting-state T2* images was performed using SPM12 (http://www.fil.ion.ucl.ac.uk/spm/). Slice-time correction, motion correction, and spatial coregistration of the resting fMRI data to the first image from each respective epoch were performed [55]. The 3D T1-weighted image of each subject was segmented into gray matter, white matter, and cerebrospinal fluid (CSF) images and was coregistered to the resting-state T2* MR image. The segmented T1 images were spatially normalized to the MNI template, and the same transform was applied to the corresponding resting-state T2* MRI images. Both images were resampled to 2 mm isotropic voxels and smoothed by a 3D Gaussian low-pass filter at 6 mm full width at half maximum (FWHM).

To remove unwanted physiological and motion effects from the resting-state T2* MR images, based on a default scheme implemented in the functional connectivity toolbox (CONN) software (CONN v.17.f) (http://web.mit.edu/swg/software.htm), denoising by way of the linear regression was conducted [56]. In other words, five principal components from white matter and CSF time series (extracted using the CompCor method [57]) and realignment parameters in six degrees of freedom (for motion correction) were regressed out. Bandpass filtering (0.008–0.09 Hz) and linear detrending were also performed [34]. Aiming to restrict further analyses only for voxels in the gray matter, a canonical gray matter mask defined in common space was applied. Finally, for each participant, all preprocessed resting-state runs were variance normalized and concatenated [13].

3.6 Preprocessing and First-Level Analysis of the Task-Based T2* Image

Preprocessing of the blood-oxygen-level-dependent (BOLD) T2* MR images was conducted using SPM12 software (https://www.fil.ion.ucl.ac.uk/spm/software/spm12/). First, images were corrected for the slice time and were realigned to the first volume of each task run for motion correction. Second, T2* images were registered to T1-weighted structural images with seven degrees of freedom, and then structural images were warped to the standard MNI space using a 12-degree affine registration. Third, T2* images were smoothed with a Gaussian kernel of 6 mm full width at half maximum (FWHM) [58, 59].

After preprocessing, first-level analyses of T2* images acquired during task performance such as the monetary incentive delay

(MID) task per participant applied a general linear model which includes event-related regressors (choice, feedback, and outcome) convolved with a hemodynamic response and trial-specific parametric regressors (choice value, feedback prediction error, and monetary outcome prediction error). In addition, a high-pass temporal filter (Fourier transform, 100 s) and motion parameters [with six degrees of freedom (DOF), estimated during preprocessing] were incorporated as regressors of no interest [32].

3.7 Construction of Seed-Based Resting-State or Task-Based Functional Connectivity

To investigate the effect of alteration in receptor (such as mGluR5) availability on functional connectivity, seed-to-ROI analysis of the preprocessed resting fMRI data was conducted using the CONN software package (https://web.conn-toolbox.org/). First, peak voxel coordinates within the clusters that showed significant between-group differences between MDD and HC in binding potential (BP_{ND}) of radio-ligand (such as [^{11}C]ABP688) were selected as seed regions for the resting-state functional connectivity. Second, seed-to-whole-brain functional connectivity network per individual was constructed using the Fisher z-transformed Pearson's correlation coefficients calculated between the time series of seed region (averaged across all voxels within the seed region) and time series of every other voxel comprising the cerebral gray matter [13]. For the seed-to-ROI functional connectivity network, correlation coefficients between the time series of seed region (averaged across all voxels within the seed region) and time series of the cortical and subcortical brain regions (predefined from the probabilistic brain atlas such as the AAL atlas) averaged across all voxels within the ROI were calculated instead [34].

To determine the directional interactions among brain regions related to the negative emotional processing, firstly, within-subject fixed-effect models for the contrast of "negative picture presentation versus neutral picture presentation" were estimated for each participant across multiple runs of task-based T2* MRI data. Secondly, group-level analyses in which all subjects are included (regardless of the diagnosis) were conducted and found clusters that showed significantly increased functional activation for the negative picture viewing compared to the viewing of neutral pictures for every participant. Thirdly, after regressing out the head motions (already uncovered in the preprocessing step of Subheading 3.6), time series of functional brain activation of these clusters (averaged per cluster found in the second stage above) were extracted from the preprocessed T2* MR images, for blocks of negative or neutral picture viewing separately [26]. Fourthly, degrees of directional influences among these clusters regarding the functional activation in response to the negative picture viewing per participant were estimated by way of the multivariate dynamical systems (MDS), a dynamic causal model that incorporates minimal priors [24].

3.8 Integrating Cross-Modal Data and Second-Level Analyses

Associations between the deficits of neurotransmitter systems and altered resting-state functional connectivity network have been explored. First, brain regions of interest (ROIs) that show statistically significant different values of binding potential (BP_{ND}, for radio-ligands such as [^{11}C]ABP688 that measure receptor availability) or distribution volume (V_T, for [^{11}C]UCB-J that quantifies synaptic density) in major depression compared to healthy controls are found. Second, resting-state seed-based functional connectivity networks between these ROIs and other brain regions are estimated. Third, correlation coefficients among the ROI's binding potential (BP_{ND}), strengths of the "ROI-to-brain regional" resting-state functional connectivity, and clinical features such as depressive symptom severity are calculated [13, 60]. For the brain white matter-based structural connectivity, associations among the number of white matter tracts that connect brain region of serotonin synthesis (raphe nucleus) to other ROIs, specific regional binding (BP_F) of the radiotracer for serotonin neurotransmission (such as radio-ligand of [^{11}C]WAY100635 for the 5-HT$_{1A}$ autoreceptor), and cortical thickness at each ROI could be explored [33].

4 How to Apply These Procedures

1. First, preprocessing of PET images and application of the T1-weighted MRI-derived ROIs (Subheading 3.2) can be done using the software of SPM12 (https://www.fil.ion.ucl.ac.uk/spm/software/spm12/), FreeSurfer (https://surfer.nmr.mgh.harvard.edu/), FSL (https://fsl.fmrib.ox.ac.uk/fsl/fslwiki), PETPVC [61], or PETSurfer [62], solely or as a part of preprocessing pipelines [63, 64]. Second, for the PET kinetic modeling (Subheadings 3.1 and 3.3), software such as the PMOD (https://www.pmod.com/web), kinfitr [65], QModeling [66], APPIAN [67], or COMKAT [68] can be applied. Third, preprocessing and white matter fiber tracking of the DWI (Subheading 3.4) can be conducted by way of the software such as FSL Diffusion Toolbox (https://fsl.fmrib.ox.ac.uk/fslcourse/lectures/practicals/fdt1/index.html) and DSI Studio (http://dsi-studio.labsolver.org/Manual). Fourth, preprocessing and calculation of resting-state functional connectivity from the resting-state T2* images (Subheadings 3.5 and 3.7) can be done using the software of CONN (https://web.conn-toolbox.org/), C-PAC (https://www.nitrc.org/projects/cpac), and REST (https://www.nitrc.org/projects/rest/), among others.

Acknowledgments

This research was funded by the Basic Science Research Program through the National Research Foundation of Korea (NRF) funded by the Ministry of Education (NRF-2017R1D1A1B03028464).

References

1. Cowen PJ, Browning M (2015) What has serotonin to do with depression? World Psychiatry 14(2):158–160

2. Maier SF, Watkins LR (2005) Stressor controllability and learned helplessness: the roles of the dorsal raphe nucleus, serotonin, and corticotropin-releasing factor. Neurosci Biobehav Rev 29(4–5):829–841

3. Ferrés-Coy A, Santana N, Castañé A et al (2013) Acute 5-HT$_1$A autoreceptor knockdown increases antidepressant responses and serotonin release in stressful conditions. Psychopharmacology 225(1):61–74

4. Savitz J, Lucki I, Drevets WC (2009) 5-HT (1A) receptor function in major depressive disorder. Prog Neurobiol 88(1):17–31

5. Fisher PM, Meltzer CC, Ziolko SK et al (2006) Capacity for 5-HT1A-mediated autoregulation predicts amygdala reactivity. Nat Neurosci 9(11):1362–1363

6. Selvaraj S, Mouchlianitis E, Faulkner P et al (2015) Presynaptic serotoninergic regulation of emotional processing: a multimodal brain imaging study. Biol Psychiatry 78(8):563–571

7. Banerjee P, Mehta M, Kanjilal B (2007) Frontiers in neuroscience the 5-HT(1A) receptor: a signaling hub linked to emotional balance. In: Chattopadhyay A (ed) Serotonin receptors in neurobiology. CRC Press/Taylor & Francis, Boca Raton, FL

8. Milak MS, DeLorenzo C, Zanderigo F et al (2010) In vivo quantification of human serotonin 1A receptor using 11C-CUMI-101, an agonist PET radiotracer. J Nucl Med 51(12): 1892–1900

9. Krishnan V, Nestler EJ (2008) The molecular neurobiology of depression. Nature 455(7215):894–902

10. Spies M, Knudsen GM, Lanzenberger R, Kasper S (2015) The serotonin transporter in psychiatric disorders: insights from PET imaging. Lancet Psychiatry 2(8):743–755

11. Wilson AA, Ginovart N, Hussey D, Meyer J, Houle S (2002) In vitro and in vivo characterisation of [11C]-DASB: a probe for in vivo measurements of the serotonin transporter by positron emission tomography. Nucl Med Biol 29(5):509–515

12. Sudhof TC (2004) The synaptic vesicle cycle. Annu Rev Neurosci 27:509–547

13. Holmes SE, Scheinost D, Finnema SJ et al (2019) Lower synaptic density is associated with depression severity and network alterations. Nat Commun 10(1):1529

14. Finnema SJ, Nabulsi NB, Mercier J et al (2018) Kinetic evaluation and test-retest reproducibility of [(11)C]UCB-J, a novel radioligand for positron emission tomography imaging of synaptic vesicle glycoprotein 2A in humans. J Cereb Blood Flow Metab 38(11):2041–2052

15. Bajjalieh SM, Frantz GD, Weimann JM, McConnell SK, Scheller RH (1994) Differential expression of synaptic vesicle protein 2 (SV2) isoforms. J Neurosci 14(9): 5223–5235

16. Nabulsi NB, Mercier J, Holden D et al (2016) Synthesis and preclinical evaluation of 11C-UCB-J as a PET tracer for imaging the synaptic vesicle glycoprotein 2A in the brain. J Nucl Med 57(5):777–784

17. Sanacora G, Treccani G, Popoli M (2012) Towards a glutamate hypothesis of depression: an emerging frontier of neuropsychopharmacology for mood disorders. Neuropharmacology 62(1):63–77

18. Kokane SS, Armant RJ, Bolaños-Guzmán CA, Perrotti LI (2020) Overlap in the neural circuitry and molecular mechanisms underlying ketamine abuse and its use as an antidepressant. Behav Brain Res 384:112548

19. Ng TH, Alloy LB, Smith DV (2019) Meta-analysis of reward processing in major depressive disorder reveals distinct abnormalities within the reward circuit. Transl Psychiatry 9(1):293

20. Kringelbach ML, Cruzat J, Cabral J et al (2020) Dynamic coupling of whole-brain neuronal and neurotransmitter systems. Proc Natl Acad Sci U S A 117(17):9566–9576

21. Duman RS, Sanacora G, Krystal JH (2019) Altered connectivity in depression: GABA and glutamate neurotransmitter deficits and reversal by novel treatments. Neuron 102(1):75–90

22. Scott J, Hidalgo-Mazzei D, Strawbridge R et al (2019) Prospective cohort study of early bio-signatures of response to lithium in bipolar-I-disorders: overview of the H2020-funded R-LiNK initiative. Int J Bipolar Disord 7(1):20

23. Parsey RV, Ogden RT, Miller JM et al (2010) Higher serotonin 1A binding in a second major depression cohort: modeling and reference region considerations. Biol Psychiatry 68(2): 170–178

24. DeLorenzo C, Kumar JS, Mann JJ, Parsey RV (2011) In vivo variation in metabotropic glutamate receptor subtype 5 binding using positron emission tomography and [11C] ABP688. J Cereb Blood Flow Metab 31(11): 2169–2180

25. DeLorenzo C, Gallezot JD, Gardus J et al (2017) In vivo variation in same-day estimates of metabotropic glutamate receptor subtype 5 binding using [(11)C]ABP688 and [(18)F] FPEB. J Cereb Blood Flow Metab 37(8): 2716–2727

26. Schneck N, Tu T, Falcone HR et al (2020) Large-scale network dynamics in neural response to emotionally negative stimuli linked to serotonin 1A binding in major depressive disorder. Mol Psychiatry 26:2393. https://doi.org/10.1038/s41380-020-0733-5

27. Hamilton JP, Sacchet MD, Hjørnevik T et al (2018) Striatal dopamine deficits predict reductions in striatal functional connectivity in major depression: a concurrent (11)C-raclopride positron emission tomography and functional magnetic resonance imaging investigation. Transl Psychiatry 8(1):264

28. Kumar JS, Prabhakaran J, Majo VJ et al (2007) Synthesis and in vivo evaluation of a novel 5-HT1A receptor agonist radioligand [O-methyl- 11C]2-(4-(4-(2-methoxyphenyl) piperazin-1-yl)butyl)-4-methyl-1,2,4-triazine-3,5(2H,4H)dione in nonhuman primates. Eur J Nucl Med Mol Imaging 34(7):1050–1060

29. Ametamey SM, Kessler LJ, Honer M et al (2006) Radiosynthesis and preclinical evaluation of 11C-ABP688 as a probe for imaging the metabotropic glutamate receptor subtype 5. J Nucl Med 47(4):698–705

30. Langer O, Någren K, Dolle F et al (1999) Precursor synthesis and radiolabelling of the dopamine D2 receptor ligand [11C]raclopride from [11C]methyl triflate. J Label Compd Radiopharm 42(12):1183–1193

31. Rahmim A, Cheng JC, Blinder S, Camborde ML, Sossi V (2005) Statistical dynamic image reconstruction in state-of-the-art high-resolution PET. Phys Med Biol 50(20): 4887–4912

32. Schneier FR, Slifstein M, Whitton AE et al (2018) Dopamine release in antidepressant-naive major depressive disorder: a multimodal [(11)C]-(+)-PHNO positron emission tomography and functional magnetic resonance imaging study. Biol Psychiatry 84(8):563–573

33. Pillai RLI, Malhotra A, Rupert DD et al (2018) Relations between cortical thickness, serotonin 1A receptor binding, and structural connectivity: a multimodal imaging study. Hum Brain Mapp 39(2):1043–1055

34. Smith GS, Kuwabara H, Gould NF et al (2021) Molecular imaging of the serotonin transporter availability and occupancy by antidepressant treatment in late-life depression. Neuropharmacology 194:108447. https://doi.org/10.1016/j.neuropharm.2021.108447

35. Finnema SJ, Nabulsi NB, Eid T et al (2016) Imaging synaptic density in the living human brain. Sci Transl Med 8(348):348–396

36. Jenkinson M, Smith S (2001) A global optimisation method for robust affine registration of brain images. Med Image Anal 5(2):143–156

37. DeLorenzo C, Klein A, Mikhno A et al (2009) A new method for assessing PET-MRI coregistration. Paper presented at the SPIE Medical Imaging, Florida, USA

38. Smith SM (2002) Fast robust automated brain extraction. Hum Brain Mapp 17(3):143–155

39. Ashburner J, Friston KJ (2005) Unified segmentation. NeuroImage 26(3):839–851

40. Holmes CJ, Hoge R, Collins L, Woods R, Toga AW, Evans AC (1998) Enhancement of MR images using registration for signal averaging. J Comput Assist Tomogr 22(2):324–333

41. Ashburner J (2007) A fast diffeomorphic image registration algorithm. NeuroImage 38(1): 95–113

42. Desikan RS, Ségonne F, Fischl B et al (2006) An automated labeling system for subdividing the human cerebral cortex on MRI scans into gyral based regions of interest. NeuroImage 31(3):968–980

43. Delorenzo C, Delaparte L, Thapa-Chhetry B, Miller J, Mann J, Parsey RV (2013) Prediction of selective serotonin reuptake inhibitor response using diffusion-weighted MRI. Front Psychiatry 4:5

44. Hirvonen J, Kajander J, Allonen T, Oikonen V, Någren K, Hietala J (2007) Measurement of serotonin 5-HT1A receptor binding using positron emission tomography and [carbonyl-(11) C]WAY-100635-considerations on the validity of cerebellum as a reference region. J Cereb Blood Flow Metab 27(1):185–195

45. Parsey RV, Arango V, Olvet DM, Oquendo MA, Van Heertum RL, John Mann J (2005)

Regional heterogeneity of 5-HT1A receptors in human cerebellum as assessed by positron emission tomography. J Cereb Blood Flow Metab 25(7):785–793

46. DeLorenzo C, Sovago J, Gardus J et al (2015) Characterization of brain mGluR5 binding in a pilot study of late-life major depressive disorder using positron emission tomography and [^{11}C] ABP688. Transl Psychiatry 5(12):e693

47. DuBois JM, Rousset OG, Rowley J et al (2016) Characterization of age/sex and the regional distribution of mGluR5 availability in the healthy human brain measured by high-resolution [(11)C]ABP688 PET. Eur J Nucl Med Mol Imaging 43(1):152–162

48. Innis RB, Cunningham VJ, Delforge J et al (2007) Consensus nomenclature for in vivo imaging of reversibly binding radioligands. J Cereb Blood Flow Metab 27(9):1533–1539

49. Lammertsma AA, Hume SP (1996) Simplified reference tissue model for PET receptor studies. NeuroImage 4(3 Pt 1):153–158

50. Wu Y, Carson RE (2002) Noise reduction in the simplified reference tissue model for neuroreceptor functional imaging. J Cereb Blood Flow Metab 22(12):1440–1452

51. Parsey RV, Slifstein M, Hwang DR et al (2000) Validation and reproducibility of measurement of 5-HT1A receptor parameters with [carbonyl-11C]WAY-100635 in humans: comparison of arterial and reference tissue input functions. J Cereb Blood Flow Metab 20(7): 1111–1133

52. Alpert NM, Badgaiyan RD, Livni E, Fischman AJ (2003) A novel method for noninvasive detection of neuromodulatory changes in specific neurotransmitter systems. NeuroImage 19(3):1049–1060

53. Liu Z, Wang Y, Gerig G et al (2010) Quality control of diffusion weighted images. Proc SPIE Int Soc Opt Eng 7628:76280j

54. Behrens TE, Berg HJ, Jbabdi S, Rushworth MF, Woolrich MW (2007) Probabilistic diffusion tractography with multiple fibre orientations: what can we gain? NeuroImage 34(1): 144–155

55. Drysdale AT, Grosenick L, Downar J et al (2017) Resting-state connectivity biomarkers define neurophysiological subtypes of depression. Nat Med 23(1):28–38

56. Whitfield-Gabrieli S, Nieto-Castanon A (2012) Conn: a functional connectivity toolbox for correlated and anticorrelated brain networks. Brain Connect 2(3):125–141

57. Behzadi Y, Restom K, Liau J, Liu TT (2007) A component based noise correction method (CompCor) for BOLD and perfusion based fMRI. NeuroImage 37(1):90–101

58. Jenkinson M, Bannister P, Brady M, Smith S (2002) Improved optimization for the robust and accurate linear registration and motion correction of brain images. NeuroImage 17(2):825–841

59. Andersson JLR, Jenkinson M, Smith S (2007) Non-linear registration, aka Spatial normalisation. FMRIB technical report TR07JA2. FMRIB Centre, Oxford

60. Kim JH, Joo YH, Son YD et al (2019) In vivo metabotropic glutamate receptor 5 availability-associated functional connectivity alterations in drug-naïve young adults with major depression. Eur Neuropsychopharmacol 29(2): 278–290

61. Thomas BA, Cuplov V, Bousse A et al (2016) PETPVC: a toolbox for performing partial volume correction techniques in positron emission tomography. Phys Med Biol 61(22): 7975–7993

62. Greve DN, Salat DH, Bowen SL et al (2016) Different partial volume correction methods lead to different conclusions: an (18)F-FDG-PET study of aging. NeuroImage 132:334–343

63. Nørgaard M, Ganz M, Svarer C et al (2019) Optimization of preprocessing strategies in Positron Emission Tomography (PET) neuroimaging: a [(11)C]DASB PET study. NeuroImage 199:466–479

64. Marcoux A, Burgos N, Bertrand A et al (2018) An automated pipeline for the analysis of PET data on the cortical surface. Front Neuroinform 12:94

65. Tjerkaski J, Cervenka S, Farde L, Matheson GJ (2020) Kinfitr - an open-source tool for reproducible PET modelling: validation and evaluation of test-retest reliability. EJNMMI Res 10(1):77

66. López-González FJ, Paredes-Pacheco J, Thurnhofer-Hemsi K et al (2019) QModeling: a multiplatform, easy-to-use and open-source toolbox for PET kinetic analysis. Neuroinformatics 17(1):103–114

67. Funck T, Larcher K, Toussaint PJ, Evans AC, Thiel A (2018) APPIAN: automated pipeline for PET image analysis. Front Neuroinform 12:64

68. Muzic RF Jr, Cornelius S (2001) COMKAT: compartment model kinetic analysis tool. J Nucl Med 42(4):636–645

Part V

Translational Interventions

Chapter 15

Translational Neuropsychopharmacology for Major Depression: Targeting Neurotransmitter Systems

Hyewon Kim, Yong-Ku Kim, and Hong Jin Jeon

Abstract

For many years, the pathology of depression had been interpreted based on psychodynamic theories. However, the accidental discovery of antidepressants has raised questions about the relationship between monoamine neurotransmitters and depression and led to numerous research studies. Currently, dozens of antidepressants are available, and the field of biological psychiatry is developing day by day. More recently, as the development of drugs targeting new neurotransmitter systems progresses rapidly, the treatment of depression is attempting to change based on new paradigms.

Key words Antidepressants, Translational neuropharmacology, Monoamine neurotransmitter, Glutamatergic modulators

1 Introduction

Major depressive disorder (MDD) is a common psychiatric disorder with significant morbidity [1]. MDD features changes in mood, cognition, and vegetative symptoms and can lead to impairments in an individual's functioning [2]. In the acute phase, one or a combination of treatment options such as pharmacotherapy, psychotherapy, and somatic therapy is used to induce the remission of major depressive episodes and restore the patient's previous functioning. According to practice guidelines, antidepressant medication is recommended as the initial treatment of choice for patients with mild to moderate depression symptoms. In addition, antidepressant medication is a major treatment option that is consistently chosen for those in need of continuation phase and maintenance treatment [3].

The currently available antidepressants mainly target the neurotransmitter system. Since the discovery of the first antidepressant in the 1950s, most antidepressants have targeted the monoamine

Yong-Ku Kim and Meysam Amidfar (eds.), *Translational Research Methods for Major Depressive Disorder*, Neuromethods, vol. 179, https://doi.org/10.1007/978-1-0716-2083-0_15,

neurotransmitter system. With the serendipitous discovery of the antidepressant effect of monoamine oxidase inhibitors (MAOIs) and tricyclic antidepressants (TCAs), a paradigm change in the treatment of depression was achieved. Moreover, biological psychiatry, which emerged along with the monoamine hypothesis, led to a flood of new drugs for depression. Until recently, guidelines for treating depression recommend antidepressants that act selectively on monoamine transporters as the initial treatment of choice. However, for the last decade, studies have shown the faster and robust antidepressant effect of glutamatergic modulators. Very recently, a glutamatergic modulator was approved by the US Food and Drug Administration (FDA) for the treatment of depression. And the indications for glutamatergic modulator use are expanding and preparing to change the paradigm of depression treatment once again.

In this chapter, we will address the discovery and development of antidepressant agents targeting the neurotransmitter system and their properties. Learning the history of drug development will not only promote understanding of the characteristics of each drug class but also help choose strategies when prescribing drugs in individual patients with depression.

2   Antidepressants Targeting Monoamine Transporters

Antidepressant use began in the 1950s when imipramine and iproniazid were used to treat depression. Until then, the only treatment for depression was electroconvulsive therapy for limited indications. Moreover, the manifestations of depression symptoms had been mainly interpreted psychodynamically and explained as results of some sort of individual's internal conflict. The other therapeutic options for depression were very limited [4]. Thus, the discovery of these two antidepressants contributed to paradigm shifts in the hypothesis of depression, neuropsychopharmacology, and treatment strategies.

The antidepressant effect of iproniazid was discovered serendipitously while being used as a treatment for patients with tuberculosis. Subsequent studies demonstrated the antidepressant effect of iproniazid and confirmed that MAOIs could be used to treat depression [5, 6]. Like iproniazid, the antidepressant effect of imipramine was also discovered by chance. Drug research on schizophrenia confirmed that imipramine, a derivative of chlorpromazine, had a mood-elevating effect instead of a neurologic effect [7].

After the discovery of their antidepressant effects, numerous clinical trials investigating MAOIs and TCAs had been conducted. However, MAOIs including iproniazid were withdrawn from use in most of the world due to hepatotoxicity [8]. But with the introduction of amitriptyline, TCAs successfully established themselves as the treatment of depression. This made psychiatrists advocate for treatment not only of the mind but also of the body and contributed to laying the foundation of biological psychiatry [4].

Clinical research has demonstrated that the mechanisms of action of TCAs included the inhibition of serotonin reuptake and showed decreased levels of cerebral serotonin in depressed patients who died from suicide [9]. This clinical evidence gave rise to the monoamine hypothesis that the dysfunction of monoamine including serotonin is the key pathophysiology in the symptoms of depression.

The monoamine hypothesis states that the deficiency of monoamine causes depression symptoms, and when monoamine supplementation with an antidepressant medication is effective, the symptoms of depression are improved. Classic antidepressants exhibit antidepressant effects by blocking more than one monoamine transporter. The representative monoamine systems associated with depression include serotonin, norepinephrine, and dopamine. Specific monoamine neurotransmitters are known to be associated with specific psychiatric manifestations. Serotonin is associated with anxiety, obsessions, and compulsions; norepinephrine is associated with alertness, energy, anxiety, attention, and interest in life; and dopamine is related to motivation, pleasure, reward, attention, and interest in life. All of these three neurotransmitters are also associated with mood [10].

Although TCAs and MAOIs were introduced to treat depression symptoms coincidentally, selective serotonin reuptake inhibitors (SSRIs) were developed by pre-planned design. Fluoxetine, the first SSRI, was developed in the process of searching for a drug that acts on the serotonin reuptake pump and does not act on a receptor that could cause adverse effects [4].

SSRIs have relative advantages in tolerance and safety but have the disadvantages of delayed response, the induction of remission in only about one-third of the patients, and non-response in about 30% of the patients [11, 12]. These flaws have led to continued research on the development of new antidepressants, and drugs with various pharmacodynamic properties including serotonin partial agonist/reuptake inhibitors (SPARIs), serotonin-norepinephrine reuptake inhibitors (SNRIs), and norepinephrine-dopamine reuptake inhibitors (NDRIs) were developed. Meanwhile, some drugs have been actively adopted in clinical settings with high efficacy, and some drugs have been withdrawn from the market due to disappointing efficacy or safety issues.

One of the important problems in the use of antidepressants targeting monoamine neurotransmitters is that although the levels of monoamine neurotransmitters in the brain rise relatively rapidly after antidepressant use, there is a delay in the clinical effect experienced by patients. These are explained by the monoamine receptor hypothesis. Chronic monoamine deficiency in depression triggers the upregulation of the monoamine receptors, and this, in turn, increases receptor sensitivity. When neurotransmitter levels suddenly increase due to the use of antidepressants, this leads to down-regulation and desensitization of the receptor, resulting in a delayed clinical effect [13].

3   Glutamatergic Modulators

The glutamatergic system recently became an important target for new antidepressant agents. Glutamate receptors are divided into ionotropic receptors and metabotropic receptors. Inotropic receptors are ligand-gated ion channels, and when an agonist such as N-methyl-D-aspartate (NMDA), kainate, or α-amino-3-hydroxy-5-methyl-4-isoxazolepropionic acid (AMPA) is bound, they open and react.

Clinical research has shown changes in glutamate/glutamine, the major excitatory neurotransmitter, and γ-aminobutyric acid (GABA), an inhibitory neurotransmitter in patients with depression. Studies have demonstrated the decreased levels of neurotransmitters in glutamatergic system in plasma and cerebrospinal fluid among people with depression [14–16]. Furthermore, studies using proton magnetic resonance spectroscopy (MRS), an in vivo imaging technique for the detection of neurochemicals, have shown a decrease in glutamate and GABA concentrations in the prefrontal cortex and anterior cingulate cortex in patients with depression [17–19]. These results, along with the need for new antidepressants due to the limitations of existing antidepressants, have led to numerous clinical studies on glutamatergic modulators over the last decade.

Ketamine is a potent NMDA antagonist and the most studied drug for its antidepressant effect among glutamatergic modulators. Ketamine is a drug that was originally used to induce anesthesia. Compared to other anesthetics, ketamine suppresses breathing less and stimulates the circulatory system, so it has been commonly used in short-term procedures that do not require muscle relaxation [20–22]. In the 1990s, studies on animal models of depression demonstrated that NMDA receptor antagonists were effective for alleviating depression symptoms [23–26]. Evidence has emerged showing that the administration of antidepressants changes the binding profiles and function of NMDA receptors and affects the expression of mRNA encoding multiple NMDA receptor subunits [27–29].

In 2000, a placebo-controlled, double-blind clinical trial demonstrated the efficacy of ketamine in patients with depression for the first time [30]. In this trial, the administration of subanesthetic intravenous (IV) doses of ketamine showed a significant reduction in depression symptoms compared to saline treatment. Since then, many clinical trials have confirmed the antidepressant effect of ketamine with consistently rapid and robust responses [31–34]. While conventional monoaminergic antidepressants take several weeks to show effects, research has shown that ketamine ameliorated the symptoms of depression within 2 h of a single infusion, and the antidepressant effect lasted more than 1 week after administration.

Clinical research also showed that the effect of ketamine was excellent in patients with treatment-resistant depression (TRD), which is defined as depression with insufficient response to standard antidepressant treatment. Ketamine was effective not only for depressed moods but also for other depression symptoms such as anhedonia, fatigue, and suicidality and led to remission in one-third of the patients with TRD [35–37]. In subsequent studies, clinical trials were performed using ketamine delivered by various routes such as intranasally, orally, or sublingually as well as by IV and showed that depression symptoms improved after the single or repeated administration of ketamine compared to placebo [38–40].

Esketamine is the S-enantiomer of ketamine. It is a nonselective and noncompetitive antagonist of the NMDA receptor with a higher affinity for the NMDA receptor than the R-enantiomer [41]. Following a multicenter, randomized, placebo-controlled trial that showed the antidepressant effect of IV esketamine [42], studies have been conducted to identify the efficacy of intranasal esketamine in patients with TRD. Studies have demonstrated both the short-term [43, 44] and maintained antidepressant effects of intranasal esketamine [45]. The FDA approved esketamine for patients with TRD in conjunction with an oral antidepressant in March 2019. And in August 2020, the FDA approved the supplemental new drug application for esketamine to treat depression symptoms in adults with MDD with acute suicidal ideation or behavior.

Other types of glutamatergic modulators are currently being investigated in clinical research. Rapastinel (GLYX-13) is a novel NMDA receptor modulator. It acts as a partial agonist of the glycine site of the NMDA receptor. An early proof-of-concept study showed that the administration of IV rapastinel reduced depression symptoms within 2 h and the effect was maintained for about 7 days in patients with MDD who had not responded to other antidepressants [46]. Although there are a limited number of clinical studies, they have demonstrated the potential of rapastinel as a novel antidepressant without psychotomimetic side effects while acting as fast as ketamine [47, 48]. AV-101 [41] and AGN-241751 [49], NMDA modulators that can be administered orally, also showed fast, robust, and sustained antidepressant effects in preclinical studies, and phase 2 clinical studies are in progress.

4   Conclusions

In this chapter, we discussed the currently available antidepressants targeting the neurotransmitter system. In recent years, we have moved away from conventional pharmacological strategies that were firmly maintained for more than three decades to a new, efficacious agent with rapid action. Meanwhile, drugs targeting other new neurotransmitter systems such as glycine site modulators, opioid receptor modulators, or modulators of the GABA receptors are undergoing verification as new antidepressants. These pathways are expected to be the cornerstone of broadening the treatment options for patients with depression and physicians who have experienced frustrations with the existing limited treatments.

References

1. Baldessarini RJ, Forte A, Selle V, Sim K, Tondo L, Undurraga J, Vazquez GH (2017) Morbidity in depressive disorders. Psychother Psychosom 86(2):65–72. https://doi.org/10.1159/000448661

2. American Psychiatric Association (2020) Diagnostic and statistical manual of mental disorders, Fifth Edition (DSM-5). APA, Washington, DC. https://www.psychiatry.org/psychiatrists/practice/dsm. Accessed 1 Sep 2020

3. American Psychiatric Association (2009) Practice guideline for the treatment of patients with major depressive disorder, 3rd edn. APA, Washington, DC

4. López-Muñoz F, Alamo C (2009) Monoaminergic neurotransmission: the history of the discovery of antidepressants from 1950s until today. Curr Pharm Des 15(14):1563–1586

5. Crane GE (1957) Iproniazid (Marsilid) phosphate: a therapeutic agent for mental disorders and debilitating diseases. Psychiatric Research Reports

6. Kline NS (1958) Clinical experience with iproniazid (Marsilid). J Clin Exp Psychopathol 19(2, Suppl. 1):72

7. Kuhn R (1958) The treatment of depressive states with G 22355 (imipramine hydrochloride). Am J Psychiatr 115(5):459–464

8. López-Munoz F, Alamo C, Juckel G, Assion H-J (2007) Half a century of antidepressant drugs: on the clinical introduction of monoamine oxidase inhibitors, tricyclics, and tetracyclics. Part I: monoamine oxidase inhibitors. J Clin Psychopharmacol 27(6):555–559

9. Shaw DM, Camps FE, Eccleston EG (1967) 5-Hydroxytryptamine in the hind-brain of depressive suicides. Br J Psychiatry 113(505):1407–1411

10. Nutt DJ (2008) Relationship of neurotransmitters to the symptoms of major depressive disorder. J Clin Psychiatry 69:4–7

11. Gumnick JF, Nemeroff CB (2000) Problems with currently available antidepressants. J Clin Psychiatry 61:5

12. Trivedi MH, Rush AJ, Wisniewski SR, Nierenberg AA, Warden D, Ritz L, Norquist G, Howland RH, Lebowitz B, McGrath PJ (2006) Evaluation of outcomes with citalopram for depression using measurement-based care in STAR* D: implications for clinical practice. Am J Psychiatr 163(1):28–40

13. Stahl SM, Stahl SM (2013) Stahl's essential psychopharmacology: neuroscientific basis and practical applications. Cambridge University Press, Cambridge

14. Altamura C, Maes M, Dai J, Meltzer HY (1995) Plasma concentrations of excitatory amino acids, serine, glycine, taurine and histidine in major depression. Eur Neuropsychopharmacol 5(Suppl):71–75. https://doi.org/10.1016/0924-977x(95)00033-l

15. Nowak G, Ordway GA, Paul IA (1995) Alterations in the N-methyl-D-aspartate (NMDA) receptor complex in the frontal cortex of suicide victims. Brain Res 675(1–2):157–164. https://doi.org/10.1016/0006-8993(95)00057-w

16. Petty F, Sherman AD (1984) Plasma GABA levels in psychiatric illness. J Affect Disord

6(2):131–138. https://doi.org/10.1016/0165-0327(84)90018-1

17. Auer DP, Pütz B, Kraft E, Lipinski B, Schill J, Holsboer F (2000) Reduced glutamate in the anterior cingulate cortex in depression: an in vivo proton magnetic resonance spectroscopy study. Biol Psychiatry 47(4):305–313. https://doi.org/10.1016/s0006-3223(99)00159-6

18. Hasler G, van der Veen JW, Tumonis T, Meyers N, Shen J, Drevets WC (2007) Reduced prefrontal glutamate/glutamine and gamma-aminobutyric acid levels in major depression determined using proton magnetic resonance spectroscopy. Arch Gen Psychiatry 64(2):193–200. https://doi.org/10.1001/archpsyc.64.2.193

19. Yildiz-Yesiloglu A, Ankerst DP (2006) Review of 1H magnetic resonance spectroscopy findings in major depressive disorder: a meta-analysis. Psychiatry Res 147(1):1–25. https://doi.org/10.1016/j.pscychresns.2005.12.004

20. Adams HA (1997) S-(+)-ketamine. Circulatory interactions during total intravenous anesthesia and analgesia-sedation. Anaesthesist 46(12):1081–1087. https://doi.org/10.1007/s001010050510

21. Heshmati F, Zeinali MB, Noroozinia H, Abbacivash R, Mahoori A (2003) Use of ketamine in severe status asthmaticus in intensive care unit. Iran J Allergy Asthma Immunol 2(4):175–180

22. Rosenbaum SB, Gupta V, Palacios JL (2020) Ketamine. StatPearls, Treasure Island, FL

23. Layer RT, Popik P, Olds T, Skolnick P (1995) Antidepressant-like actions of the polyamine site NMDA antagonist, eliprodil (SL-82.0715). Pharmacol Biochem Behav 52(3):621–627. https://doi.org/10.1016/0091-3057(95)00155-p

24. Meloni D, Gambarana C, De Montis MG, Dal Prá P, Taddei I, Tagliamonte A (1993) Dizocilpine antagonizes the effect of chronic imipramine on learned helplessness in rats. Pharmacol Biochem Behav 46(2):423–426. https://doi.org/10.1016/0091-3057(93)90374-3

25. Moryl E, Danysz W, Quack G (1993) Potential antidepressive properties of amantadine, memantine and bifemelane. Pharmacol Toxicol 72(6):394–397. https://doi.org/10.1111/j.1600-0773.1993.tb01351.x

26. Papp M, Moryl E (1994) Antidepressant activity of non-competitive and competitive NMDA receptor antagonists in a chronic mild stress model of depression. Eur J Pharmacol 263(1–2):1–7. https://doi.org/10.1016/0014-2999(94)90516-9

27. Boyer PA, Skolnick P, Fossom LH (1998) Chronic administration of imipramine and citalopram alters the expression of NMDA receptor subunit mRNAs in mouse brain. A quantitative in situ hybridization study. J Mol Neurosci 10(3):219–233. https://doi.org/10.1007/bf02761776

28. Mjellem N, Lund A, Hole K (1993) Reduction of NMDA-induced behaviour after acute and chronic administration of desipramine in mice. Neuropharmacology 32(6):591–595. https://doi.org/10.1016/0028-3908(93)90055-8

29. Paul IA, Nowak G, Layer RT, Popik P, Skolnick P (1994) Adaptation of the N-methyl-D-aspartate receptor complex following chronic antidepressant treatments. J Pharmacol Exp Ther 269(1):95–102

30. Berman RM, Cappiello A, Anand A, Oren DA, Heninger GR, Charney DS, Krystal JH (2000) Antidepressant effects of ketamine in depressed patients. Biol Psychiatry 47(4):351–354. https://doi.org/10.1016/s0006-3223(99)00230-9

31. Ionescu DF, Luckenbaugh DA, Niciu MJ, Richards EM, Zarate CA Jr (2015) A single infusion of ketamine improves depression scores in patients with anxious bipolar depression. Bipolar Disord 17(4):438–443. https://doi.org/10.1111/bdi.12277

32. Murrough JW, Wan LB, Iacoviello B, Collins KA, Solon C, Glicksberg B, Perez AM, Mathew SJ, Charney DS, Iosifescu DV, Burdick KE (2013) Neurocognitive effects of ketamine in treatment-resistant major depression: association with antidepressant response. Psychopharmacology. https://doi.org/10.1007/s00213-013-3255-x

33. Newport DJ, Carpenter LL, McDonald WM, Potash JB, Tohen M, Nemeroff CB (2015) Ketamine and other NMDA antagonists: early clinical trials and possible mechanisms in depression. Am J Psychiatry 172(10):950–966. https://doi.org/10.1176/appi.ajp.2015.15040465

34. Sos P, Klirova M, Novak T, Kohutova B, Horacek J, Palenicek T (2013) Relationship of ketamine's antidepressant and psychotomimetic effects in unipolar depression. Neuro Endocrinol Lett 34(4):287–293

35. DiazGranados N, Ibrahim LA, Brutsche NE, Ameli R, Henter ID, Luckenbaugh DA, Machado-Vieira R, Zarate CA Jr (2010) Rapid resolution of suicidal ideation after a single infusion of an N-methyl-D-aspartate antagonist in patients with treatment-resistant major depressive disorder. J Clin Psychiatry

71(12):1605–1611. https://doi.org/10.4088/JCP.09m05327blu

36. Lally N, Nugent AC, Luckenbaugh DA, Niciu MJ, Roiser JP, Zarate CA Jr (2015) Neural correlates of change in major depressive disorder anhedonia following open-label ketamine. J Psychopharmacol 29(5):596–607. https://doi.org/10.1177/0269881114568041

37. Saligan LN, Luckenbaugh DA, Slonena EE, Machado-Vieira R, Zarate CA Jr (2016) An assessment of the anti-fatigue effects of ketamine from a double-blind, placebo-controlled, crossover study in bipolar disorder. J Affect Disord 194:115–119. https://doi.org/10.1016/j.jad.2016.01.009

38. Andrade C (2017) Ketamine for depression, 4: In what dose, at what rate, by what route, for how long, and at what frequency? J Clin Psychiatry 78(7):e852–e857. https://doi.org/10.4088/JCP.17f11738

39. Andrade C (2019) Oral ketamine for depression, 1: pharmacologic considerations and clinical evidence. J Clin Psychiatry 80(2):19f12820. https://doi.org/10.4088/JCP.19f12820

40. Domany Y, Bleich-Cohen M, Tarrasch R, Meidan R, Litvak-Lazar O, Stoppleman N, Schreiber S, Bloch M, Hendler T, Sharon H (2019) Repeated oral ketamine for out-patient treatment of resistant depression: randomised, double-blind, placebo-controlled, proof-of-concept study. Br J Psychiatry 214(1):20–26. https://doi.org/10.1192/bjp.2018.196

41. Henter ID, de Sousa RT, Zarate CA Jr (2018) Glutamatergic modulators in depression. Harv Rev Psychiatry 26(6):307

42. Singh JB, Fedgchin M, Daly E, Xi L, Melman C, De Bruecker G, Tadic A, Sienaert P, Wiegand F, Manji H, Drevets WC, Van Nueten L (2016) Intravenous esketamine in adult treatment-resistant depression: a double-blind, double-randomization, placebo-controlled study. Biol Psychiatry 80(6):424–431. https://doi.org/10.1016/j.biopsych.2015.10.018

43. Fedgchin M, Trivedi M, Daly EJ, Melkote R, Lane R, Lim P, Vitagliano D, Blier P, Fava M, Liebowitz M (2019) Efficacy and safety of fixed-dose esketamine nasal spray combined with a new oral antidepressant in treatment-resistant depression: results of a randomized, double-blind, active-controlled study (TRANSFORM-1). Int J Neuropsychopharmacol 22(10):616–630

44. Popova V, Daly EJ, Trivedi M, Cooper K, Lane R, Lim P, Mazzucco C, Hough D, Thase ME, Shelton RC (2019) Efficacy and safety of flexibly dosed esketamine nasal spray combined with a newly initiated oral antidepressant in treatment-resistant depression: a randomized double-blind active-controlled study. Am J Psychiatr 176(6):428–438

45. Daly EJ, Trivedi MH, Janik A, Li H, Zhang Y, Li X, Lane R, Lim P, Duca AR, Hough D, Thase ME, Zajecka J, Winokur A, Divacka I, Fagiolini A, Cubala WJ, Bitter I, Blier P, Shelton RC, Molero P, Manji H, Drevets WC, Singh JB (2019) Efficacy of esketamine nasal spray plus oral antidepressant treatment for relapse prevention in patients with treatment-resistant depression: a randomized clinical trial. JAMA Psychiatry 76(9):893–903. https://doi.org/10.1001/jamapsychiatry.2019.1189

46. Preskorn S, Macaluso M, Mehra DO, Zammit G, Moskal JR, Burch RM (2015) Randomized proof of concept trial of GLYX-13, an N-methyl-D-aspartate receptor glycine site partial agonist, in major depressive disorder nonresponsive to a previous antidepressant agent. J Psychiatr Pract 21(2):140–149. https://doi.org/10.1097/01.pra.0000462606.17725.93

47. Ragguett RM, Rong C, Kratiuk K, McIntyre RS (2019) Rapastinel - an investigational NMDA-R modulator for major depressive disorder: evidence to date. Expert Opin Investig Drugs 28(2):113–119. https://doi.org/10.1080/13543784.2019.1559295

48. Vasilescu AN, Schweinfurth N, Borgwardt S, Gass P, Lang UE, Inta D, Eckart S (2017) Modulation of the activity of N-methyl-d-aspartate receptors as a novel treatment option for depression: current clinical evidence and therapeutic potential of rapastinel (GLYX-13). Neuropsychiatr Dis Treat 13:973–980. https://doi.org/10.2147/ndt.S119004

49. Pothula S, Liu RJ, Wu M, Sliby AN, Picciotto MR, Banerjee P, Duman RS (2020) Positive modulation of NMDA receptors by AGN-241751 exerts rapid antidepressant-like effects via excitatory neurons. Neuropsychopharmacology 46:799. https://doi.org/10.1038/s41386-020-00882-7

Chapter 16

Transcranial Direct Current Stimulation for the Treatment of Major Depressive Disorder

Lucas Borrione, Laís B. Razza, Adriano H. Moffa, and André R. Brunoni

Abstract

Transcranial direct current stimulation (tDCS) is a noninvasive brain stimulation modality, indicated for the treatment of diverse neuropsychiatric disorders, such as major depressive disorder (MDD). The procedure consists of the application of a weak electrical current to the brain, using sponge-covered and saline-humidified electrodes placed over the scalp, thereby modulating underlying neuronal networks. TDCS is considered a safe and well-tolerated technique and, due to its additional portability and low cost, can be used in home settings. We provide basic and step-by-step instructions on how to perform a tDCS session for the treatment of MDD, using a protocol that has been extensively studied and validated in previous clinical trials. We also present the neuroConn DC-Stimulator-Plus tDCS device (neuroConn, Ilmenau, Germany), with instructions for the assemblage and utilization of the equipment, and the Beam F3 method for reliable and reproducible localization of stimulation targets. General practical advice and specific challenges regarding the procedure are covered in Subheading 4.

Key words Transcranial direct current stimulation, Noninvasive brain stimulation, Major depressive disorder, Beam F3 method, Dorsolateral prefrontal cortex

1 Introduction

Transcranial direct current stimulation (tDCS) is a noninvasive brain stimulation modality, whereby a weak electrical current (generally 1–2 mA) is applied to the brain, via two electrodes placed over the scalp [1]. The current is generated by a battery-operated or rechargeable device, flows from the positive electrode (anode) to the negative electrode (cathode), and is able to modulate cortical excitability and underlying brain networks, without generating action potentials [2, 3].

Supplementary Information The online version of this chapter (https://doi.org/10.1007/978-1-0716-2083-0_16) contains supplementary material, which is available to authorized users.

Yong-Ku Kim and Meysam Amidfar (eds.), *Translational Research Methods for Major Depressive Disorder*, Neuromethods, vol. 179, https://doi.org/10.1007/978-1-0716-2083-0_16,

tDCS has been extensively studied for the treatment of major depressive disorder (MDD) and is considered moderately superior to placebo, while being deemed tolerable and safe [4, 5]. The technique is associated with few side effects, mainly itching and tingling at the stimulation site [6, 7], while the degree of the association of tDCS with manic/hypomanic episodes is small [8, 9]. TDCS has shown probable synergistic effects with sertraline [10] and cognitive interventions [11, 12], but failed to demonstrate non-inferiority to escitalopram in depressed subjects, in a large, randomized clinical trial [13].

The most recent international guideline presents anodal tDCS as definitely effective in improving depression in MDD [14]. In a very recent review, the technique was also considered with the highest quality of evidence as a treatment for MDD, together with high-frequency transcranial magnetic stimulation (TMS) [15]. Moreover, tDCS presents some advantages over TMS, such as its portability and significantly inferior cost [2]. Due to these characteristics, tDCS can be performed in home settings for the treatment of neuropsychiatric disorders, with proper training of selected patients and remote monitoring by specialized clinical staff [16, 17]. Hitherto, a few studies have been conducted to evaluate the tolerability, feasibility, and efficacy of home-use tDCS for MDD [17, 18]. This modality presents several advantages compared to tDCS application in research centers or clinics, such as lack of transportation burdens and optimization of the number of sessions, and is a feasible intervention when clinical work is rapidly disrupted, such as in pandemic crises [19]. However, the evidence of home-based tDCS is still sparse and future studies should be conducted.

Among the studies investigating the efficacy of tDCS for MDD, the montage that has been most extensively studied places the activating anode over the left dorsolateral prefrontal cortex (DLPFC) and the inhibitory cathode over the right DLPFC [20]. Among all randomized controlled trials conducted to date for MDD, all studies placed the anode over the F3 (10–20 EEG system), while the cathode varied among the F4, F8 (10–20 EEG system), or supraorbital region [4]. The rationale for this bilateral montage is based on the brain's functional interhemispheric imbalance, as previously evidenced in EEG-based studies with depressed patients [21, 22]. In these studies, the left DLPFC was shown to be hypofunctional compared to the right DLPFC, the former being the primary target for the activating anode.

Different current intensities have been investigated (1 mA to 2.5 mA), with present consensus suggesting 2 mA as an optimal value [23]. Evidence suggests that at least 10 daily 30-min sessions, over 2 weeks (excluding weekends), are necessary to induce antidepressant effects, with variable frequencies and durations of maintenance protocols (in general, twice-weekly sessions) [23, 24].

In this chapter, we will offer practical instructions directed to licenced health professionals and neuroscientists, explaining how to perform a supervised tDCS session in patients with depression, for research and clinical purposes. We have opted for a traditional protocol and montage (anode over F3 and cathode over F4), which has been successfully tested by our team in previous trials for the treatment of MDD [10, 13], using here a neuroConn DC-Stimulator-Plus device (neuroConn, Ilmenau, Germany).

We sequentially present the individual parts of the equipment and instructions on how to apply the Beam F3 method to localize stimulation targets [25] and how to assemble and utilize the tDCS device.

This chapter does not substitute the device's user manual, which should be carefully read before its operation, but rather aims at illustrating its practical application for the treatment of MDD, with step-by-step procedures, and considerations about how to handle probable challenges. Furthermore, there are commercially available solutions for home use without the need for any setups and measurements, although such systems lack flexibility for other kinds of electrode montage.

2 Materials

In this section, we will present the components of a neuroConn DC-Stimulator-Plus tDCS device (neuroConn, Ilmenau, Germany) contained in a hardshell case (Fig. 1):

1. 01 tDCS device (microprocessor constant current source).

2. 01 power outlet electrical cable (for recharging the device, as needed).

3. 02 tDCS connection cables (01 red cable—anode; 01 blue cable—cathode).

4. 04 tDCS black rubber pads [two 5 × 5 cm (25 cm^2) pads; two 7 × 5 (35 cm^2) pads].

5. 02 rubber straps (for electrode adjustment on the skull).

6. 02 plastic pins (for rubber strap adjustment)

   Additional materials (*not shown in figure 1*): 02 sponges (01 red sponge—anode; 01 blue sponge—cathode), saline for sponge humidification, paper towels for eventual excess humidification, washable colored pencil to mark stimulation targets on the skin.

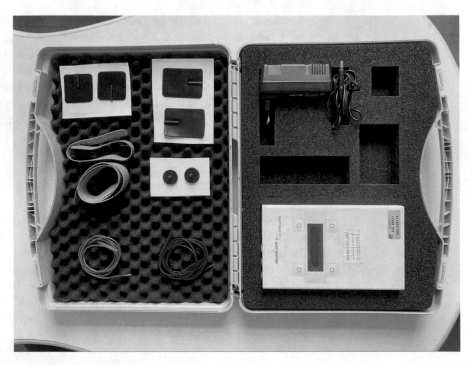

Fig. 1 Components of a Neuroconn DC-Stimulator-Plus tDCS device (neuroConn, Ilmenay, Germany) contained in a hardshell case

3   Methods

In this section, we shall cover in detail the specific steps necessary to carry out the technique. This is not necessarily a complicated procedure, but can lead to adverse effects if not done properly, or if lingering side effects are not monitored. The outlined procedure is for a traditional montage for the treatment of MDD, but the basic principles are pertinent to all tDCS applications.

3.1   Finding the Stimulation Sites (Left and Right DLPFC) Using the Beam F3 Method

As previously explained in the **Introduction** section of this chapter, the anode should be placed over the left DLPFC and the cathode over the right DLPFC. The left DLPFC corresponds to the F3 location according to the 10–20 EEG International System, while the right DLPFC corresponds to the F4 site [26].

In order to find and mark the correct stimulation sites, we recur to the **Beam F3 method** (based on the 10–20 EEG International System), which was initially devised for TMS-related research [25]. The Beam F3 method offers a free, online software (http://clinicalresearcher.org/F3/calculate.php) that was developed to locate the F3/F4 positions, using distances between four anatomical landmarks (nasion, inion, and the two ear tragi) and allowing for a fairly accurate localization of the F3 site and its contralateral correlate, F4.

The Beam F3 method also uses fewer calculations than the traditional 10–20 EEG technique, therefore being less prone to human error [25]. Studies evaluating the concordance between the Beam F3 method and a neuronavigation system suggest that the method is an acceptable alternative to locate neuronal targets [27]. To accurately calculate the F3 and F4 through Beam F3 method, we systematically present the needed steps below:

1. Head Measurements (Please Use a Tape Measure).

   (a) *Measurement 1 (nasion to inion)*: measure and take note of the distance (in centimeters) between nasion (bridge of the nose) and inion (most prominent protuberance at the posterior part of the skull); mark the midpoint with a washable colored pencil (Fig. 2).

   (b) *Measurement 2 (tragus to tragus)*: measure and take note of the distance (in centimeters) from tragus to tragus (small soft bump in front of each ear canal); mark the midpoint with a washable colored pencil (Fig. 3);

   (c) *Measurement 3 (circumference)*: measure the head circumference starting from the nasion and passing above the ears and over the inion (Fig. 4).

2. Mark the intersection of the nasion-inion plane and the tragus-tragus plane to locate the *vertex* (Fig. 5).

3. In the Beam F3 locator program (available at http://clinicalresearcher.org/F3/calculate.php), access "New Patient" and enter the three measurements obtained in item A (Fig. 6).

4. The program will output two values in cm (X and Y or Y-adjusted[1]) that are used to locate F3 and F4.

   (a) The X value must be measured from the midline (nasion) following the head circumference to the left (to find F3) or to the right (to find F4). Mark a dot along the circumference X cm from the midline (Fig. 7).

   (b) The Y-*adjusted* value is measured from the vertex to the point X on the left. Mark a dot Y-*adjusted* cm from the vertex—**this will be the F3 target**. To find F4 you should follow the same steps but along the right side (Fig. 8).

3.2 Adjusting the Electrodes Over F3 and F4

After locating the F3 and F4 targets, now one must place the electrodes over the scalp. Please follow the steps below:

1. Insert the pin leads of the electrical cables in the black rubber pads (Fig. 9);

---

[1] A recent study observed that an adjustment of 0.35 cm in the Y measurement would optimize the F3 site based on MRI-guided location [27].

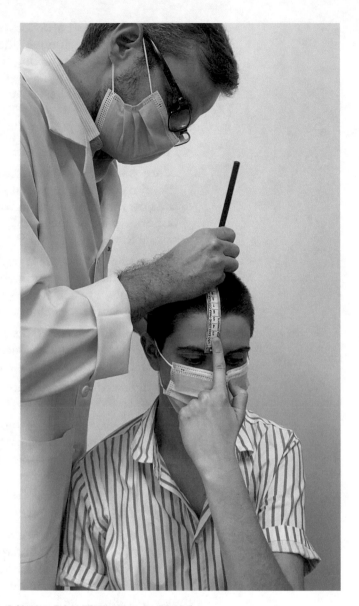

Fig. 2 Nasion-Inion Distance and midpoint

2. Insert the electrical cables (red and blue) in the tDCS device (Fig. 10);

3. Soak the sponge pads (the blue and red ones) in saline and ring out so as no fluid is dripping out (Fig. 11);

4. Place the rubber pads inside the sponges (Fig. 12);

5. Place the rubber strap and fasten it to the subject's head (tie the band at the back or use the plastic pin) (Fig. 13);

6. Place the cathode (blue cable and sponge) over the marked dot for right DLPFC (F4) and below the rubber strap (Fig. 14);

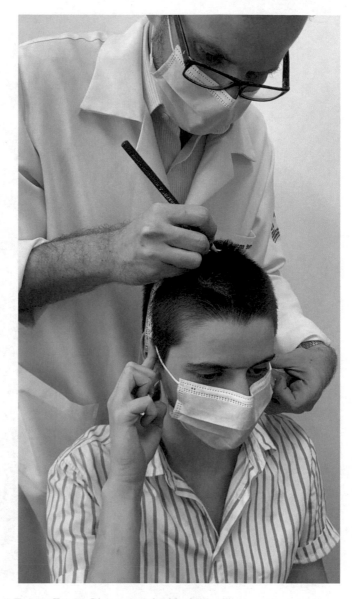

Fig. 3 Tragus-Tragus Distance and midpoint

7. Place the anode (red cable and sponge) over the marked dot for left DLPFC (F3) and below the rubber strap (Fig. 14);

8. Check the electrodes' placement to ensure they have not moved from the intended placement.

9. Make sure there is at least 7 cm between the electrodes. This step is necessary to avoid current shunting.

3.3 Turning on the tDCS Device

1. Turn the machine on by a switch at the top rear. Check the battery charge before starting the session and make sure there is enough charge to complete the tDCS session.

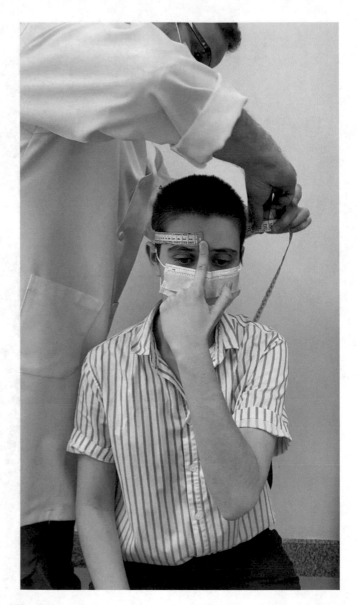

Fig. 4 Head Circumference

2. Choose a pre-programmed parameter or set a new stimulus parameter. For instance:

   (a) Current: e.g., 2000 μa = 2 mA (active).

   (b) Duration: 1800 s = 30 min.

   (c) Fade in: 30 s (active).

   (d) Fade out: 30 s (active).

   (e) Impedance: 90 kOhm.

3. After starting the session, one can see the remaining session time and the impedance alteration on the screen. If the tDCS

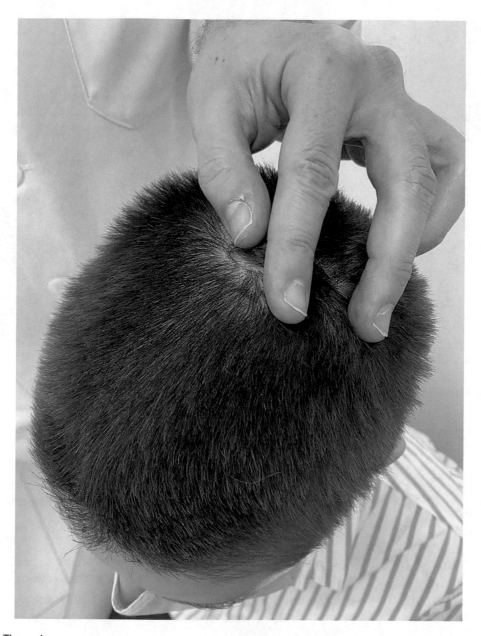

Fig. 5 The vertex

device detects high impedance, the current will be cut off automatically.

(a) In case of high impedance, the device will pause the current and will immediately set an audible warning.

(b) Check the possible reasons that are causing the high impedance. They might be lack of humidification in sponges or lack of contact of sponges with the skin (this can occur due to the amount of hair under the sponge or due to the looseness of the rubber strap, for instance).

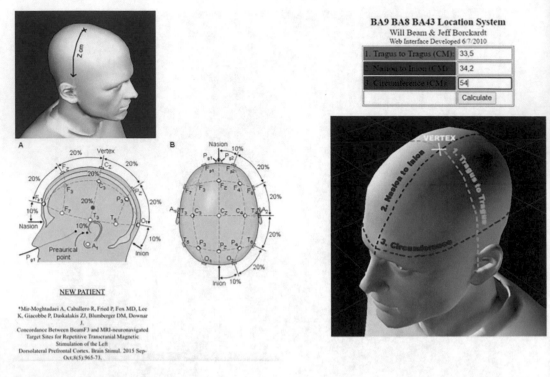

Fig. 6 Beam F3 Location Program

(c) After checking these steps, continue the tDCS session normally.

4. It is recommended that sponges are humidified every 10 min and that the impedance is monitored during the session.

5. Turn the machine off at the end of stimulation before taking off the headband and detaching electrodes.

6. Take out electrodes from sponges and discard the sponges.

7. Clean all equipment. To clean the rubber straps and electrodes after treatment, rinse with water and soap and leave to dry.

8. After the end of the tDCS session, it is highly recommended that adverse effects are accessed through a standardized questionnaire [7, 28].

4   Notes

1. Alternative ways of locating stimulation site:

   (a) "5/5.5 cm rule" (deprecated): this is one of the earliest and until some years ago the most widely used heuristics for locating the DLPFC. Although relatively simple to perform, this method was shown to suffer from significant inaccuracy, with comparisons with individual MRI

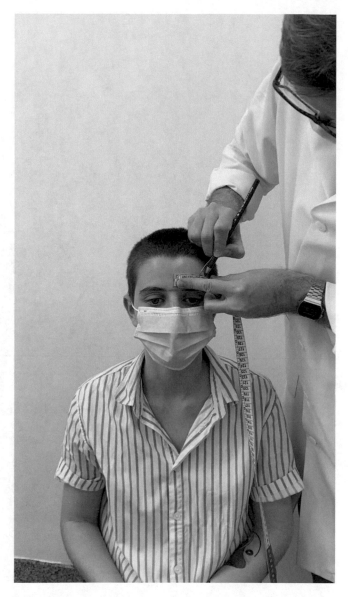

Fig. 7 The X distance

assessments showing that this approach missed the DLPFC in as many as 1/3 of patients undergoing treatment in one large TMS trial [29]. Therefore, the "5/5.5 cm rule" and its variants (6 cm or 7 cm rules) should be avoided.

(b) Neuronavigation: individualized location of the DLPFC could be improved by using MRI-guided neuronavigation to the left and right DLPFC. The main limitations of this approach are the cost and the fact that the DLPFC is a large region and there is still no consensus on its precise location [27].

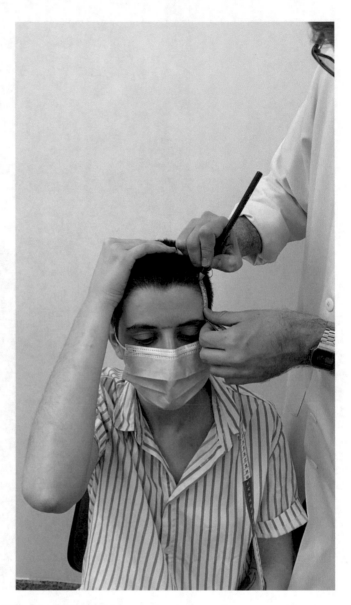

Fig. 8 The Y-adjusted distance

Targets defined with respect to the international 10–20 electroencephalography (EEG) system:

(c) EEG cap: with the use of color-coded template with electrode placement locations, it is possible to locate the F3 and F4 positions for later placement of the tDCS electrodes (Fig. 15);

To find the inion, it is easier to sense with two fingers (e.g., pointer and middle fingers) going up and down in broad

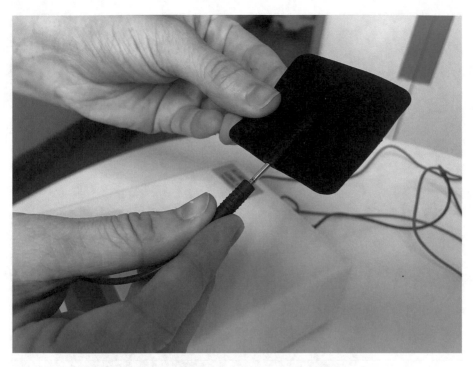

Fig. 9 Insert the pin leads of the electrical cables in the black rubber pads

movements and then, narrow them, once the inion is quite broad and you have to find the middle of that protuberance.

Don't draw just two dots when marking the midpoint of the tragus-tragus and nasion-inion because it will be difficult to work out the intercept. Draw a small line going away perpendicularly to the measurement tape.

Check with the tip of the marker placed in the **vertex** and ask the patient if that feels in the middle of the head (people are usually good to tell if it is in the middle of the head from left to right). Also, check from the front if the marker is aligned with the nose midline. Check from the side if the vertex is roughly at the ear-line (if it is too far forward or too far backward, double-check your measurements).

In case you don't have access to the "teabag" **sponge pads**, place the sponge between the electrode and the skin, and then fasten the band. Another option is to use an electrolyte-based contact, such as a conductive gel or cream between the skin and the rubber electrode.

The **cathode** can also be positioned over the F8 (10–20 EEG system) or over the supraorbital area (SOR), based on previous randomized controlled studies for MDD. For positioning the cathode over the F8, one should use an EEG cap. To apply on SOR, the cathode should be placed over the right eyebrow.

Practical safety:

Fig. 10 Insert the electrical cables (red anode and blue cathode) in the tDCS device

- Do not administer tDCS while the device is connected to the power socket.
- Do not apply tDCS to injured or irritated skin (e.g., scratched, cut, acne, scars, excessive redness/dryness).
- Attach electrodes while the device is off.
- Use an appropriate conducting solution (saline).

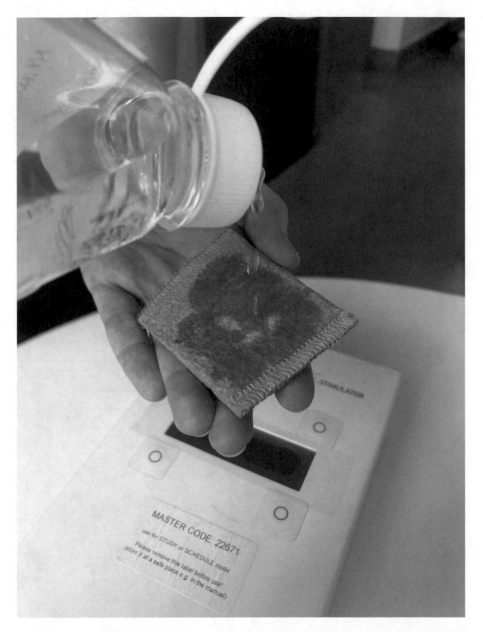

Fig. 11 Soak the sponge pads (the blue and red ones) in saline and ring out so as no fluid is dripping out

- Make adjustments for good electrode contact (add saline, tighten band, even contact between electrode and scalp).
- Ensure rubber electrodes do not come in contact with the scalp.
- Monitor side effects before, during, and after tDCS (what seem like minor side effects—e.g., gradual drying of skin—could lead to persistent redness and skin burns).

Fig. 12 Place the rubber pads inside the sponges

The **sponges** are rated for one-time use, as multiple uses cause salt deposits leading to current taking preferential current paths resulting in nonuniform current delivery across the skin. Multiple uses also lead to the stiffening of the sponges resulting in nonuniform skin contact. These issues may lead to non-tolerable stimulation. The manufacturer thus suggests one-time use which is important for patient safety, for comfort, and also for hygienic purposes.

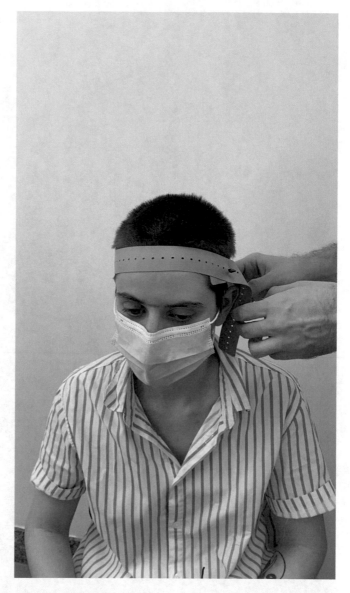

Fig. 13 Place the rubber strap and fasten it to the subject's head (tie the band at the back or use the plastic pin)

Contraindications

Do not apply tDCS in patients with cochlear implants, other metallic implants in the skull, cardiac pacemakers or implanted defibrillators, or an invasive neurostimulation device (i.e., spinal cord stimulator, vagal nerve stimulator, auricular stimulator, or deep brain stimulating electrodes).

TDCS is relatively contraindicated in patients with epilepsy, skull malformations, and certain dermatological conditions (dermatitis, psoriasis, or eczema). Special populations such as the

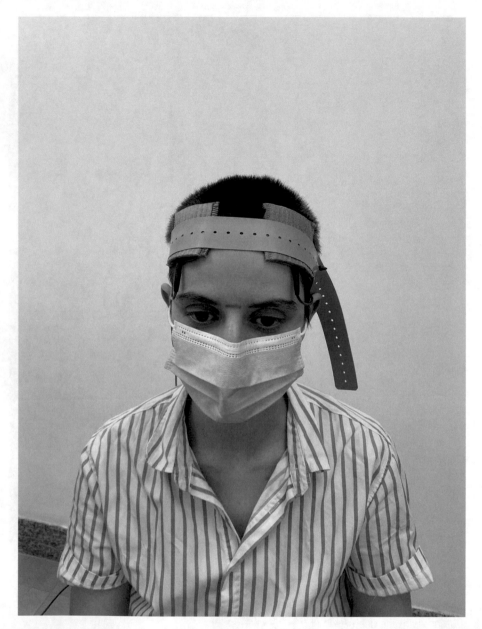

Fig. 14 Place the cathode (blue cable and sponge) over the marked dot for right DLPFC (F4) and below the rubber strap; place the anode (red cable and sponge) over the marked dot for left DLPFC (F3) and below the rubber strap

elderly, pregnant women, and children/adolescents must be carefully considered.

Special consideration and extra care should be taken in patients with MDD who experience suicidal thoughts or ideation, have a history of hypomanic/manic switches or episodes, or have been diagnosed with bipolar disorder or illicit drug use or alcohol abuse.

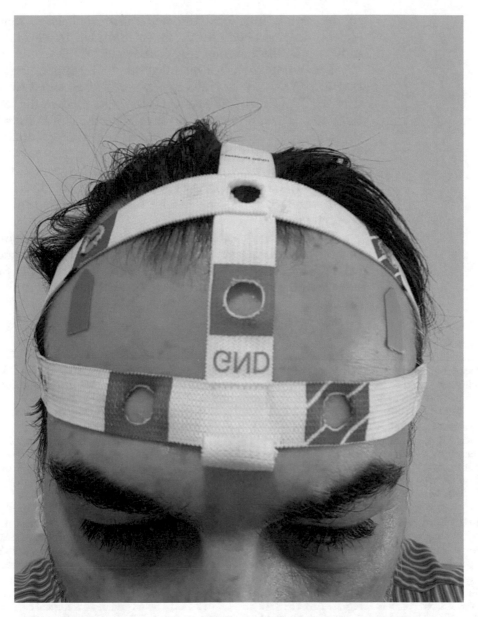

Fig. 15 EEG cap: with the use of color-coded template with electrode placement locations, it is possible to locate the F3 and F4 positions for later placement of the tDCS electrodes

5 Conclusions

tDCS is a NIBS technique able to modulate cortical excitability through a weak electrical current applied via two electrodes (anode and cathode) placed over the scalp. The tolerability and effects of the technique have been extensively evaluated for several neuropsychiatric disorders, with promising results for MDD. tDCS for MDD is typically applied bilaterally over the prefrontal cortex due to the brain's functional interhemispheric imbalance found in patients diagnosed with MDD, in which the left DLPFC is shown to be hypofunctional compared to the right DLPFC.

Commonly, the tDCS montage for MDD applies the anode over the F3 site (10–20 EEG system), while the cathode is placed over the F4 (10–20 EEG system), corresponding to the left and right dorsolateral prefrontal cortexes, respectively. It is crucial that the location of these targets is conducted by a systematic method to ensure accurate application of the electrodes. Moreover, prior knowledge regarding the operation of a tDCS device, how to perform a tDCS session, and its possible contraindications is essential to conduct a safe and quality session for both clinical and research purposes.

Therefore, in this chapter we present step-by-step instructions on how to perform a supervised tDCS session for the treatment of MDD. This systematic presentation is essential for those professionals who are interested in learning the practical application of tDCS.

References

1. Woods AJ, Antal A, Bikson M et al (2016) A technical guide to tDCS, and related non-invasive brain stimulation tools. Clin Neurophysiol 127:1031–1048

2. Borrione L, Bellini H, Razza LB et al (2020) Precision non-implantable neuromodulation therapies: a perspective for the depressed brain. Braz J Psychiatry 42(4):403–419. https://doi.org/10.1590/1516-4446-2019-0741

3. Moffa AH, Brunoni AR, Nikolin S, Loo CK (2018) Transcranial direct current stimulation in psychiatric disorders: a comprehensive review. Psychiatr Clin North Am 41:447–463

4. Razza LB, Palumbo P, Moffa AH et al (2020) A systematic review and meta-analysis on the effects of transcranial direct current stimulation in depressive episodes. Depress Anxiety 37(7): 594–608. https://doi.org/10.1002/da.23004

5. Moffa AH, Martin D, Alonzo A et al (2020) Efficacy and acceptability of transcranial direct current stimulation (tDCS) for major depressive disorder: an individual patient data meta-analysis. Prog Neuropsychopharmacol Biol Psychiatry 99:109836

6. Brunoni AR, Amadera J, Berbel B et al (2011) A systematic review on reporting and assessment of adverse effects associated with transcranial direct current stimulation. Int J Neuropsychopharmacol 14:1133–1145

7. Aparício LVM, Guarienti F, Razza LB et al (2016) A systematic review on the acceptability and tolerability of transcranial direct current stimulation treatment in neuropsychiatry trials. Brain Stimul 9:671–681

8. Berlow YA, Zandvakili A, Carpenter LL, Philip NS (2019) Transcranial direct current stimulation for unipolar depression and risk of treatment emergent mania: an updated meta-analysis. Brain Stimul 12:1066–1068

9. Brunoni AR, Moffa AH, Sampaio-Júnior B et al (2017) Treatment-emergent mania/hypomania during antidepressant treatment with transcranial direct current stimulation (tDCS): a systematic review and meta-analysis. Brain Stimul 10:260–262

10. Brunoni AR, Valiengo L, Baccaro A et al (2013) The sertraline vs. electrical current therapy for treating depression clinical study: results from a factorial, randomized, controlled trial. JAMA Psychiatry 70:383–391

11. Brunoni AR, Boggio PS, De Raedt R et al (2014) Cognitive control therapy and transcranial direct current stimulation for depression: a randomized, double-blinded, controlled trial. J Affect Disord 162:43–49

12. Segrave RA, Arnold S, Hoy K, Fitzgerald PB (2014) Concurrent cognitive control training augments the antidepressant efficacy of tDCS: a pilot study. Brain Stimul 7:325–331

13. Brunoni AR, Moffa AH, Sampaio-Junior B et al (2017) Trial of electrical direct-current therapy versus escitalopram for depression. N Engl J Med 376:2523–2533

14. Fregni F, El-Hagrassy MM, Pacheco-Barrios K et al (2021) Evidence-based guidelines and secondary meta-analysis for the use of transcranial direct current stimulation (tDCS) in neurological and psychiatric disorders. Int J Neuropsychopharmacol 24(4):256–313. https://doi.org/10.1093/ijnp/pyaa051

15. Razza LB, Afonso dos Santos L, Borrione L et al (2021) Appraising the effectiveness of electrical and magnetic brain stimulation techniques in acute major depressive episodes: an umbrella review of meta-analyses of randomized controlled trials. Braz J Psychiatry 43(5):514–524

16. Charvet LE, Kasschau M, Datta A et al (2015) Remotely-supervised transcranial direct current stimulation (tDCS) for clinical trials: guidelines for technology and protocols. Front Syst Neurosci 9:26

17. Alonzo A, Fong J, Ball N et al (2019) Pilot trial of home-administered transcranial direct current stimulation for the treatment of depression. J Affect Disord 252:475–483

18. Park J, Oh Y, Chung K et al (2019) Effect of home-based transcranial direct current stimulation (tDCS) on cognitive function in patients with mild cognitive impairment: a study protocol for a randomized, double-blind, cross-over study. Trials 20:1–9

19. Bikson M, Hanlon CA, Woods AJ et al (2020) Guidelines for TMS/tES clinical services and research through the COVID-19 pandemic. Brain Stimul 13:1124–1149

20. Borrione L, Moffa AH, Martin D et al (2018) Transcranial direct current stimulation in the acute depressive episode: a systematic review of current knowledge. J ECT 34:153–163

21. Reid SA, Duke LM, Allen JJ (1998) Resting frontal electroencephalographic asymmetry in depression: inconsistencies suggest the need to identify mediating factors. Psychophysiology 35:389–404

22. Debener S, Beauducel A, Nessler D et al (2000) Is resting anterior EEG alpha asymmetry a trait marker for depression? Neuropsychobiology 41:31–37

23. Lefaucheur J-P, Antal A, Ayache SS et al (2017) Evidence-based guidelines on the therapeutic use of transcranial direct current stimulation (tDCS). Clin Neurophysiol 128:56–92

24. Milev RV, Giacobbe P, Kennedy SH et al (2016) Canadian network for mood and anxiety treatments (CANMAT) 2016 clinical guidelines for the management of adults with major depressive disorder: Section 4. Neurostimulation treatments. Can J Psychiatry 61: 561–575

25. Beam W, Borckardt JJ, Reeves ST, George MS (2009) An efficient and accurate new method for locating the F3 position for prefrontal TMS applications. Brain Stimul 2:50–54

26. Acharya JN, Acharya VJ (2019) Overview of EEG montages and principles of localization. J Clin Neurophysiol 36:325–329

27. Mir-Moghtadaei A, Caballero R, Fried P et al (2015) Concordance between BeamF3 and MRI-neuronavigated target sites for repetitive transcranial magnetic stimulation of the left dorsolateral prefrontal cortex. Brain Stimul 8: 965–973

28. Antal A, Alekseichuk I, Bikson M et al (2017) Low intensity transcranial electric stimulation: safety, ethical, legal regulatory and application guidelines. Clin Neurophysiol 128:1774–1809

29. George MS, Lisanby SH, Avery D et al (2010) Daily left prefrontal transcranial magnetic stimulation therapy for major depressive disorder: a sham-controlled randomized trial. Arch Gen Psychiatry 67:507–516

Chapter 17

Deep Brain Stimulation for Treatment-Resistant Depression

Alexandre Paim Diaz, Brisa S. Fernandes, Valeria A. Cuellar, Joao Quevedo, Albert J. Fenoy, Marsal Sanches, and Jair C. Soares

Abstract

It is estimated that one-third of patients with depression show a lack of adequate response to different pharmacological and non-pharmacological interventions, including electroconvulsive therapy. These patients usually present with lower overall functioning and higher morbidity and mortality. Treatment-resistant depression (TRD) is also associated with substantial economic burden, increased inpatient hospitalizations, longer hospital stays, and higher healthcare costs. This book chapter discusses relevant TRD concepts, the impact of TRD, and the rationale for the use of deep brain stimulation (DBS) in the treatment of psychiatric disorders. We also present preclinical studies with DBS and animal models of depression that have helped understand the mechanisms of action of that intervention. Clinical studies have investigated different targets, like the inferior thalamic peduncle, the lateral habenular complex, the medial forebrain bundle, the nucleus accumbens, the ventral capsule/ventral striatum (the anterior limb of the internal capsule), and the subcallosal cingulate cortex. Promising results have been reported in open-label trials of DBS in TRD, with a significant reduction of depressive symptoms, tolerable adverse effects, and sustained efficacy although some of the few randomized-controlled studies have shown less encouraging results. We finalize with a discussion of candidate biomarkers for treatment response and perspectives for improving DBS's efficacy and safety to treat patients with TRD.

Key words Deep brain stimulation, Treatment-resistant depression, Brain circuitry, Brain connectivity

1 Introduction

1.1 The Concept, Epidemiology, and Impact of Treatment-Resistant Depression

Despite a lack of universal consensus on the definition of treatment-resistant depression (TRD), a patient is most commonly characterized as nonresponsive after two or more adequate treatments. Brown et al. reviewed concepts of TRD in research and clinical practice and reported that, among the 150 studies selected for review, 155 different definitions of TRD were adopted [1]. Approximately half of them defined TRD as a minimum of two treatment failures. Nevertheless, there was substantial variability across definitions with regard to the different classes of psychotropics

Yong-Ku Kim and Meysam Amidfar (eds.), *Translational Research Methods for Major Depressive Disorder*, Neuromethods, vol. 179, https://doi.org/10.1007/978-1-0716-2083-0_17,

considered, the timing of failure, and the length of treatment [1]. Some authors have proposed staging models for treatment resistance. In these models, continued lack of response to a first-line treatment indicates the need for using a second-line treatment, which would become a stage II intervention [2, 3]. For instance, the Thase and Rush Staging Method includes five stages. Stage II is similar to the usually adopted concept of TRD (failure to respond to at least two adequate trials). Stages III, IV, and V include, in addition to the previous stage features, failure to respond to an adequate trial of a tricyclic antidepressant (TCA), monoamine oxidase inhibitor (MAOI), and a course of bilateral electroconvulsive therapy (ECT), respectively [4]. Lack of response to ECT increases the overall score of the Massachusetts General Hospital Staging Method by three points versus one point per each antidepressant failure and 0.5 per each optimization strategy, augmentation, or combination [5]. Both staging methods cited clearly recognize failure to ECT as either the final or more relevant step in the sequential trials for staging TRD.

The Sequenced Treatment Alternatives to Relieve Depression (STAR*D) study is the largest open-label trial to ever evaluate treatment of depression. It included 2876 patients across four levels of treatment trials (level 1 = 2876, level 2 = 1893, level 3 = 377, and level 4 = 109), embracing monotherapies, combinations, or augmentation strategies. The findings showed that one-third of the patients who remained depressed in the four levels of treatment did not reach remission [6, 7]. With regard to the rate of TRD in the general population, Bosco-Lévy et al. estimated the annual prevalence and incidence of TRD in France. For this cohort study, which was based on a French nationwide claims database, the authors defined TRD as the use of at least three trials of different antidepressants within 3 months and the dispensing of more than two different antidepressants together or of an antidepressant with a potentiator [8]. The authors reported an annual incidence of TRD of 5.8 cases per 10,000 patients and an annual prevalence of 25.8 cases per 10,000 patients [8].

TRD has a huge individual impact in terms of psychosocial functioning [9], morbidity, and mortality. For instance, Godin et al. reported a 38% prevalence of metabolic syndrome in adults with TRD, higher than the one found among people with other psychiatric disorders [10]. Madsen et al. estimated cause-specific excess mortality and life years lost in individuals with TRD in a cohort of 8294 patients from a Danish nationwide register-based cohort study [11]. The findings pointed to adjusted and age-standardized mortality rates 34% and 39% higher for men and women with TRD, respectively, compared to individuals with non-TRD, with suicide as the highest cause-specific mortality rate ratio for both sexes [11]. These results are even more worrisome considering the already higher mortality rate associated with depression compared to the general population, from both natural and unnatural causes [12–14]. It is estimated that the 12-month prevalence and lifetime prevalence of major depression in the US population are 10.4% and 20.6%, respectively [15]. Given the epidemiological characteristics of TRD, its public health impact is substantial. Li et al. reported that, compared to non-TRD patients, those with TRD presented significantly higher inpatient hospitalizations, longer hospital stays, more emergency department visits, and increased total healthcare costs [16].

A better understanding of the pathophysiology of depression could provide insights about potential targets for effective interventions, reducing the substantial individual and familial suffering and the social and individual burden associated with TRD. In addition to genetic and epigenetic, molecular, and physiological aspects, the clarification of the brain circuitry and connectivity underlying depression, and more precisely TRD, could help identify brain regions for neuromodulation.

1.2 Deep Brain Stimulation

Deep brain stimulation (DBS) consists of the implantation of electrodes (leads) into specific brain regions guided by neuroimaging stereotactic neurosurgical techniques [17]. Stimulation is delivered by leads connected to pulse generators, which are implanted subcutaneously and contain the battery. All stimulation parameters can be set and adjusted noninvasively [17]. Figure 1 provides a representation of implanted electrodes and pulse generator in a patient. The Food and Drug Administration (FDA) has approved DBS at the ventral intermediate nucleus of the thalamus for severe tremor in Parkinson's disease (PD) in 1996 and expanded its indication in 2001, including DBS at the subthalamic nucleus (STN) or the internal segment of the globus pallidus (GPi) for advanced PD [18]. Since then, thousands of patients have benefited from the intervention [19]. PD is a neuropsychiatric disorder, characterized by motor symptoms like rest tremor, bradykinesia, muscular rigidity, and gait and postural instability [20], with a high frequency of comorbid apathy, cognitive impairment, anxiety, and depressive symptoms [21]. The basal ganglia-thalamocortical circuitry has an important role in the movement control, and a substantial death of dopaminergic neurons in the substantia nigra pars compacta (SNpc), a midbrain structure that is part of the basal ganglia circuitry, is a core aspect of PD pathophysiology [20, 22]. The hyperactivity of striatal neurons due to dopamine depletion can result in inhibition of external globus pallidus (GPe), with following disinhibition of STN and GPi, which contribute to parkinsonism [23]. DBS in the STN or GPi is associated with therapeutic modulation of this dysfunctional neurocircuitry, in addition to neurotransmitter release, increased blood flow, and neurogenesis across a neural network that goes beyond the site of stimulation [24].

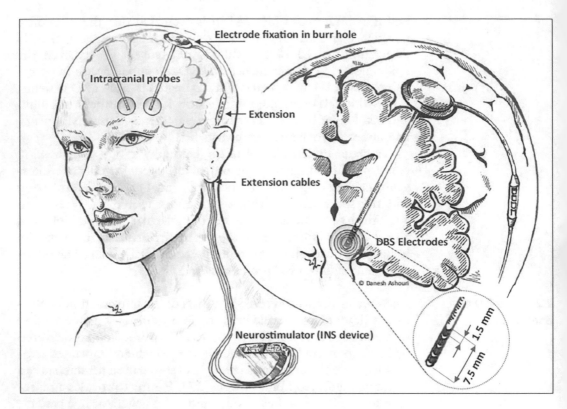

Fig. 1 Representation of implanted electrodes and pulse generator in a patient. (From Döbrössy et al., adapted from Ashouri Vajari et al. [40, 86])

1.3 Deep Brain Stimulation as a Potential Intervention for Treatment-Resistant Depression

Despite lacking a pattern of neurodegeneration similar to the one found in PD's pathophysiology, patients with depression exhibit dysfunctions in brain activity and connectivity, which raises the possibility that modulation by invasive interventions such as DBS could also result in clinical benefits. In her seminal 1999 paper, Mayberg et al. used positron emission tomography (PET) techniques to investigate the interaction between limbic and cortical regions during transient and sustained changes in affective states of people with depression and healthy controls (HC) [25]. The results showed that the experience of sadness was associated with subgenual cingulate (Brodmann area 25 [Cg25]) increased blood flow and with right dorsolateral prefrontal and inferior parietal (neocortical regions) decreased blood flow. A reverse pattern was observed with recovery, suggesting reciprocal changes involving these limbic and neocortical regions [25]. These results prompted an investigation of the benefits of DBS in the Cg25 area. Six patients with TRD (five of which had a history of previous treatment with ECT) were included in the study. The mean duration of the current depressive episode was 6 years, and the Hamilton Depression Rating Scale 17-item (HAM-D 17) score at baseline was 25 (suggesting that most participants had moderate to severe depressive symptoms) [26–28]. Four out of the six patients showed clinical response, three of which having shown dramatic improvement in depressive symptoms from baseline compared to 6-month evaluation (HAM-D 17 scores from 29 to 5, 29 to 9, and 26 to 6). The authors also reported an elevated blood flow in Cg25 at pretreatment baseline, which decreased after 3 months of stimulation, consistent with their previous findings [25, 26].

Over 160,000 patients have received DBS treatment for neurologic or non-neurologic conditions in the world [19]. In psychiatry, DBS has been extensively investigated as a potential treatment for patients with TRD. In this book chapter, we present preclinical studies focusing on the putative mechanisms underlying DBS efficacy, clinical studies and their variety of targets investigated so far, potential markers for treatment response, and adverse effects of DBS, in addition to the perspectives of deep brain stimulation for the treatment of TRD.

2   Preclinical Studies with Deep Brain Stimulation in Animal Models of Depression

2.1   Lateral Habenular Complex (LHb-c)

The habenula, comprised of two portions (lateral and medial habenula), is located in the most distal and caudal part of the epithalamus [29]. The lateral habenula (LHb) is associated with monoamine neurotransmission control and mediation of behavioral flexibility, social behavior, goal-directed processing, and aversion behavior [29]. In an animal model of treatment-resistant depression, Winter et al. pharmacologically inhibited the LHb of rats, inducing antidepressant-like behavior [30]. The antidepressant efficacy of DBS in the LHb was investigated by Meng et al. in a rat model of depression. The intervention was associated with lower depressive-like behaviors in the open-field test, in addition to increased concentration of monoamines in the brain and blood serum [31]. Phosphorylation of CaMKIIα/β and GSK3α/β in the LHb and L-type voltage-dependent calcium channel-mediated synaptic potentiation in the LHb were associated with the antidepressant effects of DBS [32, 33].

2.2   Medial Forebrain Bundle

The medial forebrain bundle (MFB) connects the main regions of the reward circuitry: the ventral tegmental area (VTA), the orbitofrontal cortex (OFC), the nucleus accumbens (NAcc), and the hypothalamus [34]. Dandekar et al. investigated DBS in the MFB of rats submitted to a chronic unpredictable stress model of depression. In addition to reversing anhedonia-like behavior, the intervention was associated with an increase in brain-derived neurotrophic factor (BDNF) levels in plasma, in cerebrospinal fluid (CSF), and in the hippocampus [35]. The same group found an increase in swimming time in the forced-swimming test associated with DBS, as well as higher protein levels of dopamine D2 receptors in the prefrontal cortex (PFC) and increased levels of dopamine transporters in the hippocampus after the stimulation of the MFB of male Wistar rats [36]. Despite the role of the MFB in connecting regions of the reward circuitry, the mechanisms of DBS in the MFB may not be related with accumbens dopamine release, one of the main neurotransmitters associated with reward-related behavior [37, 38]. Bregman et al. investigated neurotransmitter release and changes in neurocircuitry associated with DBS in the MFB of rats submitted to the forced-swimming test. In addition to the antidepressant-like effects—animals in the intervention group presented significantly lower immobility compared to the sham group—DBS was associated with an increased expression of zif268 in the prelimbic cortex, ventral tegmental area, anterior striatum, and nucleus accumbens, a gene that has a role in hippo-campal function and memory processes [39]. The authors did not find changes in dopamine and serotonin release associated with DBS at the MFB [37]. In a review, Döbrössy et al. presented experimental and clinical evidence of MFB DBS [40]. Despite some variability in animal models regarding amplitude, frequency, pulse width, and stimulus duration utilized, results show behavioral and neurobiological effects of DBS, such as reversion of depressive-like and anhedonic-like behaviors and increased dopamine release in the striatum, increased D2 receptor expression in the PFC, and increased dopamine transport in the hippocampus [40]. More recently, Thiele et al. investigated the mechanisms in which MFB DBS can reverse these reward-related behaviors using raclopride, a synthetic selective antagonist on D2 dopamine receptors, and micro-PET. The results showed that MFB DBS was able to reduce the D2 antagonism of raclopride in the striatum, as well as the depressive-like behavior associated with the synthetic compound. In addition, MFB DBS was associated with increases in mRNA coding for dopamine receptors D1 and D2 in the animals that received raclopride [41]. Other recent study evaluated how MFB DBS parameters of stimulation could affect dopamine projection toward the NAcc of Flinders sensitive line (FSL) rats. The study showed that MFB DBS was able to induce dopamine response in the NAcc of both FLS and control rats, which varied according to different pulse widths. Those results suggest distinct patterns of MFB anatomy and physiology when comparing controls and animal models of depressive-like behavior, including fiber diameter and distribution, fiber types, and dopamine neuron excitabiity [42].

2.3   Nucleus Accumbens (NAcc)

Anhedonia, a core symptom of depression [43], has been associated with reduced nucleus accumbens (NAcc) volume and decreased reward response [44]. Falowski et al. investigated the effects of NAcc DBS in an animal model of depression, with their results showing that DBS reduced open-field anxiety-like behavior and increased exploratory behavior. In addition, the intervention was associated with decreased tyrosine hydroxylase expression in the PFC and increased dendritic length in pyramidal cells in the same region, suggesting an influence of the DBS outside the site of the stimulation [45]. On the other hand, van Dijk et al. also found altered neurotransmitter levels in the PFC of rats submitted to DBS in the NAcc but in a different direction, with increases in serotonin, dopamine, and noradrenaline levels associated with the administration of the stimulation [46]. A study with 3.0 T functional magnetic resonance imaging (fMRI) identified alterations in blood oxygenation level-dependent (BOLD) signal in the brain of porcines (ten domestic male pigs) after NAcc DBS. Findings included BOLD signal changes in regions distal to the stimulation site, such as the insula, the parahippocampal cortex, the anterior prefrontal cortex, and the dorsal anterior cingulate. These results reiterate the impact of DBS in a broad network involving subcortical and cortical regions, which may underlie the efficacy of DBS in the treatment of psychiatric disorders [47]. As previously discussed, the mechanisms of action of DBS may involve a large set of biological changes, from a biochemical level to a circuitry level [24]. Mitochondrial dysfunction has been extensively described as associated with psychiatric disorders [48–50] and might modulate the efficacy of NAcc DBS in an animal model of treatment resistance to antidepressants. Kim et al. stimulated the NAcc of rats submitted to adrenocorticotropic hormone (ACTH) administration, which is associated with resistance to imipramine, and found that the intervention improved forced swim test mobility, as well as increased PFC respiratory control ratio, which represents mitochondrial function in the region, compared to sham and control groups [51].

3   Clinical Studies

In this section, the clinical evidence on the use of DBS in the treatment of TRD is discussed, according to the different stimulation sites. DBS's clinical studies and reports for TRD are according to the already investigated targets [52].

3.1   Inferior Thalamic Peduncle

The thalamic peduncles are a large bundle of fibers connecting the thalamus and cortex. The inferior thalamic peduncle (ITP) is a group of fibers that connect the thalamus pulvinar to the tip of the temporal lobe [53]. Although most of the studies on ITP DBS focused on the treatment of obsessive-compulsive disorders [54–56], this region has also been targeted for its efficacy in the management of TRD. Jimenez et al. reported a case of ITP DBS involving a 49-year-old woman with a two-decade history of recurrent depression, comorbid bulimia, and borderline personality disorder and a history of two psychiatric hospitalizations, two suicide attempts, and recurrent suicidal thoughts [57]. After surgery and during the chronic stimulation, the patient's HAM-D scores, which would range from 33 to 42 prior to the intervention, remained below 10 at most of the 2-year follow-up assessments. The global assessment functioning (GAF) changed from 20 to 80–90, before and after the chronic stimulation, respectively, without meaningful adverse events [57].

3.2 Lateral Habenular Complex (LHb-c)

Sartorius et al. reported a case of LHb DBS in a 64-year-old woman with a history of depressive episodes since the age of 18, treatment resistance during at least 9 years, two suicide attempts, and previous ECT course. After the stimulation, the patient presented sustained full remission, which was reached 4 months after the parameters were changed to higher stimulation [58]. The authors also reported a significant increase in serum BDNF levels of this patient as of week 14 following the surgery [59]. Reduced serum levels of BDNF have been consistently found in patients with depression; [60] however, it is not possible to establish a relationship between this neurotrophin level and the DBS efficacy for this case report [59].

3.3 Medial Forebrain Bundle

Denier et al. compared fractional anisotropy (FA) values of the superolateral MFB (slMFB) in acutely depressed patients with bipolar disorder (BD) with HC. The findings showed a significantly shortened tract length of the right slMFB in individuals with BD, suggesting impaired white matter microstructure in this region associated with bipolar depression [61]. In an open trial, Fenoy et al. investigated the efficacy of DBS in the MFB in ten patients with TRD. Four patients were evaluated at 26 weeks, three of them showing more than 80% decrease in the Montgomery-Asberg Depression Rating Scale (MADRS) scores from the baseline [62]. A following analysis of five patients at 1 year follow-up by the same group showed that four of the participants presented a decrease higher than 70% in the MADRS scores compared to the baseline [63]. All patients had a history of ECT treatment (range 6 to 12 times), moderate to severe depressive symptoms at baseline [64], and the average for the current depressive episode was 6 years [63]. Among the adverse events, the authors reported transient diplopia, usually vertical, which remitted after the selection of a different setting of contacts and changes in the stimulation parameters [63].

A review of DBS in the MFB for depression, including case series, case reports, and a phase I trial, showed positive results regarding short- and long-term outcomes for remission and response [40]. However, the phase I trial with a randomized-controlled onset of stimulation for 2 months after the implantation reported no significant differences between the intervention and sham groups, with both groups showing improvement in their depressive symptoms [65]. In that trial, Coenen et al. recruited 16 patients with severe depression, between 29 and 71 years of age. On average, the participants had received 19 different antidepressant medications and 20 ECT sessions. In addition to decreased depressive symptoms, the authors also reported significant improvement on quality of life and functioning at 12-month follow-up [65]. Some patients reported vision disturbances such as blurred vision and double vision that resolved after adjustments in the stimulation parameters. Transient hypomania, restlessness, transient slurred of speech, suicidal behavior, and misuse of methylphenidate were reported by one patient. Two participants progressed with severe wound healing problems [65]. Davidson et al. presented the findings of two patients with severe TRD (resistant to psychotherapeutic, pharmacological, and ECT interventions) who were treated with DBS on the superolateral branch of the MFB, a tract that contains dopaminergic fibers of the meso-cortical and meso-limbic reward pathways [63, 66]. After a follow-up of 6 months, the patients entered a crossover 2 weeks "on/off" phase of the study. Despite the absence of severe adverse events, the intervention was not associated with clinical response [66].

3.4 Nucleus Accumbens (NAcc)

Schlaepfer et al. treated three patients with TRD with NAcc DBS bilaterally [67]. The patients were a 66-year-old woman and two 37-year-old male monozygotic twins, all with treatment resistance to several trials with antidepressants (ranging from 7 to 17), psychotherapy, and ECT (number of sessions varying from 10 to 25). The authors reported improvements in depressive symptoms in all patients and a negative correlation between MADRS scores and stimulation parameters. Side effects included night sweats, sleep disturbances, agitation, and sedation [67]. In another study, additional ten patients received the same treatment, half reaching at least a 50% reduction in the depressive symptom scores [68]. Participants had a mean age or 48.6 years and a length of current depressive episode of almost 11 years, having received 21 ECT sessions on average . A history of suicide attempts was reported by three of the participants. The authors also reported a significant reduction in the Hamilton Depression Rating Scale (HDRS) scores during all follow-up evaluations. At 12-month assessment following the DBS stimulation, the HDRS total scores decreased from 32.5 (baseline) to 20.8 points. Similarly to the findings of preclinical studies, the authors found areas of metabolic changes distal to the site of stimulation, including prefrontal subregions, the posterior cingulate cortex, the subgenual cingulate region, the thalamus, and the caudate nucleus. Importantly, the participants did not present cognitive impairment during the study. Adverse effects, such as dysphagia, pain, and subjective increase in anxiety, were transient, but one participant attempted suicide and another died by suicide during the follow-up period [68, 69].

3.5 Ventral Capsule/Ventral Striatum (Anterior Limb of the Internal Capsule) VC/VS (ALIC)

The anterior limb of the internal capsule (ALIC) is a region of convergence for areas associated with emotional processing and regulation, in addition to reward-related behavior. It also has connections with the dorsolateral and ventrolateral prefrontal cortices bilaterally [70]. Nonhuman primates reared in an adverse condition presented with significant reductions in FA in the ALIC compared to animals reared under normative conditions in the study of Coplan et al. [71], and clinical studies have shown that patients with depression present significant lower FA in the ALIC compared to HC [70]. Bergfeld et al. investigated the efficacy of ALIC DBS with an open-label trial followed by a double-blind randomized crossover phase [72]. In the first phase all patients received DBS in the ALIC and the stimulation was optimized until a maximum of 52 weeks or until a stable response of at least 4 weeks could be achieved. Thereafter, patients were randomized to a double-blind crossover phase during which the stimulation was set to active or sham for 6 weeks. The sample consisted of 16 patients with TRD. Clinical characteristics of the participants included, on average, a mean age of 53.2 years, 10.8 past antidepressant medications, 68.9 past ECT sessions, 15.3 years of TRD duration, and 3.3 suicide attempts. During the open trial phase, there were significant decreases in the HAM-D and MADRS scores, 40% were classified as responders, and 20% achieved remission. Almost half of the nonresponders were considered partial responders. The depressive symptoms were significantly lower during the active DBS compared with the sham phase. Among the adverse events, one participant presented severe nausea, four attempted suicide, and two reported suicidal ideation [72]. The intervention was not associated with positive or negative cognitive outcomes in three different time points—baseline (3 weeks prior to surgery), 3 weeks after DBS, and 52 weeks following the DBS optimization phase for the intervention group, and baseline, 6 weeks following baseline and 20–24 weeks after baseline for the HC group [73]. ALIC DBS was also found to show stable efficacy over a 2-year follow-up period [74].

3.6 Subcallosal Cingulate Cortex (SCG)

The cingulate gyrus is located in the medial surface of the cerebral hemisphere, close to the corpus callosum. It is subdivided into four regions: anterior cingulate cortex (ACC), midcingulate cortex, posterior cingulate cortex, and retrosplenial cortex. In humans, the ACC includes Broadmann's areas (BA) 24, 25, 32, and 33. The subcallosal cingulate gyrus (SCG) is a subcomponent of the ACC and is related to BA25. Afferent and efferent BA25 projections include the OFC, the amygdala, the thalamus, and the hypothalamus [75], and decreased activity of this region has been associated with depression remission [25]. The efficacy of SCG DBS was investigated in a pilot randomized double-blind crossover trial with five TRD participants that presented remission during at least 3 months in the open trial phase. The results showed the importance of continued stimulation for the sustained therapeutic effects [76]. However, another randomized, double-blind, sham-controlled study evaluating the efficacy of SCG DBS in patients with TRD reported no significant differences between active and sham groups. Holtzheimer et al. randomized 90 patients from 13 different sites with participants in both groups displaying significant improvement in depressive symptoms and global functioning, despite no significant differences between the groups at 6-month follow-up [77]. In other studies that compared short-pulse- and long-pulse-width SCG DBS, despite the lack of differences in terms of efficacy between the groups, higher frequencies may be associated with better results [78, 79].

A meta-analysis and meta-regression study about the efficacy of DBS to treat TRD included 12 blinded studies that compared active intervention with sham, with sample sizes varying from 4 to 85 participants [80]. Results showed a significant reduction in depressive symptoms (HAM-D or MADRS) favoring DBS. The frequency of adverse events varied from 2% (for instance, electrode revision, postoperative *delirium*, and irritability) to 26% (headache).

4   Potential Biomarkers of Treatment Response

While relatively few studies have evaluated the clinical effects of DBS, an even lower number investigated the role of biomarkers in the prediction of response to that intervention. Given that DBS is an invasive procedure, identifying actionable and accurate biomarkers of treatment response is paramount to improving clinical outcomes and to avoiding unnecessary exposure to brain surgery [81]. Also, most trials did not include a target-engagement evaluation; thus, it is challenging to ascertain why some clinical trials yielded positive results while others did not. A tool that has been analyzed for its potential role in predicting response to DBS is quantitative electroencephalogram (qEEG). One particular qEEG method, called cordance, combines complementary information from absolute and relative power in a frequency band at a given electrode of EEG spectra, ultimately allowing the measurement of regional brain activity. Accordingly, one study analyzed the effects of subcallosal cingulate white matter (SCC) DBS in 12 people with TRD. The authors recorded resting-state EEG and showed that lower frontal theta cordance (FTC) at baseline predicted lower depression severity scores after 24 weeks. Also, greater increases in FTC in the first 4 weeks predicted greater decreases in depression severity scores at 24 weeks of follow-up, suggesting a role for FTC in predicting response to SCC DBS for TRD [82]. Another potential biomarker of interest is glutamate. Altered glutamatergic signaling in the ACC has been described in depression, and decreased levels of glutamate in the ACC have been associated with anhedonia. Accordingly, one study assessed baseline glutamate levels using proton MRS (1H-MRS) among TRD patients treated with ACC DBS. The authors reported that lower levels of glutamate in the ACC were associated with better response to ACC DBS at 6 months of follow-up [83]. In the same line, another study of SCC DBS showed that SCC brain glucose metabolism, as measured by [18F] FDG-PET, was significantly higher in responders than nonresponders [84].

5   Perspectives on the Use of Deep Brain Stimulation for Treatment-Resistant Depression

Despite the potential of DBS in the treatment of TRD and the sustained response with chronic stimulation reported by some authors, most of the available studies are open label, and the few randomized-controlled studies have limited number of participants. In addition, the intervention procedures are highly complex, limiting its use only to the treatment of severely, highly resistant individuals with depression. However, considering the clinical characteristics of the participants included in these studies, usually refractory to several pharmacological and non-pharmacological interventions, including psychotherapy and several ECT sessions, the literature is clear to suggest that DBS may represent a life-changing intervention for several patients with severely disabling symptoms. Some authors have discussed methods to improve the efficacy and safety of DBS, including electrode targeting, new materials, blinded, crossover study designs, and the selection of more homogeneous sample of patients given the high clinical and biological heterogeneity associated with classic diagnostic classifications that are used to categorize individuals with depression [81, 85, 86]. The identification of blood and neuroimaging biomarkers could also help to improve the precision of DBS treatment, which will be possible not only with an understanding of the DBS mechanisms of action, but also with a better characterization of the neurobiology associated with signs and symptoms found in depressive episodes [87]. Similarly, technological advances related to implantable pulse generator designs, biocompatibility, programming, and different stimulation methods and patterns can potentially strengthen the efficacy and safety of the stimulation.

References

1. Brown S, Rittenbach K, Cheung S, McKean G, MacMaster FP, Clement F (2019) Current and common definitions of treatment-resistant depression: findings from a systematic review and qualitative interviews. Can J Psychiatry 64(6):380–387

2. Berlim MT, Turecki G (2007) Definition, assessment, and staging of treatment-resistant refractory major depression: a review of current concepts and methods. Can J Psychiatry 52(1): 46–54

3. Akil H, Gordon J, Hen R et al (2018) Treatment resistant depression: a multi-scale, systems biology approach. Neurosci Biobehav Rev 84:272–288

4. Thase ME, Rush AJ (1997) When at first you don't succeed: sequential strategies for antidepressant nonresponders. J Clin Psychiatry 58 (Suppl 13):23–29

5. Fava M (2003) Diagnosis and definition of treatment-resistant depression. Biol Psychiatry 53(8):649–659

6. Sinyor M, Schaffer A, Levitt A (2010) The sequenced treatment alternatives to relieve depression (STAR*D) trial: a review. Can J Psychiatry 55(3):126–135

7. Rush AJ, Trivedi MH, Wisniewski SR et al (2006) Acute and longer-term outcomes in depressed outpatients requiring one or several treatment steps: a STAR*D report. Am J Psychiatry 163(11):1905–1917

8. Bosco-Levy P, Grelaud A, Blin P et al (2020) Treatment resistant depression incidence and prevalence using the French nationwide claims database. Pharmacoepidemiol Drug Saf 30(2): 169–177

9. Petersen T, Papakostas GI, Mahal Y et al (2004) Psychosocial functioning in patients with treatment resistant depression. Eur Psychiatry 19(4):196–201

10. Godin O, Bennabi D, Yrondi A et al (2019) Prevalence of metabolic syndrome and associated factors in a cohort of individuals with treatment-resistant depression: results from the FACE-DR study. J Clin Psychiatry 80(6): 19m12755

11. Madsen KB, Plana-Ripoll O, Musliner KL, Debost JP, Petersen LV, Munk-Olsen T (2020) Cause-specific life years lost in individuals with treatment-resistant depression: a Danish nationwide register-based cohort study. J Affect Disord 280(Pt A):250–257

12. Christensen GT, Maartensson S, Osler M (2017) The association between depression and mortality—a comparison of survey- and register-based measures of depression. J Affect Disord 210:111–114

13. Walker ER, McGee RE, Druss BG (2015) Mortality in mental disorders and global disease burden implications: a systematic review and meta-analysis. JAMA Psychiatry 72(4): 334–341

14. Cuijpers P, Smit F (2002) Excess mortality in depression: a meta-analysis of community studies. J Affect Disord 72(3):227–236

15. Hasin DS, Sarvet AL, Meyers JL et al (2018) Epidemiology of adult DSM-5 major depressive disorder and its specifiers in the United States. JAMA Psychiatry 75(4):336–346

16. Li G, Zhang L, DiBernardo A et al (2020) A retrospective analysis to estimate the healthcare resource utilization and cost associated with treatment-resistant depression in commercially insured US patients. PLoS One 15(9): e0238843

17. Holtzheimer PE, Mayberg HS (2011) Deep brain stimulation for psychiatric disorders. Annu Rev Neurosci 34:289–307

18. The National Institute of Neurological Disorders and Stroke (NINDS) (2019) Deep brain stimulation (DBS) for the treatment of Parkinson's disease and other movement disorders. https://www.ninds.nih.gov/About-NINDS/ Impact/NINDS-Contributions-Approved-Therapies/DBS. Accessed 8 Dec 2020

19. Lozano AM, Lipsman N, Bergman H et al (2019) Deep brain stimulation: current challenges and future directions. Nat Rev Neurol 15(3):148–160

20. Kalia LV, Lang AE (2015) Parkinson's disease. Lancet 386(9996):896–912

21. Weintraub D, Burn DJ (2011) Parkinson's disease: the quintessential neuropsychiatric disorder. Mov Disord 26(6):1022–1031

22. Sonne J, Reddy V, Beato MR (2020) Neuroanatomy, substantia Nigra. StatPearls, Treasure Island, FL

23. Wichmann T, Delong MR (2002) Neurocircuitry of Parkinson's disease. In: Davis KL, Charney D, Coyle JT, Miller LM, Nemeroff C (eds) Neuropsychopharmacology: The Fifth Generation of Progress: An Official Publication of the American College of Neuropsychopharmacology, 5th edn. Lippincott Williams & Wilkins, Philadelphia

24. Okun MS (2012) Deep-brain stimulation for Parkinson's disease. N Engl J Med 367(16): 1529–1538

25. Mayberg HS, Liotti M, Brannan SK et al (1999) Reciprocal limbic-cortical function

and negative mood: converging PET findings in depression and normal sadness. Am J Psychiatry 156(5):675–682

26. Mayberg HS, Lozano AM, Voon V et al (2005) Deep brain stimulation for treatment-resistant depression. Neuron 45(5):651–660

27. Hamilton M (1960) A rating scale for depression. J Neurol Neurosurg Psychiatry 23:56–62

28. Zimmerman M, Martinez JH, Young D, Chelminski I, Dalrymple K (2013) Severity classification on the Hamilton depression rating scale. J Affect Disord 150(2):384–388

29. Browne CA, Hammack R, Lucki I (2018) Dysregulation of the lateral Habenula in major depressive disorder. Front Synaptic Neurosci 10:46

30. Winter C, Vollmayr B, Djodari-Irani A, Klein J, Sartorius A (2011) Pharmacological inhibition of the lateral habenula improves depressive-like behavior in an animal model of treatment resistant depression. Behav Brain Res 216(1):463–465

31. Meng H, Wang Y, Huang M, Lin W, Wang S, Zhang B (2011) Chronic deep brain stimulation of the lateral habenula nucleus in a rat model of depression. Brain Res 1422:32–38

32. Kim Y, Morath B, Hu C et al (2016) Antidepressant actions of lateral habenula deep brain stimulation differentially correlate with CaMKII/GSK3/AMPK signaling locally and in the infralimbic cortex. Behav Brain Res 306:170–177

33. Zhou Q, Dong J, Xu T, Cai X (2017) Synaptic potentiation mediated by L-type voltage-dependent calcium channels mediates the anti-depressive effects of lateral habenula stimulation. Neuroscience 362:25–32

34. Bracht T, Linden D, Keedwell P (2015) A review of white matter microstructure alterations of pathways of the reward circuit in depression. J Affect Disord 187:45–53

35. Dandekar MP, Saxena A, Scaini G et al (2019) Medial forebrain bundle deep brain stimulation reverses Anhedonic-like behavior in a chronic model of depression: importance of BDNF and inflammatory cytokines. Mol Neurobiol 56(6):4364–4380

36. Dandekar MP, Luse D, Hoffmann C et al (2017) Increased dopamine receptor expression and anti-depressant response following deep brain stimulation of the medial forebrain bundle. J Affect Disord 217:80–88

37. Bregman T, Reznikov R, Diwan M et al (2015) Antidepressant-like effects of medial forebrain bundle deep brain stimulation in rats are not associated with Accumbens dopamine release. Brain Stimul 8(4):708–713

38. Berridge KC, Kringelbach ML (2015) Pleasure systems in the brain. Neuron 86(3):646–664

39. Veyrac A, Gros A, Bruel-Jungerman E et al (2013) Zif268/egr1 gene controls the selection, maturation and functional integration of adult hippocampal newborn neurons by learning. Proc Natl Acad Sci U S A 110(17):7062–7067

40. Dobrossy MD, Ramanathan C, Ashouri Vajari D, Tong Y, Schlaepfer T, Coenen VA (2021) Neuromodulation in psychiatric disorders: experimental and clinical evidence for reward and motivation network deep brain stimulation: focus on the medial forebrain bundle. Eur J Neurosci 53(1):89–113

41. Thiele S, Sorensen A, Weis J et al (2020) Deep brain stimulation of the medial forebrain bundle in a rodent model of depression: exploring dopaminergic mechanisms with Raclopride and micro-PET. Stereotact Funct Neurosurg 98(1):8–20

42. Ashouri Vajari D, Ramanathan C, Tong Y, Stieglitz T, Coenen VA, Dobrossy MD (2020) Medial forebrain bundle DBS differentially modulates dopamine release in the nucleus accumbens in a rodent model of depression. Exp Neurol 327:113224

43. American Psychiatric Association (2013) Diagnostic and statistical manual of mental disorders (DSM 5), 5th edn. American Psychiatric Association, Washington, DC

44. Wacker J, Dillon DG, Pizzagalli DA (2009) The role of the nucleus accumbens and rostral anterior cingulate cortex in anhedonia: integration of resting EEG, fMRI, and volumetric techniques. Neuroimage 46(1):327–337

45. Falowski SM, Sharan A, Reyes BA, Sikkema C, Szot P, Van Bockstaele EJ (2011) An evaluation of neuroplasticity and behavior after deep brain stimulation of the nucleus accumbens in an animal model of depression. Neurosurgery 69(6):1281–1290

46. van Dijk A, Klompmakers AA, Feenstra MG, Denys D (2012) Deep brain stimulation of the accumbens increases dopamine, serotonin, and noradrenaline in the prefrontal cortex. J Neurochem 123(6):897–903

47. Knight EJ, Min HK, Hwang SC et al (2013) Nucleus accumbens deep brain stimulation results in insula and prefrontal activation: a large animal FMRI study. PLoS One 8(2):e56640

48. Scaini G, Rezin GT, Carvalho AF, Streck EL, Berk M, Quevedo J (2016) Mitochondrial dysfunction in bipolar disorder: evidence, pathophysiology and translational implications. Neurosci Biobehav Rev 68:694–713

49. Scaini G, Fries GR, Valvassori SS et al (2017) Perturbations in the apoptotic pathway and mitochondrial network dynamics in peripheral blood mononuclear cells from bipolar disorder patients. Transl Psychiatry 7(5):e1111

50. Kuffner K, Triebelhorn J, Meindl K et al (2020) Major depressive disorder is associated with impaired mitochondrial function in skin fibroblasts. Cell 9(4):884

51. Kim Y, McGee S, Czeczor JK et al (2016) Nucleus accumbens deep-brain stimulation efficacy in ACTH-pretreated rats: alterations in mitochondrial function relate to antidepressant-like effects. Transl Psychiatry 6(6):e842

52. Galvez JF, Keser Z, Mwangi B et al (2015) The medial forebrain bundle as a deep brain stimulation target for treatment resistant depression: a review of published data. Prog Neuropsychopharmacol Biol Psychiatry 58:59–70

53. Sun C, Wang Y, Cui R et al (2018) Human thalamic-prefrontal peduncle connectivity revealed by diffusion Spectrum imaging fiber tracking. Front Neuroanat 12:24

54. Lee DJ, Dallapiazza RF, De Vloo P et al (2019) Inferior thalamic peduncle deep brain stimulation for treatment-refractory obsessive-compulsive disorder: a phase 1 pilot trial. Brain Stimul 12(2):344–352

55. Jimenez-Ponce F, Velasco-Campos F, Castro-Farfan G et al (2009) Preliminary study in patients with obsessive-compulsive disorder treated with electrical stimulation in the inferior thalamic peduncle. Neurosurgery 65(6 Suppl):203–209. discussion 209

56. Morishita T, Fayad SM, Goodman WK et al (2014) Surgical neuroanatomy and programming in deep brain stimulation for obsessive compulsive disorder. Neuromodulation 17(4):312–319. discussion 319

57. Jimenez F, Velasco F, Salin-Pascual R et al (2005) A patient with a resistant major depression disorder treated with deep brain stimulation in the inferior thalamic peduncle. Neurosurgery 57(3):585–593. discussion 585–593

58. Sartorius A, Kiening KL, Kirsch P et al (2010) Remission of major depression under deep brain stimulation of the lateral habenula in a therapy-refractory patient. Biol Psychiatry 67(2):e9–e11

59. Hoyer C, Kranaster L, Sartorius A, Hellweg R, Gass P (2012) Long-term course of brain-derived neurotrophic factor serum levels in a patient treated with deep brain stimulation of the lateral habenula. Neuropsychobiology 65(3):147–152

60. Fernandes BS, Molendijk ML, Kohler CA et al (2015) Peripheral brain-derived neurotrophic factor (BDNF) as a biomarker in bipolar disorder: a meta-analysis of 52 studies. BMC Med 13:289

61. Denier N, Walther S, Schneider C, Federspiel A, Wiest R, Bracht T (2020) Reduced tract length of the medial forebrain bundle and the anterior thalamic radiation in bipolar disorder with melancholic depression. J Affect Disord 274:8–14

62. Fenoy AJ, Schulz P, Selvaraj S et al (2016) Deep brain stimulation of the medial forebrain bundle: distinctive responses in resistant depression. J Affect Disord 203:143–151

63. Fenoy AJ, Schulz PE, Selvaraj S et al (2018) A longitudinal study on deep brain stimulation of the medial forebrain bundle for treatment-resistant depression. Transl Psychiatry 8(1):111

64. Snaith RP, Harrop FM, Newby DA, Teale C (1986) Grade scores of the Montgomery-Asberg depression and the clinical anxiety scales. Br J Psychiatry 148:599–601

65. Coenen VA, Bewernick BH, Kayser S et al (2019) Superolateral medial forebrain bundle deep brain stimulation in major depression: a gateway trial. Neuropsychopharmacology 44(7):1224–1232

66. Davidson B, Giacobbe P, Mithani K et al (2020) Lack of clinical response to deep brain stimulation of the medial forebrain bundle in depression. Brain Stimul 13(5):1268–1270

67. Schlaepfer TE, Cohen MX, Frick C et al (2008) Deep brain stimulation to reward circuitry alleviates anhedonia in refractory major depression. Neuropsychopharmacology 33(2):368–377

68. Bewernick BH, Hurlemann R, Matusch A et al (2010) Nucleus accumbens deep brain stimulation decreases ratings of depression and anxiety in treatment-resistant depression. Biol Psychiatry 67(2):110–116

69. Grubert C, Hurlemann R, Bewernick BH et al (2011) Neuropsychological safety of nucleus accumbens deep brain stimulation for major depression: effects of 12-month stimulation. World J Biol Psychiatry 12(7):516–527

70. Mithani K, Davison B, Meng Y, Lipsman N (2020) The anterior limb of the internal capsule: anatomy, function, and dysfunction. Behav Brain Res 387:112588

71. Coplan JD, Abdallah CG, Tang CY et al (2010) The role of early life stress in development of the anterior limb of the internal capsule in nonhuman primates. Neurosci Lett 480(2):93–96

72. Bergfeld IO, Mantione M, Hoogendoorn ML et al (2016) Deep brain stimulation of the ventral anterior limb of the internal capsule for treatment-resistant depression: a randomized clinical trial. JAMA Psychiatry 73(5):456–464

73. Bergfeld IO, Mantione M, Hoogendoorn MLC et al (2017) Impact of deep brain stimulation of the ventral anterior limb of the internal capsule on cognition in depression. Psychol Med 47(9):1647–1658

74. van der Wal JM, Bergfeld IO, Lok A et al (2020) Long-term deep brain stimulation of the ventral anterior limb of the internal capsule for treatment-resistant depression. J Neurol Neurosurg Psychiatry 91(2):189–195

75. Hamani C, Mayberg H, Stone S, Laxton A, Haber S, Lozano AM (2011) The subcallosal cingulate gyrus in the context of major depression. Biol Psychiatry 69(4):301–308

76. Puigdemont D, Portella M, Perez-Egea R et al (2015) A randomized double-blind crossover trial of deep brain stimulation of the subcallosal cingulate gyrus in patients with treatment-resistant depression: a pilot study of relapse prevention. J Psychiatry Neurosci 40(4):224–231

77. Holtzheimer PE, Husain MM, Lisanby SH et al (2017) Subcallosal cingulate deep brain stimulation for treatment-resistant depression: a multisite, randomised, sham-controlled trial. Lancet Psychiatry 4(11):839–849

78. Eitan R, Fontaine D, Benoit M et al (2018) One year double blind study of high vs low frequency subcallosal cingulate stimulation for depression. J Psychiatr Res 96:124–134

79. Ramasubbu R, Clark DL, Golding S et al (2020) Long versus short pulse width subcallosal cingulate stimulation for treatment-resistant depression: a randomised, double-blind, crossover trial. Lancet Psychiatry 7(1):29–40

80. Hitti FL, Yang AI, Cristancho MA, Baltuch GH (2020) Deep brain stimulation is effective for treatment-resistant depression: a meta-analysis and meta-regression. J Clin Med 9(9):2796

81. Fernandes BS, Williams LM, Steiner J, Leboyer M, Carvalho AF, Berk M (2017) The new field of 'precision psychiatry'. BMC Med 15(1):80

82. Broadway JM, Holtzheimer PE, Hilimire MR et al (2012) Frontal theta cordance predicts 6-month antidepressant response to subcallosal cingulate deep brain stimulation for treatment-resistant depression: a pilot study. Neuropsychopharmacology 37(7):1764–1772

83. Clark DL, MacMaster FP, Brown EC, Kiss ZHT, Ramasubbu R (2020) Rostral anterior cingulate glutamate predicts response to subcallosal deep brain stimulation for resistant depression. J Affect Disord 266:90–94

84. Brown EC, Clark DL, Forkert ND, Molnar CP, Kiss ZHT, Ramasubbu R (2020) Metabolic activity in subcallosal cingulate predicts response to deep brain stimulation for depression. Neuropsychopharmacology 45(10):1681–1688

85. Mosley PE, Marsh R, Carter A (2015) Deep brain stimulation for depression: scientific issues and future directions. Aust N Z J Psychiatry 49(11):967–978

86. Ashouri Vajari D, Vomero M, Erhardt JB et al (2018) Integrity assessment of a hybrid DBS probe that enables neurotransmitter detection simultaneously to electrical stimulation and recording. Micromachines (Basel) 9(10):510

87. Davidson B, Gouveia FV, Rabin JS, Giacobbe P, Lipsman N, Hamani C (2020) Deep brain stimulation for treatment-resistant depression: current status and future perspectives. Expert Rev Med Devices 17(5):371–373

Chapter 18

Serotonin Receptors and Antidepressants: Neuroimaging Findings from Preclinical and Clinical Research

Patricia A. Handschuh, Melisande E. Konadu, Benjamin Spurny-Dworak, Leo R. Silberbauer, Matej Murgas, and Rupert Lanzenberger

Abstract

Major depressive disorder is an affective disease associated with a high socioeconomic burden, social and physical impairment, as well as increased mortality rates worldwide. A multitude of in vivo neuroimaging studies has been published on alterations within neurotransmitter systems, substantiating the assumption that these changes are involved in the pathophysiology of major depression. In this context, especially alterations within the serotonergic system are known to play a central role, and the pharmacologic modulation of serotonergic neurotransmission is well accepted as one of the key elements in the treatment of depressive disorders. Throughout the last decades, special emphasis was placed on preclinical and clinical studies utilizing positron emission tomography, a nuclear imaging technique that uses radioligands to visualize certain target molecules such as neuroreceptors, transporters, as well as other proteins involved in serotonergic functioning. The chapter aims to provide a brief overview of the serotonergic neurotransmitter system, laying special emphasis on serotonin metabolism and the role of different receptors involved in serotonergic signaling pathways. In particular, we focus on the well-investigated and highly relevant inhibitory serotonin receptors $5\text{-}HT_{1A}$, $5\text{-}HT_{1B}$, and $5\text{-}HT_{1D}$. Besides, neuroimaging findings on the pharmacologically similarly important excitatory receptors $5\text{-}HT_{2A}$ and $5\text{-}HT_{2C}$ are outlined followed by a brief overview on other receptors implicated in depression: $5\text{-}HT_4$ and the $5\text{-}HT_7$. The principles of positron emission tomography are summarized, and well-established radiotracers for imaging the serotonergic system are presented. Due to the lack of highly selective radioligands, the possibilities for in vivo imaging of serotonergic neurotransmission in health and disease are somewhat limited. Notwithstanding, the most relevant findings of molecular neuroimaging studies in the context of major depressive disorder are reviewed, providing a comprehensive overview of how the availability and biodistribution of serotonergic targets might be altered in depression. For a better understanding of how pathophysiological changes are addressed in a clinical context, we outline the mechanism of action of antidepressants targeting the serotonergic system.

Key words Major depressive disorder, In vivo PET imaging, Serotonergic neurotransmission, Antidepressants

Yong-Ku Kim and Meysam Amidfar (eds.), *Translational Research Methods for Major Depressive Disorder*, Neuromethods, vol. 179, https://doi.org/10.1007/978-1-0716-2083-0_18,

1 Introduction

Over 260 million people worldwide suffer from major depressive disorder (MDD), a mental condition associated with a high socio-economic burden and an increased global mortality rate worldwide [1–3]. A depressive episode is primarily characterized by depressed mood, lack of interest, as well as significant loss of energy. Multi-center international studies on depression assume a lifetime prevalence of 10–20% [4] and the risk of suicide in depressed patients is roughly 30 times that of healthy controls [5].

The emergence and clinical manifestation of a depressive episode is based on various factors. Within the framework of a multi-factorial explanatory model, the biological underpinnings have gained importance over the last decades, and research on potentially involved neurobiological mechanisms is of enormous significance for the optimization of current and the development of future therapeutic options. These biological aspects with possibly predisposing or protective character include genetic vulnerability or resilience factors such as single nucleotide polymorphisms (SNPs) at the brain-derived neurotrophic factor (BDNF) gene [6, 7] or at ABCB1 [8]. Furthermore, polymorphisms of the tryptophan hydroxylase gene TPH1 [9] and the serotonin transporter (SERT) gene SLC6A4 were shown to influence the clinical manifestation of affective disorders [10]. Additionally, diverse biochemical disturbances found in the cerebrospinal fluid [11, 12] and neuroendocrinologic variables, e.g., dysregulation of the HPA axis [13, 14], the role of the thyroid gland [15, 16], or gonadal hormones [17–20], need to be taken into account. Moreover, inflammatory mechanisms [21] as well as volume reductions of certain brain regions (e.g., in frontal regions, the hippocampus, or the caudate nucleus [22]), partly explained by neurodegenerative and neuroplastic processes in the brain [23–25], are assumed to be involved in the pathophysiology of MDD.

With the serendipitous discovery of antidepressant effects of psychopharmacologic agents that increase the concentration of biogenic amines like serotonin (or 5-hydroxytryptamine, 5-HT) in the synaptic cleft either by inhibition of reuptake or limitation of degradation, the monoamine hypothesis and consequently the investigation of serotonergic neurotransmission have constantly gained importance. The hypothesis is based on the assumption that a relative lack of certain monoamines in the synaptic cleft facilitates the emergence of depression and other affective disorders [26]. Besides lowered monoamine levels, also the amount of degrading enzymes, such as monoamine oxidase A (MAO-A and MAO-B in the case of dopamine (DA), respectively), and pre- and postsynaptic protein expression (e.g., the SERT and 5-HT_{1A} receptors) need to be taken into account.

In this chapter, we provide an overview of serotonergic neurotransmission and its components involved in the pathophysiology of major depression. Special focus is laid on molecular neuroimaging studies performed with positron emission tomography (PET) that aim to visualize the serotonin system of healthy individuals as well as alterations of cerebral serotonergic signaling in depressed patients. Furthermore, the chapter summarizes the most relevant antidepressant agents that target the 5-HT system. In this context, the role of 5-HT receptors in the mechanism of action of antidepressants will be extensively discussed, outlining the available evidence for the influence of 5-HT receptor alterations in the pathophysiology of major depression. The chapter is divided into five sections. After the introduction (I), the major characteristics, function, and biological occurrence of serotonin will be elucidated, including its biochemical features, biosynthesis, and degradation. Moreover, the presence of serotonin in nonhuman organisms as well as its distribution, function, and mechanism of action in the human body will be depicted (II). Then, the subclasses of antidepressants influencing the 5-HT system and their particular mechanism of action will be summarized (III). This section is followed by a comprehensive summary of the preclinical and clinical application of molecular neuroimaging techniques. This passage briefly explains the basic principles of PET, followed by an overview of frequently used radiotracers and a detailed characterization of 5-HT receptor subtypes in the context of molecular brain imaging. Investigations performed with PET to quantify and visualize molecular targets within the serotonin system and to detect potentially significant alterations of their distribution and concentration in the context of major depression will be presented (IV). Lastly, we provide a brief conclusion with the most important aspects of this topic and future perspectives in the context of imaging technologies in MDD (V).

2   Serotonin—Metabolic Characteristics, Function, and Biodistribution

2.1 Serotonin Synthesis and Degradation

Serotonin belongs to the group of monoamine neurotransmitters. Due to the hydrophilic properties that hinder it from crossing the lipophilic blood-brain barrier (BBB), a compartment-targeted synthesis is required. The biosynthesis of 5-HT is an intracellular two-stage reaction. In humans, the essential amino acid L-tryptophan which serves as the precursor for serotonin has to be supplied by the intake of dietary protein. In contrast to serotonin, L-tryptophan is able to cross the BBB. In a first step, L-tryptophan is converted into the non-proteinogenic amino acid 5-hydroxytryptophan by the enzyme L-tryptophan hydroxylase which requires tetrahydrobiopterin, oxygen, NADPH, and copper or iron ions for its activity [27]. In the subsequent process,

5-hydroxytryptophan is decarboxylated to serotonin by L-aromatic amino acid decarboxylase with the cofactor vitamin B6. This results in the daily production of about 10 mg of serotonin in humans [28]. To achieve homeostasis across circulating neurotransmitters, serotonin is degraded to 5-hydroxyindoleacetic acid, largely by the enzyme monoamine oxidase A (MAO-A).

2.2 Presence of 5-HT Across Organisms

Serotonin has been reported in a variety of different life forms so far. However, it was shown to have different functions across species, ranging from growth regulation [29] in plants, overprotection from UV toxicity in higher fungi or yeast [30, 31] to neurotransmission in humans [32] as well as in other vertebrates. However, tryptophan-based serotonin synthesis is still a common characteristic among most of the abovementioned life forms. Tryptophan is one of the essential amino acids for humans that cannot be synthetized de novo. In contrast to plants, humans are lacking chloroplasts, organelles essential for tryptophan synthesis.

Although key components of 5-HT synthesis or degradation including tryptophan hydroxylase, tryptophan decarboxylase, or MAO are conserved in unicellular organisms, evidence of serotonin in prokaryotes is weak. Several bacteria were shown to be able to produce 5-HT. Especially gut bacteria like *Streptococcus, Enterococcus, Pseudomonas,* and *Escherichia* have been extensively investigated [33–36]. Serotonin synthesis by gut bacteria requires tryptophan and thus limits the availability of the essential amino acid for the host. The impact of tryptophan metabolism by the gut microbiome on human serotonergic neurotransmission has been described by several studies reporting the effects on neuropsychiatric functioning [37, 38].

As mentioned above, all animal species are dependent on exogenous tryptophan supply. Nevertheless, serotonin is found across all complexities of animal species. Even sponges—the most primitive form of animals without a nervous system—contain 5-HT [39]. As soon as primitive brains are present in animals (e.g., in *aplasia* or *C. elegans*), the spatial distribution of 5-HT changes in brain and gut regions. However, species-dependent accumulations have also been reported. While serotonergic neurons are mainly located in the mushroom bodies in insect brains [40], cerebral serotonin is mainly found in midbrain regions in higher-order animals [41–43].

2.3 Serotonin in the Human Body

The neurotransmitter 5-HT is widely distributed in the human body and regulates diverse functions in multiple organ systems, including the gastrointestinal, cardiovascular, pulmonary, endocrine, genitourinary, and central nervous system (CNS) [32]. It exerts regulatory functions in basic physiological processes as well as in neuronal networks, e.g., the sleep-wake cycle [44], sensory-motor system [45], eating behavior and appetite [46], temperature regulation [47], and pain reception [48]. Moreover, it is an essential part of neuropsychological and behavioral processes [23, 49–51, 53]. An association between serotonergic neurotransmission and mental health was already suggested in 1954 [54].

About 90% of total body serotonin is synthesized in the gastrointestinal tract, while only a small amount is produced in the CNS [55]. Within the CNS, 5-HT neurons originate from the raphe nuclei of the brainstem, which comprise a superior and inferior subdivision, consisting of nine main nuclei. 5-HT axons are projected rostrally and caudally to almost every region of the CNS [56, 57]. When looking at the various effects of 5-HT, evidence points toward a topographically complex organization of the raphe nuclei and its projections [58].

The neurotransmitter 5-HT largely acts via binding to its receptors that are classified according to their structural, pharmacological, and biochemical characteristics. At least 14 different 5-HT receptors are known, divided into families ranging from $5\text{-}HT_1$ to $5\text{-}HT_7$, exerting individual signaling pathways and comprising various isoforms [59]. All of them, except the ligand-gated $5\text{-}HT_3$ receptor, are G-protein-coupled receptors (GPCRs) located in cell membranes and consist of an extracellular, transmembrane, and intracellular segment [60]. GPCRs can be found as homo- or heterodimers but also as part of oligomers possibly leading to altered ligand pharmacology or function [60, 61]. G-proteins are intracellular heterotrimeric compounds, consisting of a G_α monomer and a $G_{\beta\gamma}$ dimer subunit. Extracellular ligand binding leads to an exchange of guanosine diphosphate (GDP) to guanosine triphosphate (GTP) at the G_α unit, followed by a separation and release of the G_α and $G_{\beta\gamma}$ complex. This is followed by downstream signaling cascades, by interplay with effector proteins, and finally by specified cellular responses [60, 62]. Many genes coding for the GPCR arise from alternative splicing or the use of alternative start codons and transcription start sites. For the G_α subunit, e.g., 53 isoforms are known on transcript level [62]. The regulation of the serotonin system is mainly driven by the $5\text{-}HT_{1A}$ and $5\text{-}HT_{1B}$ autoreceptor isoforms, as well as by the 5-HT transporter (SERT). $5\text{-}HT_{1A}$ belongs to the major inhibitory G-protein-coupled serotonin receptors [63]. It has been shown that $5\text{-}HT_{1A}$ autoreceptors control the firing rates of 5-HT neurons in the raphe nuclei [64], whereas $5\text{-}HT_{1B}$ autoreceptors terminate the release of 5-HT [65, 66]. Moreover, $5\text{-}HT_1$ receptors are assumed to be involved in the autoregulation of serotonergic neuronal development [67].

The $5\text{-}HT_1$ ($5\text{-}HT_{1A}$, $5\text{-}HT_{1B}$, $5\text{-}HT_{1D}$, $5\text{-}HT_{1E}$, $5\text{-}HT_{1F}$) receptor family [59] exerts inhibiting signals ($G_{\alpha i/o}$) via inhibition of adenylate cyclase (AC) and reduction of intracellular cyclic adenosine monophosphate (cAMP) levels [62]. The $5\text{-}HT_{1A}$ and the $5\text{-}HT_{1B}$ receptors act as presynaptic autoreceptors as well as postsynaptic heteroreceptors on non-serotonergic neurons [63]. Both receptors are distributed widely in the brain and are associated with emotion regulation [59], anxiety, and the interaction with certain antidepressants [68, 69]. The 5-HT$_2$ (5-HT$_{2A}$, 5-HT$_{2B}$, 5-HT$_{2C}$) receptors [59] exert effects via $G_{\alpha q/11}$ on phospholipase C (PLC), on inositol 1,4,5-triphosphate (IP3), and subsequently on intracellular Ca^{2+} levels [62]. The 5-HT$_{2A}$ receptor is distributed in the CNS with high densities in the cerebral cortex and claustrum, and moderate densities in the limbic system as well as basal ganglia. It is known as one of the main excitatory 5-HT receptors [70]. As summarized in detail by Celada et al., it is involved in several excitatory mechanisms throughout the brain and exerts opposing effects when compared to the 5-HT$_{1A}$ receptor, e.g., on pyramidal neurons in the medial prefrontal cortex [71]. Interestingly, lysergic acid diethylamide (LSD) and other hallucinogenic agents exhibit an agonistic effect at 5-HT$_{2A}$ receptors [59]. The 5-HT$_4$ (at least nine variants have been reported so far, 5-HT$_{4A}$–5-HT$_{4N}$), 5-HT$_6$, and 5-HT$_7$ (5-HT$_{7A}$, 5-HT$_{7B}$, 5-HT$_{7C}$, 5-HT$_{7D}$) receptor families [59] are $G_{\alpha s}$ coupled, leading to the activation of AC and an increase in cAMP levels [62]. The 5-HT$_4$ receptors are expressed in the CNS, mainly in the nigrostriatal and mesolimbic system. Hence, they are particularly involved in limbic and visuo-motor functions [59]. The 5-HT$_6$ receptor is distributed across the CNS mainly in the striatum, amygdala, nucleus accumbens, hippocampus, cortex, and olfactory tubercle [59]. It is thought to be involved in psychiatric disorders but also in cognition, including learning and memory [59, 72]. Throughout the human body, the 5-HT$_7$ receptor is mostly distributed in vessels and nonvascular smooth muscles out- and inside the brain. In the CNS the receptor is expressed in the medial thalamic nuclei and in associated limbic areas. It is assumed to be involved in the circadian rhythm, sleeping behavior, and mood [59]. The function of the 5-HT$_5$ (5-HT$_{5A}$, 5-HT$_{5B}$) family is not yet fully investigated. It either acts via $G_{\alpha i/o}$ or $G_{\alpha s}$ protein families [59]. A rat knockout model of the 5-HT$_{5A}$ receptor was linked to modified exploratory activity levels and showed changes in the effect of LSD on mice [73]. The 5-HT$_3$ (5-HT$_{3A}$, 5-HT$_{3B}$, 5-HT$_{3C}$, 5-HT$_{3D}$, and 5-HT$_{3E}$) receptor is the only ligand-gated receptor subtype. In the CNS, it is predominantly located in the dorsal vagal complex and limbic structures [59]. Additionally, it is detectable in the gastrointestinal tract, where it is involved in the control of intestinal motility and secretion [74].

The 5-HT transporter (SERT) is a monoamine transporter protein consisting of 12 transmembrane domains [75]. It belongs to the neurotransmitter-sodium symporter (NSS) family [76], and its main task is the efficient reuptake of 5-HT from the synaptic cleft and therefore the regulation of serotonergic transmission and 5-HT availability which underlies complex regulatory mechanisms [77]. SERTs are expressed with high densities in subcortical regions, including the dorsal raphe nucleus, midbrain, and thalamus. Moreover, it is found in lower concentrations in cortical regions [78]. Besides the CNS, SERTs are located in the membranes of platelets [79], in lymphocytes [80], in the gastrointestinal tract [81], as well as on pulmonary and placental membranes [82]. The short variant of the 5-HTT-linked polymorphic region was associated with vulnerability regarding affective disorders [83].

As described earlier in this chapter, MAOs are responsible for the degradation of 5-HT, catecholamines, and trace amines like β-phenylethylamines and belong to the family of mitochondrial bound enzymes [84]. The two well-known isoenzymes of MAO are referred to as MAO-A and MAO-B. The metabolization of 5-HT (and norepinephrine, respectively) is primarily conducted by MAO-A since this isoform shows a higher affinity for hydroxylated amines. Although on the genetic level the enzymes exhibit identical exon-intron organization and a peptide accordance of about 70% [85], on the protein level each enzyme reveals substrate specificity [84]. Consequently, 5-HT and norepinephrine are mainly degraded by MAO-A, whereas MAO-B largely catalyzes β-phenylethylamines such as dopamine [84]. MAO is found both extracellularly and in the cytosol. Therefore, intracellular serotonin is transported into vesicular compartments to prevent cytosolic degradation and to prepare 5-HT for the release into the synaptic cleft [56, 86, 87]. In general, MAO enzymes are both distributed widely in peripheral tissues. MAO-A is expressed predominantly in the placental tissue, while MAO-B is expressed especially in platelets [88, 89]. In the CNS, the MAO-A enzyme is mainly located in noradrenergic neurons, while MAO-B is predominantly found in serotonergic and histaminergic neurons, as well as in glial cells [90–94]. The findings of MAO-A located mainly in noradrenergic neurons are seen in contrast with the outcomes of higher 5-HT levels in MAO-A inhibited conditions, as it was investigated in MAO-A knockout mice [95], and are still a matter of debate [84].

3   Serotonin in Major Depression: Pharmacologic Agents Targeting the 5-HT System

Antidepressants interact mainly with different neurotransmitter systems by targeting receptors, transport proteins, and/or other enzymes. Based on the monoamine hypothesis, antidepressants are mainly classified according to their selectivity for these target structures. Due to their influence on certain receptors and other molecules, they are assumed to restore the physiologic balance of neurotransmitters such as 5-HT in the synaptic cleft and to modulate transmitter metabolism, receptor activity, and thus neuronal firing. Additionally, their impact on metabolic homeostasis is not limited to one transmitter system: Many antidepressants exert multimodal effects, directly or indirectly influencing various neurotransmitters, hormones, and other molecules. Moreover, combined treatment regimens might increase efficacy by addressing multiple targets on the one hand and reduce side effects on the other. However, immediate receptor modulation is only one aspect in the mode of action of antidepressants: As reviewed by Menke and Binder [96], selected antidepressant classes were shown to exert direct epigenetic and/or genetic effects. In addition, various genetic polymorphisms as well as epigenetic alterations could be linked to treatment outcome [97]. Nowadays, the knowledge on the working mechanism of well-known antidepressants goes beyond monoaminergic signaling and comprises data on, e.g., the modulation of neurotrophins like the brain-derived neurotrophic factor (BDNF) [98], thus indicating the importance of neuroplasticity and neurogenesis. Additionally, novel pharmacologic approaches involve the modulation of non-monoaminergic systems, e.g., glutamate and N-methyl-D-aspartate (NMDA) receptors [99]. As reviewed by Kraus and Kadriu, potential biomarkers, individual metabolization characteristics involving the cytochrome P450 isoenzymes, and standardized step-by-step treatment regimens are promising elements for individual treatment optimization in future [100].

Here, we will focus on the following antidepressant classes exerting direct serotonergic effects that are available for the pharmacotherapy of MDD: tricyclic antidepressants (TCA), α_2-antagonists or tetracyclic antidepressants (TeCAs), noradrenergic and specific serotonergic antidepressants (NaSSA), selective serotonin reuptake inhibitors (SSRIs), serotonin-norepinephrine reuptake inhibitors (SNRIs), serotonin 5-HT_2 antagonists and reuptake inhibitors (SARIs), glutamate modulators (GM), and reversible or irreversible monoamine oxidase inhibitors (MAOI). Also, novel antidepressants with specific multimodal mechanisms are now available, including vortioxetine and agomelatine. An overview of the mechanism of action of antidepressant classes is presented in Table 1.

3.1 Classic Tricyclic Antidepressants (TCAs)

Classic tricyclic antidepressants were developed in the 1950s. The name refers to the chemical structure of TCAs that contain three rings of atoms with varying sidechains. They exert their effect by inhibiting the reuptake of both noradrenaline (NA) and 5-HT from the synaptic cleft by blocking the SERT and the noradrenaline transporter (NET) [101]. However, they were also shown to interact with the muscarinic acetylcholine receptor, various histamine receptors, and α-adrenoceptors [102]. These interactions are known to be associated with a broad variety of side effects as summarized by Peretti et al. [103]. Additionally, many TCAs influence certain 5-HT receptors, e.g., 5-$HT_{2A/2C}$ [104] and 5-HT_7 [105], which is assumed to contribute to their antidepressant and some other clinically useful effects (e.g., antinociception) [106].

Table 1

Mechanism of action of antidepressants and binding affinity in humans as obtained from DrugCentral 2021 (http://drugcentral.org)

Substance group	TCA	SSRI	SNRI	GM	NaSSA	SARI	RIMA/ irreversible MAOI
Target	*Mechanism of action*						
Active agent	*e.g., imipramine*	*e.g., escitalopram, citalopram, fluoxetine, paroxetine, sertraline*	*e.g., duloxetine, venlafaxine, milnacipran*	*Tianeptine*	*Mirtazapine*	*Trazodone retard*	*Moclobemide, tranylcypromine*
SERT inhibition	++	+++	+++	−	−	+	−
5-HT$_2$ receptor antagonist	−	−	−[3]	?	++	+++	−
NET inhibition	++	−	+++	−	+	?	−
MAO inhibition	−	−	−	−	−	−	++
mACh receptor antagonist	+++	−[a]	−	−	−	−	−
H$_1$ receptor antagonist	+	−	−	−	++	+	−
DAT inhibition	−	−[b]	−[c]	−	−	?	−
α1 receptor antagonist	−	−	−	−	−	++	−
α2 receptor antagonist	−	−	−	?	+++	+	−
Glutamatergic neurotransmission	−	?	?	+++	−	?	?
Target	*Bioactivity value (−log[M]) in humans*						
Active agent (example)	*Imipramine*	*Fluoxetine*	*Venlafaxine*	*Tianeptine*	*Mirtazapine*	*Trazodone retard*	*Moclobemide*

(continued)

Table 1
(continued)

Substance group	TCA	SSRI	SNRI	GM	NaSSA	SARI	RIMA/ irreversible MAOI
SERT	K_d 8.85	K_d 9.09	K_d 8.05	/	/	K_d 6.80	/
5-HT$_{1A}$	K_i 5.24	/	/	/	K_d 5.30	K_i 7.02	K_d 5.30
5-HT$_{1D}$	/	/	/	/	/	K_i 6.97	/
5-HT$_{2A}$	K_i 6.82	K_i 6.61	K_i 5.65	/	K_i 6.16	K_i 7.35	/
5-HT$_{2C}$	K_i 6.80	K_i 6.40	K_i 5.70	/	K_i 7.41	K_i 6.65	/
5-HT$_4$	/	/	/	/	/	/	/
5-HT$_7$	/	/	/	/	K_i 6.58	K_i 5.75	/

[a] paroxetine → mACh receptor agonist (+)

[b] sertraline → DAT inhibition (+)

[c] venlafaxine → 5-HT$_2$ receptor antagonist (+)/DAT inhibition (+)

?: → no sufficient data available; − → no effect; + → low effect; ++ → moderate effect; +++ → strong effect; / → no sufficient data available or low binding affinity

Table 1 shows the mechanism of action of antidepressant classes as specified in Subheading 3. Furthermore, the bioactivity values of selected agents in humans are depicted. Thereby, the binding affinity (BA) of each agent for the listed serotonergic target molecules is reported (i. e., dissociation constant (K_d) or inhibition constant (K_i)). K_d measures the equilibrium between the complex (comprising a ligand and a target structure) and the dissociated components. The smaller the K_d value, the higher the BA between the ligand and the target structure. K_i, on the other hand, is a dissociation constant relating to the BA of an inhibitor to its target structure. Information on the primary mechanism of action of each pharmacologic agent as well as data on bioactivity was obtained from DrugCentral 2021, a public online drug information resource created and maintained by the Division of Translational Informatics at the University of New Mexico in collaboration with the IDG [140].

DAT: dopamine transporter; GM: glutamate modulators; MAOI: monoamine oxidase inhibitor; NaSSa: noradrenergic and specific serotonergic antidepressant; NET: norepinephrine transporter; RIMA: reversible inhibitor of monoamine oxidase A; SARI: serotonin antagonist and reuptake inhibitors; SERT: serotonin transporter; SNRI: serotonin-norepinephrine reuptake inhibitor; SSRI: selective serotonin reuptake inhibitor; TCA: tricyclic antidepressant; 5-HT: serotonin; H$_1$ receptor: histamine$_1$ receptor; mAch receptor: muscarinic acetylcholine receptor; α1: alpha-1 receptor; α2: alpha-2 receptor.

Since then, TCAs have largely been replaced by other antidepressant classes with more favorable safety and tolerability profiles; however, they are still frequently prescribed, e.g., in therapy-resistant depression [107].

3.2 Tetracyclic Antidepressants (TeCAs) and Noradrenergic and Specific Serotonergic Antidepressants (NaSSA)

Tetracyclic antidepressants were introduced in the 1970s. The antidepressant effect of mianserin, which is one of the most frequently used TeCAs, is largely exerted by the inhibition of α_2 receptors and subsequently reduces noradrenaline release from the presynapse [108]. Most TeCAs show a low affinity for the SERT, whereas NET inhibition contributes to their antidepressant effects. Furthermore, they exert weak antagonistic properties on 5-HT_{1C}, 5-HT_2, H_1, and α_1 receptors. However, the blockage of the α_2 receptors and their lower affinity for muscarinergic receptors differentiate TeCAs from TCAs [108].

Mirtazapine is seen as a further development to early tetracyclic antidepressants. The effects are based on the increased central noradrenergic and serotonergic activity. Mirtazapine is an antagonist of central presynaptic adrenoceptors, α_2 receptors. In addition, it is a potent antagonist of 5-HT_2, 5-HT_3, and histamine H_1 receptors, promoting the sedating effects of mirtazapine. The chemical structure of mirtazapine is similar to the four-ring structure of mianserin [109].

3.3 Selective Serotonin Reuptake Inhibitors (SSRIs)

The emergence of SSRIs marks a milestone in the history of psychopharmacotherapy, and this class of antidepressants belongs to the top 20 most frequently prescribed psychiatric drugs in Western countries [110]. They show a favorable safety profile, especially compared to earlier antidepressants (i.e., TCAs). They selectively block the reuptake of 5-HT into the presynaptic nerve cell by inhibiting the SERT. Thus, SSRIs elevate the amount of 5-HT in the synaptic cleft and indirectly interact with multiple, largely postsynaptic 5-HT receptors, thereby exerting their immediate (side) effects as well as their delayed antidepressant effects. As reviewed in detail by Vaswani et al., immediate effects of SSRIs are mostly adverse reactions that can be traced back to certain regions of the body where various physiologic processes are controlled. Desensitization of postsynaptic receptors is believed to alleviate these side effects over time [111]. Unfortunately, clinically significant positive effects of SSRIs are not exerted immediately. Among substances that predominantly increase extracellular 5-HT, inhibitory 5HT1_A and 5HT1_D autoreceptors are desensitized, whereby serotonergic nerve cells are disinhibited. Ultimately, increased 5-HT availability furthers postsynaptic receptor desensitization, thus contributing to the alleviation of depressive symptoms [112]. Additionally, theories on delayed drug response also involve genetic mechanisms [113] and dysfunctional membranous signaling [114].

While the exact molecular underpinnings of how the higher amount of 5-HT in the synaptic cleft exerts its antidepressant effect are still a matter of debate, imaging studies point toward connectivity changes across distinct brain regions and potentially relevant interregional differences in the presence of SERT. In this context, James et al. investigated the interregional relation of SERT availability between distinct brain areas known to be involved in MDD pathophysiology and the effect of SSRI intake over 3 weeks. Interestingly, they found a significant increase in SERT binding potential correlations across certain MDD-relevant brain regions over time, emphasizing the importance of network-based investigations within the scope of SSRI use [115].

Importantly, SSRIs show less affinity for postsynaptic adrenergic, cholinergic, histaminergic, and dopaminergic receptors than TCAs, thus lowering the risk for a variety of moderate to severe side effects and adverse events [116]. Multiple meta-analyses were performed to compare the efficacy of TCAs and SSRIs, largely concluding that both TCAs and SSRIs are similarly effective in the treatment of MDD [117]; however, SSRIs were reported to be superior in terms of efficacy and tolerability among young patients [118].

3.4 Serotonin-Norepinephrine Reuptake Inhibitors (SNRIs)

SNRIs exert a dual mode of action. Their effect is based on SERT and NET inhibition. As a result, they mediate an increase of extracellular 5-HT and NA at the synaptic cleft. The available substances (venlafaxine, milnacipran, and duloxetine) all block SERT and NET; however, duloxetine exerts a ten-fold and venlafaxine a 30-fold selectivity for 5-HT, whereas milnacipran shows equal affinity for both transporter proteins [119]. Additionally, SNRIs resulted to be effective in alleviating neuropathic and chronic pain, both in comorbid depression and in the context of other pathologic conditions [120–122].

3.5 Serotonin 5-HT2 Antagonists and Reuptake Inhibitors (SARIs)

Trazodone—the most frequently used SARI—exerts multifaceted mechanisms of action: It inhibits the G-protein-coupled 5-HT_{2A} and 5-HT_{2C} receptors. As outlined in Subheading 2, the family of 5-HT_2 receptors is primarily known for excitatory properties. In depressed individuals, insomnia, irritability, fear, inner tension, and even sexual dysfunction are attributed to these excitatory mechanisms [123]. Thus, it is assumed that the antagonistic effect on 5-HT_2 receptors mediated by SARIs reduces anxiety and inner tension and normalizes sleep architecture [124]. Furthermore, trazodone blocks the SERT and the reuptake of 5-HT into the presynapse. Thereby, SARIs increase the partially agonistic effect of the remaining 5-HT on 5-HT_{1A} receptors [125]. It is assumed that side effects associated with other antidepressants (e.g., SSRIs) can partly be traced back to the increased concentration of serotonin in the synaptic cleft and the subsequently enhanced binding to $5\text{-}HT_2$ receptors – however, the previously described inhibition of $5\text{-}HT_2$ receptors observed with trazodone could explain the more favorable side effect profile of these drugs [126]. Besides, SARIs increase the release of dopamine and NA in the prefrontal cortex as well as the synthesis of protective neurotrophic factors, e.g., BDNF [127]. Antagonistic effects at the α_1 and H_1 receptors have also been observed, further promoting sedation and anxiolysis associated with the intake of SARIs like trazodone [124].

3.6 Glutamate Modulators (GM)

The most important glutamate modulator in the context of clinical depression is tianeptine. As reviewed by McEwen et al., its atypical mechanism of action has drawn much attention [128]. Tianeptine enhances the reuptake of 5-HT into the presynapse, which challenges the monoamine hypothesis of depression to a certain extent [129]. Additionally, it induces adaptation processes and physiologic cascades, thereby promoting neuroplasticity and cellular resilience as well as exerting cytoprotective and procognitive effects [128]. These processes involve the modulation and phosphorylation of glutamate receptors in selected brain regions that are thought to be involved in the pathophysiology of depression [130].

A further pharmacologic agent influencing glutamatergic signaling is ketamine. Recently, the anesthetic was approved by the US Food and Drug Administration (FDA) for intranasal application in the treatment of therapy-resistant depression. Ketamine exerts antagonistic effects on GABA-ergic interneurons found in cortical regions and is further known to enhance excitatory glutamatergic neurotransmission in selected areas such as the prefrontal cortex [131]. Although ketamine's pharmacologic mechanisms of action represent a highly relevant research topic—also addressed by our research group [132–134]—little is known about the interaction of ketamine and serotonin receptors in particular. Nevertheless, recent molecular neuroimaging studies suggest an effect of ketamine on $5\text{-}HT_{1B}$ receptors in the brain [135].

3.7 Reversible and Irreversible Monoamine Oxidase Inhibitors (MAOI)

MAO inhibitors have been introduced into the treatment of depression in the 1950s [136]. Nowadays, several MAO inhibitors are available for therapeutic use, including reversible and irreversible as well as selective and nonselective compounds. As reviewed by Youdim et al., the antidepressant effects of MAO inhibitors are related to increased dopaminergic, noradrenergic, and serotonergic neurotransmission induced by selective MAO-A inhibition in the CNS [137]. Older, irreversible, and nonselective MAO inhibitors tend to exhibit a higher risk for hypertensive crisis induced by the intake of higher tyramine amounts. Regarding this side effect, reversible monoamine oxidase A inhibitors (RIMA) and selective MAO-B inhibitors are valued as safer options [138]. Additionally, MAO inhibitors have been strongly associated with the occurrence of the so-called serotonin syndrome, especially if used in combination with, e.g., SSRIs [139]. Therefore, pharmacological drug interactions have to be assessed carefully in clinical use.

4   Serotonin in Major Depression: Preclinical and Clinical Application of Molecular Neuroimaging Performed with Positron Emission Tomography

In addition to a great number of MRI-based neuroimaging studies conducted to investigate potentially pathologic structural [141, 142] as well as functional [143] alterations in the depressed brain and the effect of antidepressant therapy [144–146], a multitude of molecular imaging studies has been performed to analyze changes in neurotransmitter systems that might be associated with the emergence of clinical depression.

Given the good spatial resolution and high signal-to-noise ratio, PET imaging provides a highly useful method to estimate protein expression, binding behavior of certain target molecules, metabolic processes, and receptor distribution in the living human brain, especially when compared to former imaging methods like single-photon emission computed tomography (SPECT) [147]. In addition, multimodal imaging technologies with simultaneous PET/MR are considered a highly promising advancement in the field of MDD research that links molecular alterations to structural and functional changes of the brain in health and depression [148–150].

Hence, the following section provides an overview of the basic principles of PET imaging, frequently used radioligands in the context of MDD research, and PET studies performed to quantify and visualize target proteins as well as physiological processes that are involved in serotonergic neurotransmission and point toward pathological or at least predisposing changes among depressed individuals on a molecular level.

4.1   Basic Principles of PET Imaging

Positron emission tomography is a nuclear, noninvasive in vivo imaging technique used for the multidimensional visualization and quantification of molecular structures of interest and associated biochemical changes in health and disease. It is based on the (mostly intravenous) application of radioligands, i.e., radiolabeled molecules consisting of a substrate designed to bind to a specific molecule and a radioactive isotope. Following the administration, these radioligands circulate through the body and accumulate in tissue by binding to a molecular target or are metabolized (e.g., oxygen or fluorinated glucose). The decay of the isotope results in the emission of positrons, so-called antiparticles, with a positive charge and kinetic energy that collide with electrons after traveling a short distance. The unification of the positron with an electron is called annihilation and results in the emission of two photons with 511 keV energy each. These two photons are emitted at an angle of approximately 180° and subsequently are simultaneously detected outside of the body by the γ-camera of the PET scanner. The γ-camera consists of scintillation crystals that surround the body (or the head, respectively) and detect the γ-rays. If a simultaneous, pairwise detection (coincidence detection) of two photons occurs, it is assumed that the annihilation process must have happened along the intersection line between the pair of detectors. These lines are called lines of response (LORs) [151]. The detection of the coincidence photons is the first step of creating a PET image. The crystals convert the γ-ray energy into optical energy which is then transformed into an electrical signal by photomultiplier tubes [152]. The final PET image is based on the reconstruction of this electrical signal by certain mathematical operations in a process called tomographic image reconstruction. In the past, PET images used to be reconstructed using filtered back projection (FBP). Nowadays, faster and more robust ordered subset expectation maximization (OSEM) is used. Due to the use of trace amounts of radioactive substances, the radiation exposure of PET is usually low. Moreover, the substances used have a very short half-life.

4.2 Central Characteristics of Radiopharmaceuticals

A radioligand consists of a radioactive isotope and a substrate. Since radioactive isotopes have an unstable atomic nucleus and exhibit a relatively short half-life, the energy released while they decay can be used for nuclear medicine procedures as described in the past section. Most frequently, the following positron-emitting isotopes are combined with a wide range of different substrates to create a radiotracer that targets specific molecules: ^{18}F with a half-life of approx. 110 min, ^{11}C with a half-life of approx. 20 min, ^{13}N with a half-life of approx. 10 min, and ^{15}O with a half-life of approx. 2 min. Due to their limited stability and short half-life period, most of the tracers are directly produced onsite in a specialized PET center using a cyclotron. Furthermore, the selectivity and specificity of substrate for the molecule of interest remain pivotal for accurate usability in PET imaging [153].

As summarized by Pike, a suitable radiotracer for the molecular imaging of neurotransmitter systems needs to fulfill the following criteria: high affinity for and selectivity to the target molecule, sufficient ability to cross the BBB, lack of critical radiometabolites, low nonspecific binding, suitable brain pharmacokinetics concerning the half-life period, and safety regarding the administration in humans, i.e., the radioligand must be nontoxic and should not interfere with the biological system [154]. This is achieved by high radiochemical yield and specific activity which should lead to an expected occupancy of target proteins below 1–5%. Furthermore, radiotracers should not be substrates for brain efflux transporters, e.g., P-gp [155].

Besides the criteria for good tracer features, some other conditions must be guaranteed for successful PET imaging, like sufficient density of the target structure in cerebral regions of interest (ROIs) or production quality of the radioligand including adequate specific activity. Some of the major aspects that need to be taken into account are specified below.

4.2.1 Selectivity

Radiotracer selectivity is particularly important in the case of molecular targets that are widely or even ubiquitously distributed across multiple brain regions. In terms of serotonergic neurotransmission, this is true for 5-HT_{1A} [156], 5-HT_{1B} [157], or 5-HT_{2A} [158] receptors, for instance. However, it has to be considered that certain regions exhibit a relatively high density of one specific receptor subtype or target molecule. In such areas, radioligands with lower specificity can be applied for acquiring valuable binding data, provided that distribution characteristics are taken into consideration beforehand.

4.2.2 Affinity and Binding Potential

In the context of PET imaging, affinity is the strength of the binding interaction between the target molecule and the radioligand. Binding affinity is influenced by various non-covalent intermolecular interactions comprising hydrogen bonding, van der Waals forces, and ionic bonds [159]. In addition, the affinity between a ligand and its target molecule might be affected by the presence of other molecules nearby. The affinity of PET ligands usually ranges between 0.01 and 1 nM [160] and is used to calculate in vivo receptor occupancy. As summarized by Zhang and Fox, the term binding potential (BP) is defined as the product of radioligand affinity to the target receptor and receptor density (B_{max}) [161]. B_{max} defines the maximally available amount of target receptors (i.e., receptors that are not occupied by endogenous ligands). Due to the inverse relationship of binding affinity and radiotracer dissociation constant (K_D) at equilibrium, BP is calculated as the ratio of B_{max} and K_D [161, 162]. In terms of binding affinity, it should be considered that higher lipophilicity is associated with higher affinity and consequently increased nonspecific binding. Furthermore, affinity should not exceed certain predefined ranges in order to ensure reversible receptor kinetics [160, 163].

4.2.3 Lipophilicity, Passive Diffusion, and the BBB

The lipophilicity of radioligands is essential to predict the delivery of the radiotracer to the organ of interest. In the case of molecular brain imaging, moderate lipophilicity is needed in order to facilitate BBB penetration. However, BBB penetration is not only defined by the lipophilicity of the compound—since most radiotracers cross the BBB by passive diffusion, low molecular weight (<500 Da), small cross-sectional area ($<80\mathrm{\mathring{A}}^2$), low hydrogen binding capacity, and lack of formal charge are required [154, 164–167].

| 4.2.4 Specific Activity | The specific activity (SA) of a radiotracer is defined as the ratio between radioactivity and the mass of a compound [168]. SA is typically defined as mega Becquerel per microgram (MBq/μg). A sufficient SA is needed to reach adequate levels of activity without administrating too much mass. |

4.3 Quantitative Analysis of PET

PET-based technologies provide an excellent method for the quantification of receptor densities and estimation of immediate drug effects. As outlined above, the BP of target molecules (e.g., 5-HT receptors) is defined as the ratio of maximally available receptor density (B_{max}) and the radiotracer equilibrium dissociation constant (K_D) [169]. Therefore, receptor binding or binding potential (BP) will be used as terms to define receptor density in the following sections. In the context of drug challenges, the occupancy of a certain target structure by the administrated pharmacologic agent is meant to be estimated, usually by two separate PET scans performed first at baseline and then after drug challenge. Alternatively, a bolus plus constant infusion protocol can be applied [170].

4.4 Imaging Serotonin Synthesis, Transport, and Metabolization: Animal Studies

Commonly, PET tracers are firstly applied in animals before being used in human studies. A broad range of tracers has been developed so far to investigate different aspects of serotonergic metabolism and binding.

The use of labeled tryptophan as PET tracer provides insights into serotonin synthesis [171]. [^{11}C]-5-hydroxy-L-tryptophan ([^{11}C]-HTP) [172] and alpha[^{11}C]methyl-L-tryptophan ([^{11}C]-AMT) [173], radiolabeled analogs of tryptophan, were first developed and will be described in detail below. Both showed promising kinetics in animal and human applications [174, 175]. However, a study in rhesus monkeys showed a more uniform distribution of [^{11}C]AMT across the brain [176]. Moreover, [^{11}C]-HTP was proven to lack applicability for pharmacodynamic studies in rodents [177]. Soon after PET tracers for serotonin synthesis were reported, Bergstrom et al. published their development of [^{11}C] harmine, a tracer binding the enzyme monoamine oxidase A [178]. Moreover, after synthesis and degradation were successfully quantified via PET, imaging of the SERT by different tracers followed. [^{11}C]DASB or radiolabeled 3-amino-4-(2-dimethylamino-methylphenylsulfanyl)-benzonitrile was first evaluated across species in the early 2000s [179–181]. While [^{11}C]DASB is most commonly used, the tracers [^{11}C]MADAM and [^{18}F]ADAM found increasing attention over the last years [182, 183]. However, due to its high specificity, low nonspecific binding, and favorable brain kinetics, [^{11}C]DASB along with kinetic modeling is still considered the state of the art in human SERT quantification.

**4.5 Imaging
Serotonin Synthesis,
Transport, and
Metabolization in
Health and Depression:
Human Studies**

Since the primary aim of this chapter is to provide information on serotonin receptor functioning, the next paragraphs are limited to a brief insight into serotonin metabolism and its visualization by PET imaging.

*4.5.1 Serotonin
Synthesis*

As described earlier in this chapter, 5-HT is synthesized following a multistage process that starts with the conversion of L-tryptophan (Trp) to 5-HTP by L-tryptophan hydroxylase (TPH_I, TPH_{II}) [184], followed by the conversion of 5-HTP to 5-HT by aromatic amino acid decarboxylase (AADC). Since none of the mentioned enzymes is saturated at regular Trp concentrations in humans, the amount of Trp is the rate-limiting step in the process. Therefore, its quantification is used for the in vivo estimation of 5-HT synthesis measured by PET [171]. For this purpose, two radioligands could be widely established so far as mentioned in the last section: α-[^{11}C] methyltryptophan ([^{11}C]AMT) and 5-hydroxy-L-[β-11C]trypto-phan ([^{11}C]5-HTP). ([^{11}C]AMT is a radiolabeled analog of Trp. It is a substrate of TPH and can be converted to α-methylserotonin. In addition, it cannot be degraded by MAO [185]. Notwithstand-ing, limitations of [^{11}C]AMT include alternative synthesis pathways [186] and disadvantageous tracer kinetics [187]. The second radio-tracer in use, namely, ([^{11}C]5-HTP, is a radiolabeled form of 5-HTP that offers a crucial benefit: The metabolization of this radioligand is limited to the 5-HT pathway and 5-HTP is the last substrate needed for the synthesis of 5-HT [171]. However, it is difficult to produce due to complications associated with labeling 5-HTP with ^{11}C [175, 188].

In healthy humans, Chugani and colleagues reported signifi-cant regional differences of 5-HT synthesis rates throughout the brain using [^{11}C]AMT [189]. They found elevated 5-HT synthesis capacity in the thalamus, caudate nucleus, putamen, and hippocam-pus. Regarding cortical structures, the rectal gyrus of the inferior frontal lobe and the transverse temporal gyrus were found to show relatively high levels of 5-HT synthesis capacity values. In addition, they found significant gender differences with women presenting 10–20% higher rates across the brain.

Applying PET imaging with [^{11}C]AMT for the quantification of 5-HT synthesis, lower rates could be associated with neuro-chemical disturbances that potentially contribute to the develop-ment of major depressive disorder [190]. Agren and colleagues found significantly lower uptake of [^{11}C]5-HTP across the BBB in depressed patients, also pointing toward a correlation of Trp availability and depressive symptomatology [191]. Moreover, Leyton et al. found that low 5-HT synthesis rates in the prefrontal cortex might lower the threshold for suicidal behavior [192]. This could be shown by the use of [^{11}C]AMT for the measurement of regional 5-HT synthesis in the brain. Additionally, the influence of pharmacologic interventions on 5-HT synthesis could be visualized in several studies, e.g., by Berney et al. [193]. Interestingly, Neumeister and colleagues found that Trp depletion caused a transient return of depressive symptoms in subjects with remitted major depression by evaluating clinical ratings and measuring regional cerebral glucose utilization (here, [^{18}F]FDG was used) as well as tryptophan plasma concentrations [194], supporting the hypothesis that lower tryptophan levels in the brain indirectly promote the emergence of depressive symptoms by reducing serotonin synthesis rates.

4.5.2   SERT

The SERT belongs to the monoamine reuptake transporter proteins and is involved in serotonergic homeostasis by binding 5-HT in the synaptic cleft and transporting it back to the presynapse.

For the in vivo imaging of the SERT, Szabo et al. were the first to report the successful in vivo use of [^{11}C](+)McN5652, the (+) enantiomer of trans-1,2,3,5,6,10β-hexahydro-6-[4-(methylthio)-phenyl]pyrrolo-[2,1-a]-isoquinoline [195]. However, due to its high nonspecific binding, other radioligands, namely, the group of diarylsulfides, were developed. [^{11}C]MADAM (N,N-dimethyl-2-(2-amino-4-methylphenylthio)benzylamine) and [^{11}C]DASB (N,N-dimethyl-2-2-amino-4-cyanophenylthiobenzylamine) have been shown to exhibit more favorable binding characteristics with high selectivity for the SERT and lower nonspecific binding [196, 197]. [^{18}F]ADAM (N,N-dimethyl-2-(2-amino-4-(18)F-fluorophenylthio)benzylamine), on the other hand, could be established as a reasonable alternative due to its longer half-life, especially in the absence of a cyclotron. However, [^{11}C]DASB is considered the state of the art for in vivo quantification of the SERT in the human brain [198]. Recently, a novel [^{11}C]DASB bolus plus constant infusion protocol was established that allows for valid estimation of SERT binding in as little as 10–15 min of scanning [199, 200].

Among healthy individuals, high SERT expression could be found in the brainstem, basal ganglia, midbrain, and medial temporal regions [78]. As the SERT is known to be highly involved in the pathophysiology of depressive disorders and represents one of the pharmacologically most important target structures in the therapy of MDD, several in vivo imaging studies were performed to quantify and visualize SERT distribution across the brain of depressed subjects [201]. To date, most studies reported lower SERT levels associated with depression, especially in the amygdala and the midbrain, as summarized in a meta-analysis performed by Gryglewski et al. [202]. Gryglewski and colleagues analyzed imaging data (SPECT and PET) of 364 patients suffering from clinical depression. Interestingly, SERT occupancy levels of approximately 80% could be shown by Meyer and colleagues after the intake of therapeutic SSRI doses over 4 weeks, indicating that a certain plasma concentration is needed for sufficient treatment efficacy [203].

4.5.3 MAO-A

As explained in Subheading 2, the degradation of monoamines is ensured by the monoamine oxidase (MAO), present in two isoforms, namely, MAO-A and MAO-B. The degradation of 5-HT is mainly driven by MAO-A. Hence, the quantification of MAO-A as performed via PET imaging remains essential for investigating serotonin-related metabolic processes and for gaining insights into the mechanism of action of certain antidepressants (e.g., MAO inhibitors, MAOI). As reviewed by Kersemans and colleagues [204], many radiopharmaceutical compounds have been developed for this purpose so far. Among them, the highly selective [^{11}C]harmine (7-[^{11}C]Methoxy-1-methyl-9H-[3,4-b]indole)—a competitive and reversible MAO-A inhibitor—is seen as the state-of-the-art tracer for MAO-A imaging, first described by Kim et al. [205]. Moreover, other radiolabeled substances such as [^{11}C]clorgyline [206] or [^{11}C]befloxatone [207] show favorable characteristics for the visualization of pharmacologic inhibition at the MAO-A.

In support of the monoamine hypothesis, Meyer and coworkers reported elevated MAO-A levels in 17 patients suffering from major depression using [^{11}C]harmine PET [208]. In addition, elevated MAO-A binding could be detected in highly vulnerable patient populations such as perimenopausal women [209] or remitted patients treated with selective serotonin reuptake inhibitors (SSRIs) [208]. Moreover, MAO-A is implicated in non-pharmacological treatment approaches for depression. While lower MAO-A binding was recently observed following bright light therapy in seasonal affective disorder (SAD), the effect of electroconvulsive therapy on cerebral MAO-A binding is still under debate [210, 211].

4.6 Imaging 5-HT Receptors: Animal Studies

A majority of 5-HT receptors have been measured and quantified by autoradiographic or PET imaging so far. However, before 5-HT receptor distributions could be described in humans, they were extensively investigated in different animal species.

4.6.1 5-HT$_{1A}$

Several ligands for the 5-HT$_{1A}$ receptor have been published. The first discovered 5-HT$_{1A}$ tracer 8-OH-DPAT is still a widely used PET ligand [212]. Both the antagonists ([^{3}H]-WAY 100635 [213] and the agonists ([^{3}H]-ipsapirone, [^{125}I]-BH-8-MeO-N-PAT [214, 215]) followed. WAY100635 became the most commonly used tracer for the 5-HT$_{1A}$ receptor so far. A subset of new tracers was developed based on its structural skeleton, including $[^{18}F]$ FCWAY, $([^{18}F]$trans-4-fluoro-N-2-[4-(2-methoxyphenyl)piperazin-1-yl]ethyl]-N-(2 pyridyl)cyclohexanecarboxamide), 3-cis-$[^{18}F]$FCWAY [216], $[^{18}F]$MPPF, $[^{18}F]$4-(2-methoxyphenyl)-1-[2-(N-2-pyridinyl)-p-fluorobenzamido)-ethyl]piperazine [217], and $[^{18}F]$-Mefway, which were comparatively assessed across species [218, 219]. Animal studies revealed 5-HT_{1A} densities across nearly all serotonergic neurons and subsets of γ-aminobutyric acid (GABA)-ergic and glutamatergic cells found in the brainstem [220].

4.6.2   5-HT_{2A}

The 5-HT_{2A} receptor belongs, beside 5-HT_{1A}, to the most commonly investigated 5-HT receptor targets in preclinical as well as clinical research. In contrast to several other receptor subtypes, a broad range of radioligands is available for 5-HT_{2A} [221, 222]. $[^{11}C]$ketanserin was the first radiotracer for the quantification of the 5-HT_{2A} receptor [223]. However, it exhibited a low signal-to-noise ratio and suboptimal kinetics, leading to the development of numerous other PET tracers, beginning with the ^{18}F-derived antagonist $[^{18}F]$altanserin that has been extensively used across species [224, 225]. Modifications of $[^{18}F]$altanserin to increase brain uptake led to the deuterium substituted tracer $[^{18}F]$deuteroaltanserin [226] and $[^{18}F]$setoperone [227]. In addition, several carbon-derived tracers for the 5-HT_{2A} receptor were developed over the last decades. The best-described radioligands are $[^{11}C]$MDL100907 [228], $[^{11}C]$CIMBI-5, or the full 5-HT_{2A} agonist tracer $[^{11C}]$Cimbi-36 [229, 230]. A recent study showed better binding potentials of $[^{11}C]$MDL100907 compared to $[^{11}C]$ CIMBI-5 in nonhuman primates [231]. Moreover, $[^{11C}]$Cimbi-36 was shown to produce solid results in nonhuman primates making it a promising candidate for use in human studies [232]. Several radioligands were tested for the quantification of other 5-HT_2 receptors. While candidates as $[^{11}C]$WAY-163909 or $[^{11}C]$N-methylated arylazepine disappointed with moderate specific binding, $[^{18}F]$fluorophenethoxy)pyrimidine yielded promising specific binding to the 5-HT_{2C} receptors in the rat brain [233].

Although psychiatric research mainly focuses on the imaging of 5-HT_{1A} and 5-HT_{2A} receptors, other 5-HT receptor targets should not be left unmentioned. Rat studies for the quantification of the 5-HT_3 receptor used $[^3H]$-zacopride, $[^{125}I]$-zacopride [234], $[^3H]$-ICS205e930 [235], or (S)-$[^{18}F]$fesetron [236]. The agonist $[^3H]$-prucalopride and the antagonists $[^3H]$-GR113808 or $[^{125}I]$-SB201710 have been applied for the quantification of 5-HT_4 receptors across species [237–239]. Moreover, $[^{125}I]$-SB-258585 and $[^{11}C]$-GSK215083 were utilized to investigate 5-HT_6 receptor densities in marmosets, rats, and other species [240–242]. Finally, several radioligands were developed for the 5-HT_7 receptor including [^{11}C]Cimbi-806 [243], [^{11}C]Cimbi-717 [244], and [^{18}F]-2FP3 [245].

4.7 Imaging 5-HT Receptors in Health and Depression: Human Studies

The advances in PET imaging over the last decades have extended our knowledge on 5-HT receptor subtypes, their highly versatile functions in the brain, as well as their pharmacological targetability. As extensively discussed in Subheading 2, there are seven 5-HT receptor families further classified into subtypes by means of their neurochemical and neurophysiological characteristics. As reported in a multitude of studies, the distribution and functionality of 5-HT receptors throughout the brain are influenced by various factors comprising genetic (e.g., single nucleotide polymorphisms), epigenetic (e.g., stress-induced methylation), and posttranscriptional (e.g., alternative splicing) mechanisms [246], chronic malnutrition [247], inflammation, alcohol as well as aging [248, 249].

As outlined in the following part of this chapter, a large number of molecular imaging studies have been performed to elucidate the role of certain 5-HT receptor subtypes in the context of MDD pathophysiology and antidepressant therapy. Notwithstanding, it must be taken into consideration that these investigational efforts are limited by the availability of highly selective radiotracers for certain receptor subtypes. Despite their implication in MDD pathophysiology, it has yet to be answered whether the 5-HT receptors mentioned hereafter are key players in the emergence of depressive symptoms and the mechanism of action of antidepressants. Future research efforts need to focus on the development of new and the optimization of already existing radiotracers to allow for the detection and quantification of other 5-HT receptor subtypes in vivo that might play an equally or even more important role in MDD.

Thus, in the subsequent part of this chapter, well-investigated 5-HT receptor subtypes in the context of MDD that from today's point of view are thought to be highly involved in MDD pathophysiology will be characterized in detail, illustrating their distribution patterns throughout the human brain and their assumed pathophysiological implications in major depression as detected by means of PET imaging. In addition, typically used radiotracers and highly relevant literature on the effect of psychopharmacologic interventions measured with PET will be summarized. Table 2 provides an overview of all PET studies with regard to MDD that are reported in this section.

Since this chapter particularly focuses on 5-HT receptors and imaging findings on their role in the mechanism of action of antidepressants, we do not extensively report on the SERT or MAOs. Similarly, results on the potential effect of other psychopharmacologic agents (e.g., antipsychotics or mood stabilizers), electroconvulsive therapy, or psychotherapy on 5-HT receptors will be excluded as these studies would go beyond the constraints of this chapter.

Table 2

PET studies investigating serotonin receptor binding potential in major depressive disorder

Author	Investigated sample		Mean age (in years) ± SD		Radioligand	Molecular findings	Author's conclusion
	MDD	HC	MDD	HC			
5-HT receptors							
5-HT$_{1A}$							
Drevets et al., 1999	12 current MDE: – 8 MDD – 4 BD	8	35.8 ± 9.7	35.3 ± 13.5	[^{11}C]WAY-100635	– Reduced BP of 41.5% in the raphe and 26.8% in the mesiotemporal cortex in MDD compared to HC	Reduced 5-HT$_{1A}$ BP could be linked to histopathological changes within the raphe
Sargent et al., 2000	15 medication-free (1) 20 medicated (SSRI) (2)	18	(1) 37.7 ± 13.7 (2) 43.1 ± 14.8	36.4 ± 8.3	[^{11}C]WAY-100635	– Reduced mean BP of 10.8% in all investigated brain regions in unmedicated MDD compared to HC – Reduced mean BP of 11.6% in all investigated brain regions in medicated MDD compared to HC	Decreases of 5-HT$_{1A}$ BP in MDD were not altered by SSRI treatment
Meltzer et al., 2004	17 medication-free	17	71.4 ± 5.9	70.0 ± 6.7	[^{11}C]WAY-100635	– Reduced BP in dorsal raphe nuclei in MDD compared to HC	The findings support evidence of altered 5-HT$_{1A}$ receptors in MDD of elderly persons
Parsey et al., 2006a	9 remitters (1) 13 nonremitters (2) (both medication-free at baseline)	43	(1) 37.9 ± 10.4 (2) 42.9 ± 14.9	38.2 ± 15.0	[^{11}C]WAY-100635	– After 1 year of medication, nonremitters reveal higher cortical BP before treatment compared to remitters	Higher 5-HT$_{1A}$ pretreatment BP might be linked to non-remission of AD therapy in MDD
– Parsey et al., 2006b	13 AD-naïve (1) 15 AD-exposed;	43	(1) 35.8 ± 12.2 (2) 40.8 ± 13.9	38.2 ± 15.0	[^{11}C]WAY-100635	– No difference in BP between MDD compared to HC	MDD might be characterized by higher

(continued)

Table 2
(continued)

Author	Investigated sample		Mean age (in years) ± SD		Radioligand	Molecular findings	Author's conclusion
	MDD	**HC**	**MDD**	**HC**			
	medication-free 3 weeks prior to scan (2) (both during MDE)					– Increase of BP in AD-naïve MDD compared to AD-exposed MDD and HC, across all investigated regions	5-HT$_{1A}$ BP and might be long-term affected by AD exposure
Hirvonen et al., 2008	21 drug naïve	15	40.1 ± 9.0	32.6 ± 7.7	[^{11}C]WAY-100635	– Reduced BP of 9% to 25% in most brain regions in MDD compared to HC	The 5-HT$_{1A}$ receptor could contribute to the pathophysiology of MDD and might be a marker of susceptibility to MDD
Miller et al., 2009	15 MDD (remission) (1) 13 AD-naïve (current MDE) (2)	51	(1) 31.8 ± 10.9 (2) 35.9 ± 12.3	37.4 ± 14.5	[^{11}C]WAY-100635	– Increased BP in remitted MDD compared to HC – No differences between remitted MDD and AD-naïve MDD	Increased BP in remitted MDD is seen as a trait-like alteration in MDD
Lothe et al., 2012	6	18	37 ± 8.3 (range 26–48)	n.s. (range 23–48)	[^{18}F]MPPF	– Reduced BP in MDD compared to HC in all investigated regions – Increase of BP in medial orbital regions in MDD, after 1 month of SSRI treatment	Serotonergic function might be normalized by SSRI treatment
Stenkrona et al., 2013	–	4	–	n.s. (range 20–45)	[^{11}C]WAY-100635	– No differences in BP between baseline and 9 days of treatment with vortioxetine	Occupancy of the 5-HT$_{1A}$ receptor was not affected by a clinically appropriate dose of

Study	Patient group (n)	Patient age	HC (n)	HC age	Tracer	Findings	Interpretation
							vortioxetine; further research is needed
Wang et al., 2016	Meta-analysis: 10 studies (in total 218 MDD and 261 HC)				[11C]WAY-100635	– Reduced BP in the mesiotemporal cortex – Moderate reduction of BP across the raphe nuclei, hippocampus, as well as the insular, anterior, cingulate, and occipital cortice. – No relationship between symptom severity and BP	Reduced 5-HT$_{1A}$ BP and therefore serotonergic dysfunction were linked to the pathophysiology of MDD
Pillai et al., 2018	16	40.2 ± 13.9	25	40.1 ± 14.5.	[11C]WAY-100635	– Increased BP in dorsal raphe nuclei in MDD compared to HC – No difference in raphe magnus and medial raphe nucleus between MDD and HC	The combined dorsal raphe nuclei and median raphe nuclei are seen as a more sensitive and specific biomarker compared to the entire raphe, when assessing BP of the 5-HT$_{1A}$ receptor
Milak et al., 2018	9 healthy high-risk individuals (1) 30 past medication (2) 18 remitted, past medication (3)	(1) 25.5 ± 3.4 (2) 40.6 ± 13.1 (3) 34.8 ± 12.7	51	37.4 ± 14.5	[11C]WAY-100635	– Increased BP of 84.3% in midbrain raphe and 40.8% in the hippocampus in healthy high-risk individuals compared to HC	Increased BP is seen as a trait abnormality in healthy high-risk individuals and therefore seems to be familial transmitted
Langenecker et al., 2019	25 medication-free	37.8 ± 11.8	29	39.7 ± 11.0	[11C]WAY-100635	– Reduced BP of 10–20% in MDD compared to HC in predominantly temporal regions	The findings reveal evidence for measurable biomarkers in MDD
Mann et al., 2019	8 suicide attempters (>4 years) (1) 8 lifetime	(1) 39.5 ± 11.0 (2) 40.4 ± 7.2	8	40.5 ± 9.8.	[11C]WAY-100635	– No differences in BP between HC and MDD groups	No trait-like binding was detected, which may be attributed to the high

(continued)

Table 2
(continued)

Author	Investigated sample		Mean age (in years) ± SD		Radioligand	Molecular findings	Author's conclusion
	MDD	HC	MDD	HC			
	non-suicide attempters (2)					– No differences in BP between MDD suicide attempters and non-attempters	heterogeneity of nonlethal suicide attempts
Metts et al., 2019	66 medication-free (at least 2 weeks prior to scan) (1) 32 AD-naïve (2)	–	(1) 39 ± 12.6 (2) 36.4 ± 13	–	[^{11}C]WAY-100635	– No differences in BP between AD-naïve and AD-exposed MDD in cortical and subcortical brain regions	AD-dependent alterations of BP are reversed in a period of 2 weeks, when AD therapy is stopped
5-HT$_{1B}$							
Murrough et al., 2011	10 (current MDE)	10	30.8± 9.5.	30.7 ± 10.5	[^{11}C]P943	– Reduced BP of 18.7% in bilateral ventral striatum/ ventral pallidum in MDD compared to HC – Reduced BP of 16.1% in the left hemisphere and 21.1% in the right hemisphere of ventral striatum/ventral pallidum in MDD compared to HC	The findings suggest reduced action of the ventral striatum/ ventral pallidum 5-HT$_{1B}$ receptors in MDD
Nord et al., 2013	–	9	–	25 ± 4.3	[^{11}C]AZ10419369	– Increased BP after escitalopram administration in a combined region representing all projection areas, as well as in the occipital and temporal cortex	Serotonin concentrations were found to be reduced after one dose of escitalopram in serotonergic projection areas;

				– Trend to reduced BP after escitalopram administration in the raphe nuclei	these findings might support understanding the late clinical effect of SSRIs
Tiger et al., 2016	10 medication-free (current moderate MDE)	10	48 (range 24–68)	[^{11}C]AZ10419369	48 (range 25–70)	– Reduced BP of 20% in the anterior cingulate cortex, 17% in the subgenual prefrontal cortex, and 32% in the hippocampus in recurrent MDD compared to HC	The findings suggest an involvement of the anterior cingulate cortex in the pathophysiology of recurrent MDD
Tiger et al., 2020	– 30 (SSRI-resistant MDD) – 10 placebo (1) – 20 ketamine (2)	–	(1) 37.1 (2) 39.2	–	[^{11}C]AZ10419369	– Increased BP of 16.7% in the hippocampus after one ketamine infusion	The 5-HT$_{1B}$ receptor might be involved in the antidepressant effect of ketamine
5-HT$_{2A}$					
Biver et al., 1997	8 medication-free	22	48.1 ± 9.7	38.3 ± 12.0	[^{18}F]altanserin	– Reduced BP in the right posterolateral orbitofrontal cortex (Brodmann's area 47) and right anterior insular cortex in MDD compared to HC	The 5-HT$_2$ receptor might be involved in the pathophysiology of MDD in brain regions responsible for mood regulation
Meltzer et al., 1999	11 current MDE 9 with Alzheimer's disease (3 concurrent depression)	(1) 10 (2)	(1) 65 ± 5.5 (2) 69.7 ± 5.0	69.8 ± 5.0	[^{18}F]altanserin	– No differences in BP in late-life MDE compared to HC – Reduced binding in the Alzheimer's group compared to HC and MDE in various brain regions	The 5-HT$_{2A}$ receptor may be altered in a different way in late-life depression than in Alzheimer's disease
Meyer et al., 1999	14 medication-free	19	32.3 ± 6.4	31.8 ± 6.9	[^{18}F]setoperone	– No differences in MDD compared to HC	Even though no difference was found, the 5-HT$_2$ receptor could contribute to

(continued)

Table 2
(continued)

Author	Investigated sample		Mean age (in years) ± SD		Radioligand	Molecular findings	Author's conclusion
	MDD	HC	MDD	HC			
							treatment effects or might be altered in highly suicidal conditions
Attar-Lévy et al., 1999	7	7	40.0 ± 11	38 ± 10	[18F]setoperone	– Moderate reduced BP in the frontal cortex in MDD compared to HC – Reduced BP after clomipramine application in cortical regions	Clomipramine, which belongs to the group of tricyclic antidepressants, reveals a significant impact on cortical 5-HT$_2$ receptors
Yatham et al., 2000	20 medication-free	20	40.1 ± 9.5	37.2 ± 12.6	[18F]setoperone	– Reduced BP mainly in cortical regions (frontal, temporal, parietal, and occipital) in MDD compared to HC – Mean reduction of 21.7% in MDD compared to HC	5-HT$_2$ receptors might be diminished in MDD patients
Meyer et al., 2003	22 current MDE	29	31 ± 6	27 ± 5	[18F]setoperone	– Increased BP in Brodmann's area 9 (middle frontal gyrus) in MDD with extremely dysfunctional attitudes compared to HC	Low 5-HT agonism levels could causally contribute to the dysfunctional attitudes in MDD
Mintun et al., 2004	46 AD-free	29	49.6 ± 15.6	45.8 ± 15.3	[18F]altanserin	– Reduced BP of 29% in the hippocampus in MDD compared to HC	The altered serotonergic action in the hippocampus is expected to contribute

Author, year	Study population	Mean age (patients)	Mean age (HC)	Radiotracer	Study outcome	Author's conclusion
Bhagwagar et al., 2006	20 medication-free (recovered)	38.6 ± 12.1	42.6 ± 13.5	[^{11}C] MDL100,907	– Increased BP in the frontal (19%), parietal (25%), and occipital cortices (19%) in recovered MDD compared to HC	Subjects with MDD in the past might reveal increased 5-HT$_{2A}$ receptor BP ... to mood changes in MDD
Mann et al., 2019	8 suicide attempters (>4 years) (1) 8 lifetime non-suicide attempters (2)	(1) 39.5 ± 11.0 (2) 40.4 ± 7.2	40.5 ± 9.8	[^{18}F]altanserin	– No differences in MDD compared to HC – No differences between MDD suicide attempters and non-attempters	No trait-like binding was detected, which may be attributed to the high heterogeneity of nonlethal suicide attempts
5-HT$_4$ and 5-HT$_7$						
Madsen et al., 2014.	26 healthy individuals with MDD relatives	31	32 (range 20–64)	42 (range 20–86)	[^{11}C]SB207145	– Reduced BP in individuals with MDD relatives in the striatum compared to HC	Lower 5-HT$_4$ receptor BP in the striatum might be related to a higher risk of developing MDD and therefore may be part of the neurobiology underlying familial risk for depression

Table 2 shows an overview of studies using in vivo PET imaging in MDD patients or patients with depression-like symptoms. Outlined characteristics comprise data on author, publication year, study population (No. of MDD patients and HC), mean age, radiotracer specification, study outcome, and author's conclusion.
MDD: major depressive disorder; MDE: major depressive episode; BD: bipolar disorder; BP: binding potential; AD: antidepressant; HC: healthy controls; SSRI: selective serotonin reuptake inhibitor; 5-HT: serotonin.

The 5-HT$_{1A}$ receptor is the main inhibitory serotonergic receptor and the most extensively investigated subtype among all 5-HT receptors. It has been implicated in multiple neurophysiological processes such as neurocognitive functioning [250] or emotion regulation [251]. 5-HT$_{1A}$ receptors are widely distributed across the healthy brain with higher density in selected limbic areas (hippocampus CA1, subiculum, uncus) as well as the frontal, temporal, cingulate, parietal, and entorhinal cortices, whereas the basal ganglia and cerebellum were found to be significantly lower in 5-HT$_{1A}$ expression. This could be shown by Hall et al. applying an autoradiographic approach to quantify 5-HT$_{1A}$ receptors in the postmortem human brain [156]. In the midbrain raphe region, 5-HT$_{1A}$ receptors are predominantly located presynaptically on serotonergic neurons and mediate auto-inhibitory effects. In contrast, in subcortical and cortical projection areas, 5-HT$_{1A}$ heteroreceptors are located postsynaptically on non-serotonergic neurons [63, 252, 253].

In humans, in vivo 5-HT$_{1A}$ receptor distribution quantification via PET is largely performed with the 5-HT$_{1A}$ antagonist [*carbonyl*-^{11}C]WAY100635. In the 1990s, WAY100635 has been labeled in the carbonyl-^{11}C position to prevent the formation of radioactive metabolites in the brain, as it was the case while using [O-methyl-^{11}C] WAY100365 in earlier studies [254]. Due to the fast systemic metabolism of [*carbonyl*-^{11}C]WAY100635, [^{18}F]-labeled compounds were developed to optimize the metabolic profile of WAY100635, namely, [^{18}F]FCWAY ([^{18}F]trans-4-fluoro- *N*-2-[4-(2-methoxyphenyl)piperazin-1-yl]ethyl]- *N*-(2-pyridyl)cyclohexanecarboxamide) and 3-cis-[^{18}F]FCWAY. Despite their high affinity and satisfactory hippocampal to cerebellar binding ratio, 5-HT$_{1A}$ quantification in superficial brain regions was limited as a result of high bone uptake of radioactivity and defluorination of the ligand [216]. Additionally, another reversible 5-HT$_{1A}$ antagonist called [^{18}F]MPPF ([^{18}F]4-(2-methoxyphenyl)-1-[2-(*N*-2-pyridinyl)-*p*-fluorobenzamido)-ethyl]piperazine) was developed based on the structural skeleton of WAY100635, showing low nonspecific binding and simple producibility [217]. In terms of 5-HT$_{1A}$ imaging, it must be taken into account that 5-HT$_{1A}$ receptors exist either in the high- or the low-affinity state—this refers to their coupling to G-proteins: In the high-affinity state, the receptors are coupled to G-proteins; in the low-affinity state they are not [255]. This is relevant because antagonistic tracers like the aforementioned ones show similar binding affinity to 5-HT$_{1A}$ receptors at both states, while agonistic compounds (most of them based on 8-hydroxy-2-(di-*n*-propylamino) tetralin or 8-OH-DPAT) like [^{11}C]CUMI-101 ([O-methyl-^{11}C]2-(4-(4-(2-methoxyphenyl)-piperazin-1-yl)butyl)-4-methyl-1,2,4-triazine-3,5(2H,4H)dione) are known to bind 5-HT$_{1A}$ receptors with a higher affinity at the uncoupled conformation. These chemical characteristics have furthered investigations with the aim to estimate extracellular serotonin concentration and receptor occupancy of psychopharmacologic agents; however, in vivo studies in humans with $[^{11}C]$CUMI-101 yielded conflicting results [256–258]. Hence, reliable agonistic tracers to measure changes of extra-synaptic 5-HT concentration and to visualize receptor modulation by psychopharmacologic agents are still missing [252].

In the context of MDD, 5-HT$_{1A}$ receptor availability and binding affinity were shown to be reduced in multiple brain regions when compared to healthy controls; however, imaging data resulted to be somewhat conflicting with regard to the raphe nuclei. Mann et al. found no significant differences in 5-HT$_{1A}$ (and 5-HT$_{2A}$) receptor binding between healthy volunteers and depressed groups nor between healthy controls, depressed suicide attempters, and non-attempters using $[^{11}C]$WAY100635 [259]. Langenecker and colleagues reported alterations of 5-HT$_{1A}$ receptor binding when comparing depressed individuals and healthy subjects in a multimodal assessment involving PET imaging. Lowered 5HT$_{1A}$ binding was present in approximately half of the MDD group relative to the control group, especially in temporal regions [260]. Interestingly, Milak et al. investigated whether higher 5-HT$_{1A}$ binding was a biologic trait potentially transmitted to healthy high-risk offspring of MDD patients via PET imaging using $[^{11}C]$WAY100635. Across high-risk individuals, mean receptor binding potential resulted to be higher in all predefined ROIs (especially in the midbrain and hippocampus) when compared to healthy subjects [261]. Similarly, Pillai et al. found that 5-HT$_{1A}$ binding was increased in dorsal raphe nuclei across MDD patients; however, they found that differences across subtypes of the raphe nuclei must be taken into account, since the raphe magnus as well as the median raphe nuclei did not show significant differences in 5-HT$_{1A}$ binding in comparison to healthy controls [262]. In contrast, Meltzer and coworkers detected reduced binding potential in the dorsal raphe nuclei of elderly patients suffering from late-life depression [263]. Additionally, Hirvonen et al. detected reduced 5-HT$_{1A}$ receptor binding in drug-naïve-depressed individuals across multiple brain regions [264]. These results are in line with some earlier PET studies pointing toward reduced 5-HT$_{1A}$ binding across multiple regions of the depressed brain, e.g., the raphe nuclei or limbic regions [265, 266].

Finally, Wang et al. performed a meta-analysis comprising 10 studies with a total number of 218 patients with depression and 261 healthy controls and reported significantly decreased 5-HT$_{1A}$ density in the mesiotemporal cortex and smaller reductions in 5-HT$_{1A}$ binding across the raphe nuclei, hippocampus, as well as the insular, anterior, cingulate, and occipital cortices.

Although a significant association between symptom severity and 5-HT_{1A} binding had been hypothesized, no relationship was found in patients with depression [267].

Metts et al. reported no differences in [^{11}C]WAY-100635 autoreceptor binding between antidepressant-naïve and antidepressant-exposed MDD groups at least 2 weeks after discontinuation in 13 a priori defined cortical and subcortical ROIs, possibly indicating that 5-HT_{1A} downregulation reverses within 2 weeks after medication discontinuation [268]. In a PET study using [^{11}C]WAY-100635, Miller et al. could show that increased 5-HT_{1A} binding in the raphe nuclei is associated with subsequent remission in MDD patients after escitalopram treatment over 8 weeks [269]. Interestingly and partly in contrast to the latter findings, Parsey et al. demonstrated that compared to remitters, receptor binding was elevated among nonremitters after 1 year of antidepressant therapy [270]. Moreover, Parsey et al. hypothesized that antidepressant exposure might have long-term effects on 5-HT_{1A} receptor binding and found higher binding potential among antidepressant-naïve MDD patients when compared to patients after antidepressant exposure as well as to healthy controls [271]. Additionally, Lothe et al. reported dynamic changes in [^{18}F] MPPF binding potential, primarily in medial orbital regions, suggesting adaptive serotonergic mechanisms after 30 days of SSRI treatment [272].

In a longitudinal, prospective vortioxetine (Lu AA21004) study with 11 subjects examined with PET receiving either [^{11}C] MADAM (9 subjects) or [^{11}C]WAY100635 (4 subjects) at baseline, after a single dose and after 9 days of administering Lu AA21004 (up to 60 mg) for quantification of 5-HT_{1A} occupancy, Stenkrona and coworkers found no significant influence of vortioxetine treatment on 5-HT_{1A} occupancy [273].

4.7.2   5-$HT_{1B/D}$

5-HT_{1B} and 5-HT_{1D} receptors were initially considered to be species homologs of the same receptor [274]. Later on, it could be shown that these homologs were encoded by separate genes [59]. Since then, selective 5-HT_{1B} ligands could be developed for quantitative receptor imaging, e.g., AZ10419369 (5-methyl-8-(4-[^{11}C]methyl-piperazin-1-yl)-4-oxo-4H-chromene-2-carboxylic acid(4-morpholin-4-yl-phenyl)-amide). As reported by Varnas et al., [^{11}C]AZ10419369 binding was highest in the pallidum, ventral striatum, and occipital cortex. Minor binding was observed in the dorsal striatum as well as the prefrontal and temporal cortices. The lowest density was found in the thalamus and cerebellum [157]. These findings were largely in line with an autoradiographic study performed by the same group in 2001 [275]. Additionally, a high-affinity antagonist of the 5-HT_{1B} receptor called [^{11}C]P943

(R-1-[4-(2-methoxy-isopropyl)-phenyl]-3-[2-(4-methyl-pipera-zin-1-yl)benzyl]-pyrrolidin-2-one) could be established [276].

Similar to 5-HT_{1A}, 5-HT_{1B} receptors are known to either inhibit 5-HT release and modulate extracellular transmitter levels in serotonergic projection regions as autoreceptors or regulate the release of other neurotransmitters like dopamine, glutamate, acetylcholine, or GABA in their role as heteroreceptors [63]. These characteristics were reviewed in detail by Tiger and colleagues [277].

In MDD patients, 5-HT_{1B} receptor binding of $[^{11}C]P943$ was shown to be reduced in regions implicated in reward processing (ventral striatum/pallidum) [278]. Furthermore, $[^{11}C]$ AZ10419369 binding was significantly lower among drug-free MDD subjects in the anterior cingulate cortex, in the subgenual prefrontal cortex, and in selected limbic regions (e.g., the hippocampus) [279]. Additionally, $[^{11}C]AZ10419369$ was used to investigate the effects of a single dose of escitalopram on 5-HT receptors by Nord and coworkers. They found a significant increase in receptor binding along the occipital and temporal cortices and a trend toward a decrease in the raphe nuclei [280]. Moreover, Tiger et al. found that after ketamine treatment, alleviation of depressive symptoms in subjects suffering from SSRI-resistant MDD was negatively correlated with baseline 5-HT_{1B} receptor binding in the ventral striatum [135]. Although ketamine does not primarily target the serotonergic system, this finding should be taken into account since the substance is currently gaining importance in the context of therapy-resistant depression. Ketamine is an anesthetic known for its antisuicidal, antidepressant, and psychotomimetic effects at low doses [281]. It exerts inhibitory effects on cerebral NMDA receptors, modulates GABA, and interacts with the brain's opioid system [282, 283].

4.7.3 5-HT$_{2A}$

The 5-HT_{2A} receptor is known as the main excitatory receptor of the serotonin system. As outlined in Subheading 2, it is highly involved in the working mechanism of certain hallucinogenic substances (e.g., LSD) and targeted by antagonistic antidepressants as well as antipsychotic agents [284, 285]. Savli et al. reported relatively homogenous 5-HT_{2A} binding throughout the cortex with the highest values in the middle frontal, middle temporal, and angular gyri as well as the calcarine fissure, whereas the lowest binding was found in subcortical regions [286]. For the 5-HT_{2A} receptor, $[^{11}C]$ketanserin was the first 5-HT_{2A} tracer that was used for in vivo imaging in humans; however, due to unfavorable kinetic properties and low signal-to-noise ratio (as described in the context of animal studies earlier in this chapter), it was replaced by derivatives such as $[^{18}F]$altanserin [287] and $[^{18}F]$setoperone [227]; however, the interpretation of data was limited due to the formation of lipophilic metabolites and thus higher nonspecific binding in the case of $[^{18}F]$altanserin. Hence, a new bolus plus constant infusion approach had to be developed [288]. As a result of its significant affinity to dopamine D_2 receptors, the use of $[^{18}F]$ setoperone in terms of $5\text{-}HT_{2A}$ imaging remained limited to areas with low D_2 density. The first reversible antagonist with high selectivity for the $5\text{-}HT_{2A}$ was $[^{11}C]MDL100907$ ((R)-(þ)-a-(2,3-dimethoxyphenyl)-1-[2-(4-fluorophenyl)ethyl]-4-piperidine methanol), providing favorable imaging features such as high brain uptake, moderate lipophilicity, and satisfactory kinetic characteristics [228]. Later on, the ^{11}C-labeled 2-(4-bromo-2,5-dimethoxy-phenyl)-N-[(2-methoxyphenyl)methyl]ethanamine, namely, $[^{11}C]$ Cimbi-36, was developed in order to use the ability of agonistic radiotracers to estimate endogenous 5-HT release [289]; however, the compound was found to exert high affinity for $5\text{-}HT_{2C}$ receptors in selected brain regions [230].

The $5\text{-}HT_{2A}$ receptor is implicated in the pathophysiology of several neuropsychiatric conditions, e.g., schizophrenia, bipolar disorder, and major depression [290]. In the field of MDD research, PET imaging studies aiming to estimate alterations of the $5\text{-}HT_{2A}$ receptor largely provided conflicting results, with reduced [291–294], unchanged [259, 295, 296], or increased [297] receptor binding in the brain of MDD patients. However, methodological problems associated with the use of early radio-tracers as described above ($[^{11}C]$altanserin and $[^{18}F]$setoperone, respectively) must be taken into consideration. In contrast, studies performed with highly selective radiotracers such as $[^{11}C]$ MDL100907 are assumed to yield more reliable data. In a PET study investigating $5\text{-}HT_{2A}$ receptor binding potential in recovered MDD patients and the relationship between receptor binding and scores on the Dysfunctional Attitude Scale, Bhagwagar et al. found significantly higher $5\text{-}HT_{2A}$ receptor binding potential in the fron-tal, parietal, and occipital cortices across depressed individuals when compared to healthy controls. Additionally, receptor binding cor-related inversely with age in both subgroups and positively with higher levels of dysfunctional attitudes in the recovered MDD cohort [298]. The future use of $[^{11}C]$Cimbi-36 in MDD studies is seen as a promising approach to collect data on $5\text{-}HT_{2A}$ receptor binding alterations across depressed individuals. To date, the use of this highly specific full agonist in PET imaging is limited to trials for methodological purposes and pharmacologic challenges across healthy populations [299, 300].

4.7.4 5-HT$_{2C}$, 5-HT$_4$, and 5-HT$_7$

The $5\text{-}HT_{2C}$ receptor is largely limited to the central nervous system. With respect to psychopharmacologic therapy, various drugs are assumed to target the $5\text{-}HT_{2C}$ receptor exerting antago-nistic effects. In humans, the $5\text{-}HT_{2C}$ receptor is known to show higher densities across the cerebral cortex, substantia nigra, and cerebellum as shown by Abramowski via immunoblotting with antibodies [301]. However, selective radiotracers for in vivo human use have yet to be developed.

The 5-HT$_4$ receptor, albeit less extensively investigated in comparison to the aforementioned receptor subtypes, remains a promising target in the field of MDD research. Preclinical findings point toward its implication in affective disorders as well as learning and memory. Hence, 5-HT$_4$ receptor stimulation is believed to be a promising method to improve symptoms of cognitive impairment and depression. For example, Murphy et al. performed a clinical trial with prucalopride, a partial 5-HT$_4$ agonist exerting antidepressant and pro-cognitive effects in animal studies. They could show a positive effect on cognition; however, no clinical improvement of depressive symptoms could be achieved [302]. The 5-HT$_4$ receptor was shown to be densely distributed in the basal ganglia and the hippocampal formation as detected by autoradiographic analyses [303]. For imaging purposes, [^{11}C]SB207145 (8-amino-7-chloro-(N-[11C]methyl-4-piperidylmethyl)-1,4-benzodioxan-5-carboxylate) was developed and successfully used in human 5-HT receptor imaging [304]. Among healthy participants with a positive family history of major depression, the 5-HT$_4$ receptor was shown to exert reduced binding when measured with the radiotracer [^{11}C]SB207145 [305]. Additionally, Haahr et al. found a significant negative correlation between the 5-HT$_4$ receptor expression in the hippocampus and immediate recall of previously learned stimuli. Furthermore, they reported a significant negative correlation between the right hippocampal 5-HT$_4$ receptor expression and delayed recall [306].

Lastly, the 5-HT$_7$ receptor has been considered to be a promising target in the context of MDD. As known from autoradiographic approaches, the anterior thalamus and the dentate gyrus show the highest density in 5-HT$_7$ receptors, though relatively high levels were found in the brainstem nuclei including the substantia nigra, the ventral tegmental area, and the dorsal raphe nucleus [307]. In the absence of a selective radioligand for human in vivo use, no studies with in vivo PET imaging have been conducted so far. Notwithstanding prior difficulties, promising results from animal studies have been reported on, e.g., [^{11}C]Cimbi-7171 [308].

Figure 1 shows distribution patterns of the 5-HT receptor families 5-HT$_{1-7}$, the SERT, and MAO-A on the cortical surface of the human brain as predicted via mRNA expression analysis.

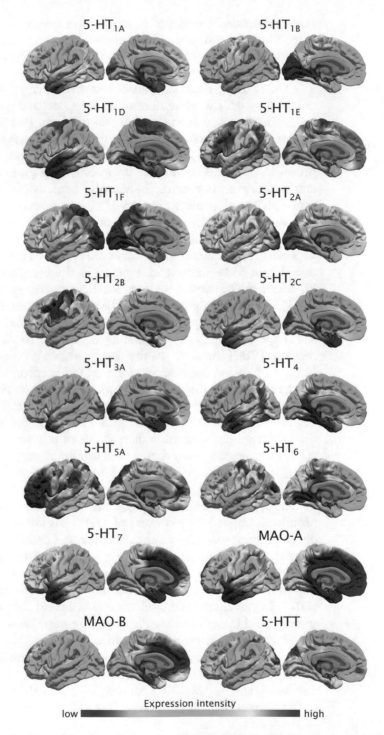

Fig. 1 Whole-brain transcriptomic patterns of genes associated with the serotonergic system. Namely, distribution patterns of the 14 serotonin receptor subtypes within the 5-HT receptor families 5-HT$_{1-7}$, the serotonin transporter (SERT), and the enzymes monoamine oxidase subtype A and B. The mRNA expression patterns are projected on the pial surface of the left hemisphere and shown from the lateral (left) and medial (right) sides. (Cortical mRNA expression images can be downloaded from our webpage https:/www.meduniwien.ac.at/neuroimaging/mRNA.html as it was performed for Fig. 1)

5   Conclusion

The human serotonin system is involved in multiple neurophysiologic processes and is implicated in the pathophysiology of several neuropsychiatric conditions. Especially in terms of major depression, the pharmacologic modulation of the SERT, 5-HT receptors, and MAO-A resulted to be a highly valuable and clinically useful treatment option. Over the course of the last decades, impressive advances in the field of in vivo molecular neuroimaging have furthered our knowledge on the mechanism of action of multiple antidepressants, e.g., SSRIs, SNRIs, TCAs, SARIs, and many others. Imaging technologies applying PET provide an exceptional opportunity to visualize certain components of the serotonin system and to quantify related alterations in the depressed brain. Using these technologies, not only pathophysiologic mechanisms on a molecular level but also therapeutic effects can be measured. As summarized in this chapter, several target structures that are involved in serotonergic neurotransmission can now be quantified with established, high-quality radiotracers. However, highly selective radioligands that meet the requirements for human use are still lacking with respect to some receptors and target molecules of the 5-HT system.

Besides radioligands for SERT and MAO-A quantification, highly selective radiotracers could be developed for visualizing certain 5-HT receptors, enabling neuroresearchers to quantify density and distribution patterns across the living human brain, both in health and depression. Table 3 *provides an* overview of up- and downregulated 5-HT receptors in the depressed brain. The 5-HT$_{1A}$ receptor belongs to one of the best characterized serotonin receptors in terms of in vivo molecular imaging. PET studies focusing on cerebral 5-HT$_{1A}$ density and distribution in depressed individuals largely point toward a downregulation in multiple brain areas, especially in cortical (e.g., the mesiotemporal cortex) and temporal regions, whereas data on the raphe nuclei remains somewhat conflicting, reporting divergent results in different subtypes of raphe nuclei. The 5-HT$_{2B}$ receptor was also found to be downregulated among depressed patients, especially in regions involved in reward processing and across selected cortical (e.g., anterior cingular and subgenual prefrontal cortices) as well as limbic areas (e.g., hippocampus). Data on the extensively investigated excitatory serotonin receptor 5-HT$_{2A}$ is rather inconclusive, particularly due to methodological problems in earlier PET studies. However, the application of [^{11}C]MDL100907 rendered more concise results, indicating higher cerebral 5-HT$_{2A}$ receptor BP in MDD patients. Additionally, in vivo human studies in a clinical context with the novel radiotracer [^{11}C]Cimbi-36 are assumed to fill certain knowledge gaps associated with the role of 5-HT$_{2A}$ in MDD

Table 3
Overview of molecular alterations of the serotonergic system in MDD as investigated by in vivo PET

5-HT receptors	
5-HT$_{1A}$	⇧
5-HT$_{1B}$	⬌
5-HT$_{2A}$	⬌
5-HT$_4$	⬌
5-HT$_7$	⬌
Transporters	
SERT	⬇
Enzymes	
MAO-A	⬆

Table 3 shows an overview of up- and downregulated 5-HT receptors in the depressed brain. Note: Up arrows indicate an increase in cerebral availability; down arrows indicate a decrease. Dark arrows indicate robust evidence; light arrows indicate preliminary evidence. Horizontal arrows indicate ambiguous evidence. MDD: major depressive disorder; SERT: serotonin transporter; MDD: major depressive episode; MAO-A: monoamine oxidase A.

pathophysiology; however, human studies are still lacking. Other serotonin receptor subtypes such as the 5-HT$_{2C}$, 5-HT$_4$, and 5-HT$_7$ represent promising candidates in the context of depression and antidepressant therapy that are likely to be involved not only in major symptoms of MDD but also in atypical depression. Moreover, subtypes of the 5-HT$_5$ and 5-HT$_6$ receptor families have not been extensively examined yet using PET imaging for lack of highly selective radiotracers with high penetration of the CNS.

Therefore, ongoing research is needed to optimize measuring technologies on the one hand and to develop suitable tracers for human use on the other. Promising research efforts as performed by our group are now focusing on multimodal imaging technologies, combining molecular PET imaging with structural and functional MRI. Furthermore, advances in the field of imaging genetics are now used by Prof. Lanzenberger et al. to show potential correlations of, e.g., mRNA expression and in vivo binding data [309] as well as selected genetic polymorphisms and 5-HT receptor binding, not only in depression [310] but also in other neuropsychiatric disorders [311]. Predicted mRNA expression patterns for genes acquired from the Allen Human Brain Atlas are accessible for download via www.meduniwien.ac.at/neuroimaging/mRNA.html. Hopefully, future investigations will pave the way for constantly improving neuroimaging methods in the field of neuropsychiatric diseases. With this in mind, researchers all around the world are working on strategies to optimize these technologies, thus improving treatment strategies and personalized therapeutic options.

References

1. Ferrari AJ, Charlson FJ, Norman RE, Patten SB, Freedman G, Murray CJ, Vos T, Whiteford HA (2013) Burden of depressive disorders by country, sex, age, and year: findings from the global burden of disease study 2010. PLoS Med 10(11):e1001547

2. Lopez AD, Murray CCJL (1998) The global burden of disease, 1990–2020. Nat Med 4(11):1241–1243. https://doi.org/10.1038/3218

3. GBD (2018) Global, regional, and national incidence, prevalence, and years lived with disability for 354 diseases and injuries for 195 countries and territories, 1990–2017: a systematic analysis for the global burden of disease study 2017. Lancet 392(10159):1789–1858. https://doi.org/10.1016/S0140-6736(18)32279-7

4. Lim GY, Tam WW, Lu Y, Ho CS, Zhang MW, Ho RC (2018) Prevalence of depression in the community from 30 countries between 1994 and 2014. Sci Rep 8(1):2861. https://doi.org/10.1038/s41598-018-21243-x

5. Harris EC, Barraclough B (1997) Suicide as an outcome for mental disorders. A meta-analysis. Br J Psychiatry 170:205–228. https://doi.org/10.1192/bjp.170.3.205

6. Schumacher J, Abou Jamra R, Becker T, Ohlraun S, Klopp N, Binder EB, Schulze TG, Deschner M, Schmäl C, Höfels S (2005) Evidence for a relationship between genetic variants at the brain-derived neurotrophic factor (BDNF) locus and major depression. Biol Psychiatry 58(4):307–314

7. Verhagen M, Van Der Meij A, Van Deurzen P, Janzing J, Arias-Vasquez A, Buitelaar J, Franke B (2010) Meta-analysis of the BDNF Val66Met polymorphism in major depressive disorder: effects of gender and ethnicity. Mol Psychiatry 15(3):260–271

8. Uhr M, Tontsch A, Namendorf C, Ripke S, Lucae S, Ising M, Dose T, Ebinger M, Rosenhagen M, Kohli M (2008) Polymorphisms in the drug transporter gene ABCB1 predict antidepressant treatment response in depression. Neuron 57(2):203–209

9. Zhang X, Gainetdinov RR, Beaulieu J-M, Sotnikova TD, Burch LH, Williams RB, Schwartz DA, Krishnan KRR, Caron MG (2005) Loss-of-function mutation in tryptophan hydroxylase-2 identified in unipolar major depression. Neuron 45(1):11–16

10. Miozzo R, Eaton WW, Joseph Bienvenu O 3rd, Samuels J, Nestadt G (2020) The serotonin transporter gene polymorphism (SLC6A4) and risk for psychiatric morbidity and comorbidity in the Baltimore ECA follow-up study. Compr Psychiatry 102:152199. https://doi.org/10.1016/j.comppsych.2020.152199

11. Ditzen C, Tang N, Jastorff AM, Teplytska L, Yassouridis A, Maccarrone G, Uhr M, Bronisch T, Miller CA, Holsboer F (2012) Cerebrospinal fluid biomarkers for major depression confirm relevance of associated pathophysiology. Neuropsychopharmacology 37(4):1013–1025

12. Heilig M, Zachrisson O, Thorsell A, Ehnvall A, Mottagui-Tabar S, Sjögren M, Åsberg M, Ekman R, Wahlestedt C, Ågren H (2004) Decreased cerebrospinal fluid neuropeptide Y (NPY) in patients with treatment refractory unipolar major depression: preliminary evidence for association with preproNPY gene polymorphism. J Psychiatr Res 38(2):113–121

13. Pariante CM, Lightman SL (2008) The HPA axis in major depression: classical theories and new developments. Trends Neurosci 31(9):464–468

14. Keller J, Gomez R, Williams G, Lembke A, Lazzeroni L, Murphy GM, Schatzberg AF (2017) HPA axis in major depression: cortisol, clinical symptomatology and genetic variation predict cognition. Mol Psychiatry 22(4):527–536

15. Hage MP, Azar ST (2012) The link between thyroid function and depression. J Thyroid Res 2012:590648

16. Bauer M, Goetz T, Glenn T, Whybrow P (2008) The thyroid-brain interaction in thyroid disorders and mood disorders. J Neuroendocrinol 20(10):1101–1114

17. Gorman JM (2006) Gender differences in depression and response to psychotropic medication. Gend Med 3(2):93–109

18. Schiller CE, Johnson SL, Abate AC, Schmidt PJ, Rubinow DR (2016) Reproductive

steroid regulation of mood and behavior. Compr Physiol 6(3):1135–1160. https://doi.org/10.1002/cphy.c150014

19. Kendler KS, Kuhn J, Prescott CA (2004) The interrelationship of neuroticism, sex, and stressful life events in the prediction of episodes of major depression. Am J Psychiatr 161(4):631–636

20. Spies M, Handschuh PA, Lanzenberger R, Kranz GS (2020) Sex and the serotonergic underpinnings of depression and migraine. Handb Clin Neurol 175:117–140

21. Miller AH, Maletic V, Raison CL (2009) Inflammation and its discontents: the role of cytokines in the pathophysiology of major depression. Biol Psychiatry 65(9):732–741

22. Koolschijn PC, van Haren NE, Lensvelt-Mulders GJ, Hulshoff Pol HE, Kahn RS (2009) Brain volume abnormalities in major depressive disorder: a meta-analysis of magnetic resonance imaging studies. Hum Brain Mapp 30(11):3719–3735. https://doi.org/10.1002/hbm.20801

23. Kraus C, Castren E, Kasper S, Lanzenberger R (2017) Serotonin and neuroplasticity - Links between molecular, functional and structural pathophysiology in depression. Neurosci Biobehav Rev 77:317–326. https://doi.org/10.1016/j.neubiorev.2017.03.007

24. Gryglewski G, Baldinger-Melich P, Seiger R, Godbersen GM, Michenthaler P, Klöbl M, Spurny B, Kautzky A, Vanicek T, Kasper S (2019) Structural changes in amygdala nuclei, hippocampal subfields and cortical thickness following electroconvulsive therapy in treatment-resistant depression: longitudinal analysis. Br J Psychiatry 214(3):159–167

25. Höflich A, Ganger S, Tik M, Hahn A, Kranz GS, Vanicek T, Spies M, Kraus C, Windischberger C, Kasper S (2017) Imaging the neuroplastic effects of ketamine with VBM and the necessity of placebo control. NeuroImage 147:198–203

26. Hirschfeld R (2000) History and evolution of the monoamine hypothesis of depression. J Clin Psychiatry 61(Suppl 6):4–6

27. Szeitz A, Bandiera SM (2018) Analysis and measurement of serotonin. Biomed Chromatogr 32(1). https://doi.org/10.1002/bmc.4135

28. Kema IP, de Vries EG, Muskiet FA (2000) Clinical chemistry of serotonin and metabolites. J Chromatogr B Biomed Sci Appl 747(1–2):33–48. https://doi.org/10.1016/s0378-4347(00)00341-8

29. Ramakrishna A, Giridhar P, Ravishankar GA (2011) Phytoserotonin: a review. Plant Signal Behav 6(6):800–809. https://doi.org/10.4161/psb.6.6.15242

30. Belenikina NS, Strakhovskaya MG, Fraikin G (1991) Near-UV activation of yeast growth. J Photochem Photobiol B 10(1–2):51–55. https://doi.org/10.1016/1011-1344(91)80211-y

31. Fraikin GY, Strakhovskaya MG, Ivanova EV, Rubin AB (1989) Near-UV activation of enzymatic conversion of 5-hydroxytryptophan to serotonin. Photochem Photobiol 49(4):475–477. https://doi.org/10.1111/j.1751-1097.1989.tb09197.x

32. Berger M, Gray JA, Roth BL (2009) The expanded biology of serotonin. Annu Rev Med 60:355–366. https://doi.org/10.1146/annurev.med.60.042307.110802

33. Hsu SC, Johansson KR, Donahue MJ (1986) The bacterial flora of the intestine of Ascaris suum and 5-hydroxytryptamine production. J Parasitol 72(4):545–549

34. Cryan JF, Dinan TG (2012) Mind-altering microorganisms: the impact of the gut microbiota on brain and behaviour. Nat Rev Neurosci 13(10):701–712. https://doi.org/10.1038/nrn3346

35. Galland L (2014) The gut microbiome and the brain. J Med Food 17(12):1261–1272. https://doi.org/10.1089/jmf.2014.7000

36. Knecht LD, O'Connor G, Mittal R, Liu XZ, Daftarian P, Deo SK, Daunert S (2016) Serotonin activates bacterial quorum sensing and enhances the virulence of Pseudomonas aeruginosa in the host. EBioMedicine 9:161–169. https://doi.org/10.1016/j.ebiom.2016.05.037

37. O'Mahony SM, Clarke G, Borre YE, Dinan TG, Cryan JF (2015) Serotonin, tryptophan metabolism and the brain-gut-microbiome axis. Behav Brain Res 277:32–48. https://doi.org/10.1016/j.bbr.2014.07.027

38. Kaur H, Bose C, Mande SS (2019) Tryptophan metabolism by gut microbiome and gut-brain-Axis: an in silico analysis. Front Neurosci 13:1365. https://doi.org/10.3389/fnins.2019.01365

39. Moroz LL, Kohn AB (2016) Independent origins of neurons and synapses: insights from ctenophores. Philos Trans R Soc Lond Ser B Biol Sci 371(1685):20150041. https://doi.org/10.1098/rstb.2015.0041

40. Blenau W, Thamm M (2011) Distribution of serotonin (5-HT) and its receptors in the insect brain with focus on the mushroom bodies: lessons from Drosophila melanogaster and Apis mellifera. Arthropod Struct Dev

40(5):381–394. https://doi.org/10.1016/j.asd.2011.01.004

41. Azmitia E, Gannon P (1983) The ultrastructural localization of serotonin immunoreactivity in myelinated and unmyelinated axons within the medial forebrain bundle of rat and monkey. J Neurosci 3(10):2083–2090

42. Jahanshahi A, Steinbusch HW, Temel Y (2013) Distribution of dopaminergic cell bodies in the median raphe nucleus of the rat brain. J Chem Neuroanat 53:60–63. https://doi.org/10.1016/j.jchemneu.2013.09.002

43. Hornung JP (2003) The human raphe nuclei and the serotonergic system. J Chem Neuroanat 26(4):331–343

44. Ursin R (2002) Serotonin and sleep. Sleep Med Rev 6(1):55–69. https://doi.org/10.1053/smrv.2001.0174

45. Jacobs BL, Fornal CA (1997) Serotonin and motor activity. Curr Opin Neurobiol 7(6):820–825. https://doi.org/10.1016/s0959-4388(97)80141-9

46. Meguid MM, Fetissov SO, Varma M, Sato T, Zhang L, Laviano A, Rossi-Fanelli F (2000) Hypothalamic dopamine and serotonin in the regulation of food intake. Nutrition 16(10):843–857. https://doi.org/10.1016/s0899-9007(00)00449-4

47. Feldberg W, Myers RD (1964) Effects on temperature of amines injected into the cerebral ventricles. A new concept of temperature regulation. J Physiol 173(2):226–231. https://doi.org/10.1113/jphysiol.1964.sp007454

48. Tenen SS (1967) The effects of p-chlorophenylalanine, a serotonin depletor, on avoidance acquisition, pain sensitivity and related behavior in the rat. Psychopharmacologia 10(3):204–219. https://doi.org/10.1007/bf00401382

49. Canli T, , Lesch K-P, (2007) Long story short: the serotonin transporter in emotion regulation and social cognition Nat Neurosci 10:1103–1109. doi:https://doi.org/10.1038/nn1964

50. Cools R, Roberts AC, Robbins TW (2008) Serotoninergic regulation of emotional and behavioural control processes. Trends Cogn Sci 12(1):31–40. https://doi.org/10.1016/j.tics.2007.10.011

51. Kranz GS, Kasper S, Lanzenberger R (2010) Reward and the serotonergic system. Neuroscience 166(4):1023–1035. https://doi.org/10.1016/j.neuroscience.2010.01.036

52. Kraus C, , Castren E, , Kasper S, , Lanzenberger R, (2017) Serotonin and neuroplasticity - Links between molecular, functional and structural pathophysiology in depression. Neurosci Biobehav Rev 77:317–326. doi:https://doi.org/10.1016/j.neubiorev.2017.03.007

53. Savitz J, Lucki I, Drevets WC (2009) 5-HT (1A) receptor function in major depressive disorder. Prog Neurobiol 88(1):17–31. https://doi.org/10.1016/j.pneurobio.2009.01.009

54. Woolley DW, Shaw E (1954) A biochemical and pharmacological suggestion about certain mental disorders. Proc Natl Acad Sci U S A 40(4):228–231. https://doi.org/10.1073/pnas.40.4.228

55. Martin AM, Young RL, Leong L, Rogers GB, Spencer NJ, Jessup CF, Keating DJ (2017) The diverse metabolic roles of peripheral serotonin. Endocrinology 158(5):1049–1063. https://doi.org/10.1210/en.2016-1839

56. Jacobs BL, Azmitia EC (1992) Structure and function of the brain serotonin system. Physiol Rev 72(1):165–229. https://doi.org/10.1152/physrev.1992.72.1.165

57. Steinbusch HW (1981) Distribution of serotonin-immunoreactivity in the central nervous system of the rat-cell bodies and terminals. Neuroscience 6(4):557–618. https://doi.org/10.1016/0306-4522(81)90146-9

58. Muzerelle A, Scotto-Lomassese S, Bernard JF, Soiza-Reilly M, Gaspar P (2016) Conditional anterograde tracing reveals distinct targeting of individual serotonin cell groups (B5-B9) to the forebrain and brainstem. Brain Struct Funct 221(1):535–561. https://doi.org/10.1007/s00429-014-0924-4

59. Hannon J, Hoyer D (2008) Molecular biology of 5-HT receptors. Behav Brain Res 195(1):198–213. https://doi.org/10.1016/j.bbr.2008.03.020

60. Oldham WM, Hamm HE (2008) Heterotrimeric G protein activation by G-protein-coupled receptors. Nat Rev Mol Cell Biol 9(1):60–71. https://doi.org/10.1038/nrm2299

61. Milligan G (2004) G protein-coupled receptor dimerization: function and ligand pharmacology. Mol Pharmacol 66(1):1–7. https://doi.org/10.1124/mol.104.000497

62. Giulietti M, Vivenzio V, Piva F, Principato G, Bellantuono C, Nardi B (2014) How much do we know about the coupling of G-proteins

to serotonin receptors? Mol Brain 7(1):49. https://doi.org/10.1186/s13041-014-0049-y

63. Barnes NM, Sharp T (1999) A review of central 5-HT receptors and their function. Neuropharmacology 38(8):1083–1152. https://doi.org/10.1016/s0028-3908(99)00010-6

64. Sprouse JS, Aghajanian GK (1987) Electrophysiological responses of serotoninergic dorsal raphe neurons to 5-HT1A and 5-HT1B agonists. Synapse 1(1):3–9. https://doi.org/10.1002/syn.890010103

65. Lanfumey L, Hamon M (2004) 5-HT1 receptors. Curr Drug Targets CNS Neurol Disord 3(1):1–10. https://doi.org/10.2174/1568007043482570

66. Sari Y (2004) Serotonin1B receptors: from protein to physiological function and behavior. Neurosci Biobehav Rev 28(6):565–582. https://doi.org/10.1016/j.neubiorev.2004.08.008

67. Rumajogee P, Vergé D, Hanoun N, Brisorgueil MJ, Hen R, Lesch KP, Hamon M, Miquel MC (2004) Adaption of the serotoninergic neuronal phenotype in the absence of 5-HT autoreceptors or the 5-HT transporter: involvement of BDNF and cAMP. Eur J Neurosci 19(4):937–944. https://doi.org/10.1111/j.0953-816x.2004.03194.x

68. Heisler LK, Chu HM, Brennan TJ, Danao JA, Bajwa P, Parsons LH, Tecott LH (1998) Elevated anxiety and antidepressant-like responses in serotonin 5-HT1A receptor mutant mice. Proc Natl Acad Sci U S A 95(25):15049–15054. https://doi.org/10.1073/pnas.95.25.15049

69. Nautiyal KM, Tritschler L, Ahmari SE, David DJ, Gardier AM, Hen R (2016) A lack of serotonin 1B autoreceptors results in decreased anxiety and depression-related behaviors. Neuropsychopharmacology 41(12):2941–2950. https://doi.org/10.1038/npp.2016.109

70. Araneda R, Andrade R (1991) 5-Hydroxytryptamine2 and 5-hydroxytryptamine 1A receptors mediate opposing responses on membrane excitability in rat association cortex. Neuroscience 40(2):399–412. https://doi.org/10.1016/0306-4522(91)90128-b

71. Celada P, Puig MV, Artigas F (2013) Serotonin modulation of cortical neurons and networks. Front Integr Neurosci 7:25. https://doi.org/10.3389/fnint.2013.00025

72. King MV, Marsden CA, Fone KC (2008) A role for the 5-HT(1A), 5-HT4 and 5-HT6 receptors in learning and memory. Trends Pharmacol Sci 29(9):482–492. https://doi.org/10.1016/j.tips.2008.07.001

73. Grailhe R, Waeber C, Dulawa SC, Hornung JP, Zhuang X, Brunner D, Geyer MA, Hen R (1999) Increased exploratory activity and altered response to LSD in mice lacking the 5-HT(5A) receptor. Neuron 22(3):581–591. https://doi.org/10.1016/s0896-6273(00)80712-6

74. Hoyer D, Hannon JP, Martin GR (2002) Molecular, pharmacological and functional diversity of 5-HT receptors. Pharmacol Biochem Behav 71(4):533–554. https://doi.org/10.1016/s0091-3057(01)00746-8

75. Coleman JA, Green EM, Gouaux E (2016) X-ray structures and mechanism of the human serotonin transporter. Nature 532(7599):334–339. https://doi.org/10.1038/nature17629

76. Kristensen AS, Andersen J, Jørgensen TN, Sørensen L, Eriksen J, Loland CJ, Strømgaard K, Gether U (2011) SLC6 neurotransmitter transporters: structure, function, and regulation. Pharmacol Rev 63(3):585–640. https://doi.org/10.1124/pr.108.000869

77. Steiner JA, Carneiro AMD, Blakely RD (2008) Going with the flow: trafficking-dependent and -independent regulation of serotonin transport. Traffic 9(9):1393–1402. https://doi.org/10.1111/j.1600-0854.2008.00757.x

78. Savli M, Bauer A, Mitterhauser M, Ding YS, Hahn A, Kroll T, Neumeister A, Haeusler D, Ungersboeck J, Henry S, Isfahani SA, Rattay F, Wadsak W, Kasper S, Lanzenberger R (2012) Normative database of the serotonergic system in healthy subjects using multitracer PET. NeuroImage 63(1):447–459. https://doi.org/10.1016/j.neuroimage.2012.07.001

79. Lesch KP, Wolozin BL, Murphy DL, Riederer P (1993) Primary structure of the human platelet serotonin uptake site: identity with the brain serotonin transporter. J Neurochem 60(6):2319–2322. https://doi.org/10.1111/j.1471-4159.1993.tb03522.x

80. Gordon J, Barnes NM (2003) Lymphocytes transport serotonin and dopamine: agony or ecstasy? Trends Immunol 24(8):438–443. https://doi.org/10.1016/s1471-4906(03)00176-5

81. Martel F (2006) Recent advances on the importance of the serotonin transporter SERT in the rat intestine. Pharmacol Res 54(2):73–76. https://doi.org/10.1016/j.phrs.2006.04.005

82. Ramamoorthy S, Bauman AL, Moore KR, Han H, Yang-Feng T, Chang AS, Ganapathy V, Blakely RD (1993) Antidepressant- and cocaine-sensitive human serotonin transporter: molecular cloning, expression, and chromosomal localization. Proc Natl Acad Sci U S A 90(6):2542–2546. https://doi.org/10.1073/pnas.90.6.2542

83. Collier DA, Stöber G, Li T, Heils A, Catalano M, Di Bella D, Arranz MJ, Murray RM, Vallada HP, Bengel D, Müller CR, Roberts GW, Smeraldi E, Kirov G, Sham P, Lesch KP (1996) A novel functional polymorphism within the promoter of the serotonin transporter gene: possible role in susceptibility to affective disorders. Mol Psychiatry 1(6): 453–460

84. Bortolato M, Chen K, Shih JC (2008) Monoamine oxidase inactivation: from pathophysiology to therapeutics. Adv Drug Deliv Rev 60(13–14):1527–1533. https://doi.org/10.1016/j.addr.2008.06.002

85. Grimsby J, Chen K, Wang LJ, Lan NC, Shih JC (1991) Human monoamine oxidase a and B genes exhibit identical exon-intron organization. Proc Natl Acad Sci U S A 88(9): 3637–3641. https://doi.org/10.1073/pnas.88.9.3637

86. Best J, Nijhout HF, Reed M (2010) Serotonin synthesis, release and reuptake in terminals: a mathematical model. Theor Biol Med Model 7:34. https://doi.org/10.1186/1742-4682-7-34

87. Ruddick JP, Evans AK, Nutt DJ, Lightman SL, Rook GA, Lowry CA (2006) Tryptophan metabolism in the central nervous system: medical implications. Expert Rev Mol Med 8(20):1–27. https://doi.org/10.1017/S1462399406000068

88. Rodríguez MJ, Saura J, Billett EE, Finch CC, Mahy N (2001) Cellular localization of monoamine oxidase a and B in human tissues outside of the central nervous system. Cell Tissue Res 304(2):215–220. https://doi.org/10.1007/s004410100361

89. Shih JC, Chen K, Ridd MJ (1999) Monoamine oxidase: from genes to behavior. Annu Rev Neurosci 22:197–217. https://doi.org/10.1146/annurev.neuro.22.1.197

90. Jahng JW, Houpt TA, Wessel TC, Chen K, Shih JC, Joh TH (1997) Localization of monoamine oxidase a and B mRNA in the rat brain by in situ hybridization. Synapse 25(1):30–36. https://doi.org/10.1002/(sici)1098-2396(199701)25:1<30::aid-syn4>3.0.co;2-g

91. Saura J, Luque JM, Cesura AM, Da Prada M, Chan-Palay V, Huber G, Löffler J, Richards JG (1994) Increased monoamine oxidase B activity in plaque-associated astrocytes of Alzheimer brains revealed by quantitative enzyme radioautography. Neuroscience 62(1):15–30. https://doi.org/10.1016/0306-4522(94)90311-5

92. Westlund KN, Denney RM, Rose RM, Abell CW (1988) Localization of distinct monoamine oxidase a and monoamine oxidase B cell populations in human brainstem. Neuroscience 25(2):439–456. https://doi.org/10.1016/0306-4522(88)90250-3

93. Willoughby J, Glover V, Sandler M (1988) Histochemical localisation of monoamine oxidase a and B in rat brain. J Neural Transm 74(1):29–42. https://doi.org/10.1007/bf01243573

94. Luque JM, Kwan SW, Abell CW, Da Prada M, Richards JG (1995) Cellular expression of mRNAs encoding monoamine oxidases a and B in the rat central nervous system. J Comp Neurol 363(4):665–680. https://doi.org/10.1002/cne.903630410

95. Cases O, Seif I, Grimsby J, Gaspar P, Chen K, Pournin S, Müller U, Aguet M, Babinet C, Shih JC et al (1995) Aggressive behavior and altered amounts of brain serotonin and norepinephrine in mice lacking MAOA. Science (New York, NY) 268(5218):1763–1766. https://doi.org/10.1126/science.7792602

96. Menke A, Binder EB (2014) Epigenetic alterations in depression and antidepressant treatment. Dialogues Clin Neurosci 16(3): 395–404. https://doi.org/10.31887/DCNS.2014.16.3/amenke

97. Kautzky A, Baldinger P, Souery D, Montgomery S, Mendlewicz J, Zohar J, Serretti A, Lanzenberger R, Kasper S (2015) The combined effect of genetic polymorphisms and clinical parameters on treatment outcome in treatment-resistant depression. Eur Neuropsychopharmacol 25(4):441–453. https://doi.org/10.1016/j.euroneuro.2015.01.001

98. Chen B, Dowlatshahi D, MacQueen GM, Wang J-F, Young LT (2001) Increased hippocampal BDNF immunoreactivity in subjects treated with antidepressant medication. Biol Psychiatry 50(4):260–265

99. Kadriu B, Greenwald M, Henter ID, Gilbert JR, Kraus C, Park LT, Zarate CA (2021) Ketamine and serotonergic psychedelics: common mechanisms underlying the effects of rapid-acting antidepressants. Int J Neuropsychopharmacol 24(1):8–21. https://doi.org/10.1093/ijnp/pyaa087

100. Kraus C, Kadriu B (2019) Prognosis and improved outcomes in major depression: a

review. Focus 9(1):127. https://doi.org/10.1038/s41398-019-0460-3

101. Tatsumi M, Groshan K, Blakely RD, Richelson E (1997) Pharmacological profile of antidepressants and related compounds at human monoamine transporters. Eur J Pharmacol 340(2–3):249–258. https://doi.org/10.1016/s0014-2999(97)01393-9

102. Gillman PK (2007) Tricyclic antidepressant pharmacology and therapeutic drug interactions updated. Br J Pharmacol 151(6):737–748. https://doi.org/10.1038/sj.bjp.0707253

103. Peretti S, Judge R, Hindmarch I (2000) Safety and tolerability considerations: tricyclic antidepressants vs. selective serotonin reuptake inhibitors. Acta Psychiatr Scand 101 (S403):17–25. https://doi.org/10.1111/j.1600-0447.2000.tb10944.x

104. Sánchez C, Hyttel J (1999) Comparison of the effects of antidepressants and their metabolites on reuptake of biogenic amines and on receptor binding. Cell Mol Neurobiol 19(4):467–489. https://doi.org/10.1023/A:1006986824213

105. Stam NJ, Roesink C, Dijcks F, Garritsen A, van Herpen A, Olijve W (1997) Human serotonin 5-HT7 receptor: cloning and pharmacological characterisation of two receptor variants. FEBS Lett 413(3):489–494. https://doi.org/10.1016/S0014-5793(97)00964-2

106. Liu J, Reid AR, Sawynok J (2013) Spinal serotonin 5-HT7 and adenosine A1 receptors, as well as peripheral adenosine A1 receptors, are involved in antinociception by systemically administered amitriptyline. Eur J Pharmacol 698(1):213–219. https://doi.org/10.1016/j.ejphar.2012.10.042

107. Olgiati P, Serretti A, Souery D, Dold M, Kasper S, Montgomery S, Zohar J, Mendlewicz J (2018) Early improvement and response to antidepressant medications in adults with major depressive disorder. Meta-analysis and study of a sample with treatment-resistant depression. J Affect Disord 227:777–786. https://doi.org/10.1016/j.jad.2017.11.004

108. Pinder RM (1991) Mianserin: pharmacological and clinical correlates. Nord Psykiatr Tidsskr 45(sup24):13–26. https://doi.org/10.3109/08039489109096678

109. Anttila SA, Leinonen EV (2001) A review of the pharmacological and clinical profile of mirtazapine. CNS Drug Rev 7(3):249–264. https://doi.org/10.1111/j.1527-3458.2001.tb00198.x

110. Fuentes AV, Pineda MD, Venkata KCN (2018) Comprehension of top 200 prescribed drugs in the US as a resource for pharmacy teaching, training and practice. Pharmacy 6(2):43

111. Vaswani M, Linda FK, Ramesh S (2003) Role of selective serotonin reuptake inhibitors in psychiatric disorders: a comprehensive review. Prog Neuro-Psychopharmacol Biol Psychiatry 27(1):85–102. https://doi.org/10.1016/S0278-5846(02)00338-X

112. Stahl SM (1998) Mechanism of action of serotonin selective reuptake inhibitors: serotonin receptors and pathways mediate therapeutic effects and side effects. J Affect Disord 51(3):215–235

113. Baudry A, Mouillet-Richard S, Schneider B, Launay JM, Kellermann O (2010) miR-16 targets the serotonin transporter: a new facet for adaptive responses to antidepressants. Science (New York, NY) 329(5998):1537–1541. https://doi.org/10.1126/science.1193692

114. Erb SJ, Schappi JM, Rasenick MM (2016) Antidepressants accumulate in lipid rafts independent of monoamine transporters to modulate redistribution of the G protein, $G_{\alpha s}$. J Biol Chem 291(38):19725–19733. https://doi.org/10.1074/jbc.M116.727263

115. James GM, Baldinger-Melich P, Philippe C, Kranz GS, Vanicek T, Hahn A, Gryglewski G, Hienert M, Spies M, Traub-Weidinger T, Mitterhauser M, Wadsak W, Hacker M, Kasper S, Lanzenberger R (2017) Effects of selective serotonin reuptake inhibitors on interregional relation of serotonin transporter availability in major depression. Front Hum Neurosci 11:48. https://doi.org/10.3389/fnhum.2017.00048

116. Peretti S, Judge R, Hindmarch I (2000) Safety and tolerability considerations: tricyclic antidepressants vs. selective serotonin reuptake inhibitors. Acta Psychiatr Scand Suppl 403:17–25. https://doi.org/10.1111/j.1600-0447.2000.tb10944.x

117. Anderson IM (2000) Selective serotonin reuptake inhibitors versus tricyclic antidepressants: a meta-analysis of efficacy and tolerability. J Affect Disord 58(1):19–36. https://doi.org/10.1016/s0165-0327(99)00092-0

118. Qin B, Zhang Y, Zhou X, Cheng P, Liu Y, Chen J, Fu Y, Luo Q, Xie P (2014) Selective serotonin reuptake inhibitors versus tricyclic antidepressants in young patients: a meta-analysis of efficacy and acceptability. Clin Ther 36(7):1087–1095.e1084. https://doi.org/10.1016/j.clinthera.2014.06.001

119. Stahl SM, Grady MM, Moret C, Briley M (2014) SNRIs: the pharmacology, clinical efficacy, and tolerability in comparison with other classes of antidepressants. CNS Spectr 10(9): 732–747. https://doi.org/10.1017/S1092852900019726

120. Lunn MP, Hughes RA, Wiffen PJ (2014) Duloxetine for treating painful neuropathy, chronic pain or fibromyalgia. Cochrane Database Syst Rev 1:CD007115

121. Lunn MP, Hughes RA, Wiffen PJ (2009) Duloxetine for treating painful neuropathy or chronic pain. Cochrane Database Syst Rev 4:CD007115

122. Barkin RL, Barkin S (2005) The role of venlafaxine and duloxetine in the treatment of depression with Decremental changes in somatic symptoms of pain, chronic pain, and the pharmacokinetics and clinical considerations of duloxetine pharmacotherapy. Am J Ther 12(5):431–438

123. Liu Y, Zhao J, Guo W (2018) Emotional roles of mono-Aminergic neurotransmitters in major depressive disorder and anxiety disorders. Front Psychol 9:2201. https://doi.org/10.3389/fpsyg.2018.02201

124. Stahl SM (2009) Mechanism of action of trazodone: a multifunctional drug. CNS Spectr 14(10):536–546. https://doi.org/10.1017/s1092852900024020

125. Odagaki Y, Toyoshima R, Yamauchi T (2005) Trazodone and its active metabolite m-chlorophenylpiperazine as partial agonists at 5-HT1A receptors assessed by [35S] GTPgammaS binding. J Psychopharmacol 19(3):235–241. https://doi.org/10.1177/0269881105051526

126. Fagiolini A, Comandini A, Dell'Osso MC, Kasper S (2012) Rediscovering trazodone for the treatment of major depressive disorder. CNS Drugs 26(12):1033–1049. https://doi.org/10.1007/s40263-012-0010-5

127. Daniele S, Zappelli E, Martini C (2015) Trazodone regulates neurotrophic/growth factors, mitogen-activated protein kinases and lactate release in human primary astrocytes. J Neuroinflammation 12(1):225. https://doi.org/10.1186/s12974-015-0446-x

128. McEwen BS, Chattarji S, Diamond DM, Jay TM, Reagan LP, Svenningsson P, Fuchs E (2010) The neurobiological properties of tianeptine (Stablon): from monoamine hypothesis to glutamatergic modulation. Mol Psychiatry 15(3):237–249. https://doi.org/10.1038/mp.2009.80

129. Mennini T, Mocaer E, Garattini S (1987) Tianeptine, a selective enhancer of serotonin uptake in rat brain. Naunyn Schmiedeberg's Arch Pharmacol 336(5):478–482. https://doi.org/10.1007/BF00169302

130. Paul IA, Skolnick P (2003) Glutamate and depression: clinical and preclinical studies. Ann N Y Acad Sci 1003(1):250–272

131. Moghaddam B, Adams B, Verma A, Daly D (1997) Activation of glutamatergic neurotransmission by ketamine: a novel step in the pathway from NMDA receptor blockade to dopaminergic and cognitive disruptions associated with the prefrontal cortex. J Neurosci 17(8):2921–2927. https://doi.org/10.1523/jneurosci.17-08-02921.1997

132. Silberbauer LR, Spurny B, Handschuh P, Klöbl M, Bednarik P, Reiter B, Ritter V, Trost P, Konadu ME, Windpassinger M, Stimpfl T, Bogner W, Lanzenberger R, Spies M (2020) Effect of ketamine on limbic GABA and glutamate: a human in vivo multivoxel magnetic resonance spectroscopy study. Front Psychiatry 11:549903. https://doi.org/10.3389/fpsyt.2020.549903

133. Höflich A, Hahn A, Küblböck M, Kranz GS, Vanicek T, Ganger S, Spies M, Windischberger C, Kasper S, Winkler D, Lanzenberger R (2017) Ketamine-dependent neuronal activation in healthy volunteers. Brain Struct Funct 222(3):1533–1542. https://doi.org/10.1007/s00429-016-1291-0

134. Spies M, James GM, Berroterán-Infante N, Ibeschitz H, Kranz GS, Unterholzner J, Godbersen M, Gryglewski G, Hienert M, Jungwirth J, Pichler V, Reiter B, Silberbauer L, Winkler D, Mitterhauser M, Stimpfl T, Hacker M, Kasper S, Lanzenberger R (2018) Assessment of ketamine binding of the serotonin transporter in humans with positron emission tomography. Int J Neuropsychopharmacol 21(2):145–153. https://doi.org/10.1093/ijnp/pyx085

135. Tiger M, Veldman ER (2020) A randomized placebo-controlled PET study of ketamine's effect on serotonin(1B) receptor binding in patients with SSRI-resistant depression. Transl Psychiatry 10(1):159. https://doi.org/10.1038/s41398-020-0844-4

136. Zisook S (1985) A clinical overview of monoamine oxidase inhibitors. Psychosomatics 26(3):240–251. https://doi.org/10.1016/S0033-3182(85)72877-0

137. Youdim MBH, Edmondson D, Tipton KF (2006) The therapeutic potential of monoamine oxidase inhibitors. Nat Rev Neurosci 7(4):295–309. https://doi.org/10.1038/nrn1883

138. Yamada M, Yasuhara H (2004) Clinical pharmacology of MAO inhibitors: safety and future. Neurotoxicology 25(1):215–221. https://doi.org/10.1016/S0161-813X(03)00097-4

139. Boyer EW, Shannon M (2005) The serotonin syndrome. N Engl J Med 352(11): 1112–1120. https://doi.org/10.1056/NEJMra041867

140. Avram S, Bologa CG, Holmes J, Bocci G, Wilson TB, Nguyen D-T, Curpan R, Halip L, Bora A, Yang JJ, Knockel J, Sirimulla S, Ursu O, Oprea TI (2020) Drug-Central 2021 supports drug discovery and repositioning. Nucleic Acids Res 49(D1): D1160–D1169. https://doi.org/10.1093/nar/gkaa997

141. Schmaal L, Hibar DP, Samann PG, Hall GB, Baune BT, Jahanshad N, Cheung JW, van Erp TGM, Bos D, Ikram MA, Vernooij MW, Niessen WJ, Tiemeier H, Hofman A, Wittfeld K, Grabe HJ, Janowitz D, Bulow R, Selonke M, Volzke H, Grotegerd D, Dannlowski U, Arolt V, Opel N, Heindel W, Kugel H, Hoehn D, Czisch M, Couvy-Duchesne B, Renteria ME, Strike LT, Wright MJ, Mills NT, de Zubicaray GI, McMahon KL, Medland SE, Martin NG, Gillespie NA, Goya-Maldonado R, Gruber O, Kramer B, Hatton SN, Lagopoulos J, Hickie IB, Frodl T, Carballedo A, Frey EM, van Velzen LS, Penninx B, van Tol MJ, van der Wee NJ, Davey CG, Harrison BJ, Mwangi B, Cao B, Soares JC, Veer IM, Walter H, Schoepf D, Zurowski B, Konrad C, Schramm E, Normann C, Schnell K, Sacchet MD, Gotlib IH, MacQueen GM, Godlewska BR, Nickson T, McIntosh AM, Papmeyer M, Whalley HC, Hall J, Sussmann JE, Li M, Walter M, Aftanas L, Brack I, Bokhan NA, Thompson PM, Veltman DJ (2017) Cortical abnormalities in adults and adolescents with major depression based on brain scans from 20 cohorts worldwide in the ENIGMA major depressive disorder working group. Mol Psychiatry 22(6):900–909. https://doi.org/10.1038/mp.2016.60

142. Schmaal L, Veltman DJ, van Erp TG, Sämann PG, Frodl T, Jahanshad N, Loehrer E, Tiemeier H (2016) Subcortical brain alterations in major depressive disorder: findings from the ENIGMA Major Depressive Disorder working group. Mol Psychiatry 21(6): 806–812. https://doi.org/10.1038/mp.2015.69

143. Kaiser RH, Andrews-Hanna JR, Wager TD, Pizzagalli DA (2015) Large-scale network dysfunction in major depressive disorder: a meta-analysis of resting-state functional connectivity. JAMA Psychiatry 72(6):603–611. https://doi.org/10.1001/jamapsychiatry.2015.0071

144. Kraus C, Seiger R, Pfabigan DM, Sladky R, Tik M, Paul K, Woletz M, Gryglewski G, Vanicek T, Komorowski A (2019) Hippocampal subfields in acute and remitted depression—an ultra-high field magnetic resonance imaging study. Int J Neuropsychopharmacol 22(8):513–522

145. Kraus C, Klöbl M, Tik M (2019) The pulvinar nucleus and antidepressant treatment: dynamic modeling of antidepressant response and remission with ultra-high field functional MRI. Mol Psychiatry 24(5):746–756. https://doi.org/10.1038/s41380-017-0009-x

146. Klöbl M, Gryglewski G, Rischka L, Godbersen GM, Unterholzner J, Reed MB, Michenthaler P, Vanicek T, Winkler-Pjrek E, Hahn A, Kasper S, Lanzenberger R (2020) Predicting antidepressant citalopram treatment response via changes in brain functional connectivity after acute intravenous challenge. Front Comput Neurosci 14:554186. https://doi.org/10.3389/fncom.2020.554186

147. Rahmim A, Zaidi H (2008) PET versus SPECT: strengths, limitations and challenges. Nucl Med Commun 29(3):193–207. https://doi.org/10.1097/MNM.0b013e3282f3a515

148. Gryglewski G, Klöbl M, Berroterán-Infante N, Rischka L, Balber T, Vanicek T, Pichler V, Kautzky A, Klebermass EM, Reed MB, Vraka C, Hienert M, James GM, Silberbauer L, Godbersen GM, Unterholzner J, Michenthaler P, Hartenbach M, Winkler-Pjrek E, Wadsak W, Mitterhauser M, Hahn A, Hacker M, Kasper S, Lanzenberger R (2019) Modeling the acute pharmacological response to selective serotonin reuptake inhibitors in human brain using simultaneous PET/MR imaging. Eur Neuropsychopharmacol 29(6):711–719. https://doi.org/10.1016/j.euroneuro.2019.04.001

149. Rischka L, Gryglewski G, Berroterán-Infante N, Rausch I, James GM, Klöbl M, Sigurdardottir H, Hartenbach M, Hahn A, Wadsak W, Mitterhauser M, Beyer T, Kasper S, Prayer D, Hacker M, Lanzenberger R (2019) Attenuation correction approaches for serotonin transporter quantification with PET/MRI. Front Physiol 10:1422. https://doi.org/10.3389/fphys.2019.01422

150. Hahn A, Gryglewski G, Nics L, Rischka L, Ganger S, Sigurdardottir H, Vraka C, Silberbauer L, Vanicek T, Kautzky A, Wadsak W, Mitterhauser M, Hartenbach M, Hacker M, Kasper S, Lanzenberger R (2018) Task-relevant brain networks identified with simultaneous PET/MR imaging of metabolism and connectivity. Brain Struct Funct 223(3):1369–1378. https://doi.org/10.1007/s00429-017-1558-0

151. Townsend D (2004) Physical principles and technology of clinical PET imaging. Ann Acad Med Singap 33(2):133–145

152. Disselhorst JA, Bezrukov I, Kolb A, Parl C, Pichler BJ (2014) Principles of PET/MR imaging. J Nucl Med 55(Supplement 2): 2S–10S

153. Wadsak W, Mitterhauser M (2010) Basics and principles of radiopharmaceuticals for PET/CT. Eur J Radiol 73(3):461–469. https://doi.org/10.1016/j.ejrad.2009.12.022

154. Pike VW (2009) PET radiotracers: crossing the blood-brain barrier and surviving metabolism. Trends Pharmacol Sci 30(8):431–440. https://doi.org/10.1016/j.tips.2009.05.005

155. Ishiwata K, Kawamura K, Yanai K, Hendrikse NH (2007) In vivo evaluation of P-glycoprotein modulation of 8 PET Radioligands used clinically. J Nucl Med 48(1): 81–87

156. Hall H, Lundkvist C, Halldin C, Farde L, Pike VW, McCarron JA, Fletcher A, Cliffe IA, Barf T, Wikström H, Sedvall G (1997) Autoradiographic localization of 5-HT1A receptors in the post-mortem human brain using [3H]WAY-100635 and [11C]way-100635. Brain Res 745(1–2):96–108. https://doi.org/10.1016/s0006-8993(96)01131-6

157. Varnas K, Nyberg S, Halldin C, Varrone A, Takano A, Karlsson P, Andersson J, McCarthy D, Smith M, Pierson ME, Soderstrom J, Farde L (2011) Quantitative analysis of [11C]AZ10419369 binding to 5-HT1B receptors in human brain. J Cereb Blood Flow Metab 31(1):113–123. https://doi.org/10.1038/jcbfm.2010.55

158. Saulin A, Savli M, Lanzenberger R (2012) Serotonin and molecular neuroimaging in humans using PET. Amino Acids 42(6): 2039–2057. https://doi.org/10.1007/s00726-011-1078-9

159. Salentin S, Haupt VJ, Daminelli S, Schroeder M (2014) Polypharmacology rescored: protein-ligand interaction profiles for remote binding site similarity assessment. Prog Biophys Mol Biol 116(2–3):174–186. https://doi.org/10.1016/j.pbiomolbio.2014.05.006

160. Laruelle M, Slifstein M, Huang Y (2003) Relationships between radiotracer properties and image quality in molecular imaging of the brain with positron emission tomography. Mol Imaging Biol 5(6):363–375. https://doi.org/10.1016/j.mibio.2003.09.009

161. Zhang Y, Fox G (2012) PET imaging for receptor occupancy: meditations on calculation and simplification. J Biomed Res 26: 69–76. https://doi.org/10.1016/S1674-8301(12)60014-1

162. Mintun MA, Raichle ME, Kilbourn MR, Wooten GF, Welch MJ (1984) A quantitative model for the in vivo assessment of drug binding sites with positron emission tomography. Ann Neurol 15(3):217–227. https://doi.org/10.1002/ana.410150302

163. Laruelle M (2000) Imaging synaptic neurotransmission with in vivo binding competition techniques: a critical review. J Cereb Blood Flow Metab 20(3):423–451. https://doi.org/10.1097/00004647-200003000-00001

164. Waterhouse RN (2003) Determination of lipophilicity and its use as a predictor of blood-brain barrier penetration of molecular imaging agents. Mol Imaging Biol 5(6): 376–389. https://doi.org/10.1016/j.mibio.2003.09.014

165. Fischer H, Gottschlich R, Seelig A (1998) Blood-brain barrier permeation: molecular parameters governing passive diffusion. J Membr Biol 165(3):201–211. https://doi.org/10.1007/s002329900434

166. Gerebtzoff G, Seelig A (2006) In silico prediction of blood-brain barrier permeation using the calculated molecular cross-sectional area as main parameter. J Chem Inf Model 46(6):2638–2650. https://doi.org/10.1021/ci0600814

167. Seelig A (2007) The role of size and charge for blood-brain barrier permeation of drugs and fatty acids. J Mol Neurosci 33(1):32–41. https://doi.org/10.1007/s12031-007-0055-y

168. Lever SZ, Fan KH, Lever JR (2017) Tactics for preclinical validation of receptor-binding radiotracers. Nucl Med Biol 44:4–30. https://doi.org/10.1016/j.nucmedbio.2016.08.015

169. Innis RB, Cunningham VJ, Delforge J, Fujita M, Gjedde A, Gunn RN, Holden J, Houle S, Huang SC, Ichise M, Iida H, Ito H, Kimura Y, Koeppe RA, Knudsen GM, Knuuti J, Lammertsma AA, Laruelle M, Logan J, Maguire RP, Mintun MA, Morris ED, Parsey R, Price JC, Slifstein M, Sossi V, Suhara T, Votaw JR, Wong DF, Carson RE (2007) Consensus nomenclature for in vivo imaging of reversibly binding radioligands. J Cereb Blood Flow Metab 27(9):1533–1539. https://doi.org/10.1038/sj.jcbfm.9600493

170. Lassen NA (1992) Neuroreceptor quantitation in vivo by the steady-state principle using constant infusion or bolus injection of radioactive tracers. J Cereb Blood Flow Metab 12(5):709–716. https://doi.org/10.1038/jcbfm.1992.101

171. Visser AK, van Waarde A, Willemsen AT, Bosker FJ, Luiten PG, den Boer JA, Kema IP, Dierckx RA (2011) Measuring serotonin synthesis: from conventional methods to PET tracers and their (pre)clinical implications. Eur J Nucl Med Mol Imaging 38(3):576–591. https://doi.org/10.1007/s00259-010-1663-2

172. Hartvig P, Lindner KJ, Tedroff J, Andersson Y, Bjurling P, Langstrom B (1992) Brain kinetics of 11 C-labelled L-tryptophan and 5-hydroxy-L-tryptophan in the rhesus monkey. A study using positron emission tomography. J Neural Transm Gen Sect 88(1):1–10. https://doi.org/10.1007/BF01245032

173. Diksic M, Nagahiro S, Sourkes TL, Yamamoto YL (1990) A new method to measure brain serotonin synthesis in vivo. I. Theory and basic data for a biological model. J Cereb Blood Flow Metab 10(1):1–12. https://doi.org/10.1038/jcbfm.1990.1

174. Muzik O, Chugani DC, Chakraborty P, Mangner T, Chugani HT (1997) Analysis of [C-11]alpha-methyl-tryptophan kinetics for the estimation of serotonin synthesis rate in vivo. J Cereb Blood Flow Metab 17(6):659–669. https://doi.org/10.1097/00004647-199706000-00007

175. Hagberg GE, Torstenson R, Marteinsdottir I, Fredrikson M, Langstrom B, Blomqvist G (2002) Kinetic compartment modeling of [11C]-5-hydroxy-L-tryptophan for positron emission tomography assessment of serotonin synthesis in human brain. J Cereb Blood Flow

Metab 22(11):1352–1366. https://doi.org/10.1097/01.WCB.0000040946.89393.9d

176. Lundquist P, Hartvig P, Blomquist G, Hammarlund-Udenaes M, Langstrom B (2007) 5-Hydroxy-L-[beta-11C]tryptophan versus alpha-[11C]methyl-L-tryptophan for positron emission tomography imaging of serotonin synthesis capacity in the rhesus monkey brain. J Cereb Blood Flow Metab 27(4):821–830. https://doi.org/10.1038/sj.jcbfm.9600381

177. Visser AK, Ramakrishnan NK, Willemsen AT, Di Gialleonardo V, de Vries EF, Kema IP, Dierckx RA, van Waarde A (2014) [(11)C]5-HTP and microPET are not suitable for pharmacodynamic studies in the rodent brain. J Cereb Blood Flow Metab 34(1):118–125. https://doi.org/10.1038/jcbfm.2013.171

178. Bergstrom M, Westerberg G, Langstrom B (1997) 11C-harmine as a tracer for monoamine oxidase a (MAO-A): in vitro and in vivo studies. Nucl Med Biol 24(4):287–293

179. Ginovart N, Wilson AA, Meyer JH, Hussey D, Houle S (2003) [11C]-DASB, a tool for in vivo measurement of SSRI-induced occupancy of the serotonin transporter: PET characterization and evaluation in cats. Synapse 47(2):123–133. https://doi.org/10.1002/syn.10155

180. Wilson AA, Ginovart N, Schmidt M, Meyer JH, Threlkeld PG, Houle S (2000) Novel radiotracers for imaging the serotonin transporter by positron emission tomography: synthesis, radiosynthesis, and in vitro and ex vivo evaluation of (11)C-labeled 2-(phenylthio)-araalkylamines. J Med Chem 43(16):3103–3110

181. Wilson AA, Jin L, Garcia A, DaSilva JN, Houle S (2001) Carbon-11 labelled cholecystokininB antagonists: radiosynthesis and evaluation in rats. Life Sci 68(11):1223–1230. https://doi.org/10.1016/s0024-3205(00)01021-3

182. Liu CT, Huang YS, Chen HC, Ma KH, Wang CH, Chiu CH, Shih JH, Kang HH, Shiue CY, Li IH (2019) Evaluation of brain SERT with 4-[(18)F]-ADAM/micro-PET and hearing protective effects of dextromethorphan in hearing loss rat model. Toxicol Appl Pharmacol 378:114604. https://doi.org/10.1016/j.taap.2019.114604

183. Halldin C, Lundberg J, Sovago J, Gulyas B, Guilloteau D, Vercouillie J, Emond P, Chalon S, Tarkiainen J, Hiltunen J, Farde L (2005) [(11)C]MADAM, a new serotonin transporter radioligand characterized in the

monkey brain by PET. Synapse 58(3): 173–183. https://doi.org/10.1002/syn. 20189

184. Walther DJ, Peter J-U, Bashammakh S, Hortnagl H, Voits M, Fink H, Bader M (2003) Synthesis of serotonin by a second tryptophan hydroxylase isoform. Science (New York, NY) 299(5603):76–76

185. Roberge AG, Missala K, Sourkes TL (1972) Alpha-methyltryptophan: effects on synthesis and degradation of serotonin in the brain. Neuropharmacology 11(2):197–209. https://doi.org/10.1016/0028-3908(72) 90092-5

186. Chugani DC, Muzik O (2000) Alpha[C-11] methyl-L-tryptophan PET maps brain serotonin synthesis and kynurenine pathway metabolism. J Cereb Blood Flow Metab 20(1):2–9. https://doi.org/10.1097/00004647-200001000-00002

187. Muzik O, Chugani DC, Chakraborty P, Mangner T, Chugani HT (1997) Analysis of [C-11]alpha-methyl-tryptophan kinetics for the estimation of serotonin synthesis rate in vivo. J Cereb Blood Flow Metab 17(6): 659–669. https://doi.org/10.1097/00004647-199706000-00007

188. Visser AKD, van Waarde A, Willemsen ATM, Bosker FJ, Luiten PGM, den Boer JA, Kema IP, Dierckx RAJO (2011) Measuring serotonin synthesis: from conventional methods to PET tracers and their (pre)clinical implications. Eur J Nucl Med Mol Imaging 38(3): 576–591. https://doi.org/10.1007/s00259-010-1663-2

189. Chugani DC, Muzik O, Chakraborty P, Mangner T, Chugani HT (1998) Human brain serotonin synthesis capacity measured in vivo with alpha-[C-11]methyl-L-tryptophan. Synapse 28(1):33–43. https://doi.org/10.1002/(sici)1098-2396(199801)28:1<33::aid-syn5>3.0.co;2-d

190. Rosa-Neto P, Diksic M, Okazawa H, Leyton M, Ghadirian N, Mzengeza S, Nakai A, Debonnel G, Blier P, Benkelfat C (2004) Measurement of brain regional alpha-[11C]methyl-L-tryptophan trapping as a measure of serotonin synthesis in medication-free patients with major depression. Arch Gen Psychiatry 61(6):556–563. https://doi.org/10.1001/archpsyc.61. 6.556

191. Agren H, Reibring L, Hartvig P, Tedroff J, Bjurling P, Hörnfeldt K, Andersson Y, Lundqvist H, Långström B (1991) Low brain uptake of L-[11C]5-hydroxytryptophan in major depression: a positron emission tomography study on patients and healthy volunteers. Acta Psychiatr Scand 83(6): 449–455. https://doi.org/10.1111/j. 1600-0447.1991.tb05574.x

192. Leyton M, Paquette V, Gravel P, Rosa-Neto P, Weston F, Diksic M, Benkelfat C (2006) α-[11C]methyl-l-tryptophan trapping in the orbital and ventral medial prefrontal cortex of suicide attempters. Eur Neuropsychopharmacol 16(3):220–223. https://doi.org/10.1016/j.euroneuro.2005.09.006

193. Berney A, Nishikawa M, Benkelfat C, Debonnel G, Gobbi G, Diksic M (2008) An index of 5-HT synthesis changes during early antidepressant treatment: alpha-[11C] methyl-L-tryptophan PET study. Neurochem Int 52(4–5):701–708. https://doi.org/10. 1016/j.neuint.2007.08.021

194. Neumeister A, Nugent AC, Waldeck T, Geraci M, Schwarz M, Bonne O, Bain EE, Luckenbaugh DA, Herscovitch P, Charney DS, Drevets WC (2004) Neural and behavioral responses to tryptophan depletion in Unmedicated patients with remitted major depressive disorder and controls. Arch Gen Psychiatry 61(8):765–773. https://doi.org/ 10.1001/archpsyc.61.8.765

195. Szabo Z, Kao PF, Scheffel U, Suehiro M, Mathews WB, Ravert HT, Musachio JL, Marenco S, Kim SE, Ricaurte GA (1995) Positron emission tomography imaging of serotonin transporters in the human brain using [11C](+) McN5652. Synapse 20(1): 37–43

196. Chalon S, Tarkiainen J, Garreau L, Hall H, Emond P, Vercouillie J, Farde L, Dasse P, Varnas K, Besnard J-C (2003) Pharmacological characterization of N, N-Dimethyl-2-(2-amino-4-methylphenyl thio) benzylamine as a ligand of the serotonin transporter with high affinity and selectivity. J Pharmacol Exp Ther 304(1):81–87

197. Wilson AA, Houle S (1999) Radiosynthesis of carbon-11 labelled N-methyl-2-(arylthio) benzylamines: potential radiotracers for the serotonin reuptake receptor. J Labelled Comp Radiopharm 42(13):1277–1288

198. Ginovart N, Wilson AA, Meyer JH, Hussey D, Houle S (2001) Positron emission tomography quantification of [11C]-DASB binding to the human serotonin transporter: modeling strategies. J Cereb Blood Flow Metab 21(11):1342–1353

199. Gryglewski G, Rischka L, Philippe C, Hahn A, James GM, Klebermass E, Hienert M, Silberbauer L, Vanicek T, Kautzky A (2017) Simple and rapid quantification of serotonin transporter binding using [11C] DASB bolus plus constant infusion. Neuro-Image 149:23–32

200. Silberbauer LR, Gryglewski G, Berroterán-Infante N, Rischka L, Vanicek T, Pichler V, Hienert M, Kautzky A, Philippe C, Godbersen GM, Vraka C, James GM, Wadsak W, Mitterhauser M, Hacker M, Kasper S, Hahn A, Lanzenberger R (2019) Serotonin Transporter Binding in the Human Brain After Pharmacological Challenge Measured Using PET and PET/MR Frontiers in molecular neuroscience. Front Mol Neurosci 12: 172. https://doi.org/10.3389/fnmol.2019. 00172

201. Spies M, Knudsen GM, Lanzenberger R, Kasper S (2015) The serotonin transporter in psychiatric disorders: insights from PET imaging. Lancet Psychiatry 2(8):743–755. https://doi.org/10.1016/s2215-0366(15) 00232-1

202. Gryglewski G, Lanzenberger R, Kranz GS, Cumming P (2014) Meta-analysis of molecular imaging of serotonin transporters in major depression. J Cereb Blood Flow Metab 34(7): 1096–1103. https://doi.org/10.1038/ jcbfm.2014.82

203. Meyer JH, Wilson AA, Sagrati S, Hussey D, Carella A, Potter WZ, Ginovart N, Spencer EP, Cheok A, Houle S (2004b) Serotonin transporter occupancy of five selective serotonin reuptake inhibitors at different doses: an [11C]DASB positron emission tomography study. Am J Psychiatry 161(5):826–835. https://doi.org/10.1176/appi.ajp.161. 5.826

204. Kersemans K, Van Laeken N, De Vos F (2013) Radiochemistry devoted to the production of monoamine oxidase (MAO-A and MAO-B) ligands for brain imaging with positron emission tomography. J Labelled Comp Radiopharm 56(3–4):78–88. https://doi. org/10.1002/jlcr.3007

205. Kim H, Sablin SO, Ramsay RR (1997) Inhibition of monoamine oxidase a by beta-carboline derivatives. Arch Biochem Biophys 337(1):137–142. https://doi.org/10.1006/ abbi.1996.9771

206. Fowler JS, MacGregor RR, Wolf AP, Arnett CD, Dewey SL, Schlyer D, Christman D, Logan J, Smith M, Sachs H et al (1987) Mapping human brain monoamine oxidase a and B with 11C-labeled suicide inactivators and PET. Science (New York, NY)

235(4787):481–485. https://doi.org/10. 1126/science.3099392

207. Bottlaender M, Dolle F, Guenther I, Roumenov D, Fuseau C, Bramoulle Y, Curet O, Jegham J, Pinquier JL, George P, Valette H (2003) Mapping the cerebral monoamine oxidase type a: positron emission tomography characterization of the reversible selective inhibitor [11C]befloxatone. J Pharmacol Exp Ther 305(2):467–473. https:// doi.org/10.1124/jpet.102.046953

208. Meyer JH, , Wilson AA, , Sagrati S, , Miler L, , Rusjan P, , Bloomfield PM, , Clark M, , Sacher J, , Voineskos AN, , Houle S, (2009) Brain monoamine oxidase a binding in major depressive disorder: relationship to selective serotonin reuptake inhibitor treatment, recovery, and recurrence Arch Gen Psychiatry. 66:1304-1312. doi:https://doi.org/10. 1001/archgenpsychiatry.2009.156

209. Rekkas PV, Wilson AA, Lee VW, Yogalingam P, Sacher J, Rusjan P, Houle S, Stewart DE, Kolla NJ, Kish S, Chiuccariello L, Meyer JH (2014) Greater monoamine oxidase a binding in perimenopausal age as measured with carbon 11-labeled harmine positron emission tomography. JAMA Psychiatry 71(8):873–879. https://doi.org/10.1001/jamapsychiatry. 2014.250

210. Spies M, James GM, Vraka C, Philippe C, Hienert M, Gryglewski G, Komorowski A, Kautzky A, Silberbauer L, Pichler V, Kranz GS, Nics L, Balber T, Baldinger-Melich P, Vanicek T, Spurny B, Winkler-Pjrek E, Wadsak W, Mitterhauser M, Hacker M, Kasper S, Lanzenberger R, Winkler D (2018) Brain monoamine oxidase a in seasonal affective disorder and treatment with bright light therapy. Transl Psychiatry 8(1): 198. https://doi.org/10.1038/s41398-018-0227-2

211. Baldinger-Melich P, Gryglewski G, Philippe C, James GM, Vraka C, Silberbauer L, Balber T, Vanicek T, Pichler V, Unterholzner J, Kranz GS, Hahn A, Winkler D, Mitterhauser M, Wadsak W, Hacker M, Kasper S, Frey R, Lanzenberger R (2019) The effect of electroconvulsive therapy on cerebral monoamine oxidase a expression in treatment-resistant depression investigated using positron emission tomography. Brain Stimul 12(3): 714–723. https://doi.org/10.1016/j.brs. 2018.12.976

212. Gozlan H, El Mestikawy S, Pichat L, Glowinski J, Hamon M (1983) Identification of presynaptic serotonin autoreceptors using a

new ligand: 3H-PAT. Nature 305(5930): 140–142. https://doi.org/10.1038/305140a0

213. Laporte AM, Lima L, Gozlan H, Hamon M (1994) Selective in vivo labelling of brain 5-HT1A receptors by [3H]WAY 100635 in the mouse. Eur J Pharmacol 271(2–3): 505–514. https://doi.org/10.1016/0014-2999(94)90812-5

214. Pazos A, Palacios JM (1985) Quantitative autoradiographic mapping of serotonin receptors in the rat brain. I. Serotonin-1 receptors. Brain Res 346(2):205–230. https://doi.org/10.1016/0006-8993(85)90856-x

215. Pazos A, Probst A, Palacios JM (1987) Serotonin receptors in the human brain--III. Autoradiographic mapping of serotonin-1 receptors. Neuroscience 21(1):97–122. https://doi.org/10.1016/0306-4522(87)90326-5

216. Lang L, Jagoda E, Schmall B, Vuong BK, Adams HR, Nelson DL, Carson RE, Eckelman WC (1999) Development of fluorine-18-labeled 5-HT1A antagonists. J Med Chem 42(9):1576–1586. https://doi.org/10.1021/jm980456f

217. Passchier J, van Waarde A, Pieterman RM, Elsinga PH, Pruim J, Hendrikse HN, Willemsen AT, Vaalburg W (2000) In vivo delineation of 5-HT1A receptors in human brain with [18F]MPPF. J Nucl Med 41(11): 1830–1835

218. Mukherjee J, Bajwa AK, Wooten DW, Hillmer AT, Pan ML, Pandey SK, Saigal N, Christian BT (2016) Comparative assessment of (18) F-Mefway as a serotonin 5-HT1A receptor PET imaging agent across species: rodents, nonhuman primates, and humans. J Comp Neurol 524(7):1457–1471. https://doi.org/10.1002/cne.23919

219. Billard T, Le Bars D, Zimmer L (2014) PET radiotracers for molecular imaging of serotonin 5-HT1A receptors. Curr Med Chem 21(1):70–81. https://doi.org/10.2174/09298673113209990215

220. Bonnavion P, Bernard JF, Hamon M, Adrien J, Fabre V (2010) Heterogeneous distribution of the serotonin 5-HT(1A) receptor mRNA in chemically identified neurons of the mouse rostral brainstem: implications for the role of serotonin in the regulation of wakefulness and REM sleep. J Comp Neurol 518(14):2744–2770. https://doi.org/10.1002/cne.22331

221. Herth MM, Knudsen GM (2015) Current radiosynthesis strategies for 5-HT2A receptor PET tracers. J Labelled Comp Radiopharm 58(7):265–273. https://doi.org/10.1002/jlcr.3288

222. L'Estrade ET, Hansen HD, Erlandsson M, Ohlsson TG, Knudsen GM, Herth MM (2018) Classics in neuroimaging: the serotonergic 2A receptor system-from discovery to modern molecular imaging. ACS Chem Neurosci 9(6):1226–1229. https://doi.org/10.1021/acschemneuro.8b00176

223. Baron JC, Samson Y, Comar D, Crouzel C, Deniker P, Agid Y (1985) In vivo study of central serotoninergic receptors in man using positron tomography. Rev Neurol 141(8–9): 537–545

224. Kroll T, Elmenhorst D, Matusch A, Wedekind F, Weisshaupt A, Beer S, Bauer A (2013) Suitability of [18F]altanserin and PET to determine 5-HT2A receptor availability in the rat brain: in vivo and in vitro validation of invasive and non-invasive kinetic models. Mol Imaging Biol 15(4):456–467. https://doi.org/10.1007/s11307-013-0621-3

225. Lemaire C, Cantineau R, Guillaume M, Plenevaux A, Christiaens L (1991) Fluorine-18-altanserin: a radioligand for the study of serotonin receptors with PET: radiolabeling and in vivo biologic behavior in rats. J Nucl Med 32(12):2266–2272

226. Staley JK, Van Dyck CH, Tan PZ, Al Tikriti M, Ramsby Q, Klump H, Ng C, Garg P, Soufer R, Baldwin RM, Innis RB (2001) Comparison of [(18)F]altanserin and [(18)F]deuteroaltanserin for PET imaging of serotonin(2A) receptors in baboon brain: pharmacological studies. Nucl Med Biol 28(3):271–279. https://doi.org/10.1016/s0969-8051(00)00212-2

227. Blin J, Pappata S, Kiyosawa M, Crouzel C, Baron JC (1988) [18F]setoperone: a new high-affinity ligand for positron emission tomography study of the serotonin-2-receptors in baboon brain in vivo. Eur J Pharmacol 147(1):73–82. https://doi.org/10.1016/0014-2999(88)90635-8

228. Ito H, Nyberg S, Halldin C, Lundkvist C, Farde L (1998) PET imaging of central 5-HT2A receptors with carbon-11-MDL 100,907. J Nucl Med 39(1):208–214

229. Finnema SJ, Stepanov V, Ettrup A, Nakao R, Amini N, Svedberg M, Lehmann C, Hansen M, Knudsen GM, Halldin C (2014) Characterization of [(11)C]Cimbi-36 as an

424   Patricia A. Handschuh et al.

agonist PET radioligand for the 5-HT(2A) and 5-HT(2C) receptors in the nonhuman primate brain. NeuroImage 84:342–353. https://doi.org/10.1016/j.neuroimage.2013.08.035

230. Ettrup A, Hansen M, Santini MA, Paine J, Gillings N, Palner M, Lehel S, Herth MM, Madsen J, Kristensen J, Begtrup M, Knudsen GM (2011) Radiosynthesis and in vivo evaluation of a series of substituted 11C-phenethylamines as 5-HT (2A) agonist PET tracers. Eur J Nucl Med Mol Imaging 38(4):681–693. https://doi.org/10.1007/s00259-010-1686-8

231. Prabhakaran J, DeLorenzo C, Zanderigo F, Knudsen GM, Gilling N, Pratap M, Jorgensen MJ, Daunais J, Kaplan JR, Parsey RV, Mann JJ, Kumar D (2019) In vivo PET imaging of [11C]CIMBI-5, a 5-HT2AR agonist radiotracer in nonhuman primates. J Pharm Pharm Sci 22(1):352–364. https://doi.org/10.18433/jpps30329

232. Yang KC, Stepanov V, Martinsson S, Ettrup A, Takano A, Knudsen GM, Halldin C, Farde L, Finnema SJ (2017) Fenfluramine reduces [11C]Cimbi-36 binding to the 5-HT2A receptor in the nonhuman primate brain. Int J Neuropsychopharmacol 20(9):683–691. https://doi.org/10.1093/ijnp/pyx051

233. Kim J, Moon BS, Lee BC, Lee HY, Kim HJ, Choo H, Pae AN, Cho YS, Min SJ (2017) A potential PET radiotracer for the 5-HT2C receptor: synthesis and in vivo evaluation of 4-(3-[(18)F]fluorophenethoxy)pyrimidine. ACS Chem Neurosci 8(5):996–1003. https://doi.org/10.1021/acschemneuro.6b00445

234. Laporte AM, Koscielniak T, Ponchant M, Verge D, Hamon M, Gozlan H (1992) Quantitative autoradiographic mapping of 5-HT3 receptors in the rat CNS using [125I]iodozacopride and [3H]zacopride as radioligands. Synapse 10(4):271–281. https://doi.org/10.1002/syn.890100402

235. Hoyer D, Neijt HC (1987) Identification of serotonin 5-HT3 recognition sites by radioligand binding in NG108-15 neuroblastomaglioma cells. Eur J Pharmacol 143(2):291–292. https://doi.org/10.1016/0014-2999(87)90547-4

236. Pithia NK, Liang C, Pan XZ, Pan ML, Mukherjee J (2016) Synthesis and evaluation of (S)-[(18)F]fesetron in the rat brain as a potential PET imaging agent for serotonin 5-HT3 receptors. Bioorg Med Chem Lett 26(8):1919–1924. https://doi.org/10.1016/j.bmcl.2016.03.018

237. Vilaro MT, Cortes R, Mengod G (2005) Serotonin 5-HT4 receptors and their mRNAs in rat and Guinea pig brain: distribution and effects of neurotoxic lesions. J Comp Neurol 484(4):418–439. https://doi.org/10.1002/cne.20447

238. Domenech T, Beleta J, Fernandez AG, Gristwood RW, Cruz Sanchez F, Tolosa E, Palacios JM (1994) Identification and characterization of serotonin 5-HT4 receptor binding sites in human brain: comparison with other mammalian species. Brain Res Mol Brain Res 21(1–2):176–180. https://doi.org/10.1016/0169-328x(94)90392-1

239. Waeber C, Sebben M, Grossman C, Javoy-Agid F, Bockaert J, Dumuis A (1993) [3H]-GR113808 labels 5-HT4 receptors in the human and Guinea-pig brain. Neuroreport 4(11):1239–1242. https://doi.org/10.1097/00001756-199309000-00007

240. Zhang X, Andren PE, Glennon RA, Svenningsson P (2011) Distribution, level, pharmacology, regulation, and signaling of 5-HT6 receptors in rats and marmosets with special reference to an experimental model of parkinsonism. J Comp Neurol 519(9):1816–1827. https://doi.org/10.1002/cne.22605

241. Parker CA, Gunn RN, Rabiner EA, Slifstein M, Comley R, Salinas C, Johnson CN, Jakobsen S, Houle S, Laruelle M, Cunningham VJ, Martarello L (2012) Radiosynthesis and characterization of 11C-GSK215083 as a PET radioligand for the 5-HT6 receptor. J Nucl Med 53(2):295–303. https://doi.org/10.2967/jnumed.111.093419

242. Colomb J, Becker G, Fieux S, Zimmer L, Billard T (2014) Syntheses, radiolabelings, and in vitro evaluations of fluorinated PET radioligands of 5-HT6 serotoninergic receptors. J Med Chem 57(9):3884–3890. https://doi.org/10.1021/jm500372e

243. Herth MM, Hansen HD, Ettrup A, Dyssegaard A, Lehel S, Kristensen J, Knudsen GM (2012) Synthesis and evaluation of [(1)(1)C]Cimbi-806 as a potential PET ligand for 5-HT(7) receptor imaging. Bioorg Med Chem 20(14):4574–4581. https://doi.org/10.1016/j.bmc.2012.05.005

244. Hansen HD, Lacivita E, Di Pilato P, Herth MM, Lehel S, Ettrup A, Andersen VL, Dyssegaard A, De Giorgio P, Perrone R, Berardi F, Colabufo NA, Niso M, Knudsen GM, Leopoldo M (2014) Synthesis, radiolabeling and in vivo evaluation of [(11)C](R)-1-[4-[2-(4-methoxyphenyl)phenyl]piperazin-1-yl]-3-(2-pyrazinyloxy)-2-propanol, a potential PET radioligand for the 5-HT(7)

receptor. Eur J Med Chem 79:152–163. https://doi.org/10.1016/j.ejmech.2014.03.066

245. Lemoine L, Andries J, Le Bars D, Billard T, Zimmer L (2011) Comparison of 4 radiolabeled antagonists for serotonin 5-HT(7) receptor neuroimaging: toward the first PET radiotracer. J Nucl Med 52(11):1811–1818. https://doi.org/10.2967/jnumed.111.089185

246. Albert PR, Le François B, Vahid-Ansari F (2019) Genetic, epigenetic and posttranscriptional mechanisms for treatment of major depression: the 5-HT1A receptor gene as a paradigm. J Psychiatry Neurosci 44(3):164–176. https://doi.org/10.1503/jpn.180209

247. Bailer UF, Frank GK, Henry SE, Price JC, Meltzer CC, Weissfeld L, Mathis CA, Drevets WC, Wagner A, Hoge J (2005) Altered brain serotonin 5-HT1A receptor binding after recovery from anorexia nervosa measured by positron emission tomography and [carbonyl11C] WAY-100635. Arch Gen Psychiatry 62(9):1032–1041

248. Dillon KA, Gross-Isseroff R, Israeli M, Biegon A (1991) Autoradiographic analysis of serotonin 5-HT1A receptor binding in the human brain postmortem: effects of age and alcohol. Brain Res 554(1–2):56–64

249. Anisman H, Merali Z, Hayley S (2008) Neurotransmitter, peptide and cytokine processes in relation to depressive disorder: comorbidity between depression and neurodegenerative disorders. Prog Neurobiol 85(1):1–74. https://doi.org/10.1016/j.pneurobio.2008.01.004

250. Ögren SO, Eriksson TM, Elvander-Tottie E, D'Addario C, Ekström JC, Svenningsson P, Meister B, Kehr J, Stiedl O (2008) The role of 5-HT1A receptors in learning and memory. Behav Brain Res 195(1):54–77

251. Selvaraj S, Mouchlianitis E, Faulkner P, Turkheimer F, Cowen PJ, Roiser JP, Howes O (2015) Presynaptic serotoninergic regulation of emotional processing: a multimodal brain imaging study. Biol Psychiatry 78(8):563–571. https://doi.org/10.1016/j.biopsych.2014.04.011

252. Silberbauer LR, James GM, Spies M, Michenthaler P, Kranz GS, Kasper S, Lanzenberger R (2020) Chapter 9 - molecular neuroimaging of the serotonergic system with positron emission tomography. In: Müller CP, Cunningham KA (eds) Handbook of behavioral neuroscience, vol 31. Elsevier, Amsterdam, pp 175–194. https://doi.org/10.1016/B978-0-444-64125-0.00009-8

253. Fink KB, Göthert M (2007) 5-HT receptor regulation of neurotransmitter release. Pharmacol Rev 59(4):360–417. https://doi.org/10.1124/pr.107.07103

254. Pike VW, McCarron JA, Lammertsma AA, Osman S, Hume SP, Sargent PA, Bench CJ, Cliffe IA, Fletcher A, Grasby PM (1996) Exquisite delineation of 5-HT1A receptors in human brain with PET and [carbonyl-11C]WAY-100635. Eur J Pharmacol 301(1):R5–R7. https://doi.org/10.1016/0014-2999(96)00079-9

255. Clawges HM, Depree KM, Parker EM, Graber SG (1997) Human 5-HT1 receptor subtypes exhibit distinct G protein coupling behaviors in membranes from Sf9 cells. Biochemistry 36(42):12930–12938. https://doi.org/10.1021/bi970112b

256. Shrestha SS, Liow JS, Lu S, Jenko K, Gladding RL, Svenningsson P, Morse CL, Zoghbi SS, Pike VW, Innis RB (2014) (11)C-CUMI-101, a PET radioligand, behaves as a serotonin 1A receptor antagonist and also binds to α(1) adrenoceptors in brain. J Nucl Med 55(1):141–146. https://doi.org/10.2967/jnumed.113.125831

257. Shrestha SS, Liow J-S, Jenko K, Ikawa M, Zoghbi SS, Innis RB (2016) The 5-HT1A receptor PET Radioligand 11C-CUMI-101 has significant binding to α1-adrenoceptors in human cerebellum, limiting its use as a reference region. J Nucl Med 57(12):1945–1948. https://doi.org/10.2967/jnumed.116.174151

258. Pinborg LH, Feng L, Haahr ME, Gillings N, Dyssegaard A, Madsen J, Svarer C, Yndgaard S, Kjaer TW, Parsey RV, Hansen HD, Ettrup A, Paulson OB, Knudsen GM (2012) No change in [11C]CUMI-101 binding to 5-HT1A receptors after intravenous citalopram in human. Synapse 66(10):880–884. https://doi.org/10.1002/syn.21579

259. Mann JJ, Metts AV, Ogden RT, Mathis CA, Rubin-Falcone H, Gong Z, Drevets WC, Zelazny J, Brent DA (2019) Quantification of 5-HT(1A) and 5-HT(2A) receptor binding in depressed suicide attempters and non-attempters. Arch Suicide Res 23(1):122–133. https://doi.org/10.1080/13811118.2017.1417185

260. Langenecker SA, Mickey BJ, Eichhammer P, Sen S, Elverman KH, Kennedy SE, Heitzeg MM, Ribeiro SM, Love TM, Hsu DT, Koeppe RA, Watson SJ, Akil H, Goldman D, Burmeister M, Zubieta JK (2019) Cognitive control as a 5-HT(1A)-based domain that is disrupted in major depressive disorder. Front

Psychol 10:691. https://doi.org/10.3389/fpsyg.2019.00691

261. Milak MS, Pantazatos S, Rashid R, Zanderigo F, DeLorenzo C, Hesselgrave N, Ogden RT, Oquendo MA, Mulhern ST, Miller JM, Burke AK, Parsey RV, Mann JJ (2018) Higher 5-HT(1A) autoreceptor binding as an endophenotype for major depressive disorder identified in high risk offspring - a pilot study. Psychiatry Res Neuroimaging 276:15–23. https://doi.org/10.1016/j.pscychresns.2018.04.002

262. Pillai RLI, Zhang M, Yang J, Boldrini M, Mann JJ, Oquendo MA, Parsey RV, DeLorenzo C (2018) Will imaging individual raphe nuclei in males with major depressive disorder enhance diagnostic sensitivity and specificity? Depress Anxiety 35(5):411–420. https://doi.org/10.1002/da.22721

263. Meltzer CC, Price JC, Mathis CA, Butters MA, Ziolko SK, Moses-Kolko E, Mazumdar S, Mulsant BH, Houck PR, Lopresti BJ, Weissfeld LA, Reynolds CF (2004) Serotonin 1A receptor binding and treatment response in late-life depression. Neuropsychopharmacology 29(12): 2258–2265. https://doi.org/10.1038/sj.npp.1300556

264. Hirvonen J, Karlsson H, Kajander J, Lepola A, Markkula J, Rasi-Hakala H, Någren K, Salminen JK, Hietala J (2008) Decreased brain serotonin 5-HT1A receptor availability in medication-naive patients with major depressive disorder: an in-vivo imaging study using PET and [carbonyl-11C]WAY-100635. Int J Neuropsychopharmacol 11(4):465–476. https://doi.org/10.1017/s1461145707008140

265. Drevets WC, Frank E, Price JC, Kupfer DJ, Holt D, Greer PJ, Huang Y, Gautier C, Mathis C (1999) PET imaging of serotonin 1A receptor binding in depression. Biol Psychiatry 46(10):1375–1387. https://doi.org/10.1016/s0006-3223(99)00189-4

266. Sargent PA, Kjaer KH, Bench CJ, Rabiner EA, Messa C, Meyer J, Gunn RN, Grasby PM, Cowen PJ (2000) Brain serotonin1A receptor binding measured by positron emission tomography with [11C]WAY-100635: effects of depression and antidepressant treatment. Arch Gen Psychiatry 57(2):174–180. https://doi.org/10.1001/archpsyc.57.2.174

267. Wang L, Zhou C, Zhu D, Wang X, Fang L, Zhong J, Mao Q, Sun L, Gong X, Xia J, Lian B, Xie P (2016) Serotonin-1A receptor alterations in depression: a meta-analysis of molecular imaging studies. BMC Psychiatry 16(1):319. https://doi.org/10.1186/s12888-016-1025-0

268. Metts AV, Rubin-Falcone H, Ogden RT, Lin X, Wilner DE, Burke AK, Sublette ME, Oquendo MA, Miller JM, Mann JJ (2019) Antidepressant medication exposure and 5-HT(1A) autoreceptor binding in major depressive disorder. Synapse 73(6):e22089. https://doi.org/10.1002/syn.22089

269. Miller JM, Brennan KG, Ogden TR, Oquendo MA, Sullivan GM, Mann JJ, Parsey RV (2009) Elevated serotonin 1A binding in remitted major depressive disorder: evidence for a trait biological abnormality. Neuropsychopharmacology 34(10):2275–2284. https://doi.org/10.1038/npp.2009.54

270. Parsey RV, Olvet DM, Oquendo MA, Huang YY, Ogden RT, Mann JJ (2006) Higher 5-HT1A receptor binding potential during a major depressive episode predicts poor treatment response: preliminary data from a naturalistic study. Neuropsychopharmacology 31(8):1745–1749. https://doi.org/10.1038/sj.npp.1300992

271. Parsey RV, Oquendo MA, Ogden RT, Olvet DM, Simpson N, Huang YY, Van Heertum RL, Arango V, Mann JJ (2006) Altered serotonin 1A binding in major depression: a [carbonyl-C-11]WAY100635 positron emission tomography study. Biol Psychiatry 59(2): 106–113. https://doi.org/10.1016/j.biopsych.2005.06.016

272. Lothe A, Saoud M, Bouvard S, Redouté J, Lerond J, Ryvlin P (2012) 5-HT1A receptor binding changes in patients with major depressive disorder before and after antidepressant treatment: a pilot [18F]MPPF positron emission tomography study. Psychiatry Res Neuroimaging 203(1):103–104. https://doi.org/10.1016/j.pscychresns.2011.09.001

273. Stenkrona P, Halldin C, Lundberg J (2013) 5-HTT and 5-HT1A receptor occupancy of the novel substance vortioxetine (Lu AA21004). A PET study in control subjects. Eur Neuropsychopharmacol 23(10): 1190–1198. https://doi.org/10.1016/j.euroneuro.2013.01.002

274. Paterson LM, Kornum BR, Nutt DJ, Pike VW, Knudsen GM (2013) 5-HT radioligands for human brain imaging with PET and SPECT. Med Res Rev 33(1):54–111. https://doi.org/10.1002/med.20245

275. Varnäs K, Hall H, Bonaventure P, Sedvall G (2001) Autoradiographic mapping of 5-HT (1B) and 5-HT(1D) receptors in the post mortem human brain using [(3)H]GR 125743. Brain Res 915(1):47–57. https://doi.org/10.1016/s0006-8993(01)02823-2

276. Gallezot JD, Nabulsi N, Neumeister A, Planeta-Wilson B, Williams WA, Singhal T, Kim S, Maguire RP, McCarthy T, Frost JJ, Huang Y, Ding YS, Carson RE (2010) Kinetic modeling of the serotonin 5-HT(1B) receptor radioligand [(11)C]P943 in humans. J Cereb Blood Flow Metab 30(1):196–210. https://doi.org/10.1038/jcbfm.2009.195

277. Tiger M, Varnäs K, Okubo Y, Lundberg J (2018) The 5-HT1B receptor - a potential target for antidepressant treatment. Psychopharmacology 235(5):1317–1334. https://doi.org/10.1007/s00213-018-4872-1

278. Murrough JW, Henry S, Hu J, Gallezot JD, Planeta-Wilson B, Neumaier JF, Neumeister A (2011) Reduced ventral striatal/ventral pallidal serotonin1B receptor binding potential in major depressive disorder. Psychopharmacology 213(2–3):547–553. https://doi.org/10.1007/s00213-010-1881-0

279. Tiger M, Farde L, Rück C, Varrone A, Forsberg A, Lindefors N, Halldin C, Lundberg J (2016) Low serotonin1B receptor binding potential in the anterior cingulate cortex in drug-free patients with recurrent major depressive disorder. Psychiatry Res Neuroimaging 253:36–42. https://doi.org/10.1016/j.pscychresns.2016.04.016

280. Nord M, Finnema SJ, Halldin C, Farde L (2013) Effect of a single dose of escitalopram on serotonin concentration in the non-human and human primate brain. Int J Neuropsychopharmacol 16(7):1577–1586. https://doi.org/10.1017/S1461145712001617

281. Price RB, Iosifescu DV, Murrough JW, Chang LC, Al Jurdi RK, Iqbal SZ, Soleimani L, Charney DS, Foulkes AL, Mathew SJ (2014) Effects of ketamine on explicit and implicit suicidal cognition: a randomized controlled trial in treatment-resistant depression. Depress Anxiety 31(4):335–343

282. Berman RM, Cappiello A, Anand A, Oren DA, Heninger GR, Charney DS, Krystal JH (2000) Antidepressant effects of ketamine in depressed patients. Biol Psychiatry 47(4): 351–354

283. Yang C, Yang J, Luo A, Hashimoto K (2019) Molecular and cellular mechanisms underlying the antidepressant effects of ketamine enantiomers and its metabolites. Transl Psychiatry 9(1):280. https://doi.org/10.1038/s41398-019-0624-1

284. Seeman P (2002) Atypical antipsychotics: mechanism of action. Can J Psychiatry 47(1):27–38

285. Van Oekelen D, Luyten WHML, Leysen JE (2003) 5-HT2A and 5-HT2C receptors and their atypical regulation properties. Life Sci 72(22):2429–2449. https://doi.org/10.1016/S0024-3205(03)00141-3

286. Savli M, Bauer A, Mitterhauser M, Ding Y-S, Hahn A, Kroll T, Neumeister A, Haeusler D, Ungersboeck J, Henry S (2012) Normative database of the serotonergic system in healthy subjects using multi-tracer PET. NeuroImage 63(1):447–459

287. Biver F, Goldman S, Luxen A, Monclus M, Forestini M, Mendlewicz J, Lotstra F (1994) Multicompartmental study of fluorine-18 altanserin binding to brain 5HT2 receptors in humans using positron emission tomography. Eur J Nucl Med 21(9):937–946

288. van Dyck CH, Tan PZ, Baldwin RM, Amici LA, Garg PK, Ng CK, Soufer R, Charney DS, Innis RB (2000) PET quantification of 5-HT2A receptors in the human brain: a constant infusion paradigm with [18F]altanserin. J Nucl Med 41(2):234–241

289. Ettrup A, da Cunha-Bang S, McMahon B, Lehel S, Dyssegaard A, Skibsted AW, Jorgensen LM, Hansen M, Baandrup AO, Bache S, Svarer C, Kristensen JL, Gillings N, Madsen J, Knudsen GM (2014) Serotonin 2A receptor agonist binding in the human brain with [(1)(1)C]Cimbi-36. J Cereb Blood Flow Metab 34(7):1188–1196. https://doi.org/10.1038/jcbfm.2014.68

290. López-Figueroa AL, Norton CS, López-Figueroa MO, Armellini-Dodel D, Burke S, Akil H, López JF, Watson SJ (2004) Serotonin 5-HT1A, 5-HT1B, and 5-HT2A receptor mRNA expression in subjects with major depression, bipolar disorder, and schizophrenia. Biol Psychiatry 55(3):225–233

291. Biver F, Wikler D, Lotstra F, Damhaut P, Goldman S, Mendlewicz J (1997) Serotonin 5-HT2 receptor imaging in major depression: focal changes in orbito-insular cortex. Br J Psychiatry 171:444–448. https://doi.org/10.1192/bjp.171.5.444

292. Attar-Lévy D, Martinot JL, Blin J, Dao-Castellana MH, Crouzel C, Mazoyer B, Poirier MF, Bourdel MC, Aymard N, Syrota A, Féline A (1999) The cortical

serotonin2 receptors studied with positron-emission tomography and [18F]-setoperone during depressive illness and antidepressant treatment with clomipramine. Biol Psychiatry 45(2):180–186. https://doi.org/10.1016/s0006-3223(98)00007-9

293. Yatham LN, Liddle PF, Shiah IS, Scarrow G, Lam RW, Adam MJ, Zis AP, Ruth TJ (2000) Brain serotonin2 receptors in major depression: a positron emission tomography study. Arch Gen Psychiatry 57(9):850–858. https://doi.org/10.1001/archpsyc.57.9.850

294. Mintun MA, Sheline YI, Moerlein SM, Vlassenko AG, Huang Y, Snyder AZ (2004) Decreased hippocampal 5-HT2A receptor binding in major depressive disorder: in vivo measurement with [18F]altanserin positron emission tomography. Biol Psychiatry 55(3):217–224

295. Meltzer CC, Price JC, Mathis CA, Greer PJ, Cantwell MN, Houck PR, Mulsant BH, Ben-Eliezer D, Lopresti B, DeKosky ST, Reynolds CF (1999) PET imaging of serotonin type 2A receptors in late-life neuropsychiatric disorders. Am J Psychiatr 156(12):1871–1878. https://doi.org/10.1176/ajp.156.12.1871

296. Meyer JH, Kapur S, Houle S, DaSilva J, Owczarek B, Brown GM, Wilson AA, Kennedy SH (1999) Prefrontal cortex 5-HT2 receptors in depression: an [18F]setoperone PET imaging study. Am J Psychiatry 156(7):1029–1034. https://doi.org/10.1176/ajp.156.7.1029

297. Meyer JH, McMain S, Kennedy SH, Korman L, Brown GM, DaSilva JN, Wilson AA, Blak T, Eynan-Harvey R, Goulding VS, Houle S, Links P (2003) Dysfunctional attitudes and 5-HT2 receptors during depression and self-harm. Am J Psychiatry 160(1):90–99. https://doi.org/10.1176/appi.ajp.160.1.90

298. Bhagwagar Z, Hinz R, Taylor M, Fancy S, Cowen P, Grasby P (2006) Increased 5-HT (2A) receptor binding in euthymic, medication-free patients recovered from depression: a positron emission study with [(11)C]MDL 100,907. Am J Psychiatry 163(9):1580–1587. https://doi.org/10.1176/ajp.2006.163.9.1580

299. Madsen MK, Fisher PM, Burmester D, Dyssegaard A, Stenbæk DS, Kristiansen S, Johansen SS, Lehel S, Linnet K, Svarer C, Erritzoe D, Ozenne B, Knudsen GM (2019) Psychedelic effects of psilocybin correlate with serotonin 2A receptor occupancy and plasma psilocin levels. Neuropsychopharmacology 44(7):1328–1334. https://doi.org/10.1038/s41386-019-0324-9

300. Erritzoe D, Ashok AH, Searle GE, Colasanti A, Turton S (2020) Serotonin release measured in the human brain: a PET study with [(11)C]CIMBI-36 and d-amphetamine challenge. Neuropsychopharmacology 45(5):804–810. https://doi.org/10.1038/s41386-019-0567-5

301. Abramowski D, Rigo M, Duc D, Hoyer D, Staufenbiel M (1995) Localization of the 5-hydroxytryptamine2C receptor protein in human and rat brain using specific antisera. Neuropharmacology 34(12):1635–1645. https://doi.org/10.1016/0028-3908(95)00138-7

302. Murphy SE, Wright LC, Browning M, Cowen PJ, Harmer CJ (2020) Role for 5-HT4 receptors in human learning and memory. Psychol Med 50(16):2722–2730

303. Varnäs K, Halldin C, Pike VW, Hall H (2003) Distribution of 5-HT4 receptors in the postmortem human brain—an autoradiographic study using [125I]SB 207710. Eur Neuropsychopharmacol 13(4):228–234. https://doi.org/10.1016/S0924-977X(03)00009-9

304. Marner L, Gillings N, Comley RA, Baaré WF, Rabiner EA, Wilson AA, Houle S, Hasselbalch SG, Svarer C, Gunn RN (2009) Kinetic modeling of 11C-SB207145 binding to 5-HT4 receptors in the human brain in vivo. J Nucl Med 50(6):900–908

305. Madsen K, Haahr MT, Marner L, Keller SH, Baaré WF, Svarer C, Hasselbalch SG, Knudsen GM (2011) Age and sex effects on 5-HT4 receptors in the human brain: a [11C] SB207145 PET study. J Cereb Blood Flow Metab 31(6):1475–1481. https://doi.org/10.1038/jcbfm.2011.11

306. Haahr ME, Fisher P, Holst K, Madsen K, Jensen CG, Marner L, Lehel S, Baaré W, Knudsen G, Hasselbalch S (2013) The 5-HT4 receptor levels in hippocampus correlates inversely with memory test performance in humans. Hum Brain Mapp 34(11):3066–3074. https://doi.org/10.1002/hbm.22123

307. Varnäs K, Thomas DR, Tupala E, Tiihonen J, Hall H (2004) Distribution of 5-HT7 receptors in the human brain: a preliminary autoradiographic study using [3H]SB-269970. Neurosci Lett '367(3):313–316. https://doi.org/10.1016/j.neulet.2004.06.025

308. Hansen HD, Herth MM, Ettrup A, Andersen VL, Lehel S, Dyssegaard A, Kristensen JL, Knudsen GM (2014) Radiosynthesis and in vivo evaluation of novel radioligands for

PET imaging of cerebral 5-HT7 receptors. J Nucl Med 55(4):640–646. https://doi.org/10.2967/jnumed.113.128983

309. Komorowski A, James G, Philippe C, Gryglewski G, Bauer A, Hienert M, Spies M, Kautzky A, Vanicek T, Hahn A (2017) Association of protein distribution and gene expression revealed by PET and post-mortem quantification in the serotonergic system of the human brain. Cereb Cortex 27(1): 117–130

310. Kautzky A, James GM, Philippe C, Baldinger-Melich P, Kraus C, Kranz GS, Vanicek T, Gryglewski G, Hartmann AM, Hahn A (2019) Epistasis of HTR1A and BDNF risk genes alters cortical 5-HT1A receptor binding: PET results link genotype to molecular phenotype in depression. Transl Psychiatry 9(1):1–10

311. Sigurdardottir HL, Kranz GS, Rami-Mark C, James GM, Vanicek T, Gryglewski G, Kautzky A, Hienert M, Traub-Weidinger T, Mitterhauser M, Wadsak W, Hacker M, Rujescu D, Kasper S, Lanzenberger R (2016) Effects of norepinephrine transporter gene variants on NET binding in ADHD and healthy controls investigated by PET. Hum Brain Mapp 37(3):884–895. https://doi.org/10.1002/hbm.23071

Chapter 19

Ketamine and Other Glutamate Receptor Antagonists As Fast-Acting Antidepressants: Evidence from Translational Research

Mu-Hong Chen, Tung-Ping Su, and Shih-Jen Tsai

Abstract

The glutamate hypothesis of depression was prominently emerging in this decade. Evidence has shown that low-dose ketamine infusion or intranasal S-ketamine spray exhibited a prominent and rapid antidepressant and antisuicidal effect. A single infusion of 0.5 mg/kg ketamine may achieve up to 70% treatment response in Caucasian patients with treatment-resistant depression (TRD), but only approximately 50% in Taiwanese patients with TRD. The BDNF release and AMPA receptor upregulation via the blockade of NMDA receptor by low-dose ketamine result in the synaptogenesis and neuroplasticity modulation, which may explain the rapid antidepressant effect of low-dose ketamine. Clinical and biological markers, such as depression severity, body mass index, subjective feeling during infusion, and prefrontal cortex (PFC) and anterior cingulate cortex (ACC) functioning, may predict the treatment response of low-dose ketamine infusion. The 18F-FDG-PET studies found that a short activation in the PFC engendered by ketamine infusion may work as a kindler, facilitating the persistent increase in glucose metabolism in the dorsal ACC, which may further explain the outcome that the antidepressant effects of a single infusion of ketamine may be approximately 2 weeks. In recent years, whether several potential glutamate receptor antagonists/modulators, including R-ketamine, ketamine metabolites (i.e., hydroxynorketamine), lanicemine, and D-cycloserine, may also have a fast-acting antidepressant effect is still being studied. However, the exact neuromechanisms of the rapid antidepressant and antisuicidal effects of low-dose ketamine and intranasal S-ketamine spray in the TRD may go beyond the current evidence, which needs further investigation.

Key words Ketamine, Treatment-resistant depression, Glutamate, Fast-acting antidepressant, Biomarkers

1 Pandemic-Related Increase in Depression over Prevalence in Recent Decades

Depression, with an estimated lifetime prevalence of 10–25% among women and 5–12% among men, is a major chronic mental illness and has been a leading cause of disease burden since 2015 [1–4]. The point prevalence of depression has been significantly higher in women (approximately 20%) and particularly higher in

Yong-Ku Kim and Meysam Amidfar (eds.), *Translational Research Methods for Major Depressive Disorder*, Neuromethods, vol. 179, https://doi.org/10.1007/978-1-0716-2083-0_19,

countries with a medium human development index (up to 30%) [2]. The global disease burden of depression, measured by disability-adjusted life years, increased by up to approximately 40% in the past decade [4]. The National Survey on Drug Use and Health in the United States reported that 12-month depression prevalence increased significantly from 2005 (6.6%) to 2015 (7.4%), especially among the youngest and oldest age groups, non-Hispanic white persons, the lowest income groups, and the highest education and income groups [5]. The phenomenon of significantly increased depression prevalence is more prominent in developing countries [6]. In India, for example, the contribution of mental disorders, especially depressive disorder, to the total disease burden has doubled between 1990 and 2017 [6]. In Taiwan, the prevalence of common mental disorders also doubled from 11.5% in 1990 to 23.8% in 2010, correlating with increases in national rates of unemployment, divorce, and suicide [7].

In 2020, the year of the COVID-19 outbreak, the prevalence of depression symptoms reported in the United States was more than threefold higher than in the years preceding the COVID-19 pandemic [8]. Furthermore, a meta-analysis of 12 studies from China, India, Vietnam, Italy, the United Kingdom, and Denmark demonstrated that compared with a global estimated prevalence of depression of 3.44% in 2017, the pooled prevalence of 25% appeared to be seven times greater in the period of the COVID-19 outbreak [8, 9].

2   Glutamate System Modulation as a Potential Treatment Target for Rapid Antidepressant Effect

Since the first tricyclic antidepressant imipramine was discovered in 1951 and later introduced for medical use in 1957, depression treatment research has been dominated by work associated with the monoamine hypothesis regarding serotonin, norepinephrine, and dopamine and the development of related new drugs, such as selective serotonin reuptake inhibitors, serotonin norepinephrine reuptake inhibitors, and norepinephrine dopamine reuptake inhibitors [10–12]. However, traditional antidepressants targeting reuptake inhibitors always take several weeks to achieve their optimal antidepressant effects, which may be partially due to the characteristics of the corresponding binding receptors, including 5-hydroxytryptamine (5-HT) receptors, dopamine receptors, and norepinephrine receptors (α, β) [13–15]. All 5-HT receptors ($5\text{-}HT_{1-7}$, with the exception of $5\text{-}HT_3$), dopamine receptors (D_{1-5}), and norepinephrine receptors (α_{1-2}, β_{1-3}) are a class of G protein-coupled receptors, which produce intracellular action through downstream signal transduction and secondary messenger

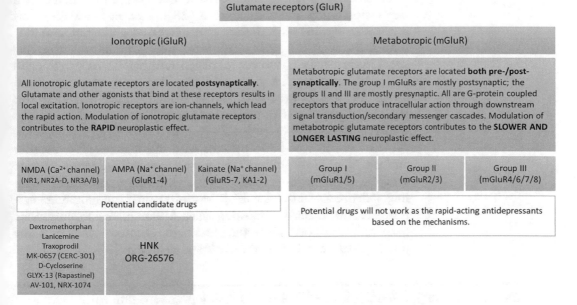

Fig. 1 Glutamate receptor subtypes and potential novel rapid-acting antidepressants

cascades [13, 16, 17]. Based on these molecular characteristics, the modulation of G protein-coupled receptors contributes to this slower-achieved but longer-lasting effect [13, 14].

Compared with 5-HT receptors, dopamine receptors, and nor-epinephrine receptors, predominantly G protein-coupled receptors, glutamate receptors, and gamma-aminobutyric acid (GABA) receptors are ligand-gated ion channels (ionotropic receptors) or G protein-coupled receptors (metabotropic receptors; Fig. 1) [13, 18–20]. Modulation of ionotropic receptors resulting in the local excitation and the generation of action potential with the influxes of sodium or calcium ions contributes to the faster effect compared with metabotropic receptors, resulting in intracellular secondary messenger cascades [20–22]. Ionotropic receptors of glutamate receptors include N-methyl-D-aspartate (NMDA), α-amino-3-hydroxyl-5-methyl-4-isoxazoleproprionic acid (AMPA), and kainate receptors and are located postsynaptically [20, 21]. Metabotropic receptors of glutamate receptors include three groups: group I (mGluR1/5), group II (mGluR2/3), and group III (mGluR4/6/7/8) [20, 21]. The group I mGluRs are mostly postsynaptic, whereas groups II and III have predominantly presynaptic localization [20, 21].

The glutamate hypothesis of depression has come to prominence [23–27]. A meta-analysis study of 1180 patients with depression and 1066 healthy controls that examined the levels of glutamate using proton magnetic resonance spectroscopy (^{1}H-MRS) identified significant decreases in glutamate and glutamine levels within the medial frontal cortex in patients with depression compared with controls [23]. Another ^{1}H-MRS meta-analysis study further indicated that decreased glutamate and glutamine levels in the prefrontal cortex were correlated with treatment severity (i.e., number of failed antidepressant treatments) [28]. Previous studies have also provided evidence of reduced glutamate metabolite levels in the frontal cortex and cingulate regions of patients with major depressive disorder in the midst of a current depressive episode [29–31]. In addition, a genome-wide association study of 4346 patients with major depressive disorder and 4430 controls found that genes involved in glutamatergic synaptic neurotransmission were significantly associated with major depressive disorder [32]. Modulation of the glutamate system—more specifically, boosting synaptic glutamate concentrations that further act on ionotropic glutamate receptors—may be a key potential strategy for rapid depression treatment [13, 18, 21].

3   Ketamine: Prototype Medication with a Rapid Antidepressant and Antisuicidal Effect

3.1   Ketamine Introduction

Ketamine's history begins with phencyclidine [33, 34]. Ketamine was first synthesized in 1962 by Calvin L. Stevens and was known by the developmental code name CI-581; it was approved by the US Food and Drug Administration (US FDA) in 1970 for anesthetic use both in adult and pediatric surgery and was first given to American soldiers during the Vietnam War [33, 35–37]. Ketamine is a racemic mixture compound (R,S-ketamine) containing equal parts of S-ketamine and R-ketamine [38]. Ketamine undergoes extensive metabolism, initially through nitrogen demethylation to norketamine by the cytochrome P450 liver enzymes CYP2B6 and CYP3A4 [38–40]. Furthermore, norketamine is metabolized to the hydroxynorketamines (HNKs) and dehydronorketamine (DHNK) [38–40]. When administering intravenous 0.5 mg/kg ketamine, the average time for ketamine metabolites to reach peak plasma concentration after infusion was estimated to be approximately 1.33 h for norketamine and 3.83 h for DHNK and HNK [41].

In early 2000, ketamine, as a NMDA antagonist, was rediscovered to be potentially effective for depression in a small-scale clinical trial of seven patients with major depressive disorder who completed 2 test days that involved intravenous treatment with ketamine hydrochloride (0.5 mg/kg) or saline solutions under randomized double-blind conditions [42]. S-ketamine was introduced for medical (anesthetic) use in 1997 and was reported to have a remarkably smooth emergence period, a profound postoperative analgesic effect, and a less psychotomimetic phenomenon and to be associated with a more rapid recovery of cerebral functions in comparison to racemic ketamine [43]. S-ketamine was approved for treatment-resistant depression in 2019 and additionally approved for the short-term treatment of suicidal thoughts in 2020 by the US FDA.

3.2 Clinical Trials of Low-Dose Ketamine Infusion in Treatment-Resistant Depression and Suicide

Evidence has shown that low-dose ketamine infusion or intranasal S-ketamine spray exhibits prominent and rapid antidepressant and antisuicidal effects [44–48]. A recent meta-analysis study of 24 clinical trials representing 1877 participants demonstrated that racemic ketamine, relative to S-ketamine, demonstrated greater overall response and remission rates as well as fewer dropouts [47]. Unexpectedly, our clinical trial of 71 Taiwanese patients with treatment-resistant depression only achieved an approximately 50% response rate after a single ketamine infusion, which was much lower than in trials of Caucasian patients, which reported up to a 70% response rate [49–51]. By analyzing the serum ketamine and norketamine levels in our patients, we found that both ketamine (115.86 ng/mL vs. 204.13 ng/mL) and norketamine (33.39 ng/mL vs. 55.52 ng/mL) levels at 40 min after ketamine infusion were much lower in Taiwanese patients than in Caucasian patients [51, 52]. Estimation of the pooled prevalence of the brain-derived neurotrophic factor (BDNF) Val66Met polymorphism demonstrated that the frequency of the Val allele was approximately 80% in Caucasians but was much lower (<50%) in Han Chinese and in Taiwanese, which may partially explain the difference in the treatment response to low-dose ketamine infusion between Caucasians and Taiwanese [52–54].

In addition, previous studies have suggested that the duration of the antisuicidal effects of ketamine infusion may last no longer than 1 week, whereas the antidepressant effects of ketamine infusion may last for approximately 2 weeks [49, 55]. A meta-analysis of 10 randomized placebo-controlled trials involving 167 patients with major depression, bipolar depression, or posttraumatic stress disorder reported that ketamine rapidly (within 1 day) and significantly reduced suicidal ideation both in terms of clinician-administered and self-reported outcome measures [56]. Effect sizes were moderate to large (Cohen's $d = 0.48$–0.85) at all time points following administration (days 1–7) [56]. Wilkinson et al. further suggested that the effects of ketamine on reducing suicidal ideation were partially independent of its effects on mood [56].

Finally, increasing evidence suggests that repeated ketamine infusion or intranasal S-ketamine spray may sustain the antidepressant and antisuicidal effects of ketamine and S-ketamine among patients with treatment-resistant depression [57–63]. A clinical trial of 41 patients with treatment-resistant depression who were randomized to a single infusion of ketamine versus midazolam (an active placebo control) that was followed with a course of six open-label ketamine infusions thrice weekly over 2 weeks found that 59% of participants met response criteria after repeated infusions, with a median of three infusions required to achieve response [61]. The clinical trials of add-on 4-week intranasal S-ketamine spray twice weekly versus placebo spray demonstrated that changes in depressive (Montgomery-Asberg Depression Rating Scale, MADRS) and suicidal (Clinical Global Impression-Severity of Suicidality-revised) symptoms with S-ketamine plus antidepressant were significantly greater than the results achieved with an antidepressant plus placebo at day 28. Likewise, clinically meaningful improvements were observed in the esketamine plus antidepressant arm at earlier time points [62–65].

3.3 Molecular Mechanisms of the Rapid Antidepressant Effect of Ketamine

Two hypotheses, the disinhibition hypothesis of GABA interneuron (the "go" hypothesis) and the direct inhibition hypothesis in pyramidal glutamate neuron (the "stop" hypothesis), were proposed to explain the rapid antidepressant and antisuicidal effects of low-dose ketamine infusion (Fig. 2) [13, 18, 21, 66–69].

The "go" hypothesis suggests that the increased synaptic glutamate release is due to (mechanism 1) the disinhibition effect of GABA interneurons, which are inhibited by the blockade of the NMDA receptor (NMDAR) by ketamine, an NMDA antagonist [13, 18, 21, 66, 69]. The intrasynaptic glutamates bind to and activate postsynaptic Na^+-dependent AMPA receptors (AMPAR), which results in (mechanism 2) depolarization that further activates voltage-gated calcium channels [13, 18, 21, 66, 69]. Ca^{2+} triggers the activity-dependent synaptic release of BDNF, which then acts on the surface receptor tropomyosin receptor kinase B (TrkB) and further leads to two major downstream signaling cascades (MEK-ERK and PI3K-Akt) and finally converges at the mammalian target of rapamycin (mTOR) [13, 18, 21, 66, 69]. mTOR regulates cell growth, cell proliferation, cell motility, cell survival, protein synthesis, autophagy, and transcription [70, 71]. In the mTOR-dependent antidepressant effect of ketamine, mTOR activation leads to disinhibition of synaptic protein translation, including the newly synthesized AMPARs and other synaptic components that are inserted into the postsynaptic density [13, 18, 21, 66, 69].

The "stop" hypothesis suggests that (mechanism 3) the direct inhibition of postsynaptic NMDARs acts on the pyramidal neurons, which at rest keep eukaryotic elongation factor 2 (eEF2) phosphorylated and inhibit BDNF synaptic translation [13, 18, 21, 66, 69]. The direct blockade of postsynaptic NMDARs by ketamine affects the neuronal NMDAR-mediating spontaneous excitatory transmission, which results in the desuppression of BDNF translation that then contributes to changes in synaptic plasticity [13, 18, 21, 66, 69]. Increased levels of synaptic BDNF bind to TrkB, which further activates the mTOR pathway [13, 18, 21, 66, 69]. In addition, (mechanism 4) ketamine selectively blocks extrasynaptic GluN2B-containing NMDARs, which are tonically

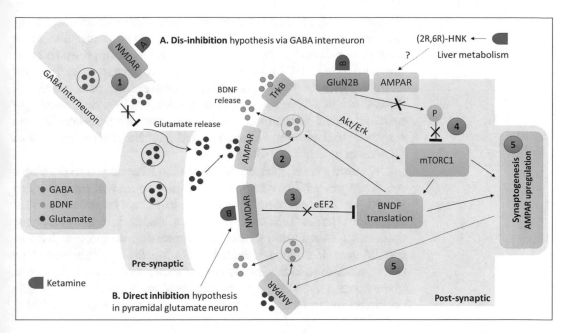

Fig. 2 Hypothesized mechanisms of ketamine as a rapid-acting antidepressant

activated by low levels of ambient glutamate regulated by the excitatory amino acid transporter 2 located on astrocytes [13, 18, 21, 66, 69]. Inhibition of the extrasynaptic GluN2B-NMDARs desuppresses mTOR pathway function, which in turn induces protein synthesis (i.e., AMPARs) and increases postsynaptic density through synaptogenesis [13, 18, 21, 66, 69]. Lower synaptic density has been reported to contribute to the severity of depression in regions associated with affective processing (the prefrontal cortex, anterior cingulate cortex (ACC), and hippocampus) [72].

Both the disinhibition hypothesis and the direct inhibition hypothesis of the rapid antidepressant effect of ketamine involve AMPAR and BDNF [13, 18, 21, 66, 69, 73]. Briefly, BDNF release and AMPAR upregulation via the blockade of NMDAR triggered by low-dose ketamine result in (mechanism 5) synaptogenesis and neuroplasticity modulation via AMPAR upregulation, which may play crucial roles in ketamine's rapid antidepressant and antisuicidal effects.

4 Predicting Treatment Response to Low-Dose Ketamine Infusion

4.1 Clinical Markers of Depression

Using a single marker to predict the treatment response to ketamine infusion is always difficult because potential predictors, such as depression severity and the comorbidity of an anxiety disorder, commonly occur together and their collinearity interferes with regression analysis, although several clinical and demographic characteristics, such as concomitant comorbidity of anxiety disorder, high body mass index (BMI), and positive family history of alcohol use disorder, may be related to reduced response to ketamine infusion [74–77]. However, counterintuitively, several clinical characteristics of treatment-resistant depression, such as age at disease onset, severity of depression, duration of a depressive episode, and level of treatment resistance, may not be related to the response to ketamine infusion based on the clinical trials with Caucasian patients [78–80].

We used a classification and regression tree analysis to develop a predicted probability composed of various potential predictors to investigate the treatment response to low-dose ketamine infusion and found that in Taiwanese patients, a longer duration of current episode (>24 months) and severe depression symptoms ($\geq$24 in the 17-item Hamilton Depression Rating Scale) were related to poor response to ketamine infusion; factors such as younger age at disease onset (<30 years) and obesity/overweight (BMI $\geq$ 25) predicted treatment response to low-dose ketamine infusion [74].

4.2 Clinical Markers During Infusion

Although dissociative symptoms commonly appear during ketamine infusion, few studies have investigated the subjective feelings experienced during ketamine infusion in relation to the antidepressant effect and treatment outcome when administering low-dose ketamine [81, 82]. However, whether the dissociative effects of ketamine may be related to the antidepressant and antisuicidal effects of ketamine remains unknown [82, 83]. Using the 5-Dimensional Altered States of Consciousness Rating Scale to assess subjective feelings during ketamine infusion, Aust et al. demonstrated that ketamine-induced anxiety during infusion had a negative effect on the antidepressant efficacy of ketamine [81].

In addition to ketamine-related negative feelings (i.e., anxiety, anxious ego disintegration) during infusion, the feeling of happiness during ketamine infusion has been rarely mentioned but is often observed [84]. Specifically, happiness is a subjective experience of positive affect and is often called subjective well-being or emotional well-being [85]. The lack of positive affect is likely to be more common and is usually seen in different stages of major depressive disorder, including the acute, chronic, and remitted phases [86, 87]. Our clinical trial employed the visual analogue scale for happiness to assess happiness during infusion and found that subjective happiness during ketamine infusion, which was not related to the psychotomimetic effect of ketamine, predicted the antidepressant effect of both 0.5 mg/kg and 0.2 mg/kg ketamine infusion over time [88]. The effect size of infusion response type (happiness vs. nonhappiness) for treatment response was 0.49, indicating a moderate predictive power. Furthermore, subjective happiness that occurred during ketamine infusion persisted up to day 14 postinfusion both in 0.5 mg/kg and 0.2 mg/kg ketamine infusion groups [88].

4.3 Neuroimaging Biomarkers

Based on the ^{1}H-MRS studies of significant decreases in glutamate and glutamine levels in the frontal cortex and cingulate regions and of the lower glutamate and glutamine response in the ventromedial prefrontal cortex with improved antidepressant response ketamine treatment [29–31, 89], it is reasonable to assess whether the baseline brain structure and neural functioning involved with the frontal cortex and cingulate regions may predict the treatment response to low-dose ketamine infusion [90–94].

Voxel-based morphometry of 33 depressed patients acquired and observed both before and 24 h after a single ketamine infusion demonstrated that the greater baseline volume of the bilateral rostral anterior cingulate cortex significantly predicts rapid symptom reduction of depression [93]. A recent diffusion magnetic resonance imaging study further indicated that higher preinfusion fractional anisotropy in the left cingulum bundle (projecting from the cingulate gyrus to the entorhinal cortex) and the left superior longitudinal fasciculus (projecting from the frontal lobe through the operculum to the posterior end of the lateral sulcus and then radiating to the occipital and temporal lobes) was associated with greater depression symptom improvement 24 h after ketamine infusion [92]. Both we and Mkrtchian et al. have reported that baseline hypoconnectivity of the bilateral superior frontal cortex to the executive region of the striatum was associated with a greater reduction of depression symptoms after ketamine infusion and that, further, ketamine increased the frontostriatal functional connectivity among patients with treatment-resistant depression [90, 94]. Finally, Gärtner et al. demonstrated that low baseline functional connectivity between the PFC and the subgenual cingulate cortex predicted the treatment outcome to ketamine infusion; moreover, they also found that functional connectivity increases between the right lateral PFC and the subgenual ACC (sgACC) after ketamine infusion, and this connectivity is positively linked to treatment response [91].

5 Neurobiological Mechanisms of the Rapid Antidepressant and Antisuicidal Effect of Ketamine

The half-life values of ketamine and its active metabolites, norketamine and dehydronorketamine, are approximately 3, 5, and 7 h, respectively [95]. Several ^{18}F-FDG-PET studies have been conducted to assess the brain activation at different time points before and after ketamine infusion, including the baseline before infusion and 40 min, 120 min, and 24 h after infusion [96–99]. Lally et al. and Nugent et al. evaluated the brain glucose metabolism prior to low-dose ketamine infusion and at 120 min after infusion using ^{18}F-FDG-PET scanning among patients with treatment-resistant depression and demonstrated that the overall depression improvement, as measured by MADRS, was related to the higher activity in the sgACC and reduced anhedonia was associated with increased glucose metabolism in the hippocampus and dorsal ACC (dACC) [98, 99]. Our randomized, placebo-controlled PET clinical trial assessed the brain glucose metabolism prior to infusion and 40 min and 24 h after infusion in severely depressed patients and revealed that glucose metabolism of the PFC, supplementary motor area (SMA), and dACC in patients provided with the low-dose ketamine infusion were higher than those in the control group at 40-min postinfusion [96]. Furthermore, we found that the activation in the PFC disappeared on scans 24 h after infusion and reported that only the activation in the SMA and dACC persisted 24 h after infusion [97]. Our findings support the aforementioned evidence of no elevated glucose metabolism in the PFC 120 min after ketamine infusion [98, 99]. In addition, a recent ^{13}C MRS study examining glutamate neurotransmission in the frontal cortex 45 min after infusion observed that ketamine increased prefrontal glutamate-glutamine cycling, as indicated by a 13% increase in ^{13}C glutamine enrichment compared with the placebo [100]. The imaging findings may reflect the molecular results that the transient surge (40–60 min after ketamine infusion) of synaptic glutamate activates the PFC functioning (i.e., BDNF release, AMPAR upregulation, increased synaptic density), which further achieves the rapid antidepressant effect [13, 18, 21, 66–69, 96, 97, 100].

By putting the preceding pieces of evidence together, we hypothesize that the effect of ketamine infusion on PFC activation rapidly occurs within 1 h, probably owing to the glutamate surge in the PFC, and then rapidly disappears approximately 2 h later. A short activation in the PFC engendered by ketamine infusion may be a trigger, facilitating the persistent increase in glucose metabolism in the SMA and dACC; therefore, the PFC can still consider to play a key role in improving treatment-resistant depression (Fig. 3) [96, 97, 100].

6   Other Potential Glutamate Receptor Antagonists As Fast-Acting Antidepressants

6.1 Ketamine Metabolites

The ketamine metabolite HNK has recently been suggested to be an ideal antidepressant for treating animal models of depression. However, its effects and mechanisms are subject to debate [101–107]. Studies have suggested that (2S,6S; 2R,6R)-HNK exhibited low binding affinities for NMDAR but may have greater binding affinity for AMPAR, further triggering excitatory postsynaptic potentials and BDNF expression [107, 108].

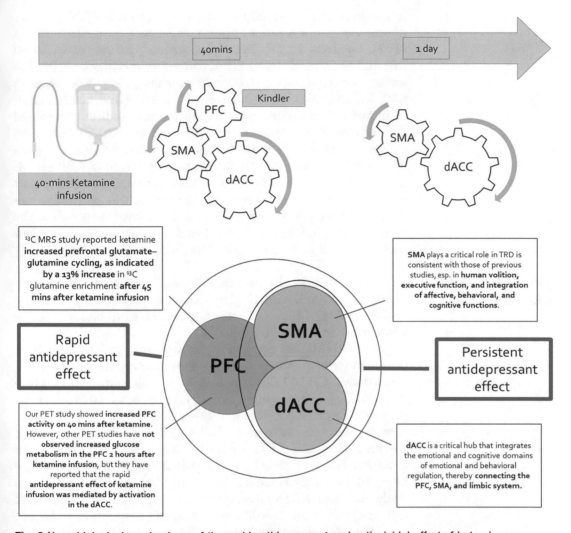

Fig. 3 Neurobiological mechanisms of the rapid antidepressant and antisuicidal effect of ketamine

A recent preclinical study of a mouse model of depression induced by chronic corticosterone (CORT) injection suggested that a single administration of S-norketamine and (2S,6S)-HNK dose-dependently reduced enhanced immobility at 30 min after injection in chronic CORT-treated mice, whereas R-norketamine or (2R,6R)-HNK did not [105]. Chen et al. further reported that (2S,6S)-HNK and (2R,6R)-HNK protected against distinct stress-induced behaviors in both male and female mice, with (2S,6S)-HNK attenuating learned fear in male mice and (2R,6R)-HNK preventing stress-induced depressive-like behavior in both sexes [102]. However, a human clinical trial reported that in patients with treatment-resistant depression who received a single ketamine infusion, ketamine concentration positively predicted a distal anti-depressant response at day 11 after infusion, and an inverse relationship was observed between (2S,6S;2R,6R)-HNK concentration and antidepressant response at 3 and 7 days after infusion [104]. Contrary to preclinical observations, Farmer et al. found that higher (2S,6S;2R,6R)-HNK concentrations may be related to weaker antidepressant response in humans [104]. The negative relationship between higher levels of (2R,6R)-HNK and less clinical improvement in depression and suicide has also been reported in the previous human study [109]. However, the above results were collected as indirect findings based on the low-dose ketamine infusion studies instead of originating from a direct (2R,6R)-HNK clinical trial [103]. Abdallah et al. commented that given the urgent need for novel rapid-acting antidepressants, the preclinical evidence, and the potential for reduced abuse liability as well as the enhanced side effect profile compared with ketamine, it remains useful to invest in the HNK pipeline [103].

6.2 Other Glutamate Receptor Modulators

Based on the molecular characteristics of glutamate receptors, candidate compounds targeting the ionotropic receptors, including NMDAR, AMPAR, and kainate receptors, may exhibit rapid-acting antidepressant effects [13, 18, 76].

Several NMDAR antagonists/modulators, such as dextromethorphan (plus quinidine or bupropion), lanicemine (AZD6765), traxoprodil (CP-101,606), MK-0657 (CERC-301), D-cycloserine, GLYX-13 (rapastinel), AV-101, and NRX-1074, may be potential future rapid-acting antidepressants [13, 18, 76]. Our clinical trial of D-cycloserine (250 mg/day for 2 days, 500 mg/day for 2 days, and 750 mg/day for 3 days to 1000 mg/day for 5 weeks) failed to reveal any additional antidepressant effect of D-cycloserine after ketamine infusion [110]. A recent preclinical study reported dextromethorphan (32 mg/kg) produced acute (30 min) antidepressant-like effects in tail suspension tests, but treatment with dextromethorphan (32 mg/kg) alone or in combination with CYP2D6 enzyme inhibitor quinidine (32 mg/kg) failed to produce rapid antidepressant-like effects in the novelty-induced hypophagia test [111]. A phase IIa open-label clinical trial of the 10-week treatment of dextromethorphan and quinidine in patients with treatment-resistant depression found that a significant change in MADRS was observed after a treatment period of at least 4 weeks, which did not work as a rapid-acting antidepressant [112]. A randomized, placebo-controlled, crossover study of 22 patients with treatment-resistant depression found that a single infusion of lanicemine 150 mg resulted in a 32% response rate compared with 15% for placebo [113]. However, a clinical trial of 302 patients with treatment-resistant depression who were randomized to 15 double-blind intravenous infusions of adjunctive lanicemine 50 mg, lanicemine 100 mg, or saline over a 12-week course found that neither dose was superior to placebo in reducing depressive symptoms at the end of the study [114]. A cross-over study with five patients with treatment-resistant depression who were randomized to receive either MK-0657 monotherapy (4–8 mg/day) or placebo for 12 days revealed that a significant improvement in depressive symptoms measured by the Hamilton Depression Rating Scale and Beck Depression Inventory but not MADRS (primary outcome) was observed as early as day 5 in patients receiving MK-0657 compared with those receiving placebo [115]. A second attempt at a phase II trial in 2016 also found that MK-0657 failed to demonstrate efficacy against depression (Clin301-203; NCT02459236). ORG-26576 is an AMPA receptor and positive allosteric modulator and was examined in a phase II study where the dose of 100–600 mg b.i.d. was safe and tolerable in depressed patients throughout the 28-day treatment period [116, 117].

Finally, several NMDAR antagonists/modulators and AMPAR agonists/modulators, such as memantine, lamotrigine, and riluzole, are available in current clinical practice [118, 119]. However, none exhibit the rapid-acting antidepressant effects of ketamine and S-ketamine. The exact neuromechanisms of the rapid antidepressant and antisuicidal effects of low-dose ketamine and intranasal S-ketamine spray in treatment-resistant depression may go beyond the current evidence; further investigation is needed. If researchers and clinicians cannot comprehensively elucidate the definite neuromechanisms of low-dose ketamine and S-ketamine as rapid-acting antidepressants, it will be impossible to definitively produce a second ketamine or a totally novel rapid-acting antidepressant.

7   Conclusion

This century is proving to be difficult for many. With an increasing rate of depression and suicide worldwide, both in developed and developing countries, the public anticipates an optimal and efficient treatment strategy for depression and suicide as well as a prompt response from clinical practice experts and the public health field. Low-dose ketamine infusion and intranasal S-ketamine, which have been shown to be efficacious as the first rapid-acting antidepressant and antisuicidal drug in the history of depression treatment and clinical psychiatry, may equip clinicians and public health providers with the means to conquer the growing depression and suicide crisis in the near future. However, the definite neuromechanisms of ketamine and S-ketamine against depression and suicide remain unclear. The long-term safety and the potential risk of abuse of ketamine and S-ketamine still need further investigation. In addition, the psychotomimetic and dissociative effects also confound the universal use of ketamine and S-ketamine in depression treatment. However, we most certainly continue to anticipate a novel rapid-acting antidepressant is developed without such adverse effects as soon as possible, which would mean a drug with the benefits of ketamine and none of its detrimental effects. Before we have such a strong and safe bullet against depression and suicide, further preclinical and clinical studies will be necessary to clearly and comprehensively elucidate the exact neuromechanisms of the rapid antidepressant and antisuicidal effects of ketamine and S-ketamine.

Acknowledgements

The authors thank Mr I-Fan Hu, MA (Courtauld Institute of Art, University of London; National Taiwan University) for his friendship and support. Mr Hu declares no conflicts of interest. *Funding source*: The study was supported by grant from Taipei Veterans General Hospital (V106B-020, V107B-010, V107C-181, V108B-012), Yen Tjing Ling Medical Foundation (CI-110-30), and the Ministry of Science and Technology, Taiwan (107-2314-B-075-063-MY3, 108-2314-B-075 -037). The funding source had no role in any process of our study.

References

1. Moore JD, Bona JR (2001) Depression and dysthymia. Med Clin North Am 85:631–644
2. Lim GY, Tam WW, Lu Y, Ho CS, Zhang MW, Ho RC (2018) Prevalence of depression in the community from 30 countries between 1994 and 2014. Sci Rep 8:2861
3. Charlson F, van Ommeren M, Flaxman A, Cornett J, Whiteford H, Saxena S (2019) New WHO prevalence estimates of mental disorders in conflict settings: a systematic review and meta-analysis. Lancet 394:240–248
4. Ferrari AJ, Charlson FJ, Norman RE, Patten SB, Freedman G, Murray CJ, Vos T, Whiteford HA (2013) Burden of depressive disorders by country, sex, age, and year: findings from the global burden of disease study 2010. PLoS Med 10:e1001547
5. Weinberger AH, Gbedemah M, Martinez AM, Nash D, Galea S, Goodwin RD (2018) Trends in depression prevalence in the USA from 2005 to 2015: widening disparities in vulnerable groups. Psychol Med 48:1308–1315
6. India State-Level Disease Burden Initiative Mental Disorders Collaborators (2020) The burden of mental disorders across the states of India: the global burden of disease study 1990-2017. Lancet Psychiatry 7:148–161
7. Fu TS, Lee CS, Gunnell D, Lee WC, Cheng AT (2013) Changing trends in the prevalence of common mental disorders in Taiwan: a 20-year repeated cross-sectional survey. Lancet 381:235–241
8. Ettman CK, Abdalla SM, Cohen GH, Sampson L, Vivier PM, Galea S (2020) Prevalence of depression symptoms in US adults before and during the COVID-19 pandemic. JAMA Netw Open 3:e2019686
9. Bueno-Notivol J, Gracia-Garcia P, Olaya B, Lasheras I, Lopez-Anton R, Santabarbara J (2021) Prevalence of depression during the COVID-19 outbreak: a meta-analysis of community-based studies. Int J Clin Health Psychol 21:100196
10. Lopez-Munoz F, Alamo C (2009) Monoaminergic neurotransmission: the history of the discovery of antidepressants from 1950s until today. Curr Pharm Des 15:1563–1586
11. Wong DT, Perry KW, Bymaster FP (2005) Case history: the discovery of fluoxetine hydrochloride (Prozac). Nat Rev Drug Discov 4:764–774
12. Hirschfeld RM (2000) History and evolution of the monoamine hypothesis of depression. J Clin Psychiatry 61(Suppl 6):4–6
13. Harmer CJ, Duman RS, Cowen PJ (2017) How do antidepressants work? New

perspectives for refining future treatment approaches. Lancet Psychiatry 4:409–418

14. Harmer CJ, Goodwin GM, Cowen PJ (2009) Why do antidepressants take so long to work? A cognitive neuropsychological model of antidepressant drug action. Br J Psychiatry 195: 102–108

15. Taylor C, Fricker AD, Devi LA, Gomes I (2005) Mechanisms of action of antidepressants: from neurotransmitter systems to signaling pathways. Cell Signal 17:549–557

16. Koelle MR (2018) Neurotransmitter signaling through heterotrimeric G proteins: insights from studies in C. elegans. WormBook 2018:1–52

17. Trumpp-Kallmeyer S, Hoflack J, Bruinvels A, Hibert M (1992) Modeling of G-protein-coupled receptors: application to dopamine, adrenaline, serotonin, acetylcholine, and mammalian opsin receptors. J Med Chem 35:3448–3462

18. Kadriu B, Musazzi L, Henter ID, Graves M, Popoli M, Zarate CA Jr (2019) Glutamatergic neurotransmission: pathway to developing novel rapid-acting antidepressant treatments. Int J Neuropsychopharmacol 22:119–135

19. Pinheiro PS, Mulle C (2008) Presynaptic glutamate receptors: physiological functions and mechanisms of action. Nat Rev Neurosci 9: 423–436

20. Niciu MJ, Kelmendi B, Sanacora G (2012) Overview of glutamatergic neurotransmission in the nervous system. Pharmacol Biochem Behav 100:656–664

21. Park M, Niciu MJ, Zarate CA Jr (2015) Novel glutamatergic treatments for severe mood disorders. Curr Behav Neurosci Rep 2:198–208

22. Kannampalli P, Sengupta JN (2015) Role of principal ionotropic and metabotropic receptors in visceral pain. J Neurogastroenterol Motil 21:147–158

23. Moriguchi S, Takamiya A, Noda Y, Horita N, Wada M, Tsugawa S, Plitman E, Sano Y, Tarumi R, ElSalhy M, Katayama N, Ogyu K, Miyazaki T, Kishimoto T, Graff-Guerrero A, Meyer JH, Blumberger DM, Daskalakis ZJ, Mimura M, Nakajima S (2019) Glutamatergic neurometabolite levels in major depressive disorder: a systematic review and meta-analysis of proton magnetic resonance spectroscopy studies. Mol Psychiatry 24:952–964

24. Sanacora G, Treccani G, Popoli M (2012) Towards a glutamate hypothesis of depression: an emerging frontier of neuropsychopharmacology for mood disorders. Neuropharmacology 62:63–77

25. Li CT, Yang KC, Lin WC (2018) Glutamatergic dysfunction and glutamatergic compounds for major psychiatric disorders: evidence from clinical neuroimaging studies. Front Psychiatry 9:767

26. Abdallah CG, Jiang L, De Feyter HM, Fasula M, Krystal JH, Rothman DL, Mason GF, Sanacora G (2014) Glutamate metabolism in major depressive disorder. Am J Psychiatry 171:1320–1327

27. Lener MS, Niciu MJ, Ballard ED, Park M, Park LT, Nugent AC, Zarate CA Jr (2017) Glutamate and gamma-aminobutyric acid Systems in the Pathophysiology of major depression and antidepressant response to ketamine. Biol Psychiatry 81:886–897

28. Arnone D, Mumuni AN, Jauhar S, Condon B, Cavanagh J (2015) Indirect evidence of selective glial involvement in glutamate-based mechanisms of mood regulation in depression: meta-analysis of absolute prefrontal neuro-metabolic concentrations. Eur Neuropsychopharmacol 25:1109–1117

29. Auer DP, Putz B, Kraft E, Lipinski B, Schill J, Holsboer F (2000) Reduced glutamate in the anterior cingulate cortex in depression: an in vivo proton magnetic resonance spectroscopy study. Biol Psychiatry 47:305–313

30. Michael N, Erfurth A, Ohrmann P, Arolt V, Heindel W, Pfleiderer B (2003) Metabolic changes within the left dorsolateral prefrontal cortex occurring with electroconvulsive therapy in patients with treatment resistant unipolar depression. Psychol Med 33:1277–1284

31. Hasler G, van der Veen JW, Tumonis T, Meyers N, Shen J, Drevets WC (2007) Reduced prefrontal glutamate/glutamine and gamma-aminobutyric acid levels in major depression determined using proton magnetic resonance spectroscopy. Arch Gen Psychiatry 64:193–200

32. Lee PH, Perlis RH, Jung JY, Byrne EM, Rueckert E, Siburian R, Haddad S, Mayerfeld CE, Heath AC, Pergadia ML, Madden PA, Boomsma DI, Penninx BW, Sklar P, Martin NG, Wray NR, Purcell SM, Smoller JW (2012) Multi-locus genome-wide association analysis supports the role of glutamatergic synaptic transmission in the etiology of major depressive disorder. Transl Psychiatry 2:e184

33. Maddox VH, Godefroi EF, Parcell RF (1965) The synthesis of phencyclidine and other 1-Arylcyclohexylamines. J Med Chem 8: 230–235

34. Mion G, Villevieille T (2013) Ketamine pharmacology: an update (pharmacodynamics and

molecular aspects, recent findings). CNS Neurosci Ther 19:370–380

35. Kurdi MS, Theerth KA, Deva RS (2014) Ketamine: current applications in anesthesia, pain, and critical care. Anesth Essays Res 8:283–290

36. Li L, Vlisides PE (2016) Ketamine: 50 years of modulating the mind. Front Hum Neurosci 10:612

37. Domino EF (2010) Taming the ketamine tiger. 1965. Anesthesiology 113:678–684

38. Zanos P, Moaddel R, Morris PJ, Riggs LM, Highland JN, Georgiou P, Pereira EFR, Albuquerque EX, Thomas CJ, Zarate CA Jr, Gould TD (2018) Ketamine and ketamine metabolite pharmacology: insights into therapeutic mechanisms. Pharmacol Rev 70:621–660

39. Kharasch ED, Labroo R (1992) Metabolism of ketamine stereoisomers by human liver microsomes. Anesthesiology 77:1201–1207

40. Hijazi Y, Boulieu R (2002) Contribution of CYP3A4, CYP2B6, and CYP2C9 isoforms to N-demethylation of ketamine in human liver microsomes. Drug Metab Dispos 30:853–858

41. Zhao X, Venkata SL, Moaddel R, Luckenbaugh DA, Brutsche NE, Ibrahim L, Zarate CA Jr, Mager DE, Wainer IW (2012) Simultaneous population pharmacokinetic modelling of ketamine and three major metabolites in patients with treatment-resistant bipolar depression. Br J Clin Pharmacol 74:304–314

42. Berman RM, Cappiello A, Anand A, Oren DA, Heninger GR, Charney DS, Krystal JH (2000) Antidepressant effects of ketamine in depressed patients. Biol Psychiatry 47:351–354

43. Himmelseher S, Pfenninger E (1998) The clinical use of S-(+)-ketamine--a determination of its place. Anasthesiol Intensivmed Notfallmed Schmerzther 33:764–770

44. Han Y, Chen J, Zou D, Zheng P, Li Q, Wang H, Li P, Zhou X, Zhang Y, Liu Y, Xie P (2016) Efficacy of ketamine in the rapid treatment of major depressive disorder: a meta-analysis of randomized, double-blind, placebo-controlled studies. Neuropsychiatr Dis Treat 12:2859–2867

45. McGirr A, Berlim MT, Bond DJ, Fleck MP, Yatham LN, Lam RW (2015) A systematic review and meta-analysis of randomized, double-blind, placebo-controlled trials of ketamine in the rapid treatment of major depressive episodes. Psychol Med 45:693–704

46. Coyle CM, Laws KR (2015) The use of ketamine as an antidepressant: a systematic review and meta-analysis. Hum Psychopharmacol 30:152–163

47. Bahji A, Vazquez GH, Zarate CA Jr (2021) Comparative efficacy of racemic ketamine and esketamine for depression: a systematic review and meta-analysis. J Affect Disord 278:542–555

48. Papakostas GI, Salloum NC, Hock RS, Jha MK, Murrough JW, Mathew SJ, Iosifescu DV, Fava M (2020) Efficacy of esketamine augmentation in major depressive disorder: a meta-analysis. J Clin Psychiatry 81:19r12889

49. Murrough JW, Iosifescu DV, Chang LC, Al Jurdi RK, Green CE, Perez AM, Iqbal S, Pillemer S, Foulkes A, Shah A, Charney DS, Mathew SJ (2013) Antidepressant efficacy of ketamine in treatment-resistant major depression: a two-site randomized controlled trial. Am J Psychiatry 170:1134–1142

50. Wan LB, Levitch CF, Perez AM, Brallier JW, Iosifescu DV, Chang LC, Foulkes A, Mathew SJ, Charney DS, Murrough JW (2015) Ketamine safety and tolerability in clinical trials for treatment-resistant depression. J Clin Psychiatry 76:247–252

51. Zarate CA Jr, Brutsche N, Laje G, Luckenbaugh DA, Venkata SL, Ramamoorthy A, Moaddel R, Wainer IW (2012) Relationship of ketamine's plasma metabolites with response, diagnosis, and side effects in major depression. Biol Psychiatry 72:331–338

52. Su TP, Chen MH, Li CT, Lin WC, Hong CJ, Gueorguieva R, Tu PC, Bai YM, Cheng CM, Krystal JH (2017) Dose-related effects of adjunctive ketamine in Taiwanese patients with treatment-resistant depression. Neuropsychopharmacology 42(13):2482–2492

53. Gratacos M, Gonzalez JR, Mercader JM, de Cid R, Urretavizcaya M, Estivill X (2007) Brain-derived neurotrophic factor Val66Met and psychiatric disorders: meta-analysis of case-control studies confirm association to substance-related disorders, eating disorders, and schizophrenia. Biol Psychiatry 61:911–922

54. Chen MH, Lin WC, Wu HJ, Cheng CM, Li CT, Hong CJ, Tu PC, Bai YM, Tsai SJ, Su TP (2019) Antisuicidal effect, BDNF Val66Met polymorphism, and low-dose ketamine infusion: reanalysis of adjunctive ketamine study of Taiwanese patients with treatment-resistant depression (AKSTP-TRD). J Affect Disord 251:162–169

55. Murrough JW, Soleimani L, DeWilde KE, Collins KA, Lapidus KA, Iacoviello BM, Lener M, Kautz M, Kim J, Stern JB, Price RB, Perez AM, Brallier JW, Rodriguez GJ, Goodman WK, Iosifescu DV, Charney DS

(2015) Ketamine for rapid reduction of suicidal ideation: a randomized controlled trial. Psychol Med 45:3571–3580

56. Wilkinson ST, Ballard ED, Bloch MH, Mathew SJ, Murrough JW, Feder A, Sos P, Wang G, Zarate CA Jr, Sanacora G (2018) The effect of a single dose of intravenous ketamine on suicidal ideation: a systematic review and individual participant data meta-analysis. Am J Psychiatry 175:150–158

57. McIntyre RS, Rodrigues NB, Lee Y, Lipsitz O, Subramaniapillai M, Gill H, Nasri F, Majeed A, Lui LMW, Senyk O, Phan L, Carvalho IP, Siegel A, Mansur RB, Brietzke E, Kratiuk K, Arekapudi AK, Abrishami A, Chau EH, Szpejda W, Rosenblat JD (2020) The effectiveness of repeated intravenous ketamine on depressive symptoms, suicidal ideation and functional disability in adults with major depressive disorder and bipolar disorder: results from the Canadian rapid treatment Center of Excellence. J Affect Disord 274:903–910

58. Lipsitz O, McIntyre RS, Rodrigues NB, Kaster TS, Cha DS, Brietzke E, Gill H, Nasri F, Lin K, Subramaniapillai M, Kratiuk K, Teopiz K, Lui LMW, Lee Y, Ho R, Shekotikhina M, Mansur RB, Rosenblat JD (2020) Early symptomatic improvements as a predictor of response to repeated-dose intravenous ketamine: results from the Canadian rapid treatment Center of Excellence. Prog Neuro-Psychopharmacol Biol Psychiatry 105:110126

59. Shiroma PR, Thuras P, Wels J, Albott CS, Erbes C, Tye S, Lim KO (2020) A randomized, double-blind, active placebo-controlled study of efficacy, safety, and durability of repeated vs single subanesthetic ketamine for treatment-resistant depression. Transl Psychiatry 10:206

60. Phillips JL, Norris S, Talbot J, Hatchard T, Ortiz A, Birmingham M, Owoeye O, Batten LA, Blier P (2020) Single and repeated ketamine infusions for reduction of suicidal ideation in treatment-resistant depression. Neuropsychopharmacology 45:606–612

61. Phillips JL, Norris S, Talbot J, Birmingham M, Hatchard T, Ortiz A, Owoeye O, Batten LA, Blier P (2019) Single, repeated, and maintenance ketamine infusions for treatment-resistant depression: a randomized controlled trial. Am J Psychiatry 176:401–409

62. Popova V, Daly EJ, Trivedi M, Cooper K, Lane R, Lim P, Mazzucco C, Hough D, Thase ME, Shelton RC, Molero P, Vieta E, Bajbouj M, Manji H, Drevets WC, Singh JB

(2019) Efficacy and safety of flexibly dosed Esketamine nasal spray combined with a newly initiated Oral antidepressant in treatment-resistant depression: a randomized double-blind active-controlled study. Am J Psychiatry 176:428–438

63. Daly EJ, Singh JB, Fedgchin M, Cooper K, Lim P, Shelton RC, Thase ME, Winokur A, Van Nueten L, Manji H, Drevets WC (2018) Efficacy and safety of intranasal Esketamine adjunctive to Oral antidepressant therapy in treatment-resistant depression: a randomized clinical trial. JAMA Psychiatry 75:139–148

64. Fu DJ, Ionescu DF, Li X, Lane R, Lim P, Sanacora G, Hough D, Manji H, Drevets WC, Canuso CM (2020) Esketamine nasal spray for rapid reduction of major depressive disorder symptoms in patients who have active suicidal ideation with intent: double-blind, randomized study (ASPIRE I). J Clin Psychiatry 81:19m13191

65. Ionescu DF, Fu DJ, Qiu X, Lane R, Lim P, Kasper S, Hough D, Drevets WC, Manji H, Canuso CM (2020) Esketamine nasal spray for rapid reduction of depressive symptoms in patients with major depressive disorder who have active suicide ideation with intent: results of a phase 3, double-blind, randomized study (ASPIRE II). Int J Neuropsychopharmacol 24(1):22–31

66. Aleksandrova LR, Wang YT, Phillips AG (2017) Hydroxynorketamine: implications for the NMDA receptor hypothesis of Ketamine's antidepressant action. Chronic Stress (Thousand Oaks) 1:2470547017743511

67. Lener MS, Kadriu B, Zarate CA Jr (2017) Ketamine and beyond: investigations into the potential of glutamatergic agents to treat depression. Drugs 77:381–401

68. Zanos P, Gould TD (2018) Intracellular signaling pathways involved in (S)- and (R)-ketamine antidepressant actions. Biol Psychiatry 83:2–4

69. Miller OH, Moran JT, Hall BJ (2016) Two cellular hypotheses explaining the initiation of ketamine's antidepressant actions: direct inhibition and disinhibition. Neuropharmacology 100:17–26

70. Lipton JO, Sahin M (2014) The neurology of mTOR. Neuron 84:275–291

71. Hay N, Sonenberg N (2004) Upstream and downstream of mTOR. Genes Dev 18: 1926–1945

72. Holmes SE, Scheinost D, Finnema SJ, Naganawa M, Davis MT, DellaGioia N, Nabulsi N, Matuskey D, Angarita GA, Pietrzak RH, Duman RS, Sanacora G, Krystal JH,

Carson RE, Esterlis I (2019) Lower synaptic density is associated with depression severity and network alterations. Nat Commun 10: 1529

73. Groc L, Choquet D (2020) Linking glutamate receptor movements and synapse function. Science 368:eaay4631

74. Chen MH, Wu HJ, Li CT, Lin WC, Bai YM, Tsai SJ, Hong CJ, Tu PC, Cheng CM, Su TP (2020) Using classification and regression tree modelling to investigate treatment response to a single low-dose ketamine infusion: post hoc pooled analyses of randomized placebo-controlled and open-label trials. J Affect Disord 281:865–871

75. Chen MH, Lin WC, Wu HJ, Bai YM, Li CT, Tsai SJ, Hong CJ, Tu PC, Cheng CM, Su TP (2020) Efficacy of low-dose ketamine infusion in anxious vs nonanxious depression: revisiting the adjunctive ketamine study of Taiwanese patients with treatment-resistant depression. CNS Spectr 26(4):362–367

76. Niciu MJ, Luckenbaugh DA, Ionescu DF, Guevara S, Machado-Vieira R, Richards EM, Brutsche NE, Nolan NM, Zarate CA Jr (2014) Clinical predictors of ketamine response in treatment-resistant major depression. J Clin Psychiatry 75:e417–e423

77. Rong C, Park C, Rosenblat JD, Subramaniapillai M, Zuckerman H, Fus D, Lee YL, Pan Z, Brietzke E, Mansur RB, Cha DS, Lui LMW, McIntyre RS (2018) Predictors of response to ketamine in treatment resistant major depressive disorder and bipolar disorder. Int J Environ Res Public Health 15: 771

78. Li CT, Chen MH, Juan CH, Huang HH, Chen LF, Hsieh JC, Tu PC, Bai YM, Tsai SJ, Lee YC, Su TP (2014) Efficacy of prefrontal theta-burst stimulation in refractory depression: a randomized sham-controlled study. Brain 137:2088–2098

79. Khan A, Leventhal RM, Khan SR, Brown WA (2002) Severity of depression and response to antidepressants and placebo: an analysis of the Food and Drug Administration database. J Clin Psychopharmacol 22:40–45

80. Naudet F, Maria AS, Falissard B (2011) Antidepressant response in major depressive disorder: a meta-regression comparison of randomized controlled trials and observational studies. PLoS One 6:e20811

81. Aust S, Gartner M, Basso L, Otte C, Wingenfeld K, Chae WR, Heuser-Collier I, Regen F, Cosma NC, van Hall F, Grimm S, Bajbouj M (2019) Anxiety during ketamine infusions is associated with negative treatment responses in major depressive disorder. Eur Neuropsychopharmacol 29:529–538

82. Pennybaker SJ, Niciu MJ, Luckenbaugh DA, Zarate CA (2017) Symptomatology and predictors of antidepressant efficacy in extended responders to a single ketamine infusion. J Affect Disord 208:560–566

83. Luckenbaugh DA, Niciu MJ, Ionescu DF, Nolan NM, Richards EM, Brutsche NE, Guevara S, Zarate CA (2014) Do the dissociative side effects of ketamine mediate its antidepressant effects? J Affect Disord 159: 56–61

84. Gaydos SJ, Kelley AM, Grandizio CM, Athy JR, Walters PL (2015) Comparison of the effects of ketamine and morphine on performance of representative military tasks. J Emerg Med 48:313–324

85. Medvedev ON, Landhuis CE (2018) Exploring constructs of Well-being, happiness and quality of life. PeerJ 6:e4903

86. Nutt D, Demyttenaere K, Janka Z, Aarre T, Bourin M, Canonico PL, Carrasco JL, Stahl S (2007) The other face of depression, reduced positive affect: the role of catecholamines in causation and cure. J Psychopharmacol 21: 461–471

87. Watson D, Clark LA, Carey G (1988) Positive and negative affectivity and their relation to anxiety and depressive disorders. J Abnorm Psychol 97:346–353

88. Chen MH, Lin WC, Wu HJ, Bai YM, Li CT, Tsai SJ, Hong CJ, Tu PC, Cheng CM, Su TP (2020) Happiness during low-dose ketamine infusion predicts treatment response: Reexploring the adjunctive ketamine study of Taiwanese patients with treatment-resistant depression. J Clin Psychiatry 81:20m13232

89. Milak MS, Rashid R, Dong Z, Kegeles LS, Grunebaum MF, Ogden RT, Lin X, Mulhern ST, Suckow RF, Cooper TB, Keilp JG, Mao X, Shungu DC, Mann JJ (2020) Assessment of relationship of ketamine dose with magnetic resonance spectroscopy of Glx and GABA responses in adults with major depression: a randomized clinical trial. JAMA Netw Open 3:e2013211

90. Chen MH, Chang WC, Lin WC, Tu PC, Li CT, Bai YM, Tsai SJ, Huang WS, Su TP (2020) Functional dysconnectivity of frontal cortex to striatum predicts ketamine infusion response in treatment-resistant depression. Int J Neuropsychopharmacol 23(12): 791–798

91. Gartner M, Aust S, Bajbouj M, Fan Y, Wingenfeld K, Otte C, Heuser-Collier I, Boker H, Hattenschwiler J, Seifritz E,

Grimm S, Scheidegger M (2019) Functional connectivity between prefrontal cortex and subgenual cingulate predicts antidepressant effects of ketamine. Eur Neuropsychopharmacol 29:501–508

92. Sydnor VJ, Lyall AE, Cetin-Karayumak S, Cheung JC, Felicione JM, Akeju O, Shenton ME, Deckersbach T, Ionescu DF, Pasternak O, Cusin C, Kubicki M (2020) Studying pre-treatment and ketamine-induced changes in white matter microstructure in the context of ketamine's antidepressant effects. Transl Psychiatry 10:432

93. Herrera-Melendez A, Stippl A, Aust S, Scheidegger M, Seifritz E, Heuser-Collier I, Otte C, Bajbouj M, Grimm S, Gartner M (2020) Gray matter volume of rostral anterior cingulate cortex predicts rapid antidepressant response to ketamine. Eur Neuropsychopharmacol 43:63–70

94. Mkrtchian A, Evans JW, Kraus C, Yuan P, Kadriu B, Nugent AC, Roiser JP, Zarate CA Jr (2020) Ketamine modulates fronto-striatal circuitry in depressed and healthy individuals. Mol Psychiatry 26(7):3292–3301

95. Hijazi Y, Bodonian C, Bolon M, Salord F, Boulieu R (2003) Pharmacokinetics and haemodynamics of ketamine in intensive care patients with brain or spinal cord injury. Br J Anaesth 90:155–160

96. Li CT, Chen MH, Lin WC, Hong CJ, Yang BH, Liu RS, Tu PC, Su TP (2016) The effects of low-dose ketamine on the prefrontal cortex and amygdala in treatment-resistant depression: a randomized controlled study. Hum Brain Mapp 37:1080–1090

97. Chen MH, Li CT, Lin WC, Hong CJ, Tu PC, Bai YM, Cheng CM, Su TP (2018) Persistent antidepressant effect of low-dose ketamine and activation in the supplementary motor area and anterior cingulate cortex in treatment-resistant depression: a randomized control study. J Affect Disord 225:709–714

98. Nugent AC, Diazgranados N, Carlson PJ, Ibrahim L, Luckenbaugh DA, Brutsche N, Herscovitch P, Drevets WC, Zarate CA Jr (2014) Neural correlates of rapid antidepressant response to ketamine in bipolar disorder. Bipolar Disord 16:119–128

99. Lally N, Nugent AC, Luckenbaugh DA, Niciu MJ, Roiser JP, Zarate CA Jr (2015) Neural correlates of change in major depressive disorder anhedonia following open-label ketamine. J Psychopharmacol 29:596–607

100. Abdallah CG, De Feyter HM, Averill LA, Jiang L, Averill CL, Chowdhury GMI, Purohit P, de Graaf RA, Esterlis I, Juchem C, Pittman BP, Krystal JH, Rothman DL, Sanacora G, Mason GF (2018) The effects of ketamine on prefrontal glutamate neurotransmission in healthy and depressed subjects. Neuropsychopharmacology 43:2154–2160

101. Chou D (2020) Brain-derived neurotrophic factor in the ventrolateral periaqueductal gray contributes to (2R,6R)-hydroxynorketamine-mediated actions. Neuropharmacology 170:108068

102. Chen BK, Luna VM, LaGamma CT, Xu X, Deng SX, Suckow RF, Cooper TB, Shah A, Brachman RA, Mendez-David I, David DJ, Gardier AM, Landry DW, Denny CA (2020) Sex-specific neurobiological actions of prophylactic (R,S)-ketamine, (2R,6R)-hydroxynorketamine, and (2S,6S)-hydroxynorketamine. Neuropsychopharmacology 45:1545–1556

103. Abdallah CG (2020) (2R,6R)-Hydroxynorketamine (HNK) plasma level predicts poor antidepressant response: is this the end of the HNK pipeline? Neuropsychopharmacology 45:1245–1246

104. Farmer CA, Gilbert JR, Moaddel R, George J, Adeojo L, Lovett J, Nugent AC, Kadriu B, Yuan P, Gould TD, Park LT, Zarate CA Jr (2020) Ketamine metabolites, clinical response, and gamma power in a randomized, placebo-controlled, crossover trial for treatment-resistant major depression. Neuropsychopharmacology 45:1398–1404

105. Yokoyama R, Higuchi M, Tanabe W, Tsukada S, Naito M, Yamaguchi T, Chen L, Kasai A, Seiriki K, Nakazawa T, Nakagawa S, Hashimoto K, Hashimoto H, Ago Y (2020) (S)-norketamine and (2S,6S)-hydroxynorketamine exert potent antidepressant-like effects in a chronic corticosterone-induced mouse model of depression. Pharmacol Biochem Behav 191:172876

106. Hare BD, Pothula S, DiLeone RJ, Duman RS (2020) Ketamine increases vmPFC activity: effects of (R)- and (S)-stereoisomers and (2R,6R)-hydroxynorketamine metabolite. Neuropharmacology 166:107947

107. Zanos P, Moaddel R, Morris PJ, Georgiou P, Fischell J, Elmer GI, Alkondon M, Yuan P, Pribut HJ, Singh NS, Dossou KS, Fang Y, Huang XP, Mayo CL, Wainer IW, Albuquerque EX, Thompson SM, Thomas CJ, Zarate CA Jr, Gould TD (2016) NMDAR inhibition-independent antidepressant actions of ketamine metabolites. Nature 533:481–486

108. Moaddel R, Abdrakhmanova G, Kozak J, Jozwiak K, Toll L, Jimenez L, Rosenberg A,

Tran T, Xiao Y, Zarate CA, Wainer IW (2013) Sub-anesthetic concentrations of (R,S)-ketamine metabolites inhibit acetylcholine-evoked currents in alpha7 nicotinic acetylcholine receptors. Eur J Pharmacol 698:228–234

109. Grunebaum MF, Galfalvy HC, Choo TH, Parris MS, Burke AK, Suckow RF, Cooper TB, Mann JJ (2019) Ketamine metabolite pilot study in a suicidal depression trial. J Psychiatr Res 117:129–134

110. Chen MH, Cheng CM, Gueorguieva R, Lin WC, Li CT, Hong CJ, Tu PC, Bai YM, Tsai SJ, Krystal JH, Su TP (2019) Maintenance of antidepressant and antisuicidal effects by D-cycloserine among patients with treatment-resistant depression who responded to low-dose ketamine infusion: a double-blind randomized placebo-control study. Neuropsychopharmacology 44: 2112–2118

111. Saavedra JS, Garrett PI, Honeycutt SC, Peterson AM, White JW, Hillhouse TM (2020) Assessment of the rapid and sustained antidepressant-like effects of dextromethorphan in mice. Pharmacol Biochem Behav 197:173003

112. Murrough JW, Wade E, Sayed S, Ahle G, Kiraly DD, Welch A, Collins KA, Soleimani L, Iosifescu DV, Charney DS (2017) Dextromethorphan/quinidine pharmacotherapy in patients with treatment resistant depression: a proof of concept clinical trial. J Affect Disord 218:277–283

113. Zarate CA Jr, Mathews D, Ibrahim L, Chaves JF, Marquardt C, Ukoh I, Jolkovsky L, Brutsche NE, Smith MA, Luckenbaugh DA (2013) A randomized trial of a low-trapping nonselective N-methyl-D-aspartate channel blocker in major depression. Biol Psychiatry 74:257–264

114. Sanacora G, Johnson MR, Khan A, Atkinson SD, Riesenberg RR, Schronen JP, Burke MA, Zajecka JM, Barra L, Su HL, Posener JA, Bui KH, Quirk MC, Piser TM, Mathew SJ, Pathak S (2017) Adjunctive Lanicemine (AZD6765) in patients with major depressive disorder and history of inadequate response to antidepressants: a randomized, Placebo-Controlled Study. Neuropsychopharmacology 42:844–853

115. Ibrahim L, Diaz Granados N, Jolkovsky L, Brutsche N, Luckenbaugh DA, Herring WJ, Potter WZ, Zarate CA Jr (2012) A randomized, placebo-controlled, crossover pilot trial of the oral selective NR2B antagonist MK-0657 in patients with treatment-resistant major depressive disorder. J Clin Psychopharmacol 32:551–557

116. Nations KR, Bursi R, Dogterom P, Ereshefsky L, Gertsik L, Mant T, Schipper J (2012) Maximum tolerated dose evaluation of the AMPA modulator org 26576 in healthy volunteers and depressed patients: a summary and method analysis of bridging research in support of phase II dose selection. Drugs R D 12:127–139

117. Nations KR, Dogterom P, Bursi R, Schipper J, Greenwald S, Zraket D, Gertsik L, Johnstone J, Lee A, Pande Y, Ruigt G, Ereshefsky L (2012) Examination of org 26576, an AMPA receptor positive allosteric modulator, in patients diagnosed with major depressive disorder: an exploratory, randomized, double-blind, placebo-controlled trial. J Psychopharmacol 26: 1525–1539

118. Du J, Suzuki K, Wei Y, Wang Y, Blumenthal R, Chen Z, Falke C, Zarate CA Jr, Manji HK (2007) The anticonvulsants lamotrigine, riluzole, and valproate differentially regulate AMPA receptor membrane localization: relationship to clinical effects in mood disorders. Neuropsychopharmacology 32:793–802

119. Kishi T, Matsunaga S, Iwata N (2017) A meta-analysis of Memantine for depression. J Alzheimers Dis 57:113–121

Chapter 20

Antidepressants, Sexual Behavior, and Translational Models for Male Sexual Dysfunction: Development of Animal Models, Pharmacology, and Genetics

Jocelien D. A. Olivier, Josien Janssen, Tommy Pattij, Stephen De Prêtre, and Berend Olivier

Abstract

The discovery and development of the first generations of antidepressants in the last century, the tricyclic antidepressants and serotonin reuptake blockers, were a breakthrough in the pharmacological treatment of major depression. Along with the antidepressant activity came the sexual side effects, which contributed considerably to the high level of stopping treatment. In the subsequent search for new and better antidepressants, early detection of potential sexual side effects is of paramount importance, hence the need for predictive animal models. Sexual behavior of the male rat has been frequently used to detect inhibiting effects of psychotropic drugs. We developed a male rat model of sexual behavior that mirrored the human profile of antidepressants: sexual inhibitory effects only after chronic but not after acute administration. We extensively describe the methodology and the model and show the profile of various antidepressants and other psychotropics in male rat sexual behavior. To generate male rats for our experiments, we employ large cohorts of male Wistar rats that are trained once weekly in 30-min tests with estrous females for at least 4–7 times. During this training each individual rat develops its own stable sexual phenotype, being between 0 and 5 ejaculations per 30-min test. Such a (endo)phenotype appears very stable over time and animals can be used repeatedly for pharmacological experiments for over a year, providing an ideal intra-male experimental model. For testing the effects of drugs (e.g., antidepressants) on sexual behavior, we standardly use rats with stable ejaculation numbers of 2–3 per test, providing a model that is sensitive to both sexual-stimulating (prosexual) and sexual-inhibiting effects, and are able to dissect acute and chronic effects of drugs. The effects of various drugs tested in this model over the last decades are given.

By using the large cohort approach and sexual training, we discovered that the number of rats with 0, 1, 2, 3, 4, or 5 ejaculations (E) per test shows a bell-shaped distribution. Relatively few rats have either 0 or 1 E or 4 or 5 E/test, whereas those with 2 or 3 E/test are much more abundant. Based on the similarity of rat ejaculation number distribution with that of ejaculation latency distribution in human males, we postulate that fast ejaculating rats (4 or 5 E/test) are a translational model for premature ejaculation, whereas slow or not ejaculating rats (0 or 1 E/test) could model an-ejaculation or delayed ejaculation in men. Several pharmacological experiments are described supporting the use of these translational endophenotypic models of normal, slow, and fast ejaculating rats. The importance of the serotonergic system and in particular the role of 5-HT$_{1A}$ receptors in male sexual behavior is highlighted. The serotonin transporter knockout rat illustrates the influence of genetic modifications in male rat sexual behavior. This model

Yong-Ku Kim and Meysam Amidfar (eds.), *Translational Research Methods for Major Depressive Disorder*, Neuromethods, vol. 179, https://doi.org/10.1007/978-1-0716-2083-0_20,
© The Author(s), under exclusive license to Springer Science+Business Media, LLC, part of Springer Nature 2022

displays a comparable sexual phenotype as chronically SSRI-treated rats. Such a genetic model may be useful in detecting underlying mechanisms of sexual dysfunctions (like delayed ejaculation) but may also contribute to the study of critically involved neurochemical systems. Finally, testing drugs with multimodal mechanisms of action in such genetic models might unravel new mechanisms involved, finally contributing to better treatments.

Key words Antidepressants, Sexual behavior, Premature ejaculation, Delayed ejaculation, Rat model, Cohort studies, Endophenotype, Pharmacology, Genetics, Serotonin

1 Antidepressants in Human Major Depression: Side Effects on Sexual Behavior

Two extensive studies were published in 2011 about the prevalence and associated disability, including associate disease burden and financial cost, of brain disorders in Europe [1–3]. In psychiatric disorders, anxiety disorders with 14% and major depression with 7% had the highest 12-month prevalence. Major depression is a severe brain disorder associated with long-term disability and very low quality of life.

Over the last century, the search for antidepressant medicines gradually emerged, but it lasted till the 1950s and 1960s before the first antidepressants were found. Early antidepressants like imipramine and the irreversible monoamine oxidase inhibitors (MAOI) were serendipitously discovered. The "accidental" discoveries triggered intensive research and led to a series of new antidepressants, including the tricyclics (TCA, e.g., imipramine, nortriptyline, amitriptyline, and clomipramine) and a series of both reversible and irreversible MAOIs. TCAs and MAOIs, although still clinically available, are not anymore first-line medications, mainly because of their often severe side effects. The ongoing research in the 1960s and 1970s revealed that TCAs block monoamine transporters (reuptake carriers) mainly for serotonin (5-HT) and noradrenaline (NA) or NE (norepinephrine) to a varying extent. Some TCAs are preferential serotonin transporter (SERT) blockers (clomipramine, amitriptyline), whereas others are preferential noradrenaline transporter (NET) blockers (desipramine, maprotiline) or mixed SERT/NET blockers (doxepin, imipramine). TCAs concomitantly block various neurotransmitter receptors, in particular muscarinic cholinergic, H_1 histaminergic, and α_1-adrenoreceptors. These additional mechanisms cause considerable and unwanted side effects, including sedation, dry mouth, and constipation.

In line with the early hypothesis that low serotonin and/or noradrenaline levels or activity in the CNS is causing or associated with depressive mood, several new drugs were developed, including the selective serotonin reuptake inhibitors (SSRIs, e.g., fluoxetine, fluvoxamine, sertraline, paroxetine, (es)citalopram), selective noradrenaline reuptake inhibitors (NRIs, e.g., reboxetine, atomoxetine), and mixed 5-HT/NA reuptake blockers (SNRIs, e.g., venlafaxine, duloxetine). Several other antidepressants with different mechanisms of action have also been developed, but in general SSRIs are the drug of choice and first-line in the treatment of major depression [4].

The most disturbing side effects of SSRIs and some TCAs (e.g., clomipramine) are on sexual functions like libido, orgasm, and arousal problems. In contrast to most other side effects that are prominent in the first treatment phase, and often diminish or disappear upon drug continuation (e.g., nausea, dizziness), sexual side effects do not disappear or diminish and are often leading to drug discontinuation.

Major depression itself is associated on its own with a 50–70% enhanced risk of sexual dysfunction, and prescribing antidepressants with associated sexual side effects strongly increases the risk on noncompliance or discontinuation of drug-taking [5]. The relationship between depression, its pharmacological treatment, and sexual dysfunction is highly complex. There is a high comorbidity between depression and sexual dysfunction in both male and female patients [6]. Although successful treatment of depression may improve some aspects of sexual function (especially libido), most antidepressants induce sexual dysfunction, especially ejaculation [7]. Consequently, it is often difficult or impossible to distinguish whether the disease causes a certain sexual dysfunction, by pharmacological treatment or other factors. If a depressed patient upon pharmacological treatment complains about sexual dysfunction, this could indicate either a nonresponse to the treatment or a drug side effect [8].

2   Sexual Side Effects of Antidepressants (SSRIs) Are Therapeutic in Human Males: Premature Ejaculation

Premature ejaculation (PE) has emerged the last couple of decades as an ejaculation disorder or ejaculation "complaint" [9]. In the latest version of DSM (DSM-5), premature ejaculation has been defined as persistent or recurrent ejaculation with minimal sexual stimulation before, on, or shortly after penetration and before the person wishes it. The disturbance causes marked distress or interpersonal difficulties. Moreover, the PE is not exclusively due to the direct effects of a substance or withdrawal of a substance. This definition became inadequate over the years in particular when drugs had to be investigated and consequently the International Society for Sexual Medicine (ISSM) has redefined PE based on evidence-based features [10]. The ISSM defined lifelong PE as an ejaculation within 1 min of vaginal penetration in the majority of sexual encounters, without the ability to delay ejaculation, and with associated negative personal consequences. Waldinger [9] makes a distinction in the description of PE as either a "disorder" or a "complaint." He estimates the prevalence of PE as a complaint in the order of 15–30%, whereas the prevalence as "disorder" seems to be much lower (2.3%). He explains this by showing that there are four different PE subtypes, viz., lifelong PE (LPE), acquired PE (aPE), natural variable PE, and premature (P)-like ejaculatory dysfunction, distinguished on the basis of the IELTs. LPE and aPE might represent the PE "disorder" type and occur with a relatively low prevalence. Waldinger [9] hypothesizes that the latter two types have a biological substrate including genetic correlates, whereas the natural variable PE and P-like ejaculatory dysfunction might be considered either a normal variation of ejaculatory performance or caused by psychological, cultural, or interrelationship factors and are considered not caused by neurobiological or genetic factors.

3   SSRIs Have Inhibitory Effects on the Ejaculation Latency in Men with Premature Ejaculation

Treatment by selective serotonin reuptake inhibitors (SSRIs) has led to a paradigm shift in the understanding and treatment of PE [11].

Although anecdotal reports were present on a possible inhibitory role of serotonergic antidepressants (mainly SSRIs) on premature ejaculation, the early placebo-controlled studies of various SSRIs performed by Waldinger formed a stable basis for extensive "off-label" use of SSRIs for treatment of PE. The tricyclic antidepressant clomipramine, the most "serotonergic" of the tricyclics (TCAs) and clinically available before the SSRIs were introduced (from 1984 on), was the treatment of choice for PE, although only a few double-blind placebo-controlled studies have been performed [12–15].

Development of the SSRIs provided a new approach to delay ejaculation. Waldinger et al. [11] were the first in establishing delayed ejaculation times in men with PE and very short ejaculation latencies (<40 s). Paroxetine (40 mg/day) resulted in a delayed latency time of 7.5 min after 3 weeks of treatment. A later study [16] in a larger population, using timing with a clock and applying 20 mg or 40 mg/day paroxetine, confirmed these earlier findings. In a subsequent study [17], fluoxetine (20 mg/day), fluvoxamine (100 mg/day), paroxetine (20 mg/day), and sertraline (50 mg/day) were tested in a double-blind, placebo-controlled study over 6 weeks using a stopwatch method to determine the intravaginal ejaculation latency time (IELT). All groups had a basal IELT of approx. 10–20 s at the start of the study and placebo treatment did not affect the IELT. At single doses, akin to effective antidepressant doses, paroxetine, fluoxetine, and sertraline increased IELT considerably, whereas fluvoxamine had modest effects. Other studies (*see* [18]) confirmed this SSRI-induced elongation of the IELT [10]. Dapoxetine, a rapid-acting and short half-life SSRI, introduced as having on-demand effects on PE [19], is the only SSRI specifically introduced into the market for premature ejaculation. There is some dispute about the effectiveness of dapoxetine [20, 21], also probably reflected in its limited use [22].

There seem to be different sexual inhibitory properties of the various SSRIs both in healthy men [23–26] and in rats [27]. De Jong et al. [28] postulated that the differential inhibitory effects of the various SSRIs might be due to differences in desensitization of 5-HT_{1A} receptors located on oxytocinergic neurons. The serotonergic system seems intrinsically involved in the regulation of sexual behavior [29, 30] and seems also to play an important role in premature ejaculation, based on pharmacological [31, 32] and genetic [33–36] data. The serotonergic system is widespread and reaches practically all parts of the brain including the spinal cord [37]. A big unknown question is whether premature ejaculation is related to dysfunctions in the serotonergic systems in the brain and spinal cord and to what other neurotransmitter systems the serotonergic system talks during the performance of the (premature) ejaculation and associated behavioral and physiological events [23, 30].

In this contribution we focus on the "hard core" PE with a neurobiological and genetic substrate as defined by Waldinger [38] as that enables biological research into animal models. First, we shortly describe the neurobiological and anatomical processes involved in the ejaculatory process in order to delineate possible entrances toward animal models of ejaculation and ejaculation disorders.

4 Sequential Organization of Sexual Behavior in the Rat: From Initiation to Ejaculation—Methodological and Theoretical Issues

The neurobiological and peripheral mechanism involved in the overall ejaculation process in humans and mammals is rather complicated involving many neural mechanisms in the central and peripheral nervous system.

Veening and Coolen [39] introduced an intriguing funnel model of male sexual behavior in the rat (*see* Fig. 1). Before starting sexual behavior a male rat determines whether the environment is safe and the female is in estrus. In the funnel model, the male is in phase 1, the scanning and initiation phase. The male progresses to the appetitive or precopulatory phase where it actively explores the female [40], by approaching, following, and genital sniffing,

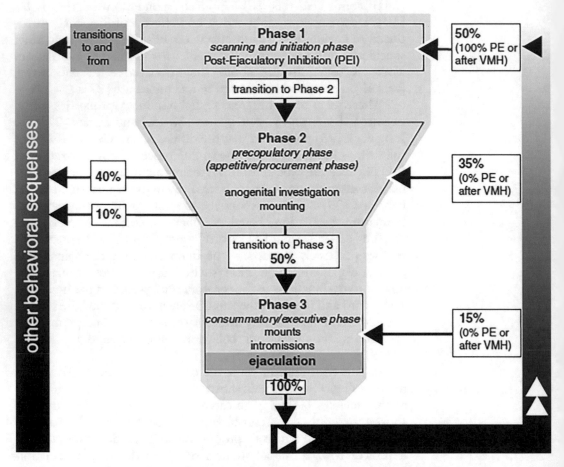

Fig. 1 The "funnel model" of sexual behavior [39]. The sequential analysis of all behavioral transitions indicates that the possibility to choose for another transition than the regularly preferred one is gradually decreasing during the progression of the behavioral sequence (funnel model). During the male sexual sequence, in the initiation phase (phase 1), many transitions are possible leading to or away from sexual activities, but after mounting at least 50% of the transitions are followed by intromission and after a few intromissions the chance on ejaculation increases to 100%. In phase 2 the male explores the female, and after establishment of her behavioral status (estrus), the male starts mounting which leads to transition to phase 3, the consummatory or executive phase, where the male vigorously mounts and intromits, finally leading to ejaculation. After ejaculation in the post-ejaculatory inhibitory interval, the animal returns to phase 1. After this post-ejaculatory interval, the sequence is repeated and repeated

presumably obtaining olfactory information for initiation of the copulatory phase. If the female is in estrus, she will respond by proceptive behavior, including darting, hopping, and ear wiggling. This probably triggers the male to initiate mounting. In the funnel model, this is the transition from phase 2 (precopulatory phase) to phase 3, the consummatory/executive phase. A series of mounts and intromissions will follow, each typically followed (particularly the intromissions) by short bouts of genital grooming, finally leading to ejaculation. Ejaculation is followed by an extended period (minutes) of genital grooming and sexual inactivity, during which the animal is unable to restart sexual activities (refractory period), the post-ejaculatory interval (PEI). Because the refractory male stays very attentive to the environment, PEI has been included in phase 1. After the PEI, the male restarts the sexual cycle and, if let free for several hours, may reach up to 5–8 ejaculatory series [41, 42]. Such males finally reach sexual satiety during which sexual behavior may be inhibited for several days to 1 week [43, 44].

The funnel model explains the sequential organization of male rat sexual behavior and indicates that the normal sexual sequence can be easily disrupted in phase 1; however, progressing to phase 2 but particularly in phase 3, it becomes more difficult to interrupt the regular sequence. Figure 2 gives a schematic overview of the main brain areas and connections involved in sexual behavior and ejaculation [39, 45, 46]. Ejaculation and its associated orgasm are

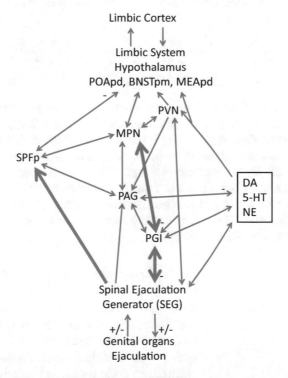

Fig. 2 "Ejaculation control." Different brain areas and connections are shown controlling the performance of male sexual behavior. DA, dopamine; 5-HT, serotonin; NE, norepinephrine; MPN, medial preoptic nucleus; PGI, paragigantocellular nucleus; PAG, periaqueductal gray; POApd, posterodorsal preoptic area; BNSTpm, posteromedial part of the bed nucleus of the stria terminalis; MEApd, posterodorsal medial amygdaloid nucleus; SPFp, parvocellular part of the subparafascicular nucleus; PVN, paraventricular nucleus of the hypothalamus

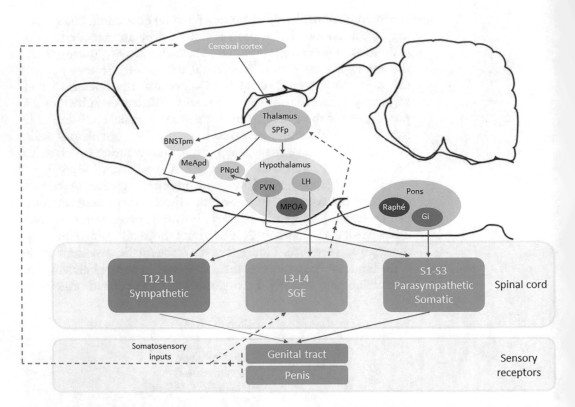

Fig. 3 Diagram of the CNS network controlling ejaculation. Brain nuclei forming the brain circuitry of ejaculation are indicated with their connections to spinal nuclei involved in the control of ejaculation. Somatosensory afferents originating in the genitals and terminating in the CNS are indicated. Abbreviations: BNSTpm, posteromedial part of the bed nucleus of the stria terminalis; Gi, gigantocellular nuclei; LH, lateral hypothalamus; MeApd, posterodorsal part of the medial amygdala; MPOA, medial preoptic area; PNpd, posterodorsal part of the preoptic nucleus; PVN, paraventricular nucleus; SGE, spinal generator of ejaculation; SPFp, parvicellular part of the subparafascicular nucleus. T12–L1: spinal level from sections thoracic 1 to lumbar 1; L3–L4: spinal level from sections lumbar 3 to lumbar 4; S1–S3: spinal level from sections sacral 1 to 3; SPG: spinal ejaculation generator

the final consummatory phase of the sexual behavior cycle. Ejaculation is a reflex comprising sensory receptors and areas, afferent pathways, cerebral sensory areas, cerebral motoric areas, spinal motor centers, and efferent pathways (Fig. 3). Moreover, integrative areas in the brain probably influence several aspects of this cycle, e.g., merging psychological and motivational factors. The ejaculatory cycle seems to be under control of strong dopaminergic and serotonergic neurotransmission although various other systems (e.g., cholinergic, GABA-ergic, adrenergic, and oxytocinergic) play a role too.

Ejaculation involves three basic mechanisms: emission, ejection, and orgasm. Emission of spermatozoa from the testes and seminal vesicles and prostate is induced by sympathetic motor neurons in the thoracolumbar intermediolateral cell column

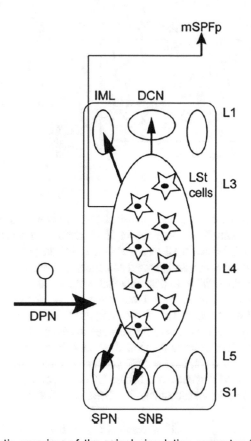

Fig. 4 Schematic overview of the spinal ejaculation generator (SPG). Sensory inputs from the dorsal penile nerve (DPN) trigger activation of lumbar spinothalamic (LSt) cells in spinal levels L3–4. LSt cell axons project to spinal areas involved in the coordinated autonomic and motor outflow controlling emission and expulsion: sympathetic preganglionic cells in the intermediolateral cell column (IML) and dorsocentral nucleus (DCN), and parasympathetic preganglionic cells in the sacral parasympathetic nucleus of the bulbocavernosus (SNB). It is hypothesized that LSt cells trigger ejaculation via release of neurotransmitters in these target areas. In addition, LSt cells project to the medial subdivision of the parvocellular subparafascicular thalamic nucleus (mSPFp). This spinothalamic projection is hypothesized to contribute to ejaculation-related reward and inhibition of subsequent mating behavior (from Veening and Coolen 2014)

(IML) and parasympathetic motor neurons in the sacral parasympathetic nucleus (SPN) (Fig. 4). Somatic motor neurons in the dorsolateral and dorsomedial ventral horn of the lumbosacral spinal cord cause rhythmic contractions of the striated ischiocavernosus and bulbospongiosus muscles in the pelvic floor leading to forceful expulsion of semen from the urethra [47]. The motor neurons involved in ejaculation are triggered by sensory input from the genitals and reaches the dorsal horns and dorsal gray commissure of the lumbosacral spinal cord via the dorsal penile nerve (Fig. 3).

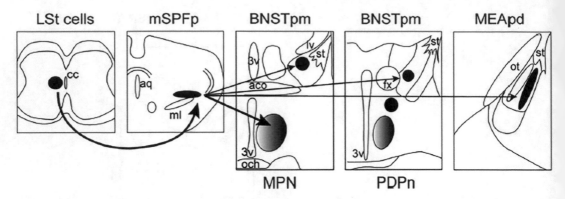

Fig. 5 Schematic overview of the spino-thalamic-forebrain pathway for relay of ejaculation-related signals in the male rat. Black-filled areas depict the locations of ejaculation-specific cFos expression in lumbar spinothalamic (LSt) cells in L3–4 spinal levels, medial subdivision of the parvocellular subparafascicular thalamic nucleus (mSPFp), anterior and posterior subdivisions of the posteromedial part of the bed nucleus of the stria terminalis (BNSTpm), posterodorsal preoptic nucleus (PDPn), and lateral zone of the posterodorsal medial amygdala (MEApd). Neural activation in the medial preoptic nucleus (MPN) is induced by all consummatory elements of sexual behavior, including ejaculation, and is depicted as graded shades of gray. Arrows illustrate the connections between the ejaculation-activated areas and flow of ejaculation-related signals; thick arrows illustrate demonstrated connections, while thinned arrows show connections between the brain areas in general, but have not yet been established for the ejaculation-activated areas specifically. Based on references in Veening and Coolen (2014). cc, central canal; aq, aqueduct; aco, anterior commissure; och, optic chiasm; 3v, third ventricle; lv, lateral ventricle; st, stria terminalis; fx, fornix; ot, optic tract

The relay of genitosensory input to ejaculatory motor output occurs at the level of the spinal cord in the so-called lumbar spinothalamic cells (LSt cells) [48]. Galaninergic fibers originating from these LSt cells project to all areas in the spinal cord containing motor neurons involved in ejaculation, and selective lesioning of these LSt cells eliminates ejaculation without affecting other parameters of sexual behavior [48]. Activation of the LSt cells seems to play an important role in the ejaculation threshold. Serotonin and serotonergic fibers are present in all the areas of the spinal cord involved in ejaculation, including LSt cells, and might be influencing ejaculatory processes at this spinal level [47]. Cerebral sensory and motor areas are involved in all phases of sexual behavior, including ejaculation. However, the question is whether there are "specific sexual" sensory or motor areas or that depending on the necessary conditions such areas are recruited on demand. There seem to be brain areas that are specifically activated when rats ejaculate. It is suggested that there exist a specific "ejaculation" network underlying male sexual behavior (Fig. 5), including areas in the posteromedial part of the bed nucleus of the stria terminalis (BNSTpm), the posterodorsal preoptic nucleus (PDPn), the posterodorsal medial amygdaloid nucleus (MeApd), and the parvocellular part of the subparafascicular thalamus (SPFp) [49]. Although their precise role is yet to be determined, they seem to relay their genital sensory information to the medial preoptic area (MPOA), a pivotal area in the control of male sexual behavior.

The above-described brain areas are part of a larger network that regulates male sexual behavior and ejaculation. This network (Fig. 2) includes the MPOA, the paraventricular nucleus of the hypothalamus (PVN), and the paragigantocellular nucleus (nPGi). In rats MPOA and PVN exert excitatory influences on ejaculation, whereas the nPGi exerts inhibitory influence. The MPOA is involved in modulation of emission and expulsion phases of ejaculation and integrates the genital and other sensory inputs. The MPOA does not project directly into the lumbosacral spinal cord, but modulates several relay stations that directly innervate the cord (including PVN, PAG, and nPGi). The PVN projects both to the nPGi directly and also directly to the spinal autonomic and pudendal nerves. The nPGi seems a vital center in the control of ejaculation as it exerts a tonic inhibition on ejaculation. Between the limbic system, controlling behavior, and the spinal ejaculation generator (SEG), controlling ejaculation, a variety of brain areas are involved in the proper performance and inhibition of response. Many of these brain regions are reciprocally connected, and for only a few of them the excitatory or inhibitory effects have been shown unequivocally. The control mechanisms are apparently extensive and complicated. Two "main control lines" appear to be most important, however, one descending and the other ascending. The inhibitory control of the spinal ejaculation generator appears to be located in the lateral paragigantocellular nucleus (PGl) of the caudal brainstem. The medial preoptic nucleus (MPN) is able to inhibit the inhibitory PGl effects, thereby releasing the ejaculation pattern. This "motor-control line" (MPN-PGl-SEG) is based on inhibitory control mechanisms, modulated by the other brain areas. The other "main line" contains information about the ejaculatory process and ascends among others to the SPFp. From there the information is spread to a number of hypothalamic and limbic brain areas, like MPN, posterodorsal medial amygdala (MEApd), posteromedial part of the bed nucleus of the stria terminalis (BNSTpm), and the posterodorsal preoptic nucleus (POApd). Several findings suggest that these messages contain some kind of "satiety signal" and play a role in the temporary inhibition of SEG mechanisms, in the phase of "post-ejaculatory inhibition." Due to the reciprocal relations between these areas, it is yet completely impossible to define their contribution more specifically.

The paraventricular hypothalamic nucleus (PVN) is extensively involved in autonomic control mechanisms and may convey the deteriorating effects of stress on copulation. About the effects of the mesencephalic periaqueductal gray (PAG), its role in female sexual behavior (the "lordosis response") is well known. Its role in masculine sexual behavior is less clear, but its extensive set of relations with the ejaculation-controlling brain areas is remarkable.

Finally, monoaminergic systems (dopamine (DA), serotonin (5-HT), and noradrenaline (NA)) have been indicated at the right side of the scheme in Fig. 2. They are able to influence a variety of aspects of sexual behavior and ejaculation, at all levels from the limbic system to the SEG. Their excitatory or inhibitory actions have not been specified in the scheme. In addition, the effects of numerous neuropeptides involved in the control of sexual activities, like galanin, oxytocin, or β-endorphin, have not been included in the present scheme.

Allard et al. [50] have formulated several putative hypotheses that might underlie rapid or premature ejaculation. These authors explain ejaculation as a spinal reflex controlled by the spinal ejaculation generator (SEG) that is modulated both by pelvic sensory information and descending input from inhibitory and excitatory areas in the brainstem and hypothalamus. Higher cortico-limbic brain areas are involved in sexual arousal integrating psychological and voluntary aspects of "libido." Lower areas and pathways correspond to the autonomic control of ejaculation, not under voluntary control. During sexual intercourse the higher cortico-limbic areas modulate (inhibit or activate) inhibitory and excitatory areas resp., shifting the supraspinal tone on the SEG from inhibitory to excitatory. Under normal satisfying conditions there is a subtle balance in all these processes leading to ejaculation (*see* Fig. 4a in [50]). Premature ejaculation might, in these authors's hypothesis, (a) be caused by too high stimulation of the SEG, either by peripheral or central activation; another possibility (b) is that PE is due to impairment in descending inhibitory tone. Alternatively (c), a "constitutive" hypersensitivity of the SEG might be causative. In these alternatives disturbances at higher integrating areas (like in the hypothalamus or limbic areas) or at lower integrating areas (supraspinal or spinal) might be involved. As serotonin is clearly involved in aspects of sexual behavior, including ejaculation, a role for a serotonergic disturbance in PE might be possible. On the other hand, if serotonergic imbalance does not underlie the PE disturbances, it still is possible that manipulating the serotonergic tone (e.g., with an SSRI) restores the balance somewhere in the system. Although it has been suggested that the process might happen at the level of the descending inhibitory serotonergic projections from the brainstem to the SEG [24, 51], this is far from proved.

5   Male Rat Sexual Behavior: A Translational Model for Human Sexual Behavior and its Dysfunctions—Methods and Applications

Animals, and we restrict us in this overview mainly to rats, can be used to model all aspects present in sexual behavior and its underlying physiology and pharmacology. Although clear differences are present between humans and animals in the whole sexual area (e.g., hormonal dependence, performance of sexual behavior, higher cerebral importance (*see* [52, 53] for an extensive elaboration and comparison)), there is however a lot of comparability between sexual behavior in rats and humans, particularly the mechanisms in the brain, spinal cord, and autonomic regulation, anatomically, physiologically, and pharmacologically [45, 46, 54]. This makes it feasible to use animal models of sexual behavior to predict putative mechanisms and functions in humans and its pathology. Studies can be performed using physiological preparations that can be used to study at organ level how certain mechanisms are locally organized or respond to certain experimental maneuvers. Physiological experiments on erection or ejaculation can be performed under anesthesia and can help to study, e.g., the effects of drugs on such preparations (cf. [54]). However, the present chapter will focus on intact animals and under free behaving conditions. Under such, preferably naturalistic, conditions, sexual behaviors are displayed, representing its associated full repertoire of behaviors, sounds, and temporal sequences. Sexual behavior between a male rat and a female rat in estrus (Fig. 6), in an appropriate environment, is often used to study various aspects of the underlying mechanisms or its neuropharmacology [55].

Figures 7 and 8 show in a schematic way the interaction between an experienced male rat and a female rat in behavioral estrus during an encounter (e.g., during 30 min) in a rather large cage. Under such conditions the male rat drives the interaction and a number of ejaculations may occur (e.g., up to five ejaculations in a 30-min encounter). Sexual behavior runs in a cascade of events [52, 56]. After some (urogenital) investigation of the female by the male, the male starts mounting to which the female responds with a lordosis posture. Depending upon the sexual experience of the male, some more mounts and intromissions will follow ending in an ejaculation. After some time the male restarts this cycle, ending again in an ejaculation. On average sexually experienced male rats ejaculate 2–3 times in a 30-min encounter.

mount   intromission   ejaculation

Fig. 6 A picture of sexual behavior postures, viz., mounting, intromission, and ejaculation [239]

Sexual Behavior Parameters

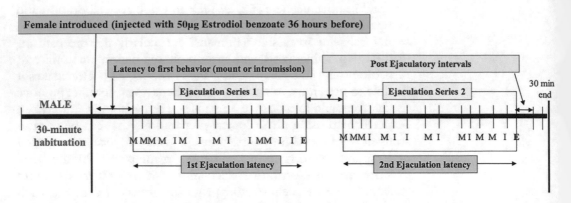

Fig. 7 Sexual behavior parameters of a rat displaying two ejaculations in a 30-min test with data broken down into the first and second ejaculatory series (*M* mount, *I* intromission, *E* ejaculation)

Apart from these behavioral sequences, male rats also produce ultrasonic sounds during mating, the so-called 22-kHz and 50-kHz calls [57] which are produced during different phases of sexual cycles and play important roles in the social interactions between the male and female [58–60]. Typically, 22-kHz calls (which fall into durations between 300 and 200 ms and frequencies between 18 and 32 kHZ) are associated with post-ejaculatory interval where a male displays a refractory period before resuming copulatory behavior [61, 62]. The role of the males' 50-kHz calls is less clear [60, 63].

In these sequences one can distinguish various phases, including appetitive, precopulatory, consummatory, and post-consummatory or preparatory phases. During the appetitive phase male rats show behavioral activation, including grooming, motor activation, and genital investigation. This is followed by a precopulatory phase in which animals become more aroused and activated and pursue the female. In the consummatory phase the male mounts, intromits, and ejaculates. The post-consummatory or preparatory phase is quite similar to the precopulatory phase in that animals are preparing for a next round. During this phase, males produce the so-called post-ejaculatory sounds of approx. 20–30 kHz, reflecting an absolute refractory period in which the male cannot resume sexual activities. A fully proceptive and receptive female (often induced by estrogen/progesterone injections) is completely in synchrony with the behavior of the male showing appetitive, precopulatory (proceptive behavior like hops, darts, ear wiggling, and pacing), and consummatory (receptive) behavior (lordosis). During the post-ejaculatory period of the male, the female abstains from solicitation, suggesting that the ejaculations induced a state of post-consummatory (preparatory) behavior in

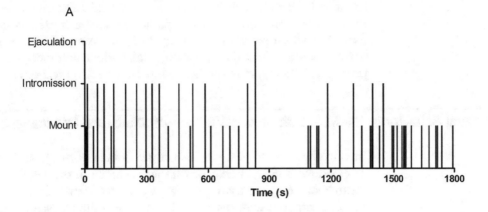

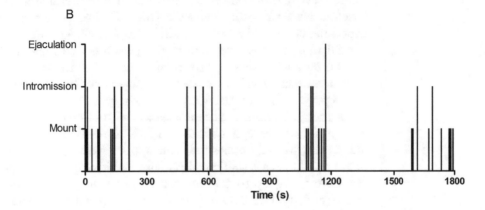

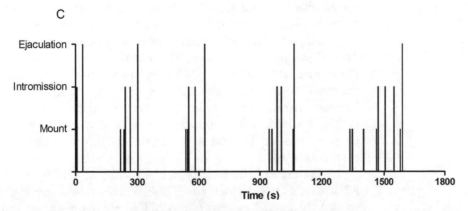

Fig. 8 Temporal distribution of mounts, intromissions, and ejaculation in three representative, sexually well-trained male Wistar rats during sexual behavior with an estrous female over an experimental period of 1800 s (30 min). (**a**) shows a rat that ejaculates once, (**b**) a rat with three ejaculations, and (**c**) a rat with five ejaculations. Vertical lines either depict one unit (mount), two units (intromission), or three units (ejaculation)

the female as well probably induced by the 22-kHz calls from the male. This male-female pairing paradigm is the most frequently used behavioral paradigm to study all kinds of aspects of rat male sexual behavior, including physiological, endocrinological, genetic, pharmacological, neurochemical, and neuroanatomical ones.

6   Sexual Behavior in the Male Rat: Cohort Studies—Methods and Implications

In our pharmacological studies on the effects of drugs and medicines on male sexual behavior that were run from the 1980s of last century on [64], we used quite large cohorts of male rats (80–120 per experiment) and discovered that the animals needed to be trained several times on sexual performance before reaching a stable baseline level of sexual behavior [65–69]. For pharmacological experiments we used rats with a stable number (2, 3) of ejaculations per 30 min, because we wanted the possibility to measure both decreasing and increasing effects of drugs. Over the years we run large male sexual behavior studies (cohorts of 90–120 males are tested per experiment) using outbred Wistar males. During a sexual encounter of 30 min, a male rat interacts with a hormonally primed, proceptive, and receptive female and his sexual behavior is scored. Standardly, we train males four times, once weekly, and the sexual behavior of males is stable and remains stable over the next period (we have tested up to 6 months afterward), suggesting a lifelong stable (endo)phenotype. It appears that the ejaculation frequencies are distributed following a bell-shaped curve. In over 2000 males tested this way, we found that the number of "fast" ejaculating animals (4–5 ejaculations/test) fluctuates between 3% and 10% depending on the cohort and over time. Similarly, the number of "slow" (sluggish) ejaculators (0–1 ejaculation/test) also fluctuates over time and cohort (3–20%) and accordingly, the number of "normal" ejaculating rats (2–3 ejaculations/test) fluctuates in between these numbers. In most cohorts the number of "fast" (rapid) ejaculators is around 10%, and consequently large cohorts of rats have to be used to generate enough experimental "fast ejaculating" rats.

As an example, in a very recent experiment (2020), a cohort of 32 male rats was trained for 7 consecutive weeks (*see* Fig. 9). At week 1 the majority of rats (almost 70%) did not ejaculate, but upon subsequent weekly testing, the number of ejaculating animals increased and at week 4 a more or less sexual stable pattern was present. Only two animals did not ejaculate at all, although during subsequent testing this number increased to four. The two non-ejaculating animals did not ejaculate once during all these seven tests and even did not mount or intromit, whereas the other two animals mounted and intromitted but did not ejaculate.

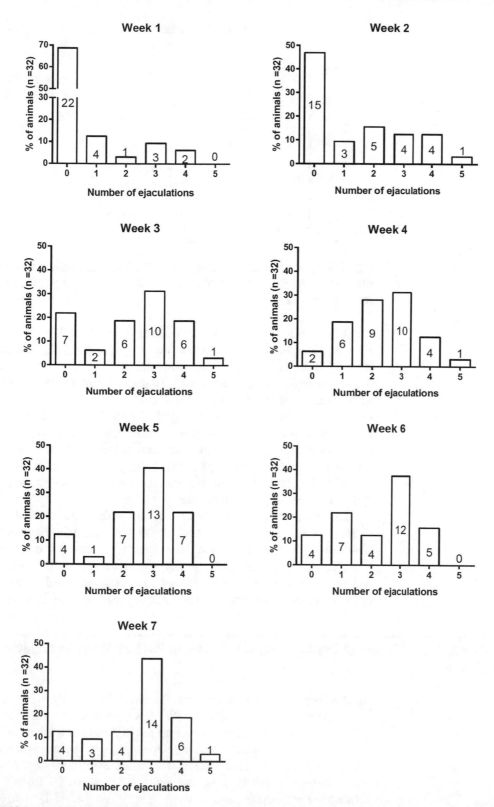

Fig. 9 Male rats ($N = 32$) are weekly trained with a sexually receptive and proceptive female during a 30-min test. The distribution of rats that ejaculate 0 to 5 times during each test is portrayed as % of total, and also the absolute number of animals per category is indicated (number in bars)

Ejaculation latency

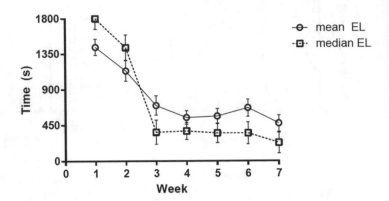

Fig. 10 The mean and median latencies from the first sexual act (either mount or intromission) till the first ejaculation are given for the 7 consecutive weekly sexual tests in 32 male rats

The stability of sexual behavior in this group of 32 animals can also be illustrated by the time the first ejaculation (ejaculation latency) is reached, either measured from the start of the test or measured from the first sexual act (either mount or intromission, whichever comes first). Figure 10 shows that from test 3 on, these values are quite stable, illustrating that after 3–4 training tests the sexual behavior of these animals is stable and remains stable over time. We have run tests of male rat sexual behavior for more than 8 months, and using weekly interval testing of 1, the level of sexual behavior (measured in the number of ejaculations/test and first ejaculation latency) was constant.

In this particular experiment 16 animals were needed for a pharmacological experiment and we selected animals with a stable baseline of 3–4 ejaculations. In subsequent sections the use of low-frequency and high-frequency ejaculators as models of resp. delayed or ejaculating rats (0–1 ejaculation/test) and premature ejaculation (high number of ejaculations/test) will be illustrated.

7   Methods and Models of Normal Sexual Behavior in the Rat: Pharmacological Experiments

In this paper we focus on the "normal" ejaculating males for psychopharmacological studies. We reason that their level of sexual behavior allows detection of both inhibitory and stimulatory effects of drugs. Antidepressants have a delayed onset of therapeutic action [70]. Although some side effects of antidepressants emerge rapidly (e.g., nausea, dizziness), sexual side effects seem to mirror the antidepressant profile and emerge over time [71–73]. An ideal animal model predicting sexual side effects should follow such a time course: acutely no or marginal effects, with an inhibitory effect on sexual behavior occurring after subchronic (1 week) or chronic (2 weeks) treatment. If the model were also able to detect prosexual activities of psychotropics, this would further support its use in predicting their putative influence on sexual behavior in humans. Therefore, we designed an experimental drug test that consists of 14 days of daily treatment with (several doses of) an experimental compound (mostly) followed by a washout test 1 week after stopping the last treatment. We measure the sexual performance of all rats after acute, 8 days (subchronic), 15 days (chronic), and 22 days (washout) of treatment.

Using this experimental design, we tested various drugs which are either clinically used as antidepressants, viz., paroxetine, citalopram and escitalopram, (selective serotonin reuptake inhibitors, SSRIs), venlafaxine (serotonin-noradrenalin reuptake inhibitor, SNRI), bupropion (dopamine and noradrenaline reuptake inhibitor, DNRI), buspirone (5-HT_{1A} receptor agonist), vortioxetine (an SSRI with additional various serotonergic receptor modulatory properties), vilazodone (an SSRI with 5-HT_{1A} receptor agonistic properties), tramadol (a μ-opiate receptor agonist and serotonin/noradrenaline reuptake inhibitor), as well as putative antidepressants in development for clinical use (DOV-216,303, a triple (5-HT, NA, DA) reuptake inhibitor, TRI) and S32006 (5-HT_{2C} receptor antagonist)) [74]. Additionally, we added studies on antidepressants with various mechanisms to this review.

Antidepressants in general, and specifically SSRIs (paroxetine, fluoxetine, (es)citalopram, fluvoxamine, sertraline, sertraline), are notorious for their sexual side effects [17, 23, 75–77]. Although venlafaxine has been promoted as having less sexual side effects, emerging data indicate no substantial difference from the SSRIs [78, 79]. Bupropion, lacking effects on the serotonergic transporter (SERT), has been suggested to have few sexual side effects, and even to favor sexual function [80–82]. Buspirone, a partial 5-HT_{1A} receptor agonist (but also a dopamine D_2 receptor antagonist), is mainly used in anxious patients but also possesses antidepressant properties [70, 83–85] and has not been associated with sexual side effects [86, 87]. Further, in animal studies, prosexual effects of buspirone have been reported [88–90]. The putative antidepressant, DOV-216,303, a triple monoaminergic reuptake blocker [91, 92], may have less or no sexual side effects because of the dopaminergic stimulatory profile inherent in its mechanism of action. 5-HT_{2C} receptor antagonists like S32006 display an antidepressant profile in preclinical models and elevate extracellular levels of DA and NA but not 5-HT [74, 93–96]. They have been claimed to possess stimulatory sexual effects or at least not to compromise sexual function [70, 93, 97].

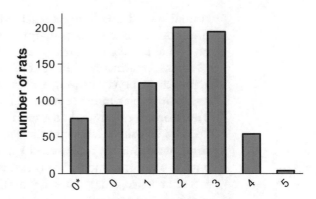

Fig. 11 Distribution of ejaculation frequencies of sexually trained male Wistar rats in the fourth training (total n = 766, obtained from 7 experiments, gathered from year 2004 to 2009). Based on the fourth mating test, these male rats can be classified with stable copulatory behaviors as "normal" (2 to 3 ejaculations/ 30 min), "sluggish" (0 to 1 ejaculation/30 min), and "fast" (4 to 5 ejaculations/ 30 min) ejaculating. The 0* represents "real zeros," i.e., never reached an ejaculation during any of the training sessions

Figure 11 presents cohort studies performed over a period of almost 6 years. All rats were similarly trained, and in all cohorts we found the typical bell-shaped distribution of what we call "sexual endophenotypes." In all separate studies, for our pharmacological experiments, we were able to select at least 60 stable "normal" sexually performing (2–3 ejaculations/30 min) rats. These animals displayed stable sexual behavior during the individual experiments. All control groups (Fig. 12) showed a relatively constant level during successive vehicle tests (acute; subchronic, chronic, and washout), emphasizing the stability of individual sexual behavior. Clearly, individual rats do not further improve their sexual performance with increasing sexual experience or over time, supporting the notion of fixed individual "sexual endophenotypes." An additional finding is the stability of sexual behavior over the years. During these 6 years of experiments, and already before that [67, 98], we have always found—sometimes with variability in the course of a year (seasonal variation)—the same level of sexual performance within the cohorts. In general, the number of "slow" ejaculators is larger than the number of "fast" ejaculators [69], but we always have at least 60 animals in a cohort that, upon training, exhibit at least 2 ejaculations per 30-min test. Such a paradigm of sexual behavior is ideally suited to test the effects of psychotropic drugs, because both inhibitory and stimulatory effects can be detected. Moreover, the paradigm tests the sexual behavior of individual rats both after acute and (sub)chronic administration, paralleling the putative onset of action of the drugs tested. In our paradigm, we use the SSRI paroxetine as the standard reference drug and 10 mg/kg as the reference dose: this always has inhibitory

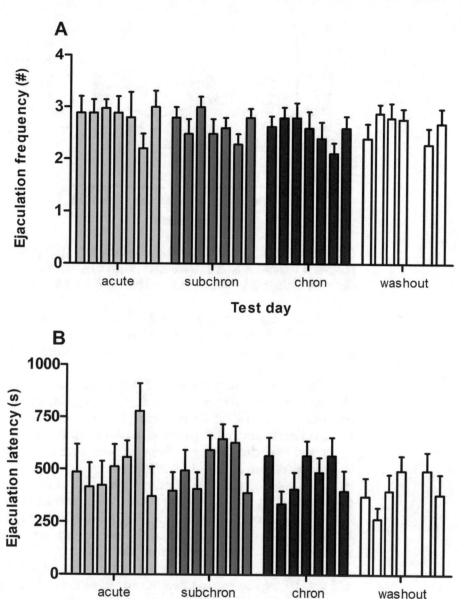

Fig. 12 Ejaculation frequencies of the vehicle-treated animals over seven experiments. In each experiment, normal "endophenotypic" male Wistar rats are treated with the vehicle chronically and tested in a 30-min session with an estrous female rat. For each treatment day the experiments 1 to 7 are listed from left to right. Latency to first ejaculation of the vehicle-treated animals over seven experiments. In exp. 5 no washout experiment was performed

effects after (sub)chronic dosing. In the seven experiments according to the protocol described in Chan et al. [99], over a period of 5 years with different routes of administration and slightly different vehicles. During each of the testing periods, the vehicle-treated

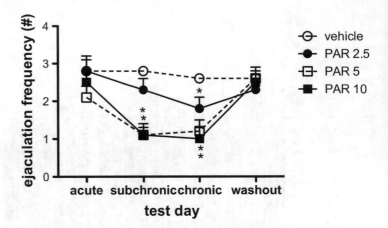

Fig. 13 Effects of chronic administration of paroxetine (0, 2.5, 5, and 10 mg/kg, PO) over 4 treatments days on ejaculation frequency of normal "endophenoty-pic" male Wistar rats in a 30-min sex test with an estrous female rat

animals did not differ significantly in ejaculation frequency and latency to first ejaculation (Fig. 12; *see* also [100]). No significant differences were found between the vehicle treatments for these important behaviors.

Figure 13 shows the effect of paroxetine on the number of ejaculations on the 4 experimental days. During acute treatment of paroxetine, none of the doses tested (2.5, 5, and 10 mg/kg, PO) had an effect in any of the parameters examined. Paroxetine significantly reduced the number of ejaculations after subchronic and chronic treatment at 5 and 10 mg/kg doses, whereas it inhibited it at 2.5 mg/kg only chronically. One week after cessation of treatment, all groups had returned to their normal baseline. At mid and high doses (5 and 10 mg/kg), paroxetine strongly inhibited sexual behaviors in many of the parameters examined after subchronic and chronic treatments (Table 1).

After subchronic and particularly after chronic administration of 5 and 10 mg/kg paroxetine, the intromission latency, the ejaculatory latency, and the first post-ejaculatory latencies were enhanced in line with the strong sexually inhibitory effects of paroxetine. The number of animals that reached at least one ejaculation also decreased considerably, making comparisons of the second ejaculation series awkward. The low dose of paroxetine (2.5 mg/kg) only had effects on the ejaculation frequency and the latency to first ejaculation after chronic administration. Figure 14 clearly illustrates the suppressing effects of two SSRIs (paroxetine and citalopram) on the distribution of ejaculations per test, in contrast to vilazodone, which even has some stimulatory effects [101]. Paroxetine and most other SSRIs [17, 51, 77] have strong inhibitory effects on sexual behavior of humans, both in healthy individuals [26] and in depressed patients [51, 102–105]. Paroxetine seems to have sexual

Table 1
Effects of paroxetine (2.5, 5, and 10 mg/kg, PO) on various sexual behavior parameters of normal "endophenotypic" male Wistar rats in a 30-min sex test with an estrous female rat. *EF* ejaculation frequency, *ML* latency to first mount, *IL* latency to first intromission, *MF* first series mount frequency, *IF* first series intromission frequency, *CE* first series copulatory efficiency, *EL* latency to first ejaculation, and *PEI* first series post-ejaculatory interval

Parameter	Vehicle	PAR 2.5	PAR 5	PAR 10	ANOVA
Acute treatment					
EF	2.8 ± 0.3	2.8 ± 0.4	2.1 ± 0.3	2.5 ± 0.3	0.451
ML	39.9 ± 21.8	8.2 ± 3.1	7.6 ± 3.3	28.3 ± 9.5	0.466
IL	132.6 ± 98.5	62.3 ± 19.4	89.3 ± 54.5	46.9 ± 11.8	0.793
MF	11.6 ± 2.9	15.0 ± 4.8	19.4 ± 4.1	17.7 ± 4.8	0.540
IF	9.1 ± 2.5	3.6 ± 0.9	8.6 ± 1.0	5.9 ± 0.7	0.214
CE	49.9 ± 5.7	28.0 ± 8.8	38.0 ± 6.3	34.9 ± 5.1	0.098
EL	508.7 ± 123.2	477.9 ± 177.2	661.7 ± 111.9	595.8 ± 138.2	0.832
PEI	462.6 ± 115.4	489.8 ± 187.7	330.0 ± 37.1	548.8 ± 145.8	0.741
Subchronic treatment					
EF	2.8 ± 0.1	2.3 ± 0.3	1.1 ± 0.3[a]	1.1 ± 0.2[a]	0.000
ML	17.8 ± 5.1	12.6 ± 6.9	10.6 ± 3.1	11.2 ± 3.4	0.681
IL	46.1 ± 14.6	36.1 ± 12.6	265.2 ± 193.4	558.9 ± 173.0[a]	0.005
MF	12.5 ± 3.8	15.0 ± 3.3	33.4 ± 8.9[a]	29.9 ± 4.3[a]	0.005
IF	6.2 ± 0.5	9.3 ± 1.6	9.1 ± 2.1	5.4 ± 1.0	0.420
CE	44.0 ± 3.8	41.0 ± 5.1	25.2 ± 7.3[a]	17.0 ± 4.0[a]	0.000
EL	395.3 ± 67.0	593.1 ± 106.9	1340.5 ± 136.6[a]	1066.4 ± 108.1[a]	0.000
PEI	311.4 ± 8.5	339.9 ± 22.5	994.5 ± 255.5[a]	595.3 ± 124.2	0.000
Chronic treatment					
EF	2.6 ± 0.1	1.8 ± 0.3[a]	1.2 ± 0.3[a]	1.0 ± 0.2[a]	0.000
ML	23.0 ± 15.2	32.2 ± 15.8	34.5 ± 14.7	129.0 ± 58.4	0.109
IL	13.9 ± 3.4	62.5 ± 29.7	451.0 ± 231.3[a]	266.0 ± 91.7	0.009
MF	10.4 ± 1.8	17.6 ± 3.3	25.8 ± 4.1[a]	25.2 ± 4.6[a]	0.002
IF	6.7 ± 0.6	10.3 ± 0.9	6.9 ± 1.5	5.4 ± 1.0	0.012
CE	46.3 ± 4.9	46.7 ± 6.4	25.6 ± 7.1	22.9 ± 4.0[a]	0.001
EL	306.1 ± 36.1	789.6 ± 142.7[a]	1125.6 ± 155.8[a]	1062.6 ± 127.6[a]	0.000
PEI	309.3 ± 11.0	565.1 ± 166.9	1060.5 ± 246.9[a]	745.2 ± 172.0	0.003

(continued)

Table 1
(continued)

Parameter	Vehicle	PAR 2.5	PAR 5	PAR 10	ANOVA
Washout					
EF	2.6 ± 0.3	2.3 ± 0.3	2.6 ± 0.2	2.5 ± 0.2	0.922
ML	24.7 ± 12.0	6.4 ± 1.6	12.1 ± 6.5	100 ± 3.6	0.368
IL	46.0 ± 19.9	99.7 ± 50.6	70.3 ± 31.7	83.8 ± 55.0	0.830
MF	19.9 ± 3.1	18.4 ± 3.8	23.5 ± 4.4	12.8 ± 2.2	0.135
IF	8.3 ± 0.9	7.6 ± 1.3	5.8 ± 0.7	7.1 ± 1.1	0.381
CE	34.4 ± 4.6	32.7 ± 5.9	24.7 ± 4.5	39.1 ± 5.7	0.316
EL	533.9 ± 63.4	586.4 ± 63.4	583.6 ± 115.5	470.3 ± 83.0	0.763
PEI	336.1 ± 23.3	334.9 ± 27.3	433.3 ± 124.9	408.9 ± 83.8	0.719

[a]$p < 0.05$ significance when compared to the vehicle in Post Hoc analyses (Bonferroni)

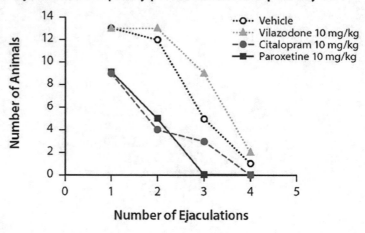

Fig. 14 The number of rats ($N = 14$/group) that had 1, 2, 3, or 4 ejaculations during a 30-min sexual behavior test on day 14 after testing vehicle, vilazodone (10 mg/kg), citalopram (10 mg/kg), and paroxetine (10 mg/kg)

inhibitory effects after at least 1 week [11, 25, 72] or several weeks (in depressed patients). In our rat paradigm, paroxetine has a dose-dependent inhibitory effect on sexual behavior as can best be seen by its effects on the number of ejaculations/test (Fig. 12). The 5 and 10 mg/kg doses seem to be equipotent, whereas the 2.5 mg/kg dose only inhibits sexual behavior after 2 weeks of administration. Remarkably, acute administration of paroxetine does not reliably inhibit sexual behavior. This parallels the human situation where at least 1 week of administration is needed to affect sexual behavior [4, 71]. One week after cessation of treatment, the sexual performance of all paroxetine-treated groups returned to the control level, showing the absence of irreversible or delayed effects. In humans, no data are available as to how fast the sexual side effects of SSRIs wane after cessation of treatment, but data in healthy males with premature ejaculations suggest that after 6 weeks of treatment, sexual inhibitory effects subside within a week [106]. In this aspect, these sex and effects seem to be relatively independent of the antidepressant effects of SSRIs which take time to emerge, but after long-term and effective treatment, depression seems to recur only (if at all) after a long period following cessation of treatment [107, 108]. This might also indicate that the sexual inhibitory effects of SSRIs in depressed patients could be blocked by pharmacological treatment independent of therapeutic effects. Venlafaxine is a serotonin-noradrenaline reuptake inhibitor (SNRI) and its effects on sexual behavior may be different from pure SSRIs. Although at lower doses, venlafaxine did not affect sexual behavior, it had a comparable sexual inhibitory profile as paroxetine at the highest dose tested (40 mg/kg), suggesting that it may also compromise sexual performance in humans. Human data [78, 79] show that venlafaxine, at doses that exert antidepressant efficacy, induces comparable sexual side effects to SSRIs including fluoxetine and paroxetine. Venlafaxine most potently increases 5-HT levels and only at higher doses acts as a noradrenaline reuptake blocker [70, 94, 109]. This might explain its profile in our rat paradigm (Table 2) consistent with sexual side effects of venlafaxine in humans [110]. On the other hand, the limited data available on venlafaxine in premature ejaculation does not suggest that it has a strong sexual-inhibiting effect in healthy men [111, 112]. Bupropion, a noradrenaline-dopamine reuptake inhibitor [113], exerted a sexual stimulatory profile in our paradigm (Table 2), particularly after acute dosing. The number of ejaculations/test increased considerably after acute administration of 15 mg/kg. After (sub)-chronic dosing, there was still some sexual-stimulating activity, although less than after acute dosing. The effect of bupropion was only found in an increase in ejaculations; all other parameters showed a trend toward prosexual effects but this never reached significance. This suggests that the relatively weak blockade of the dopamine transporter (DAT) with bupropion [74] might not be strong enough to lead to permanent sexual stimulatory activity and that some tolerance for prosexual activity had occurred. In a similar experiment [114], the mixed D_2/D_3 dopamine receptor agonist apomorphine showed a comparable behavioral profile, although revealing a somewhat stronger prosexual profile than bupropion (see later). This illustrates that dopaminergic mechanisms may exert prosexual activity that potentially antagonizes inhibitory effects induced by SSRIs. Human data on bupropion show that it is not itself associated with sexual inhibitory effects [80] and that it can

Table 2
Summary of effects of various drugs on male sexual behavior after acute, subchronic, or chronic treatment. ○: no effect; ↓: inhibition; ↑: stimulation; nt: not tested. SSRI: selective serotonergic reuptake inhibitor; SNRI: serotonergic and noradrenergic reuptake inhibitor; NDRI: noradrenergic and dopaminergic reuptake inhibitor; TCA: tricyclic antidepressant; TRI: triple monoaminergic reuptake inhibitor; R: receptor. Bold printed drugs have been tested in our laboratories

Drug	Mechanism of action	Rat acute	Rat chronic	Human sexual behavior
Paroxetine	SSRI	○	↓	↓
Clomipramine	TCA	○	↓	↓
(Es)citalopram	SSRI	○	↓	↓
Fluoxetine	SSRI	○	↓	↓
Fluvoxamine	SSRI	○	○↓	○↓
Sertraline	SSRI	○	↓	↓
Dapoxetine	SSRI	○	nt	nt
Venlafaxine	SNRI	○	↓	↓
Buspirone	5-HT$_{1A}$R agonist	↑	○	○
Bupropion	NDRI	↑	○	↑
DOV216,303	TRI	○	↓	nt
Vilazodone	SSRI+5-HT$_{1A}$R agonist	○	○	○
Hypidone	SSRI+5-HT$_{1A}$R agonist	○	○	nt
Vortioxetine	SSRI+ 5-HT$_{1A,1B,1D,7}$ R agonist; 5-HT$_3$R antagonist	○	○	○
Tramadol	SNRI+μ-Opioid R agonist	○↓	nt	↓
Mirtazapine	NaSSA: α$_2$R antagonist; 5-HT$_{2A, 2C}$; 5-HT$_3$R antagonist	↓	↓	↑
Apomorphine	Dopamine D$_{2,1,4}$ R agonist	↑	○	↑
Flesinoxan	5-HT$_{1A}$ R agonist	↑	○↑	○
CP-94253	5-HT$_{1B}$ R agonist	○	↓	↓
Ondansetron	5-HT$_3$ R antagonist	○	○	nt

alleviate SSRI-induced sexual side effects [70, 82, 115]. DOV216,303 is a triple monoaminergic reuptake inhibitor [91, 116] that blocks NE, 5-HT, and DA reuptake: it was designed to combine the antidepressant efficacy of SNRIs with dopamine reuptake blockade, thereby reversing several side effects, including sexual ones [91, 92]. DOV216,303 has been tested in animal paradigms of depression and shown antidepressant efficacy

[91, 117, 118]. Pilot studies in depressed patients have seen some efficacy of DOV216,303 in depression and significantly reduced HAM-D scores similar to citalopram [91]. At doses that exert antidepressant effects in animal models of depression, DOV216,305 showed no effects on male rat sexual behavior (Table 2). This suggests that dopaminergic stimulation (via DAT blockade) antagonizes inhibitory effects on sexual behavior induced by SERT blockade. A putative contribution and/or interaction between SERT and noradrenaline transporter (NET) blockade is possible, but requires exploration. Buspirone, a partial 5-HT_{1A} receptor agonist and weak dopamine D_2 receptor antagonist, is a clinically effective anxiolytic with antidepressant properties [70, 119]. 5-HT_{1A} receptor agonists, including buspirone, have prosexual activity in rats upon acute administration [28, 47, 67]. However, to our knowledge chronic administration studies of sexual behavior have not been performed. The present data in rats (Table 2) suggest that low doses of buspirone, which are also antidepressant in animal depression paradigms, have mild prosexual activities. Human data on the sexual side effects of buspirone are scant and do not suggest inhibitory activity [84]. Most human studies that have examined buspirone as a putative treatment for sexual side effects induced by SSRIs suggest that buspirone is modestly beneficial [86, 120]. These human data seem to concur with our data: either no effect or mild stimulatory activity.

S32006 is a benzourea derivative and 5-HT_{2C} receptor antagonist, and it has shown antidepressant and anxiolytic properties in various rodent behavioral models after acute and chronic administration [74]. At doses that display antidepressant activity, S32006 does not have any effect on sexual behavior (Table 2). Because 5-HT_{2C} receptor agonists clearly inhibit the sexual behavior of rats [121], it was thought that an antagonist may have some prosexual activity: though this was not the case, interaction studies with SSRIs would be of interest [93, 94].

Vortioxetine is a more recently developed multimodal antidepressant that exerts potent SSRI activity but also has significant effects on various serotonergic receptors including 5-$HT_{1A, \ 1B}$ agonistic and 5-$HT_{1D, \ 3 \ and \ 7}$ antagonistic activity [122]. In our rat model (Table 2) vortioxetine did not affect sexual behavior at doses of 1 and 10 mg/kg/day orally [123]. The low vortioxetine dose led to approx. 50% occupancy of the SERTs, whereas the 10-mg/kg dose occupies around 90% of the SERTs, comparable to human vortioxetine dose (20 mg) that exerts antidepressant activity [122]. The animal data correspond to human data where vortioxetine does not exert sexual-inhibiting activity, neither in depressed patients [124] nor in healthy, nondepressed people [125].

Vilazodone is an antidepressant with an SSRI and partial 5-HT$_{1A}$ receptor agonistic mechanism of action [126]. Early studies [127] indicated that vilazodone had relatively low sexual side effects comparable to placebo treatment that was later confirmed in an extensive phase-IV clinical trial [128]. In a later study in healthy (nondepressed) adults, vilazodone also did not exert a significant effect on sexual functioning, whereas paroxetine did [129].

In our rat model (Table 2) it can be seen that vilazodone does not exert any effect on sexual behavior, neither after acute nor (sub)chronic dosing, whereas the reference SSRIs citalopram and paroxetine significantly inhibited sexual behavior after (sub)chronic administration [101]. This can be illustrated by portraying the distribution of the number of ejaculations per group after 14 days of treatment (Fig. 14), where in the vilazodone-treated group ($N = 14$ animals), more animals with higher number of ejaculations occur than in the citalopram and paroxetine groups.

Hypidone, in (pre)clinical development, has a comparable mechanistic profile as vilazodone and has a comparable behavioral profile with regard to its antidepressant and lack of sexual side effects [130].

Tramadol is, based on μ-opioid agonistic mechanism, mainly used for pain relief, but has also considerable serotonergic and noradrenergic reuptake inhibiting properties [131]. This activity profile suggests antidepressant properties, which has been confirmed [132]. Tramadol has also, as an off-label application, sexual inhibitory effects on sexual behavior [133, 134], notably in premature ejaculation, comparable to the SSRIs [26]. Tramadol's potential risk on opioid addiction remains a serious problem [135]. Tramadol has acute sexual inhibitory effects in our rat model but at relatively high doses [136]. We have not tested tramadol chronically (Table 2), but its acute inhibitory action supports its acute effects in premature ejaculation ([137]; see later in PE section).

Mirtazapine is an antidepressant that increases the synaptic concentrations of both serotonin and noradrenaline; moreover, it is a receptor antagonist at 5-HT$_{2A/2C}$ and 5-HT$_3$ receptors [138]. Mirtazapine belongs to a category of antidepressants called NaSSAs [139]. It potently antagonizes central α$_2$ adrenergic auto- and heteroreceptors, resulting in enhanced release of noradrenaline (via inhibition of α$_2$ autoreceptors) and serotonin, due to blockade of inhibitory heteroreceptors on serotonergic terminals [140]. The preclinical data on mirtazapine [141] are somewhat conflicting with both inhibitory and stimulatory effects on various parameters, but its overall inhibitory effects prevail. Some clinical studies indicate that in SSRI-treated depressed patients that suffered from sexual complaints, adding mirtazapine restored sexual function [142, 143].

Apomorphine, a nonselective dopamine $D_{2,1,3}$ receptor agonist [144], has been introduced clinically as an on-demand treatment for erectile dysfunction that leads to severe sexual complaints [144]. It has been known for decades that dopamine plays an important role in the control of male sexual behavior, associated with the involvement of hypothalamic and mesolimbic DA systems [40, 145]. The data on the effects of apomorphine on male rat sexual behavior is complicated and involves motivational, sexual experience and sexual satiation aspects [146]. Most reports have tested apomorphine only after acute administration; Olivier et al. [114] have tested apomorphine after acute and chronic (2 weeks) administration (Table 2). Acutely, apomorphine (0.4 mg/kg SC) enhanced the number of ejaculations and reduced the number of mounts and intromissions. Chronically, the effects were less clear, only the mount intromission frequencies were lower, but the post-ejaculatory interval latency was increased. This profile indicates apomorphine's on-demand character and illustrates the usability and predictability of our rat model even for erectile dysfunctions.

Flesinoxan, a potent and selective $5\text{-}HT_{1A}$ receptor agonist [147], that has been in development for anxiety disorders and major depression [29] was tested at a dose (2.5 mg/kg SC, twice daily) leading to an adequate $5\text{-}HT_{1A}$ receptor occupancy [148] in male rat sexual behavior [123]. Acutely flesinoxan had a prosexual effect (shortened latency to first ejaculation and tendency to enhanced number of ejaculations). Subchronic and chronic administration had no clear prosexual effects (Table 2). Earlier acute studies already showed the acute prosexual effects of flesinoxan [42, 149] but no other chronic studies have been performed. These treatment-length-dependent effects of $5\text{-}HT_{1A}$ receptor stimulation suggest involvement of receptor desensitization [150]. This profile seems in line with the effects seen with buspirone in our rat model: acutely stimulation, (sub)chronically no effects. The downregulation of $5\text{-}HT_{1A}$ receptors after chronic administration of flesinoxan might underlie its failure to reach clinical applications.

Studies on vortioxetine [123] were performed investigating the mechanisms underlying its lack of sexual side effects in our rat model (Table 2). We tested a selective $5\text{-}HT_{1B}$ receptor agonist (CP-94253) and a $5\text{-}HT_3$ receptor antagonist, ondansetron, both after acute and chronic administration. Both compounds were tested at doses that occupied 80% and 60%, respectively, of the receptors involved [148] comparable to those receptor occupancies after giving vortioxetine at active antidepressant doses.

CP-94253 reduced the number of ejaculations after subchronic and chronic dosing and increased the ejaculation to first ejaculation after chronic treatment, suggesting inhibitory effects on male rat sexual behavior (Table 2). This profile is somewhat similar to that of

SSRIs, and it cannot be excluded that 5-HT_{1B} receptor stimulation is part of or causing the inhibitory action of SSRIs after chronic administration.

Ondansetron lacks, at a dose occupying approx. 60% of brain 5-HT_3 receptors, any effect on male sexual behavior in the rat (Table 2), confirming earlier data after acute administration of 5-HT_3 receptor antagonists [151]. Probably, 5-HT_3 receptors do not play a (important) role in the modulation of male sexual behavior.

7.1 Conclusions

The male rat paradigm used here to examine the inhibitory and stimulatory effects of antidepressant and other drugs on sexual behavior of male rats relates well to their known and predicted effects in humans. In line with clinical experience, marked blockade of SERTs (paroxetine, clomipramine, fluoxetine, citalopram, escitalopram, sertraline, fluvoxamine, and venlafaxine (a mixed SERT and NET blocker)) interfered with male sexual behavior, in contrast to the DAT/NET blocker bupropion. Typically, the inhibitory action on sexual behavior emerges after chronic administration comparable to the situation in human patients.

Furthermore, other drugs that primarily increase brain levels of DA and NA vs. 5-HT (the partial 5-HT_{1A} receptor agonist, buspirone; the triple monoaminergic reuptake inhibitor, DOV216,303; and the 5-HT_{2C} receptor antagonist, S32006) exerted no detrimental influence or a mild stimulatory effect on sexual performance, and they are predicted to have little or no sexual side effects in men. It is suggested that blockade of DATs or NATs, as well as 5-HT_{2C} receptor blockade, would be a useful avenue to clinically explore for the reduction of SERT-mediated sexual inhibition. Blocking DATs might be particularly useful to overcome the sexual inhibition due to disruption of SERTs. These possibilities also justify investigation employing the present experimental model and combinations of drugs. More generally, exploration of the present paradigm, if possible in parallel with therapeutic investigations, should provide important insights into the influence of antidepressants and other classes of psychotropic agent on male sexual function.

8   Methods and Models of Premature Ejaculation

If the central finding in human PE is a strong reduction of the ejaculation latency after vaginal penetration, an animal model should at least have a comparable outcome to have predictive and possibly construct validity. Several experimental techniques have been exploited over the years to induce faster copulating and ejaculating animals, in particular rats. The resulting models might be used to study what the underlying mechanisms of rapid or premature ejaculations are or screen for drugs that alleviate

PE. One could distinguish (electrical/chemical) lesion or stimulation models, pharmacological models, behavioral models, and genetic models. SSRIs, as they display clearly ejaculation latency enhancing effects in humans, should be tested in such models and, akin to humans, induce lengthened ejaculation latencies.

9 Lesion and Electrical Stimulation

Sexual behavior in rats has been the subject of studies for many years, and already in the 1960s and 1970s of the last century, mechanisms in the brain involved in male sexual behavior were investigated. Already in the 1960s electrical stimulation and lesion experiments indicated hypothalamic areas where sexual behavior could be affected, in particular the medial preoptic area (MPOA). Bilateral MPOA lesions can completely wipe out sexual behaviors [152, 153], whereas electrical stimulation can facilitate rat copulatory behavior [154, 155]. Lesions of the nucleus paragigantocellularis (nPGi) in the ventral medulla of male rats facilitated ejaculation in rats, confirming the hypothesis that the nPGi plays a key role in the descending inhibition to genital reflexes [156, 157]. Both paradigms could be considered putative models of premature ejaculation. Unfortunately, no studies with SSRIs have been performed in the MPOA stimulation model. An acute high dose of fluoxetine (20 mg/kg, IP) in nPGi-lesioned rats showed that fluoxetine inhibited sexual behaviors in the sham-lesioned group and had no effects in the nPGi-lesioned group, suggesting a serotonergic mediated inhibitory effect on sexual behavior in the nPGi under normal conditions [158]. Lesions of the ventrolateral periaqueductal gray (vlPAG), the source of serotonin in the nPGi (which expresses various types of serotonergic receptors), result in disinhibition of ejaculation (reduced ejaculation latency and enhanced number of ejaculations) [159]. These latter authors suggest that the serotonergic vlPAG-nPGi pathway contributes an important regulatory mechanism and is involved in the SSRI-induced inhibition of ejaculation. Disruption of the dorsal raphe serotonergic system by either electrolytic lesions or neurotoxic 5,7-DHT lesions shortened the ejaculatory latency and the post-ejaculatory refractory period, indicating fastened ejaculation [156]. It is likely that the lesions of the dorsal raphe nucleus in the latter study (partly) also had damaged the vlPAG, confirming the influence of serotonergic tone in the nPGi for the inhibition of ejaculation and other sexual activities and indicating a putative site of action of the inhibitory influence of SSRIs on these activities.

Therefore, an mPOA-PAG-nPGi-spinal cord pathway system might be crucial in the expression of ejaculation and associated behaviors and a serotonergic mechanism seems clearly involved. No further pharmacological studies have been performed on the PGi-lesioned male rat.

10   Pharmacological Models and Methods in PE

In the last 40 years many investigators have illustrated the important role of serotonin in sexual behavior [31, 32]. Depleting 5-HT in the CNS, e.g., by para-chlorophenylalanine (p-CPA), an inhibitor of 5-HT synthesis, facilitates ejaculation and increases the number of ejaculations [160]. Also localized injection of the neurotoxin 5,7-DHT, decreasing central 5-HT levels, facilitates sexual behavior, suggesting that central 5-HT neurotransmission has an inhibitory action on ejaculatory and other sexual activities [161]. However, p-CPA administration leads to temporary 5-HT depletion and is accompanied by serious side effects, making extensive experiment difficult if not impossible. No sexual behavior studies have been performed on (chronic) SSRI treatment of p-CPA-treated male rats, and such a model seems not that attractive to find new medications for sexual dysfunctions.

It is feasible to directly facilitate sexual behavior including ejaculation by administration of serotonergic and dopaminergic ligands. The prototype serotonin agonist used is 8-OH-DPAT, a full and potent agonist on $5\text{-}HT_{1A}$ receptors, but also other $5\text{-}HT_{1A}$ receptor agonists can be used to facilitate male sexual behavior in rats [42, 160, 162–164].

8-OH-DPAT (racemic mixture or (+)-enantiomer is used), administered shortly before a 30-min encounter, dose-dependently affects male sexual behavior. Ejaculation latencies decrease, the number of ejaculations per session increases, and the number of mounts and intromissions per ejaculation session decreases [163]. The overall picture reflects a strongly facilitated male sexual behavior. Some animals may ejaculate upon the first mount/intromission, strongly reflecting the extremely fast PE in some human males. It could be suggested that $5\text{-}HT_{1A}$ receptor stimulation facilitates sexual behavior by lowering the ejaculation threshold and as such would constitute an animal model of PE. Chronic treatment with an SSRI (paroxetine) that delays ejaculation reduced the facilitation of ejaculation induced by 8-OH-DPAT in rats [165, 166], indicating a role for downregulated $5\text{-}HT_{1A}$ receptors in the sexual side effects of SSRIs on sexual behavior. Because $5\text{-}HT_{1A}$ receptors are located both presynaptically (autoreceptors on serotonergic neurons) and postsynaptically (heteroreceptors on non-serotonergic neurons), it would be important to unravel whether pre- or postsynaptically located $5\text{-}HT_{1A}$ receptors are involved in the prosexual effects of $5\text{-}HT_{1A}$ receptor agonists. We tested the acute effects of the biased receptor agonists F-13714, a preferential $5\text{-}HT_{1A}$ autoreceptor agonist, and F-15599, a preferential $5\text{-}HT_{1A}$ heteroreceptor agonist, on sexual behavior in trained and sexually stable male rat [167]. Both compounds had prosexual activity, although the potency of F-13714 was much higher compared to F-15599. Another 5-HT$_{1A}$ receptor ligand, a mixed 5-HT$_{1A}$ autoreceptor agonist/heteroreceptor antagonist S15535, did not affect sexual behavior at all. These results do not further unravel the precise role of different 5-HT$_{1A}$ receptor pools in male sexual behavior but add to the notion of a probably very complex interaction between presynaptic and postsynaptic serotonergic mechanisms in the regulation of male sexual behavior. Of course, direct 5-HT$_{1A}$ receptor antagonists like WAY100,635 antagonize 5-HT$_{1A}$ receptor agonist-induced prosexual activity [164, 168]. Unpublished data (Olivier et al. –Utrecht University) have shown that the prosexual effects of F-13714 and F-15599 could be antagonized by the 5-HT$_{1A}$ receptor antagonist WAY-100,635. Although not conclusive, some data suggest that postsynaptic 5-HT$_{1A}$ receptors mediate this prosexual activity of 5-HT$_{1A}$ receptor stimulation [169]. Nevertheless, the applicability of 5-HT$_{1A}$ receptor-induced stimulation as a translational model in the study of PE is still attractive for acute testing of inhibitory drugs.

Brain dopamine plays an important role in the control of male sexual behavior, in which the hypothalamic and mesolimbic dopaminergic systems play a major role [40, 170]. The mesolimbic system is activated by natural rewards including copulation [171]. The hypothalamic system (including mPOA and others) is directly involved in the execution of sexual activities [172].

A second class of drugs, dopaminergic agonists, is also able to facilitate sexual behavior including ejaculation [161, 170, 173]. The prototypic nonselective dopamine receptor agonist apomorphine (but other more selective agonists as well) is able to reduce the ejaculation threshold both after acute and chronic administration [114]. Recent evidence suggests that it is particularly activation of dopaminergic D$_2$ receptors that may cause this effect [146]. However, the effects of 5-HT$_{1A}$ receptor activation are qualitatively different from that of dopamine D$_2$ receptor activation; the latter reflects more reduction of mounts and intromissions and induction of penile erections but not so much decreases in ejaculation latencies as seen after serotonergic manipulation [114]. Although several other mechanisms (e.g., oxytocin, GABA, noradrenaline) have been probed pharmacologically, no clear stimulatory profiles have emerged that may lead to an animal prosexual model. Therefore, 5-HT$_{1A}$ receptor activation seems the preferred pharmacologically induced model of PE.

11   Behavioral Models and Methods in PE

We hypothesized that the ejaculation latency in human males is part of a biological variation with a genetic component, whereas the distribution of the IELT follows a "Gaussian" distribution. According to this hypothesis, some men will have throughout their life fast ejaculations, others more normal, whereas some men display retarded or even no ejaculations. Waldinger et al. [174, 175] found some empirical evidence for this hypothesis in a stopwatch-assessed intravaginal ejaculation study in a random cohort of 491 men from 5 countries with a median IELT of 5.4 min (range 0.55–44.1 min).

Based on this hypothesis and anecdotal observations in our research over many years, we investigated whether biological variations in sexual behavior do exist in male rats. We run over the years large male sexual behavior studies (cohorts of 90–120 males are tested) using outbred Wistar males. During a sexual encounter of 30 min, a male rat interacts with a hormonally primed, proceptive, and receptive female and his sexual behavior is scored. Standardly, we train males four times, once weekly, and the sexual behavior of males is stable and remains stable over the next period (we have tested up to 6 months afterward), suggesting a lifelong stable phenotype. It appears that the ejaculation frequencies are distributed following a bell-shaped curve. In over 2000 males tested in this way, we found that the number of "fast" ejaculating animals (4–5 ejaculations/test) fluctuated between 2% and 20% depending on the cohort and over time. Similarly, the number of "slow" (sluggish) ejaculators (0–1 ejaculation/test) fluctuates more over time and cohort (10–60%), and accordingly, the number of "normal" ejaculating rats (2–3 ejaculations/test) fluctuates between 35% and 80%. In most cohorts the number of "fast" (rapid) ejaculators is around 10%, and consequently large cohorts of rats have to be used to generate enough experimental "fast ejaculating" rats. This data has been portrayed in a theoretical distribution of the number of ejaculations and first ejaculation latency (Fig. 15).

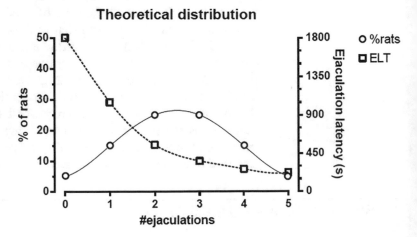

Fig. 15 Theoretical distribution (% of total) of animals according to the number of ejaculations per test and the ejaculation latency per test. Data are based on more than 2500 male rats tested in our standard sexual behavior test of 30 min in sexually trained animals

We postulated that these "rapid" ejaculating rats might be the rat equivalent of human PE. Studying the sexual behavior of selected groups of "sluggish," "normal," and "fast" ejaculating rats showed that the ejaculation latency of "fast" rats was extremely shorter than that of both the "normal" and "sluggish" ones. Concomitantly, the mount, but not the intromission frequency, was inversely related to the EL (sluggish>normal>fast). The post-ejaculatory interval (from the first ejaculation till the next mount or intromission (whichever came first)) was not different between the three phenotypes although this could not really be measured in the "slow" animals. The three sexual "phenotypes" were also studied in various behavioral paradigms to judge whether the differences in sexual behavior were overall present, but it appeared that they were restricted to their sexual behavior [67]. This nicely concurs with the human PE phenotype that also does not coincide with other behavioral or psychiatric features [18, 38].

We determined whether the three phenotypes were sensitive to the prosexual activity of a rather high dose of 8-OH-DPAT (0.8 mg/kg SC). Although the "rapid" ejaculators already had a high starting level of sexual activity, it appeared that the drug still was able to further stimulate the sexual behavior, probably toward an absolute ceiling of sexual activities possible in half an hour testing [67]. The number of ejaculations was enhanced, the ejaculation latency decreased, and the mount frequency dramatically decreased; in this case many animals immediately ejaculated upon the first mount. When we tested the animals 1 week later without any treatment, all animals had resumed their original phenotype. This strongly suggests that the (endo)phenotypes in sexual behavior reflect a neurobiological rather than psychological or experiential background.

Thus far no pharmacological experiments or investigations into brain mechanisms or changes therein have been performed on "fast" ejaculating rats. Although studies have been performed comparing sexually naïve versus experienced rats, no studies are known into the underlying neurobiological substrates of the "fast" ejaculating rats. Although their serotonergic system (at least 5-HT_{1A} receptors) seems intact, nothing is known about the controlling mechanisms of ejaculation processes in these rats. Whether (part of) serotonergic systems are different in "fast" versus normal/slow ejaculating rats is unknown and subject to future research. Although we never explicitly tested the effects of SSRIs on these "fast" ejaculating rats, we have some data of the effects of paroxetine on male sexual behavior in relatively fast ejaculating rats (animals with three and four ejaculations/test). Paroxetine has no inhibiting effects on sexual behavior after acute administration but strongly inhibits it after chronic (14 days/once daily) administration, fulfilling one criterion needed for a model to be used for PE: sensitive to the same therapy as their human counterpart. This makes it a very attractive model although with limitations: many animals have to be tested in a very labor-intensive manner in order to get a relatively low number of "fast" ejaculating animals. The stability of the (endo)phenotype is a very strong point for the model; animals can be retested many times and can be used for chronic administration. Many studies have to be performed to investigate brain mechanisms involved in the PE animals. Moreover, neuroendocrine, physiological, and genetic studies have to be designed to further develop this promising animal model of premature ejaculation.

12 In Search for the (Endo)Phenotype in Male Rat Ejaculatory Behavior

Over the last decades our group (Olivier and coworkers) has found that male rats display a certain (endo)phenotype in their male sexual behavior toward female rats in estrus. We repeatedly found that large cohorts of male rats, tested once a week for 30 min against an estrous female, develop a rather stable ejaculatory pattern [67–69]. Upon their first sexual encounter with an estrous female, male rats generally display a low level of sexual activity, expressed in a low number of animals ejaculating. However, upon repeated weekly testing the sexual behavior often improves (higher number of ejaculations; shorter ejaculation latency) and stabilizes over time. In our drug studies, we generally use rats that have been trained 4–6 times and have relatively stable levels of sexual behavior, at least during the last two tests. Remarkably, all rats have a quite stable sexual (endo)phenotype, typically displaying 0 to 5 ejaculations per 30 min. In large cohorts of rats tested this way, a typical bell-shaped distribution in the number of ejaculations displayed by the animals is found [67–69]. Concomitantly, the latency to the first ejaculation of these rats follows a linear decreasing distribution (*see* Fig. 15).

Apparently, sexually naïve rats upon their first encounter with an estrous female have to learn and get experience with this new situation, which takes some time but clearly each individual, upon repeated testing, develops its "own" sexual (endo)phenotype, ranging from zero to five ejaculations in a 30-min test.

13 Can We Change the (Endo)Phenotype by Pharmacological Intervention?

In an attempt to investigate the level of ejaculations, whether it is mainly determined via learning (experience) or is really "endogenous," we did some piloting where we tried to manipulate the level of sexual behavior by pharmacological stimulation using the 5-HT$_{1A}$ receptor agonist 8-OH-DPAT. 5-HT$_{1A}$ receptor agonists stimulate male sexual behavior in the rat [42]. In an earlier study [67] we selected male rats, tested six times in weekly sexual behavior tests with an estrous female, and after training created three groups of rats: sluggish (<1 ejaculation/test), normal (1–2 ejaculations/test), and rapid or fast ejaculators (≥3 ejaculations/test). Immediately after the training, all animals were treated randomly over a 2-week period with vehicle or 8-OH-DPAT (0.8 mg/kg SC). 8-OH-DPAT increased the number of ejaculations in all three groups to almost the same (high) level, accompanied by strong reduction in the latency to the first ejaculation. Testing all three groups 1 week later after the experiments without any drug treatment revealed that all three groups returned to their original (endo)phenotype. These results clearly indicate that animals with a low basal sexual behavior (0 or 1 ejaculation/test) can be changed in high-frequency performers, but that this pharmacological boost does not lead to permanent changes in their endogenous sexual behavior.

Because these studies were performed in sexually well-trained rats, we were interested to see whether treatment with (±)-8-OH-DPAT at the very first sexual training test in sexually naïve rats would change the development of the various phenotypes (notably faster/higher number of ejaculations) or did not affect future behavior at all.

Table 3 shows the experimental design. Two groups of male rats ($N = 24$ rats/group) were tested over 7 weeks (1 sexual behavior test for 30 min each week).

Group 1 was injected with vehicle at test 1 and with ±8-OH-DPAT (0.8 mg/kg SC) on test 6. Group 2 started at test 1 with ±8-OH-DPAT (0.8 mg/kg SC) and received ±8-OH-DPAT (0.8 mg/kg SC) also at test 6. On tests 1, 4, 6, and 7, sexual behavior was recorded. All animals were also sexually trained in weeks 2 and 3. On weeks 4 and 7 no injections were given. The data are shown in Fig. 16. We found that 8-OH-DPAT did not robustly stimulate sexual behavior at the very first test (T1), probably due to interfering behavioral and physiological effects by this relatively high dose of 8-OH-DPAT, including 5-HT$_{1A}$ receptor-mediated serotonergic behaviors, like flat body posture, forepaw treading, lower lip retraction, and others [176, 177]. At test 4 (at T2 and T3 animals have been sexually trained), no difference is noted between

Table 3
Experimental design of development of sexual (endo)phenotypes in two groups of male rats ($N = 24$ rats/group) tested once weekly over 7 weeks. Experimental treatments: *Veh* vehicle (0 mg/kg SC), *DPAT* ±8-OH-DPAT (0.8 mg/kg SC)

Tests (weeks)	1	2	3	4	5	6	7
Group 1	Veh	–	–	–	–	DPAT	–
Group 2	DPAT	–	–	–	–	DPAT	–

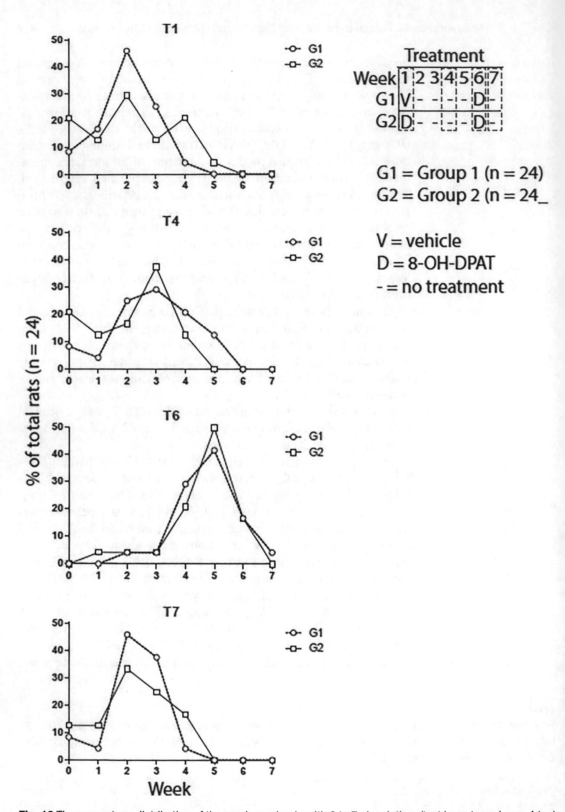

Fig. 16 The percentage distribution of the number animals with 0 to 7 ejaculations/test is portrayed over 4 test days during 7 subsequent weekly sexual behavior tests of 30 min in two groups of male rats (Group 1 (G1) and Group 2 (G2)), each consisting of 24 animals. In week 1 (T1), G1 received (±)-8-OH-DPAT (0.8 mg/kg SC); G2 received vehicle. In week 4 (T4) animals were scored without being injected. In week 6 (T6), both groups received (±)-OH-DPAT (0.8 mg/kg SC) and in week 7 (T7) animals were not injected but scored

Table 4
The mean ± SEM of the first ejaculation latency in the experiment with two groups of rats (similar to the previous table)

Treatment Group 1	*Group 1* mean (± SEM) ejaculation latency (sec)	Treatment Group 2	*Group 2* mean (± SEM) ejaculation latency (sec)	Test week
Vehicle	758 ± 83	±8-OH-DPAT	208 ± 45.3	T1
None	436 ± 101	None	719 ± 119	T4
8-OH-DPAT	24.5 ± 6.6	±8-OH-DPAT	70 ± 29.6	T6
None	413 ± 91.8	None	661 ± 91.8	T7

the two groups. At test 6 both groups received 8-OH-DPAT (0.8 mg/kg SC) and showed similar prosexual responses: a clear right shift occurred in the number of animals with higher number of ejaculations, confirming the earlier findings of Pattij et al. [67]. One week later (T7), the two groups were back to "normal" (cf. T4), again indicating that stimulation of sexual behavior, either early in sexually naïve animals or in well-trained ones, does not change the "basal" (or endo)phenotype of male rats in sexual behavior.

A comparable picture arises in the first ejaculation latency data (Table 4); 8-OH-DPAT shortens the EL, particularly at T6, and EL returns to a basal level at T7 (and T4), again confirming the robustness of the sexual (endo)phenotype.

To further substantiate the difference in ±8-OH-DPAT-induced effects in sexually naïve versus sexually experienced male rats, 5 groups of animals ($N = 25$ per group) were tested at training tests 1 and 7 with vehicle, or a series of ±8-OH-DPAT (0.1, 0.2, 0.4, and 0.8 mg/kg SC). Figure 17 shows that ±8-OH-DPAT does not induce prosexual activity (EF and EL) in sexually naïve rats, whereas in sexually trained rats (having six training tests before testing on week 7), even low doses of ±8-OH-DPAT (from 0.2 mg/kg on) stimulate sexual behavior, suggesting that establishment of a robust sexual phenotype more or less "protects" against disturbing side effects of the drug.

The data of all these experiments clearly indicate that individual male rats have a stable but "own" sexual behavior "level" ((endo)-phenotype) that in general is learned over a couple of successive sex tests (once a week). Each animal reaches a certain level of ejaculations per test that ranges for individual rats from 0 to 5 ejaculations/test. A small percentage (<5–10%) of animals never ejaculates, although basically they are able to do so, e.g., after stimulation with an appropriate dose of a 5-HT_{1A} receptor agonist (e.g., 8-OH-DPAT or others). These "sluggish" males may represent an animal model of human delayed ejaculation or an

Effect of 8-OH-DPAT in sexually naive animals

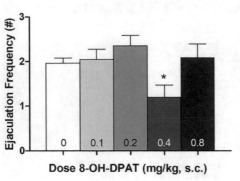 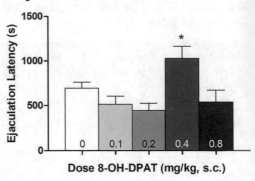

Effect of 8-OH-DPAT in sexually experienced animals

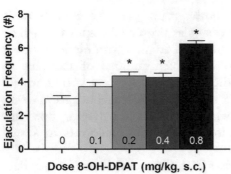 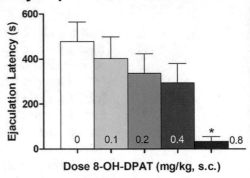

Fig. 17 Five groups of rats ($N = 25$ per group) were trained weekly on sexual behavior in a 30-min test with an estrous female. Rats obtained either vehicle or one dose of ($\pm$)-8-OH-DPAT (0.1, 0.2, 0.4 or 0.8 mg/kg SC) on test 1(naïve) and on test 7 (experienced). The mean number of ejaculations (+SEM) and the latency to the first ejaculation (+SEM) are shown. * = $p < 0.05$ compared to vehicle treatment

ejaculation [41, 67, 68]. Another small percentage (<5–10%) has a high number of ejaculations (>4) after training and might be a model for human lifelong premature ejaculation. The remaining animals (1–3 ejaculations/test) are considered as "normal" ejaculators and model normal human ejaculation. This latter group has been extensively studied in our research on the effects of various drug classes (in particular antidepressants) on male rat sexual behavior. This animal model has a high predictive validity for drug effects on human sexual behavior and can be used to predict whether newly developed, innovative drugs have sexual side effects in patients [32].

14 Genetic Models and Methods in Sexual Behavior in the Rat

Basal strain differences in the level of sexual behavior in rats seem obvious although no extensive and systematic studies have been performed to find putative genetic substrates. Different research groups have used several strains of rats including inbred and outbred strains. A recent study [178] compared the sexual behaviors of three strains, the inbred Fischer F344 and Lewis strains and the outbred Sprague-Dawley strain [179]. Remarkably, in quite extensive tests over five successive weekly sessions, Lewis rats did not ejaculate at all during the 50-min tests, whereas Fischer F344 rats improved their sexual behavior considerably over time, getting a plateau around four ejaculations/test, whereas Sprague-Dawley rats showed a constant level of sexual behavior, around one ejaculation/test. Accordingly, Lewis rats showed low levels of mounts and intromissions. All these behavioral differences were not dependent when the males of either strain were coupled to females of all three strains, indicating that external stimuli associated with the female do not contribute to the sexual phenotypes of the males. It would be interesting to study the effects of the 5-HT_{1A} receptor agonist 8-OH-DPAT in these three strains.

The group of Overstreet [180, 181] developed rat lines selectively bred for different thermal responses to the 5-HT_{1A} receptor agonist 8-OH-DPAT: the high-sensitive (HDS) and low-sensitive (LDS) lines. HDS animals respond with a strong hypothermia to 8-OH-DPAT, the LSD rats do not. These rat lines respond differently in various behavioral tests, including the social interaction anxiety paradigm and the forced-swim depression test. These HDS/LDS differences seem associated with differences in 5-HT_{1A} receptor numbers in the forebrain (HDS > LDS = RDS; [182]. Sura et al. [178] investigated the sexual behavior of sexually naïve and experienced male rats of HDS, LDS, and RDS (randomly bred) rat lines. Naïve HDS rats had lower sexual activities compared to LDS and RDS rats. Upon getting experience the HDS group improved ejaculatory behavior, comparable to that of LDS rats. 8-OH-DPAT (0.05 mg/kg SC) facilitated ejaculatory behavior in HSD and LSD rats, although more strongly in the LDS rats. The 5-HT_{1A} receptor antagonist WAY100,635 had no behavioral effects in either group. Sura et al. [178] conclude an important role of 5-HT_{1A} receptors in the forebrain in the initiation of sexual behavior when rats are sexually inexperienced.

Schijven et al. [183] investigated male rat sexual behavior in two inbred rat lines, the Flinders sensitive line (FSL) and Flinders insensitive line (FIL or FRL), derived from the outbred Sprague-Dawley strain [184]. The FSL rats are considered as an animal genetic model of depression [184] and have been used extensively over the last 30 years. FSL rats exhibit several behavioral and physiological symptoms strongly resembling human depression symptoms. FSL rats, compared to FIL rats, show several abnormalities in the serotonergic system, suggestive of a reduced 5-HT turnover and desensitized 5-HT receptors. Schijven et al. [183] investigated male rat sexual behavior in the FSL and FIL rats and focused on the status of the 5-HT_{1A} receptor. All rats were trained once weekly for 30 min with a sexually receptive and proceptive female to a sexually stable (experienced) level. On subsequent weekly tests, they received saline at week 7 and low doses of $(\pm)$-8-OH-DPAT (0.01 and 0.03 mg/kg SC) in weeks 8 and 9, resp., followed by WAY100,635 (0.1 mg/kg IP) + 8-OH-DPAT (0.03 mg/kg SC) in week 10 and again saline in week 11. FSL rats have higher ejaculation frequencies than FIL rats, which does not fit with a more depressive phenotype. FRL rats were more sensitive to 8-OH-DPAT measured in the ejaculation latencies and intromission frequencies, suggesting that the blunted response to 8-OH-DPAT in the FSL rats might be due to lower 5-HT_{1A} receptor density. Several "depressive-like" behaviors seen in the FSL rats could be reversed by chronic, but not acute, administration of SSRIs [185, 186]. It would be interesting to see whether a similar pattern would emerge after acute and chronic SSRIs in FSL rats.

However, we are not aware of studies in different rat strains that have looked into the existence of (endo)phenotypes, which is not expected in inbred, but probably could be present in other outbred strains like Sprague-Dawley. It would be advised that experimental animals should be studied on their basal sexual performance before using them in, e.g., pharmacological or hormonal studies as the outcome may depend on their basal endogenous level of sexual drive.

The possibility to generate transgenics, knockouts, and knockdowns is particularly developed in mice and possibilities emerge to study the effects of such genetic modifications in mice. Moreover, clear differences in sexual behavior between various mouse strains exist, enabling the study of genetic influences on male sexual behavior. The disadvantage of mice is the large variability within individual mice and the unpredictability of sexual behavior over time [187]. For this reason hardly any research has been performed on brain physiology or pharmacology of mouse sexual behavior. An attempt was tried using chromosomal substitution strains based on two strains of mice (C57Bl/6J and A/J). A/J mice have more sexual abnormalities than C57Bl/6J mice, and making consomic strains using the A/J as donor and the C57Bl/6J as host strain searching for quantitative loci involved in sexual functions and disorders seems feasible. There was some evidence that loci on certain chromosomes influenced the sexual behavior of some consomic strains [69].

Recently, gene knockout rats have been developed, creating the possibility to study the effects of selective knockout of a certain gene on sexual behavior. We studied the sexual behavior of male Wistar rats that lacked the serotonin transporter (SERT), an important regulator of serotonergic neurotransmission and the target of an important class of antidepressants, the SSRIs. In this case, constitutive lacking of the SERT led to reduction in basal sexual behavior [55, 168], which reflects comparable effects after chronic SSRI treatment.

In a more recent experiment we replicated and extended these findings [188]. Figure 18 shows data on training of three male genotype groups, homozygous SERT$^{+/+}$, heterozygous SERT$^{+/-}$, and homozygous SERT$^{-/-}$ male rats that were trained for 6 weeks in weekly tests of 30 min with a receptive female rat. Rats need approx. 3–4 sexual tests to establish a stable sexual (endo)-phenotype. Wildtype (SERT$^{+/+}$) and heterozygous knockout (SERT$^{-/-}$) male rats do not differ on any measure of sexual behavior [55, 188], illustrated here in the number of ejaculations and the first ejaculation latency per test. From test (week) 2 on, homozygous SERT$^{-/-}$ rats display reduced sexual behavior as evidenced in

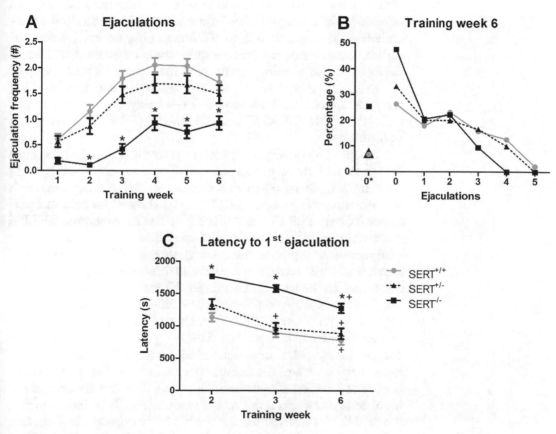

Fig. 18 (a) The mean number of ejaculations (± SEM) for the three genotype groups (SERT$^{+/+}$ (N = 95), SERT$^{+/-}$ (N = 60), and SERT$^{-/-}$ (N = 63)) of male rats is shown for the successive 6 weekly training tests of 30 min with an estrous female. * = significant difference from the SERT$^{+/+}$ and SERT$^{+/-}$ groups. (b) The distribution of the number of animals with particular phenotypes (0 to 5 ejaculations/test) in the last (6th) training test. 0* depicts the animals that never ejaculated during all the training tests (they are also included in the 0 ejaculation group). (c) The latencies to the first ejaculation for the three genotype groups are shown during training weeks 2, 3, and 6. *$p < 0.05$ compared to SERT$^{+/-}$ and SERT$^{+/+}$ groups; +$p < 0.05$ compared to week 2

the number of ejaculations and the first ejaculation latencies. Typically, as shown in Fig. 18b, SERT$^{-/-}$ rats had fewer ejaculations per test showing a leftward shift of the distribution.

In humans, all SSRIs delay or completely inhibit ejaculation, associated with libido and arousal reductions [189]. Chronic SSRIs at effective antidepressant doses result in minimal 80% blockade of the SERTs [190]. SERT$^{+/-}$ rats, which have a 50% reduced SERT level and function [191], have no disturbance in sexual behavior, whereas SERT$^{-/-}$ rats (100% lack of SERTs) show decreased (male) sexual behaviors. The basal level of sexual behavior of SERT$^{-/-}$ rats is comparable to that of male rats chronically treated with SSRIs [24, 192]. It seems plausible that a critical number (threshold) of available and functional 5-HT transporters is necessary to enable normal male rat sexual behavior. Acute SSRI administration leading to >80% SERT occupation does not lead to sexual behavior inhibition [66], but chronic administration does apparently lead to adaptations in 5-HT functioning underlying the disturbed sexual behavior. There is quite some evidence that changes (adaptations) in 5-HT$_{1A}$ receptors in the brain may underlie (part of) these changes, and we studied this by applying a 5-HT$_{1A}$ receptor agonist and antagonist to sexually trained and stable male rats with (SERT$^{+/+}$) and without (SERT$^{-/-}$) serotonin transporters.

In our early (at Utrecht university) studies [55, 168], 90 male Wistar rats of three genotypes (SERT$^{+/+}$, SERT$^{+/-}$, SERT$^{-/-}$) were trained up to seven times (once weekly for 30 min) with sexually receptive females. SERT$^{-/-}$ rats showed lower sexual performance than SERT$^{+/+}$ and SERT$^{+/-}$. The heterozygous SERT$^{+/-}$ rats did not differ from the wildtype SERT$^{+/+}$ rats at any behavioral measure. Typically, the three genotypes did not differ in the number of tests needed to stabilize their sexual behavior (approx. our tests). In the later (at Groningen University) studies [32, 167, 188], we essentially replicated these findings as can be seen in Fig. 18. SERT$^{+/+}$ and SERT$^{+/-}$ male rats were completely similar in their sexual behavior. The SERT$^{-/-}$ males had significantly reduced sexual behavior compared to the SERT$^{+/+}$ and $^{+/-}$ genotypes as evidenced in the number of ejaculations (Fig. 18a) and the latency to the first ejaculation (Fig. 18c). The distribution of the number of ejaculations per test at week 6 (Fig. 18b) indicates that the SERT$^{-/-}$ genotype has a left-shifted distribution, showing that almost 50% of the animals did not ejaculate. An independent research group [193], although using very few animals, essentially replicated our findings on male sexual behavior.

We consider the low sexual behavior level of male SERT$^{-/-}$ rats as a model of sexual disturbances induced by chronic administration of SSRIs in men [32, 194]. For that reason the SERT-KO rat is attractive to better understand the mechanisms in the brain that are involved in sexual disturbances after SSRIs, but also to detect the influence of non-SSRI drugs on sexual behavior. Moreover, testing drugs with multimodal mechanisms including serotonin reuptake blockade (e.g., NSRIs, vilazodone, vortioxetine, tramadol) might unravel the contribution of the non-SSRI mechanisms in a drug.

Serotonergic activation of sexual behavior in male rats leads to activation of 5-HT_{1A} receptors, based on the prosexual effects of 5-HT_{1A} receptor agonists [42]. Coadministration of a 5-HT_{1A} receptor antagonist and an SSRI leads to a facilitation of the SSRI-induced inhibition of male sexual behavior, indicative for the role of 5-HT_{1A} receptors in the modulation of male rat sexual behavior [195, 196]. The 5-HT_{1A} receptor agonist $(\pm)\text{-8-OH-}$DPAT dose-dependently $(0, 0.01, 0.1, \text{and } 1 \text{ mg/kg SC})$ increased the number of ejaculations and decreased the first ejaculation frequency (EF) in both the wildtype ($SERT^{+/+}$) and homozygous knockout ($SERT^{-/-}$) rats [168] (Fig. 19 top panels). The 5-HT_{1A} receptor antagonist WAY100,635 in a wide dose range did not affect the sexual behavior (EF and first ejaculation latency (EL)) in the wildtypes (except at a high dose of 1 mg/kg, probably due to non-serotonergic mechanisms). In contrast, WAY100,635 strongly and dose-dependently inhibited EF and enhanced EL in the $SERT^{-/-}$ rats (Fig. 19 middle panels). Combining a fixed effective dose of WAY100.635 (0.1 mg/kg) with a dose range of 8-OH-DPAT (Fig. 19 bottom panels), it is found that WAY100.635 completely antagonized the prosexual effects of 8-OH-DPAT in the $SERT^{+/+}$ rats and (partially) antagonized the prosexual effects of 8-OH-DPAT in the $SERT^{-/-}$ rats. Based on these findings we postulated that complete absence of SERTs leads to changes in certain 5-HT_{1A} receptor pools with different sensitivity levels. One of these pools mediates the prosexual effects of 5-HT_{1A} receptor stimulation and is not desensitized. The other receptor pool, mediating the inhibitory effects of 5-HT_{1A} receptor antagonists, is sensitized in $SERT^{-/-}$ rats [168]. Summarizing, two populations of 5-HT_{1A} receptors are involved in male rat sexual behavior. For normal sexual functioning, activation of a certain 5-HT_{1A} receptor pool is needed; this pool is desensitized in $SERT^{-/-}$ rats. The prosexual effects of 5-HT_{1A} receptor agonists are mediated via a different pool and are not desensitized in $SERT^{-/-}$ rats.

In a further attempt to unravel the different 5-HT_{1A} receptor targets (pools) in male sexual behavior, we studied the sexual behavior effects of two biased 5-HT_{1A} receptor agonists, F-13714, a preferential 5-HT_{1A} autoreceptor agonist, and F-15599, a preferential 5-HT_{1A} heteroreceptor agonist [197, 198] in male rat sexual behavior of well-trained and sexually stable $SERT^{+/+}$ and $SERT^{-/-}$ rats. Both biased agonists had prosexual effects in $SERT^{+/+}$ and $SERT^{-/-}$ rats; F-13714 was much more potent than F-15599. Compared to $SERT^{+/+}$ rats, the

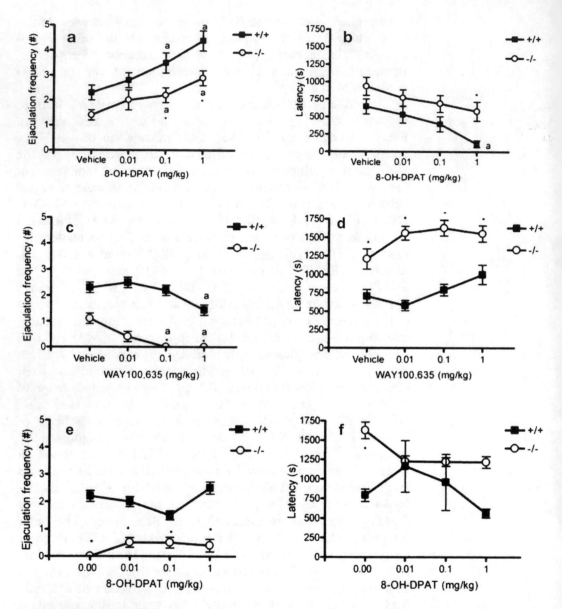

Fig. 19 Effects of the 5-HT$_{1A}$ receptor agonist 8-OH-DPAT, 5-HT$_{1A}$ receptor antagonist WAY100,635, and their interaction are shown on two measures of sexual behavior in male SERT$^{+/+}$ and SERT$^{-/-}$ rats, viz., the mean number of ejaculations (EF) per test ($\pm$SEM) and the latency to the first ejaculation (EL in seconds: $\pm$SEM). (**a** and **b**) Show the effects of ($\pm$)-8-OH-DPAT in a dose range of 0, 0.01, 0.1, and 1 mg/kg (SC). (**c** and **d**) Show the effects of WAY100,635 (0, 0.01, 0.1, and 1 mg/kg, IP) on EF and EL, and (**e** and **f**) show the interaction of a fixed dose of WAY100,635 (0.1 mg/kg, IP) with a dose range of ($\pm$)-8-OH-DPAT (0, 0.01, 0.1, and 1.0 mg/kg, SC)

F-13714 dose-response curve in SERT$^{-/-}$ rats had shifted to the right, which did not occur in F-15599 [167]. A mixed 5-HT$_{1A}$ autoreceptor agonist/heteroreceptor antagonist (S-15535) did not influence male sexual behavior in either genotype or in groups of

SERT$^{+/+}$ and SERT$^{-/-}$ rats selected on low sexual behavior levels [167]. It is up to now unclear how pre- and postsynaptic 5-HT$_{1A}$ receptor mechanisms are precisely involved in the modulation of male rat sexual behavior.

Along the multimodal mechanism line approach, we tested tramadol, a worldwide used painkiller, producing antinociception by activation of μ-opioid receptors [199]. Tramadol is a racemic mixture of two enantiomers [200]; the (+)-enantiomer and its metabolite ((+)-M1) are selective agonists of the μ-opioid receptor and selective serotonin reuptake inhibitors (SSRIs). The (−)-enantiomer and its (−)-M2 metabolite are noradrenaline reuptake inhibitors [131]. Tramadol has, like other SSRIs, inhibitory effects in premature ejaculation in humans [134, 201]. Tramadol at high doses inhibits sexual behavior in wildtype male rats and we postulated this effect due to its SSRI mechanism [136].

Tramadol had no acute inhibiting effects on male rat sexual behavior in SERT$^{+/+}$, SERT$^{+/-}$, and SERT$^{-/-}$ rats [188]. WAY100,635 (0, 0.1, 0.3, and 1 mg/kg IP) had no effects on male rat sexual behavior in SERT$^{+/+}$ rats, but strongly inhibited sexual behavior in SERT$^{-/-}$ rats, confirming the earlier finding of Chan et al. [168]. Combining tramadol (20 mg/kg IP) with WAY 100,635 (0.3 mg/kg IP) reduced sexual behavior in SERT$^{+/+}$ rats, but completely abolished it in SERT$^{-/-}$ rats, confirming the important role of the SSRI mechanism in tramadol [188].

Recently, a dopamine transporter (DAT)-knockout rat has been developed by silencing the gene encoding DAT by using zinc finger nuclease technology [202–204]. DAT-KO rats develop normally but have lower body weights than heterozygous (DAT$^{+/-}$) and wildtype (DAT$^{+/+}$) rats. They display locomotor hyperactivity and have various behavioral and cognitive abnormalities (*see* [205] for a review). The three genotypes were tested on male sexual behavior and obtained five sequential 60-min sex tests at 3-day intervals with an estrous female [205]. Intriguingly, DAT-KO male rats seemed to more rapidly acquire stable sexual activity levels; upon reaching stable sexual behavior levels, DAT-KO males had higher levels of sexual behavior than DAT$^{+/-}$ and DAT$^{+/+}$ rats: more ejaculations/ test, shorter ejaculation latency, higher intromission ratio, and lower mount and intromission frequencies. The sexual phenotype of the DAT$^{+/+}$ male rat makes it an interesting model of "enhanced" sexual behavior, comparable to that induced by pharmacological blocking of the DAT by, e.g., cocaine or selective DAT inhibitors [206, 207].

15   Delayed Ejaculation

Ejaculatory dysfunctions in men occur frequently, are often complex, and may arise from a combination of neurobiological, physiological, pharmacological, and psychological factors. There is clearly variability in the ejaculation time of males, and there is no clinical consensus about a "normal" ejaculation time in normal partners/relations nor what is considered as unacceptably short or long [18, 38]. A stopwatch study measuring the intravaginal ejaculation latency time (IELT) from almost 500 males in 5 different countries demonstrated an IELT continuum [174, 175]. The shape of this distribution was positively skewed, with a median IELT of 5.4 min (range 33 s till 44 min). The data showed both extremely fast (<1 min) ejaculating men but also (extremely) delayed ejaculating (DE; >30 min) ones in this cohort, illustrating the variability in the ejaculation latency of healthy men. Clinically, often a distinction between lifelong delayed ejaculation (llDE) and acquired DE (aDE) is made. The prevalence of llDE in the general population is estimated to be up to 1%, whereas aDE is estimated at 3–4% [208–211]. Although DE is often considered to be caused by psychological problems, it has increasingly become clear that neurobiological and neuropharmacological mechanisms play a big role in ejaculation (and orgasmic) processes in the brain and spinal cord [24, 106]. Waldinger and Olivier formulated a dynamic neurobiological theory on the distribution of ejaculation times in men [67, 106, 212]. According to this "ejaculation distribution theory" (EDT), llDE and premature ejaculation (PE) are part of a normal biological variability of IELT in men. Waldinger and Olivier [106] postulated that any random sample of men is likely to include some men with early (or premature) ejaculation, some other always experiencing (suffering from) delayed or even an ejaculation, while the majority have "normal" or "average" IELTs. In this theory, llDE is considered primarily a neurobiological variant; this may secondarily lead to psychological or psychosocial distress. The earlier described stopwatch study in 491 healthy males in 5 countries [174] and later a positive replication [175], focusing on IELTs, clearly support this "ejaculation distribution theory."

Pharmacotherapy for delayed ejaculation. DSM-5 defines DE as a marked delay in ejaculation or a marked infrequency or absence of ejaculation on almost all occasions (75–100% of time) of partnered sexual activity without the individual wanting the delay, persisting for at least 6 months and causing significant distress to the person. The pathophysiology of llDE is multifactorial, including (neuro)-biological and psychosocial factors [213], and treatment is often problematic [213–215] and includes pharmacotherapy. Delayed ejaculation or anejaculation is often caused by side effects of various drugs [213, 216]. Up till now no adequate pharmaco-therapeutic approach is present and neither drug discovery nor development for DE seems underway.

15.1 The Involvement of the Serotonergic System in Delayed Ejaculation

The serotonergic (5-HT) system in the brain plays an important modulatory role in many aspects of male sexual behavior [29, 31, 32, 194]. Apart from the serotonin transporter (inhibited by selective serotonin reuptake inhibitors), 14 different 5-HT receptors exist and are known to be involved in many physiological and behavioral mechanisms in the brain [217, 218]. One of these 5-HT receptors, the 5-HT_{1A} receptor, plays an important role in the regulation of serotonergic cell firing in the raphe nuclei, the origin of serotonergic cells in the brain. Application of a 5-HT_{1A} receptor agonist stimulates the so-called 5-HT_{1A} autoreceptors on the 5-HT cell bodies, thereby blocking the cell firing which leads to a decrease of 5-HT release at all postsynaptically located 5-HT receptors (14 different types). At the same time, 5-HT_{1A} receptor agonists also stimulate postsynaptic 5-HT_{1A} heteroreceptors, thereby specifically activating (or inhibiting) non-serotonergic neural systems that are supplied with 5-HT_{1A} receptors. It is unknown however whether dysfunctions in the serotonergic system, including 5-HT_{1A} receptors, underlie delayed ejaculation. Apart from 5-HT_{1A} receptors, activation of 5-HT_{2C} receptors also leads to pro-ejaculatory effects [219]. This stimulatory profile is behaviorally quite different from that of 5-HT_{1A} receptor agonists: no increase in the number of ejaculations, an increase in the onset of the first mount or intromission in the first ejaculation series, and a decrease in the number of intromissions in the first and second ejaculation series. These changes concur however with a strong reduction in ambulation, suggestive of associated sedation. Further research into the role of 5-HT_{2C} receptors is needed to unravel its potential role in (delayed) ejaculation.

It is long known that standard 5-HT_{1A} receptor agonists like 8-OH-DPAT have prosexual activity in the male rat [42]. 8-OH-DPAT shortens the IELT in rats and increases the number of ejaculations/30 min. However, 8-OH-DPAT cannot be used in humans, and other selective 5-HT_{1A} receptor agonists that have been developed and tested for depression and anxiety have not been tested in humans in any sexual dysfunction, let alone llDE [220]. The only 5-HT_{1A} receptor agonist available for human use is buspirone (for depression/anxiety), but buspirone is a partial 5-HT_{1A} receptor agonist, has an interfering functional metabolite (side effects), and is also a dopamine D_2 receptor antagonist responsible for buspirone's sedative effects at higher doses. So far, it remains unknown whether activation of autoreceptors or (certain pools of) heteroreceptors contributes to the sexual-stimulating activity of nonselective (with regard to auto- or heteroreceptor) 5-HT_{1A} receptor agonists. This is essential knowledge to pave the way for drug targets treating delayed ejaculation disorder.

15.2 Animal Research into Ejaculatory Behavior

In our research we developed a translational model in rats that has a high predictability for human male sexual behavior, including premature, normal, and delayed ejaculation [41, 67]. This rat model has a high similarity profile with regard to the effects of antidepressants, psychostimulants, and other psychoactive drugs in human sexual behavior and its dysfunctions [69, 100] and has been used to predict whether new antidepressants (vortioxetine, vilazodone) might exert sexual inhibitory effects in depressed patients [101, 123, 221].

16 Methods and Models: Model of Delayed Ejaculation in Rats

When large populations of male rats are tested on their sexual activity during a number of successive, weekly tests (4–6 tests) against an estrous female, they develop over time a very stable sexual behavior that can be qualified as either "slow," "normal," or "fast," as characterized by the number of ejaculations per test [29, 41, 67, 99]. We have trained over 2000 male rats this way, and ejaculation frequencies (rats ejaculate during our standard 30-min test from 0 to 5 ejaculations/30 min) follow a kind of bell-shaped, Gaussian distribution with approx. 5–10% displaying "hyposexual" (0 ejaculation) and approx. 5–10% "hypersexual" behavior (4–5 ejaculations/30 min). The remaining animals display "normal" ejaculations, between one and three ejaculations/test. A similar, but then reversed, picture emerged when the ejaculation latency (the rat-IELT (rIELT)) was portrayed. The higher the number of ejaculations, the shorter the rIELTs. We postulated that in male rats, like in human males, a biological ejaculation time distribution is found and propose that rats that do not (or seldomly) ejaculate, after extensive sexual training, are comparable to human males with llDE.

We have used several strategies to create low sexual behavior levels in male rats, e.g., using sexually naïve rats or rats that after extensive training display very limited or no sexual activity (sexually sluggish rats) [55, 222]. Figure 20 shows data of a study we performed in the 1990s on the effects of four 5-HT$_{1A}$ receptor agonists on sexual behavior of sexually naïve rats during a 15-min sex test [222]. Under these conditions rats display very low sexual activity, clearly reflected in the very low number of ejaculations. Several 5-HT$_{1A}$ receptor agonists (8-OH-DPAT, flesinoxan, buspirone, and ipsapirone) dose-dependently increased the number of ejaculations and decreased the ejaculation latency. Buspirone at 10 mg/kg did not facilitate sexual behavior probably due to sedation.

In sexually extensively trained (six tests) male rats, we used "sluggish" rats as a model of delayed ejaculation [41, 67, 68]. From 100 trained male rats, 12 rats were selected with a

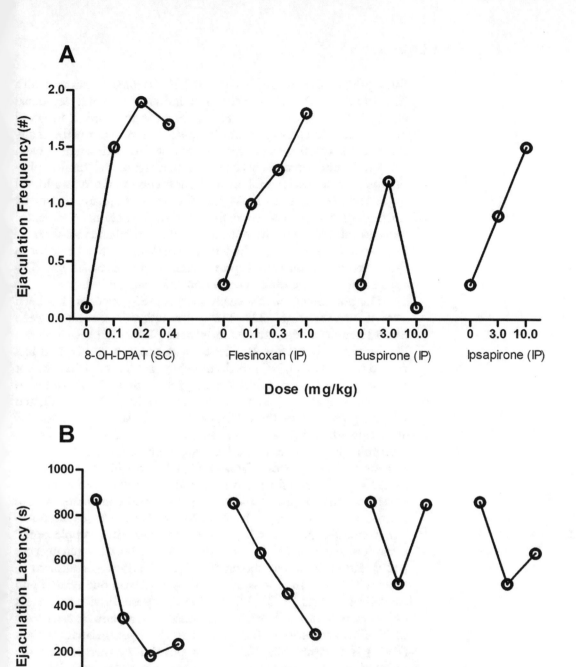

Fig. 20 Effects of four 5-HT$_{1A}$ receptor agonists on male sexual behavior in sexually naïve rats. For each dose-response study, a group of 60 naïve male Wistar rats was used. Rats were randomized to one dose of a specific drug, injected 30 min before a test of 15-min duration against a female in behavioral estrus. The top panel **a** shows the effects of ($\pm$)-8-OH-DPAT (0, 0.1, 0.2, and 0.4 mg/kg SC), flesinoxan (0, 0.1, 0.3, and 1.0 mg/kg IP), buspirone (0, 3, and 10 mg/kg IP), and ipsapirone (0, 3, and 10 mg/kg IP). The bottom panel **b** shows the effect on the first ejaculation latency. * indicates significant difference ($p < 0.05$) from vehicle treatment

"sluggish" profile (mean EF 0.2 ± 0.1 E/30 min). Treatment with ($\pm$)-8-OH-DPAT (0.8 mg/kg SC) induced a strong prosexual effect, indicating that the 5-HT_{1A} receptor mechanism involved in the ejaculation process is intact. Testing 1 week later without any treatment showed that all animals returned to their original basal "sluggish" phenotype. This indicates that the sexual "endo" phenotype of male rats is a biological phenomenon that can be affected by external (e.g., pharmacological) factors, but apparently not by learning or experience. Apart from the differences in basal sexual behavior, sluggish rats did not differ from normal and fast ejaculating rats in locomotor activity and approach-avoidance behavior in the elevated-plus maze and apomorphine-induced stereotypy [67], suggesting no large difference in dopamine susceptibility.

The phenomenon of sexually sluggish (SS) male rats has been known for a long time [223–225]. Although a distinction between SS and non-copulating (NC) rats is made [226, 227] in that SS rats display sexual behaviors but fail to ejaculate within a certain test period (30–60 min), NC rats do not show any sexual behavior. For the purpose of this review we categorize these two groups as "sexually sluggish" (SS) rats. Various studies have investigated whether, apart from the differences in male sexual behavior, SS and copulating (C) rats differ. Similar to C males, SS males can discriminate between neutral odors, male urine, and between estrous and non-estrous females [228]. No difference was also found in preference for receptive females or bedding of estrous or anestrous females [227]. NC and SS males had comparable serum testosterone and estradiol levels [226], whereas supplemental testosterone did not promote sexual behavior and supplemental estradiol-suppressed ejaculation in both phenotypes [229]. Recently, it was shown that NC and SS males differed in certain olfactory functions, including sex-pheromone-related and food-related cues [230]. Whether this hyposensitivity to certain olfactory stimuli is related to difference in sympathetic sensitivity or NMDA receptors in the paraventricular hypothalamic nucleus (PVN) is unclear [231, 232]. The possibility emerges that the reduced sexual function in male SS rats has certain physiological causes, different from reproductive or endocrinological functioning, apart from the 5-HT_{1A} receptor agonists that stimulate sexual behavior in SS males [67] and sexually naïve males (Fig. 19). Several pharmacological studies have tested various mechanisms in SS or SS-like (naïve) rats. Anandamide, an endocannabinoid, induces ejaculations in SS rats [233] through activation of CB_1 receptors [234]. Anandamide targets the ejaculation threshold by decreasing the ejaculation latency [235]. Studies on the effects of D_2 and D_3 dopaminergic agonists [236] in SS rats did not reveal clear stimulating effects on ejaculation, whereas it did in normal ejaculating rats. Some suggestions that sildenafil, a phosphodiesterase-5 inhibitor, might also facilitate ejaculation, possibly via a dopaminergic mechanism, has been made but further research is necessary to unravel peripheral versus central effects [237].

The area of pharmacological treatment of delayed ejaculation and an-ejaculation in men is clearly in its infancy, and not much progress has been made in the last decade, although a possible 5-HT$_{1A}$ approach seems promising.

17   Conclusions and Discussion

Although the field of sexual behavior paradigms in animals, particularly in rats, is a relatively well-investigated area, no animal models of premature ejaculation have emerged thus far. Although it is feasible to induce fast ejaculation in rats via brain lesions, electrical brain stimulation, or pharmacological means, the resulting paradigm does not create ideal animal models for PE, in particular because they are technically difficult to perform and no therapeutic drugs have been tested in such paradigms.

A new animal model in rats, the spontaneous "fast" ejaculating rat that can be selected from a normal population of outbred Wistar rats, seems to have several features that makes it a promising model to study animal PE but also delayed ejaculation (*see* also [238]). However, a considerable amount of scientific research into its physiological, neurochemical, and genetic background has to be performed to further strengthen its predictive and construct validity for human PE disturbance.

References

1. Gustavsson A, Svensson M, Jacobi F, Allgulander C, Alonso J, Beghi E, Dodel R, Ekman M, Faravelli C, Fratiglioni L, Gannon B, Jones DH, Jennum P, Jordanova A, Jönsson L, Karampampa K, Knapp M, Kobelt G, Kurth T, Lieb R, Linde M, Ljungcrantz C, Maercker A, Melin B, Moscarelli M, Musayev A, Norwood F, Preisig M, Pugliatti M, Rehm J, Salvador-Carulla L, Schlehofer B, Simon R, Steinhausen HC, Stovner LJ, Vallat JM, Van den Bergh P, van Os J, Vos P, Xu W, Wittchen HU, Jönsson B, Olesen J (2011) Cost of disorders of the brain in Europe 2010. Eur Neuropsychopharmacol 21(10):718–779. https://doi.org/10.1016/j.euroneuro.2011.08.008

2. Nutt DJ (2011) The full cost and burden of disorders of the brain in Europe exposed for the first time. Eur Neuropsychopharmacol 21(10):715–717. https://doi.org/10.1016/j.euroneuro.2011.09.005

3. Wittchen HU, Jacobi F, Rehm J, Gustavsson A, Svensson M, Jönsson B, Olesen J, Allgulander C, Alonso J, Faravelli C, Fratiglioni L, Jennum P, Lieb R, Maercker A, van Os J, Preisig M, Salvador-Carulla L, Simon R, Steinhausen HC (2011) The size and burden of mental disorders and other disorders of the brain in Europe 2010. Eur Neuropsychopharmacol 21(9):655–679. https://doi.org/10.1016/j.euroneuro.2011.07.018

4. Balon R (2006) SSRI-associated sexual dysfunction. Am J Psychiatry 163:1504–1509

5. Chokka PR, Hankey JR (2018) Assessment and management of sexual dysfunction in the context of depression. Ther Adv Psychopharmacol 8(1):13–23. https://doi.org/10.1177/2045125317720642

6. Makhlouf A, Kparker A, Niederberger CS (2007) Depression and erectile dysfunction. Urol Clin North Am 34(4):565–574. https://doi.org/10.1016/j.ucl.2007.08.009

7. Segraves RT, Balon R (2014) Antidepressant-induced sexual dysfunction in men. Pharmacol Biochem Behav 121:132–137. https://doi.org/10.1016/j.pbb.2013.11.003

8. Lahon K, Shetty HM, Paramel A, Sharma G (2011) Sexual dysfunction with the use of antidepressants in a tertiary care mental health setting - a retrospective case series. J Pharmacol Pharmacother 2(2):128–131. https://doi.org/10.4103/0976-500X.81913

9. Waldinger MD (2011) Toward evidence-based genetic research on lifelong premature ejaculation: a critical evaluation of methodology. Korean J. Urol 52(1):1–8. https://doi.org/10.4111/kju.2011.52.1.1

10. Althof SE, Mcmahon CG, Waldinger MD, Serefoglu EC, Shindel AW, Adaikan PG, Becher E, Dean J, Giuliano F, Hellstrom WJG, Giraldi A, Glina S, Incrocci L, Jannini E, Mccabe M, Parish S, Rowland D, Segraves RT, Sharlip I, Torres LO (2014) An update of the International Society of Sexual Medicine's guidelines for the diagnosis and treatment of premature ejaculation (PE). Sex. Med 2(2):60–90. https://doi.org/10.1002/sm2.28

11. Waldinger MD, Zwinderman AH, Schweitzer DH, Olivier B (2004) Relevance of methodological design for the interpretation of efficacy of drug treatment of premature ejaculation: a systematic review and meta-analysis. Int J Impot Res 16(4):369–381. https://doi.org/10.1038/sj.ijir.3901172

12. Goodman RE (1980) An assessment of clomipramine (Anafranil®) in the treatment of premature ejaculation. J Int Med Res 8(S3):53–59

13. Girgis SM, El-Haggar S, El-Hermouzy S (1982) A double-blind trial of clomipramine in premature ejaculation. Andrologia 14(4):364–368. https://doi.org/10.1111/j.1439-0272.1982.tb02278.x

14. Segraves RT, Saran A, Segraves K, Maguire E (1993) Clomipramine versus placebo in the treatment of premature ejaculation: a pilot study. J Sex Marital Ther 19(3):198–200. https://doi.org/10.1080/00926239308404904

15. Althof SE, Levine SB, Corty EW, Risen CB, Stern EB, Kurit DM (1995) A double-blind crossover trial of clomipramine for rapid ejaculation in 15 couples. J Clin Psychiatry 56(9):402–407

16. Waldinger MD, Hengeveld MW, Zwinderman AH (1994) Paroxetine treatment of premature ejaculation: a double-blind, randomized, placebo-controlled study. Am J Psychiatry 151(9):1377–1379. https://doi.org/10.1176/ajp.151.9.1377

17. Waldinger MD, Hengeveld MW, Zwinderman AH, Olivier B (1998) Effect of SSRI antidepressants on ejaculation. J Clin Psychopharmacol 18:274–281

18. Waldinger MD (2005) Lifelong premature ejaculation: current debate on epidemiology and SSRI treatment. World J Urol 23:102–108

19. Pryor JL, Althof SE, Steidle C, Rosen RC, Hellstrom WJ, Shabsigh R, Miloslavsky M, Kell S (2006) Efficacy and tolerability of dapoxetine in treatment of premature ejaculation: an integrated analysis of two double-blind, randomised controlled trials. Lancet 368(9539):929–937. https://doi.org/10.1016/S0140-6736(06)69373-2

20. Waldinger MD, Schweitzer DH (2008) Premature ejaculation and pharmaceutical company-based medicine: the dapoxetine case. J Sex Med 5(4):966–997. https://doi.org/10.1111/j.1743-6109.2008.00633.x

21. Waldinger MD, Schweitzer DH, Olivier B (2006) Dapoxetine treatment of premature ejaculation. Lancet 368:1869

22. Park HJ, Park NC, Kim TN, Baek SR, Lee KM, Choe S (2017) Discontinuation of Dapoxetine treatment in patients with premature ejaculation: a 2-year prospective observational study. Sex Med 5:e99–e105

23. Waldinger MD, Berendsen HHG, Blok BFM, Olivier B, Holstege G (1998) Premature ejaculation and serotonergic antidepressants-induced delayed ejaculation: the involvement of the serotonergic system. Behav Brain Res 92(2):111–118. https://doi.org/10.1016/S0166-4328(97)00183-6

24. Waldinger MD (2002) The neurobiological approach to premature ejaculation. J Urol 168(6):2359–2367. https://doi.org/10.1016/S0022-5347(05)64146-8

25. Waldinger MD, Zwinderman AH, Olivier B (2003) Antidepressants and ejaculation: a double-blind, randomized, fixed-dose study with mirtazapine and paroxetine. J Clin Psychopharmacol 23(5):467–470. https://doi.org/10.1097/01.jcp.0000088904.24613.e4

26. Waldinger MD, Olivier B (2004) Utility of selective serotonin reuptake inhibitors in premature ejaculation. Curr Opin Investig Drugs 5(7):743–747

27. Waldinger MD, Van de Plas A, Pattij T, Van Oorschot R, Coolen LM, Veening JG, Olivier B (2002) The selective serotonin re-uptake inhibitors fluvoxamine and paroxetine differ

in sexual inhibitory effects after chronic treatment. Psychopharmacology 160(3): 283–289. https://doi.org/10.1007/s00213-001-0980-3

28. de Jong TR, Veening JG, Olivier B, Waldinger MD (2007) Oxytocin involvement in SSRI-induced delayed ejaculation: a review of animal studies. J Sex Med 4(1):14–28. https://doi.org/10.1111/j.1743-6109.2006.00394.x

29. Olivier B (2015) Serotonin: a never-ending story. Eur J Pharmacol 753:2–18. https://doi.org/10.1016/j.ejphar.2014.10.031

30. Rubio-Casillas A, Rodríguez-Quintero CM, Rodríguez-Manzo G, Fernández-Guasti A (2015) Unraveling the modulatory actions of serotonin on male rat sexual responses. Neurosci Biobehav Rev 55:234–246. https://doi.org/10.1016/j.neubiorev.2015.05.003

31. Olivier J, Esquivel Franco D, Waldinger M, Olivier B (2019) In: Tricklebank E, Daly M (eds) The serotonin system: history, neuropharmacology, and pathology. Elsevier, Academic press, London, pp 117–132

32. Olivier J, Olivier B (2019) Antidepressants and sexual dysfunction; translational aspects. Curr Sex Heal Rep 11:156–166

33. Janssen PKC, Van Schaik R, Zwinderman AH, Olivier B, Waldinger MD (2014) The 5-HT1A receptor C(1019)G polymorphism influences the intravaginal ejaculation latency time in Dutch Caucasian men with lifelong premature ejaculation. Pharmacol Biochem Behav 121:184–188. https://doi.org/10.1016/j.pbb.2014.01.004

34. Janssen PKC, Schaik R, Olivier B, Waldinger MD (2014) The 5-HT2C receptor gene Cys23Ser polymorphism influences the intravaginal ejaculation latency time in Dutch Caucasian men with lifelong premature ejaculation. Asian J Androl 16(4):607–610. https://doi.org/10.4103/1008-682X.126371

35. Janssen PKC, Zwinderman AH, Olivier B, Waldinger MD (2014) Serotonin transporter promoter region (5-HTTLPR) polymorphism is not associated with paroxetine-induced ejaculation delay in Dutch men with lifelong premature ejaculation. Korean J Urol 55(2):129–123. https://doi.org/10.4111/kju.2014.55.2.129

36. Janssen PKC, Bakker SC, Réthelyi J, Zwinderman AH, Touw DJ, Olivier B, Waldinger MD (2009) Serotonin transporter promoter region (5-HTTLPR) polymorphism is associated with the intravaginal ejaculation latency time in Dutch men with lifelong premature

ejaculation. J Sex Med 6(1):276–284. https://doi.org/10.1111/j.1743-6109.2008.01033.x

37. Flaive A, Fougère M, van der Zouwen CI, Ryczko D (2020) Serotonergic modulation of locomotor activity from basal vertebrates to mammals. Front Neural Circuits 14:–590299. https://doi.org/10.3389/fncir.2020.590299

38. Waldinger MD (2016) The pathophysiology of lifelong premature ejaculation. Transl Androl Urol 5(4):424–433. https://doi.org/10.21037/tau.2016.06.04

39. Veening JG, Coolen LM (2014) Neural mechanisms of sexual behavior in the male rat: emphasis on ejaculation-related circuits. Pharmacol Biochem Behav 121: 170–183. https://doi.org/10.1016/j.pbb.2013.12.017

40. Pfaus JG (2009) Pathways of sexual desire. J Sex Med 6(6):1506–1533. https://doi.org/10.1111/j.1743-6109.2009.01309.x

41. Pattij T, Olivier B, Waldinger M (2005) Animal models of ejaculatory behavior. Curr Pharm Des 11(31):4069–4077. https://doi.org/10.2174/138161205774913363

42. Snoeren EMS, Veening JG, Olivier B, Oosting RS (2014) Serotonin 1A receptors and sexual behavior in male rats: a review. Pharmacol Biochem Behav 121:102–114. https://doi.org/10.1016/j.pbb.2013.11.007

43. Larsson K (1956) Conditioning and sexual behavior in the male albino rat. In: Elmgren J (ed) Acta Psychologica Gothoburgensia I. Almqvist & Wiksell, Stockholm pp. 1–269

44. Fernández-Guasti A, Rodríguez-Manzo G (2003) Pharmacological and physiological aspects of sexual exhaustion in male rats. Scand J Psychol 44(3):257–263. https://doi.org/10.1111/1467-9450.00343

45. Giuliano F, Clément P (2005) Neuroanatomy and physiology of ejaculation. Annu Rev Sex Res 16:190–216. https://doi.org/10.1080/10532528.2005.10559833

46. Giuliano F, Clément P (2005) Physiology of ejaculation: emphasis on serotonergic control. Eur Urol 48(3):408–417. https://doi.org/10.1016/j.eururo.2005.05.017

47. de Jong TR, Veening JG, Waldinger MD, Cools AR, Olivier B (2006) Serotonin and the neurobiology of the ejaculatory threshold. Neurosci Biobehav Rev 30(7):893–907. https://doi.org/10.1016/j.neubiorev.2006.01.001

48. Truitt WA, Coolen LM (2002) Identification of a potential ejaculation generator in the

spinal cord. Science 297(5586):1566–1569. https://doi.org/10.1126/science.1073885

49. Coolen LM, Allard J, Truitt WA, McKenna KE (2004) Central regulation of ejaculation. Physiol Behav 83(2):203–215. https://doi.org/10.1016/j.physbeh.2004.08.023

50. Allard J, Truitt WA, McKenna KE, Coolen LM (2005) Spinal cord control of ejaculation. World J Urol 23(2):119–126. https://doi.org/10.1007/s00345-004-0494-9

51. Gregorian RS, Golden KA, Bahce A, Goodman C, Kwong WJ, Khan ZM, De Bittner MR, Vanier MC (2002) Antidepressant-induced sexual dysfunction. Ann Pharmacother 36(10):1577–1589. https://doi.org/10.1345/aph.1A195

52. Ågmo A (1997) Male rat sexual behavior. Brain Res Protocol 1(2):203–209. https://doi.org/10.1016/S1385-299X(96)00036-0

53. Ågmo A (2007) Functional and Dysfunctional Sexual Behavior. Elsevier, Amsterdam

54. Giuliano F, Clément P (2012) Pharmacology for the treatment of premature ejaculation. Pharmacol Rev 64(3):621–644. https://doi.org/10.1124/pr.111.004952

55. Olivier B, Chan JSW, Snoeren EM, Olivier JDA, Veening JG, Vinkers CH, Waldinger MD, Oosting RS (2011) Differences in sexual behaviour in male and female rodents: Role of serotonin. Curr Top Behav Neurosci 8:15–36

56. Heijkoop R, Huijgens PT, Snoeren EMS (2018) Assessment of sexual behavior in rats: the potentials and pitfalls. Behav Brain Res 352:70–80. https://doi.org/10.1016/j.bbr.2017.10.029

57. Hernandez C, Sabin M, Riede T (2017) Rats concatenate 22 kHz and 50 kHz calls into a single utterance. J Exp Biol 220(Pt 5):814–821. https://doi.org/10.1242/jeb.151720

58. Brudzynski SM (2009) Communication of adult rats by ultrasonic vocalization: biological, sociobiological, and neuroscience approaches. ILAR J 50(1):43–50. https://doi.org/10.1093/ilar.50.1.43

59. Willadsen M, Seffer D, Schwarting RKW, Wöhr M (2014) Rodent ultrasonic communication: male prosocial 50-khz ultrasonic vocalizations elicit social approach behavior in female rats (rattus norvegicus). J Comp Psychol 128(1):56–64. https://doi.org/10.1037/a0034778

60. Snoeren EMS, Ågmo A (2014) The incentive value of males' 50-khz ultrasonic vocalizations for female rats (rattus norvegicus). J Comp Psychol 28(1):40–55. https://doi.org/10.1037/a0033204

61. Barfield RJ, Geyer LA (1972) Sexual behavior: ultrasonic postejaculatory song of the male rat. Science 176(4041):1349–1350. https://doi.org/10.1126/science.176.4041.1349

62. Barfield RJ, Geyer LA (1975) The ultrasonic postejaculatory vocalization and postejaculatory refractory period of the male rat. J Comp Physiol Psychol 88(2):723–734. https://doi.org/10.1037/h0076435

63. Ågmo A, Snoeren EMS (2015) Silent or vocalizing rats copulate in a similar manner. PLoS One 10(12):e0144164. https://doi.org/10.1371/journal.pone.0144164

64. Olivier B, Mos J (1988) Effects of psychotropic drugs on sexual behaviour in male rats. In: Olivier B, Mos J (eds) Depression, anxiety and aggression: preclinical and clinical interfaces. Medidact, Houten, pp 121–134

65. Mos J, Olivier B, Bloetjes BK, Poth M (1990) Drug-induced facilitation of sexual behavior in the male rat: behavioural and pharmacological aspects. In: Slob MJ, Baum AK (eds) Psychoneuroendocrinology of growth and development. Medicom, Rotterdam, pp 221–232

66. Mos J, Mollet I, Tolboom JTBM, Waldinger MD, Olivier B (1999) A comparison of the effects of different serotonin reuptake blockers on sexual behaviour of the male rat. Eur Neuropsychopharmacol 9(1-2):123–135. https://doi.org/10.1016/S0924-977X(98)00015-7

67. Pattij T, De Jong TR, Uitterdijk A, Waldinger MD, Veening JG, Cools AR, Van Der Graaf PH, Olivier B (2005) Individual differences in male rat ejaculatory behaviour: searching for models to study ejaculation disorders. Eur J Neurosci 22(3):724–734. https://doi.org/10.1111/j.1460-9568.2005.04252.x

68. Olivier B, Chan JSW, Pattij T, De Jong TR, Oosting RS, Veening JG, Waldinger MD (2006) Psychopharmacology of male rat sexual behavior: modeling human sexual dysfunctions? Int J Impot Res 18(Suppl 1):S14–S23. https://doi.org/10.1038/sj.ijir.3901330

69. Chan JSW, Olivier B, de Jong TR, Snoeren EMS, Kooijman E, van Hasselt FN, Limpens JHW, Kas MJH, Waldinger MD, Oosting RS (2008) Translational research into sexual disorders: pharmacology and genomics. Eur J Pharmacol 585(2-3):426–435. https://doi.org/10.1016/j.ejphar.2008.02.098

70. Millan MJ (2006) Multi-target strategies for the improved treatment of depressive states: conceptual foundations and neuronal substrates, drug discovery and therapeutic

application. Pharmacol Ther 110(2): 135–370. https://doi.org/10.1016/j. pharmthera.2005.11.006

71. Waldinger MD, Zwinderman AH, Olivier B (2001) Antidepressants and ejaculation: a double-blind, randomized, placebo-controlled, fixed-dose study with paroxetine, sertraline, and nefazodone. J Clin Psychopharmacol 21(3):293–297. https://doi.org/10.1097/00004714-200106000-00007

72. Waldinger MD, Zwinderman AH, Olivier B (2004) On-demand treatment of premature ejaculation with clomipramine and paroxetine: a randomized, double-blind fixed-dose study with stopwatch assessment. Eur Urol 46(4):510–515. https://doi.org/10.1016/j. eururo.2004.05.005

73. Waldinger MD, Scweitzer DH, Olivier B (2005) On- demand SSRI treatment of premature ejaculation: Pharmacodynamic limitations for relevant ejaculation delay and consequent solutions. J Sex Med 2(1): 121–131. https://doi.org/10.1111/j. 1743-6109.2005.20112.x

74. Dekeyne A, Mannoury La Cour C, Gobert A, Brocco M, Lejeune F, Serres F, Sharp T, Daszuta A, Soumier A, Papp M, Rivet JM, Flik G, Cremers TI, Muller O, Lavielle G, Millan MJ (2008) S32006, a novel 5-HT2C receptor antagonist displaying broad-based antidepressant and anxiolytic properties in rodent models. Psychopharmacology 199(4):549–568. https://doi.org/10.1007/s00213-008-1177-9

75. Dorevitch A, Davis H (1994) Fluvoxamine-associated sexual dysfunction. Ann Pharmacother 28(7-8):872–874. https://doi.org/10.1177/106002809402800709

76. Hsu JH, Shen WW (1995) Male sexual side effects associated with antidepressants: a descriptive clinical study of 32 patients. Int J Psychiatry Med 25(2):191–201. https://doi.org/10.2190/1DHU-Y7L7-9GKG-V7WV

77. Montejo-Gonzalez AL, Liorca G, Izquierdo JA, Ledesma A, Bousono M, Calcedo A, Carrasco JL, Ciudad J, Daniel E, de la Gandara J, Derecho J, Franco M, Gomez MJ, Macias JA, Martin T, Perez V, Sanchez JM, Sanchez S, Vicens E (1997) SSRI-induced sexual dysfunction: fluoxetine, paroxetine, sertraline, and fluvoxamine in a prospective, multicenter, and descriptive clinical study of 344 patients. J Sex Marital Ther 23(3):176–194. https://doi.org/10.1080/00926239708403923

78. Lee KU, Lee YM, Nam JM, Lee HK, Kweon YS, Lee CT, Jun TY (2010) Antidepressant-induced sexual dysfunction among newer antidepressants in a naturalistic setting.

Psychiatry Investig 7(1):55–59. https://doi.org/10.4306/pi.2010.7.1.55

79. Kennedy SH, Eisfeld BS, Dickens SE, Bacchiochi JR, Bagby RM (2000) Antidepressant-induced sexual dysfunction during treatment with moclobemide, paroxetine, sertraline, and venlafaxine. J Clin Psychiatry 61(4):276–281. https://doi.org/10.4088/JCP.v61n0406

80. Nieuwstraten CE, Dolovich LR (2001) Bupropion versus selective serotonin-reuptake inhibitors for treatment of depression. Ann Pharmacother 35(12):1608–1613. https://doi.org/10.1345/aph.1A099

81. Clayton AH, Pradko JF, Croft HA, Brendan Montano C, Leadbetter RA, Bolden-Watson-C, Bass KI, Donahue RMJ, Jamerson BD, Metz A (2002) Prevalence of sexual dysfunction among newer antidepressants. J Clin Psychiatry 63(4):357–366. https://doi.org/10.4088/JCP.v63n0414

82. Kennedy SH, McCann SM, Masellis M, McIntyre RS, Raskin J, McKay G, Baker GB (2002) Combining bupropion SR with venlafaxine, paroxetine, or fluoxetine: a preliminary report on pharmacokinetic, therapeutic, and sexual dysfunction effects. J Clin Psychiatry 63(3):181–186. https://doi.org/10.4088/JCP.v63n0302

83. Gobert A, Rivet JM, Cistarelli L, Melon C, Millan MJ (1999) Buspirone modulates basal and fluoxetine-stimulated dialysate levels of dopamine, noradrenaline and serotonin in the frontal cortex of freely moving rats: activation of serotonin(1A) receptors and blockade of α2-adrenergic receptors underlie its actions. Neuroscience 93(4):1251–1262. https://doi.org/10.1016/S0306-4522(99)00211-0

84. Rickels K, Rynn M (2002) Pharmacotherapy of generalized anxiety disorder. J Clin Psychiatry 63(Suppl 1):9–16

85. Gitlin M (2003) Sexual dysfunction with psychotropic drugs. Expert Opin Pharmacother 4(12):2259–2269. https://doi.org/10.1517/14656566.4.12.2259

86. Landén M, Eriksson E, Ågren H, Fahlén T (1999) Effect of buspirone on sexual dysfunction in depressed patients treated with selective serotonin reuptake inhibitors. J Clin Psychopharmacol 19(3):268–271. https://doi.org/10.1097/00004714-199906000-00012

87. Landén M, Högberg P, Thase ME (2005) Incidence of sexual side effects in refractory depression during treatment with citalopram or paroxetine. J Clin Psychiatry 66(1):

100–106. https://doi.org/10.4088/JCP.v66n0114

88. Mathes CW, Smith ER, Popa BR, Davidson JM (1990) Effects of intrathecal and systemic administration of buspirone on genital reflexes and mating behavior in male rats. Pharmacol Biochem Behav 36(1):63–68. https://doi.org/10.1016/0091-3057(90)90126-3

89. Uphouse L, Caldarola-Pastuszka M, Montanez S (1992) Intracerebral actions of the 5-HT1A agonists, 8-OH-DPAT and buspirone and of the 5-HT1A partial agonist/antagonist, NAN-190, on female sexual behavior. Neuropharmacology 31(10):969–981. https://doi.org/10.1016/0028-3908(92)90097-9

90. Fernández-Guasti A, Roldán-Roldán G, Larsson K (1991) Anxiolytics reverse the acceleration of ejaculation resulting from enforced Intercopulatory intervals in rats. Behav Neurosci 105(2):230–240. https://doi.org/10.1037/0735-7044.105.2.230

91. Skolnick P, Krieter P, Tizzano J, Basile A, Popik P, Czobor P, Lippa A (2006) Preclinical and clinical pharmacology of DOV 216,303, a "triple" reuptake inhibitor. CNS Drug Rev 12(2):123–134. https://doi.org/10.1111/j.1527-3458.2006.00123.x

92. Millan MJ (2009) Dual- and triple-acting agents for treating Core and co-morbid symptoms of major depression: novel concepts, new drugs. Neurotherapeutics 6(1):53–77. https://doi.org/10.1016/j.nurt.2008.10.039

93. Millan MJ (2005) Serotonin 5-HT2C receptors as a target for the treatment of depressive and anxious states: focus on novel therapeutic strategies. Therapie 60(5):441–460. https://doi.org/10.2515/therapie:2005065

94. Millan MJ, Brocco M, Gobert A, Dekeyne A (2005) Anxiolytic properties of agomelatine, an antidepressant with melatoninergic and serotonergic properties: role of 5-HT2C receptor blockade. Psychopharmacology 177(4):448–458. https://doi.org/10.1007/s00213-004-1962-z

95. Nic Dhonnchadha BA, Ripoll N, Clénet F, Hascoët M, Bourin M (2005) Implication of 5-HT2 receptor subtypes in the mechanism of action of antidepressants in the four plates test. Psychopharmacology 179(2):418–429. https://doi.org/10.1007/s00213-004-2044-y

96. Zupancic M, Guilleminault C (2006) Agomelatine: a preliminary review of a new antidepressant. CNS Drugs 20(12):981–992. https://doi.org/10.2165/00023210-200620120-00003

97. Foreman MM, Fuller RW, Nelson DL, Calligaro DO, Kurz KD, Misner JW, Garbrecht WL, Parli CJ (1992) Preclinical studies on LY237733, a potent and selective serotonergic antagonist. J Pharmacol Exp Ther 260(1):51–57

98. Waldinger MD, Olivier B (2005) Animal models of premature and retarded ejaculation. World J Urol 23(2):115–118. https://doi.org/10.1007/s00345-004-0493-x

99. Chan JSW, Waldinger MD, Olivier B, Oosting RS (2010) Drug-induced sexual dysfunction in rats. Curr Protoc Neurosci. Chapter 9:Unit 9.34. https://doi.org/10.1002/0471142301.ns0934s53

100. Bijlsma EY, Chan JSW, Olivier B, Veening JG, Millan MJ, Waldinger MD, Oosting RS (2014) Sexual side effects of serotonergic antidepressants: mediated by inhibition of serotonin on central dopamine release? Pharmacol Biochem Behav. https://doi.org/10.1016/j.pbb.2013.10.004

101. Oosting RS, Chan JS, Olivier B, Banerjee P, Choi YK, Tarazi F (2016) Differential effects of vilazodone versus citalopram and paroxetine on sexual behaviors and serotonin transporter and receptors in male rats. Psychopharmacology 233(6):1025–1034. https://doi.org/10.1007/s00213-015-4198-1

102. Rothschild AJ (1995) Selective serotonin reuptake inhibitor-induced sexual dysfunction: efficacy of a drug holiday. Am J Psychiatry 152(10):1514–1516. https://doi.org/10.1176/ajp.152.10.1514

103. Bobes J, González MP, Bascarán MT, Clayton A, Garcia M, Rico-Villade Moros F, Banús S (2002) Evaluating changes in sexual functioning in depressed patients: sensitivity to change of the CSFQ. J Sex Marital Ther 28(2):93–103. https://doi.org/10.1080/00926230252851852

104. Baldwin D, Hutchison J, Donaldson K, Shaw B, Smithers A (2008) Selective serotonin re-uptake inhibitor treatment-emergent sexual dysfunction: randomized double-blind placebo-controlled parallel-group fixed-dose study of a potential adjuvant compound, VML-670. J Psychopharmacol 22(1):55–63. https://doi.org/10.1177/0269881107078490

105. Gartlehner G, Thieda P, Hansen RA, Gaynes BN, DeVeaugh-Geiss A, Krebs EE, Lohr KN (2008) Comparative risk for harms of second-generation antidepressants: a systematic review and meta-analysis. Drug Saf 31(10):

851–865. https://doi.org/10.2165/00002018-200831100-00004

106. Waldinger MD, Olivier B (1998) Selective serotonin reuptake inhibitor-induced sexual dysfunction: clinical and research considerations. Int Clin Psychopharmacol 13(Suppl 6): S27–S33. https://doi.org/10.1097/00004850-199802002-00005

107. Remick RA (2002) Diagnosis and management of depression in primary care: a clinical update and review. CMAJ 167:1253–1260

108. Manning C, Marr J (2003) "Real-life burden of depression" surveys - GP and patient perspectives on treatment and management of recurrent depression. Curr Med Res Opin 19(6):526–531. https://doi.org/10.1185/030079903125002117

109. Stahl SM (1998) Mechanism of action of serotonin selective reuptake inhibitors. Serotonin receptors and pathways mediate therapeutic effects and side effects. J Affect Disord 51(3):215–235. https://doi.org/10.1016/S0165-0327(98)00221-3

110. Shelton RC (2019) Serotonin and norepinephrine reuptake inhibitors. In: Handbook of experimental pharmacology. Springer Nature, Switzerland, pp 145–180

111. Kiliç S, Ergin H, Baydinç YC (2005) Venlafaxine extended release for the treatment of patients with premature ejaculation: a pilot, single-blind, placebo-controlled, fixed-dose crossover study on short-term administration of an antidepressant drug. Int J Androl 28(1): 47–52. https://doi.org/10.1111/j.1365-2605.2005.00507.x

112. Safarinejad MR (2008) Safety and efficacy of venlafaxine in the treatment of premature ejaculation: a double-blind, placebo-controlled, fixed-dose, randomised study. Andrologia 40(1):49–55. https://doi.org/10.1111/j.1439-0272.2008.00813.x

113. Stahl SM, Pradko JF, Haight BR, Modell JG, Rockett CB, Learned-Coughlin S (2004) A review of the neuropharmacology of bupropion, a dual norepinephrine and dopamine reuptake inhibitor. Prim Care Companion J Clin Psychiatry 6(4):159–166. https://doi.org/10.4088/pcc.v06n0403

114. Olivier JDA, de Jong TR, Jos Dederen P, van Oorschot R, Heeren D, Pattij T, Waldinger MD, Coolen LM, Cools AR, Olivier B, Veening JG (2007) Effects of acute and chronic apomorphine on sex behavior and copulation-induced neural activation in the male rat. Eur J Pharmacol 576(1-3):61–76. https://doi.org/10.1016/j.ejphar.2007.08.019

115. Fatemi SH, Emamian ES, Kist DA (1999) Venlafaxine and bupropion combination therapy in a case of treatment- resistant depression. Ann Pharmacother 33(6):701–703. https://doi.org/10.1345/aph.18249

116. Skolnick P, Popik P, Janowsky A, Beer B, Lippa AS (2003) Antidepressant-like actions of DOV 21,947: a "triple" reuptake inhibitor. Eur J Pharmacol 461(2-3):99–104. https://doi.org/10.1016/S0014-2999(03)01310-4

117. Paterson NE, Balci F, Campbell U, Olivier BE, Hanania T (2011) The triple reuptake inhibitor DOV216,303 exhibits limited antidepressant-like properties in the differential reinforcement of low-rate 72-second responding assay, likely due to dopamine reuptake inhibition. J Psychopharmacol 25: 1357–1364

118. Breuer ME, Chan JSW, Oosting RS, Groenink L, Korte SM, Campbell U, Schreiber R, Hanania T, Snoeren EMS, Waldinger M, Olivier B (2008) The triple monoaminergic reuptake inhibitor DOV 216,303 has antidepressant effects in the rat olfactory bulbectomy model and lacks sexual side effects. Eur Neuropsychopharmacol 18(12):908–916. https://doi.org/10.1016/j.euroneuro.2008.07.011

119. Millan MJ, Gobert A, Lejeune F, Dekeyne A, Newman-Tancredi A, Pasteau V, Rivet JM, Cussac D (2003) The novel melatonin agonist agomelatine (S20098) is an antagonist at 5-hydroxytryptamine2C receptors, blockade of which enhances the activity of frontocortical dopaminergic and adrenergic pathways. J Pharmacol Exp Ther 306(3):954–956. https://doi.org/10.1124/jpet.103.051797

120. Michelson D, Bancroft J, Targum S, Yongman K, Tepner R (2000) Female sexual dysfunction associated with antidepressant administration: a randomized, placebo-controlled study of pharmacologic intervention. Am J Psychiatry 157(2):239–243. https://doi.org/10.1176/appi.ajp.157.2.239

121. Padoin MJ, Lucion AB (1995) The effect of testosterone and DOI (1-(2,5-dimethoxy-4-iodophenyl)-2-aminopropane) on male sexual behavior of rats. Eur J Pharmacol 277:1–6

122. Sanchez C, Asin KE, Artigas F (2015) Vortioxetine, a novel antidepressant with multimodal activity: review of preclinical and clinical data. Pharmacol Ther 145:43–57. https://doi.org/10.1016/j.pharmthera.2014.07.001

123. Li Y, Pehrson AL, Oosting RS, Gulinello M, Olivier B, Sanchez C (2017) A study of time- and sex-dependent effects of vortioxetine on

rat sexual behavior: possible roles of direct receptor modulation. Neuropharmacology 121:89–99. https://doi.org/10.1016/j.neuropharm.2017.04.017

124. Jacobsen PL, Mahableshwarkar AR, Chen Y, Chrones L, Clayton AH (2015) Effect of vortioxetine vs. escitalopram on sexual functioning in adults with well-treated major depressive disorder experiencing ssri-induced sexual dysfunction. J Sex Med 12(10):2036–2048. https://doi.org/10.1111/jsm.12980

125. Jacobsen P, Zhong W, Nomikos G, Clayton A (2019) Paroxetine, but not Vortioxetine, impairs sexual functioning compared with placebo in healthy adults: a randomized, controlled trial. J Sex Med 16(10):1638–1649. https://doi.org/10.1016/j.jsxm.2019.06.018

126. Dawson LA, Watson JM (2009) Vilazodone: a 5-HT1A receptor agonist/serotonin transporter inhibitor for the treatment of affective disorders. CNS Neurosci Ther 15(2):107–117. https://doi.org/10.1111/j.1755-5949.2008.00067.x

127. Clayton AH, Kennedy SH, Edwards JB, Gallipoli S, Reed CR (2013) The effect of vilazodone on sexual function during the treatment of major depressive disorder. J Sex Med 10(10):2465–2476. https://doi.org/10.1111/jsm.12004

128. Clayton AH, Gommoll C, Chen D, Nunez R, Mathews M (2015) Sexual dysfunction during treatment of major depressive disorder with vilazodone, citalopram, or placebo: results from a phase IV clinical trial. Int Clin Psychopharmacol 30(4):216–223. https://doi.org/10.1097/YIC.0000000000000075

129. Clayton AH, Durgam S, Li D, Chen C, Chen L, Mathews M, Gommoll CP, Szegedi A (2017) Effects of vilazodone on sexual functioning in healthy adults. Int Clin Psychopharmacol 32:27–35

130. Zhang L-M, Wang X-Y, Zhao N, Wang Y-L, Hu X-X, Ran Y-H, Liu Y-Q, Zhang Y-Z, Yang R-F, Li Y-F (2017) Neurochemical and behavioural effects of hypidone hydrochloride (YL-0919): a novel combined selective 5-HT reuptake inhibitor and partial 5-HT 1A agonist. Br J Pharmacol 174:769–780

131. Matthiesen T, Wöhrmann T, Coogan TP, Uragg H (1998) The experimental toxicology of tramadol: an overview. Toxicol Lett 95(1):63–71. https://doi.org/10.1016/S0378-4274(98)00023-X

132. Rojas-Corrales MO, Gibert-Rahola J, Micó JA (1998) Tramadol induces antidepressant-type effects in mice. Life Sci 63(12): PL175–PL180. https://doi.org/10.1016/S0024-3205(98)00369-5

133. Eassa BI, El-Shazly MA (2013) Safety and efficacy of tramadol hydrochloride on treatment of premature ejaculation. Asian J Androl 15(1):138–142. https://doi.org/10.1038/aja.2012.96

134. Yang L, Qian S, Liu H, Liu L, Pu C, Han P, Wei Q (2013) Role of tramadol in premature ejaculation: a systematic review and meta-analysis. Urol Int 91(2):197–205. https://doi.org/10.1159/000348826

135. Waldinger MD (2018) Drug treatment options for premature ejaculation. Expert Opin Pharmacother 19(10):1077–1085. https://doi.org/10.1080/14656566.2018.1494725

136. Olivier J, Esquivel Franco DC, Oosting R, Waldinger M, Sarnyai Z, Olivier B (2017) Tramadol: effects on sexual behavior in male rats are mainly caused by its 5-HT reuptake blocking effects. Neuropharmacology 116:50–58. https://doi.org/10.1016/j.neuropharm.2016.11.020

137. Abdel-Hamid IA, Andersson KE, Waldinger MD, Anis TH (2016) Tramadol abuse and sexual function. Sex Med Rev 4(3):235–246. https://doi.org/10.1016/j.sxmr.2015.10.014

138. Smith WT, Glaudin V, Panagides J, Gilvary E (1990) Mirtazapine vs. amitriptyline vs. Placebo in the treatment of major depressive disorder. Psychopharmacol Bull 26(2):191–196

139. Kent JM (2000) SNaRIs, NaSSAs, and NaRIs: new agents for the treatment of depression. Lancet 355(9207):911–918. https://doi.org/10.1016/S0140-6736(99)11381-3

140. De Boer T, Ruigt GSF, Berendsen HHG (1995) The α2-selective adrenoceptor antagonist org 3770 (mirtazapine, Remeron®) enhances noradrenergic and serotonergic transmission. Hum Psychopharmacol Clin Exp 10(Suppl 2):S107–S118. https://doi.org/10.1002/hup.470100805

141. Benelli A, Frigeri C, Bertolini A, Genedani S (2004) Influence of mirtazapine on the sexual behavior of male rats. Psychopharmacology 171(3):250–258. https://doi.org/10.1007/s00213-003-1591-y

142. Boyarsky BK, Haque W, Rouleau MR, Hirschfeld RMA (1999) Sexual functioning in depressed outpatients taking mirtazapine. Depress. Anxiety 9(4):175–179. https://doi.org/10.1002/(SICI)1520-6394(1999)9:4<175::AID-DA5>3.0.CO;2-0

143. Farah A (1999) Relief of SSRI-induced sexual dysfunction with mirtazapine treatment. J Clin Psychiatry 60(4):260–261. https://doi.org/10.4088/JCP.v60n0412a

144. Auffret M, Drapier S, Véris M (2019) New tricks for an old dog: a repurposing approach of apomorphine. Eur J Pharmacol 843:66–79

145. Hull EM, Lorrain DS, Du J, Matuszewich L, Lumley LA, Putnam SK, Moses J (1999) Hormone-neurotransmitter interactions in the control of sexual behavior. Behav Brain Res 105(1):105–116. https://doi.org/10.1016/S0166-4328(99)00086-8

146. Guadarrama-Bazante IL, Canseco-Alba A, Rodríguez-Manzo G (2014) Dopamine receptors play distinct roles in sexual behavior expression of rats with a different sexual motivational tone. Behav Pharmacol 25(7):684–694. https://doi.org/10.1097/FBP.0000000000000086

147. Olivier B, Soudijn W, Van Wijngaarden I (1999) The 5-HT(1A) receptor and its ligands: structure and function. Prog Drug Res 52:103–165. https://doi.org/10.1007/978-3-0348-8730-4_3

148. du Jardin KG, Jensen JB, Sanchez C, Pehrson AL (2014) Vortioxetine dose-dependently reverses 5-HT depletion-induced deficits in spatial working and object recognition memory: a potential role for 5-HT1A receptor agonism and 5-HT3 receptor antagonism. Eur Neuropsychopharmacol 24(1):160–171. https://doi.org/10.1016/j.euroneuro.2013.07.001

149. Ahlenius S, Larsson K, Wijkström A (1991) Behavioral and biochemical effects of the 5-HT1A receptor agonists flesinoxan and 8-OH-DPAT in the rat. Eur J Pharmacol 200(2-3):259–266. https://doi.org/10.1016/0014-2999(91)90580-J

150. Hensler JG (2003) Regulation of 5-HT1A receptor function in brain following agonist or antidepressant administration. Life Sci 72(15):1665–1682. https://doi.org/10.1016/S0024-3205(02)02482-7

151. Tanco SA, Watson NV, Gorzalka BB (1993) Lack of effects of 5-HT3 antagonists on normal and morphine-attenuated sexual behaviours in female and male rats. Experientia 49(3):238–241. https://doi.org/10.1007/BF01923532

152. Larsson K, Heimer L (1964) Mating behaviour of male rats after lesions in the preoptic area. Nature 202:413–414. https://doi.org/10.1038/202413a0

153. Christensen LW, Nance DM, Gorski RA (1977) Effects of hypothalamic and preoptic lesions on reproductive behavior in male rats. Brain Res Bull 2(2):137–141. https://doi.org/10.1016/0361-9230(77)90010-7

154. Van Dis H, Larsson K (1971) Induction of sexual arousal in the castrated male rat by intracranial stimulation. Physiol Behav 6(1):85–86. https://doi.org/10.1016/0031-9384(71)90021-7

155. Malsbury CW (1971) Facilitation of male rat copulatory behavior by electrical stimulation of the medial preoptic area. Physiol Behav 7(6):797–805. https://doi.org/10.1016/0031-9384(71)90042-4

156. Normandin JJ, Murphy AZ (2011) Excitotoxic lesions of the nucleus paragigantocellularis facilitate male sexual behavior but attenuate female sexual behavior in rats. Neuroscience 175:212–223. https://doi.org/10.1016/j.neuroscience.2010.11.030

157. Yells DP, Hendricks SE, Prendergast MA (1992) Lesions of the nucleus paragigantocellularis: effects on mating behavior in male rats. Brain Res 596(1-2):73–79. https://doi.org/10.1016/0006-8993(92)91534-L

158. Yells DP, Prendergast MA, Hendricks SE, Nakamura M (1994) Fluoxetine-induced inhibition of male rat copulatory behavior: modification by lesions of the nucleus paragigantocellularis. Pharmacol Biochem Behav 49(1):121–127. https://doi.org/10.1016/0091-3057(94)90465-0

159. Normandin JJ, Murphy AZ (2011) Serotonergic lesions of the periaqueductal gray, a primary source of serotonin to the nucleus paragigantocellularis, facilitate sexual behavior in male rats. Pharmacol Biochem Behav 98(3):369–375. https://doi.org/10.1016/j.pbb.2011.01.024

160. Uphouse L, Guptarak J (2010) Serotonin and sexual behavior. In: Müller C, Jacobs B (eds) Handbook of behavioral neurobiology of serotonin, vol 21. Elsevier, Amsterdam, pp 347–365

161. Bitran D, Hull EM (1987) Pharmacological analysis of male rat sexual behavior. Neurosci Biobehav Rev 11(4):365–389. https://doi.org/10.1016/S0149-7634(87)80008-8

162. Haensel SM, Mos J, Olivier B, Slob AK (1991) Sex behavior of male and female wistar rats affected by the serotonin agonist 8-OH-DPAT. Pharmacol Biochem Behav 40(2):221–228. https://doi.org/10.1016/0091-3057(91)90543-B

163. Ahlenius S, Larsson K, Svensson L, Hjorth S, Carlsson A, Lindberg P, Wikström H, Sanchez D, Arvidsson LE, Hacksell U, Nilsson JLG (1981) Effects of a new type of 5-HT

receptor agonist on male rat sexual behavior. Pharmacol Biochem Behav 15(5):785–792. https://doi.org/10.1016/0091-3057(81) 90023-X

164. Ahlenius S, Larsson K (1997) Specific involvement of central 5-HT(1A) receptors in the mediation of male rat ejaculatory behavior. Neurochem Res 22(8): 1065–1070. https://doi.org/10.1023/A:1022443413745

165. De Jong TR, Pattij T, Veening JG, Dederen PJWC, Waldinger MD, Cools AR, Olivier B (2005) Effects of chronic paroxetine pretreatment on (±)-8-hydroxy-2-(di-n-propyl-amino)tetralin induced c-fos expression following sexual behavior. Neuroscience 134(4):1351–1361. https://doi.org/10.1016/j.neuroscience.2005.05.012

166. De Jong TR, Pattij T, Veening JG, Waldinger MD, Cools AR, Olivier B (2005) Effects of chronic selective serotonin reuptake inhibitors on 8-OH-DPAT-induced facilitation of ejaculation in rats: comparison of fluvoxamine and paroxetine. Psychopharmacology 179(2): 509–515. https://doi.org/10.1007/s00213-005-2186-6

167. Esquivel-Franco DC, de Boer SF, Waldinger M, Olivier B, Olivier JDA (2020) Pharmacological studies on the role of 5-HT1A receptors in male sexual behavior of wildtype and serotonin transporter knockout rats. Front Behav Neurosci 14:40. https://doi.org/10.3389/fnbeh.2020.00040

168. Chan JSW, Snoeren EMS, Cuppen E, Waldinger MD, Olivier B, Oosting RS (2011) The serotonin transporter plays an important role in male sexual behavior: a study in serotonin transporter knockout rats. J Sex Med 8(1): 97–108. https://doi.org/10.1111/j.1743-6109.2010.01961.x

169. Fernández-Guasti A, Escalante A (1991) Role of presynaptic serotonergic receptors on the mechanism of action of 5-HT1A and 5-HT1B agonists on masculine sexual behaviour: physiological and pharmacological implications. J Neural Transm 85(2): 95–107. https://doi.org/10.1007/BF01244702

170. Hull EM, Muschamp JW, Sato S (2004) Dopamine and serotonin: influences on male sexual behavior. Physiol Behav 83(2): 291–307. https://doi.org/10.1016/j.physbeh.2004.08.018

171. Wise RA, Bozarth MA (1985) Brain mechanisms of drug reward and euphoria. Psychiatr Med 3(4):445–460

172. Hull EM, Dominguez JM (2007) Sexual behavior in male rodents. Horm Behav 52(1):45–55. https://doi.org/10.1016/j.yhbeh.2007.03.030

173. Melis MR, Argiolas A (1995) Dopamine and sexual behavior. Neurosci Biobehav Rev 19(1):19–38. https://doi.org/10.1016/0149-7634(94)00020-2

174. Waldinger MD, Quinn P, Dilleen M, Mundayat R, Schweitzer DH, Boolell M (2005) A multinational population survey of intravaginal ejaculation latency time. J Sex Med 2(4):492–497. https://doi.org/10.1111/j.1743-6109.2005.00070.x

175. Waldinger MD, McIntosh J, Schweitzer DH (2009) A five-nation survey to assess the distribution of the intravaginal ejaculatory latency time among the general male population. J Sex Med 6(10):2888–2895. https://doi.org/10.1111/j.1743-6109.2009.01392.x

176. Berendsen HHG, Jenck F, Broekkamp CLE (1989) Selective activation of 5HT1A receptors induces lower lip retraction in the rat. Pharmacol Biochem Behav 33(4):821–827. https://doi.org/10.1016/0091-3057(89) 90477-2

177. Berendsen HHG, Broekkamp CLE, Van Delft AML (1990) Antagonism of 8-OH-DPAT-induced behaviour in rats. Eur J Pharmacol 187(1):97–103. https://doi.org/10.1016/0014-2999(90)90344-6

178. Sura A, Overstreet DH, Marson L (2001) Selectively bred male rat lines differ in naïve and experienced sexual behavior. Physiol Behav 72(1-2):13–20. https://doi.org/10.1016/S0031-9384(00)00300-0

179. Hurwitz ZE, Riley AL (2011) The differential expression of male sexual behavior in the Lewis, Fischer and Sprague-Dawley rat strains. Learn Behav 39(1):36–45. https://doi.org/10.3758/s13420-010-0006-2

180. Overstreet DH, Rezvani AH, Pucilowski O, Gause L, Janowsky DS (1994) Rapid selection for serotonin-1a sensitivity in rats. Psychiatr Genet 4(1):57–62. https://doi.org/10.1097/00041444-199421000-00008

181. Overstreet DH, Rezvani AM, Knapp DJ, Crews FT, Janowsky DS (1996) Further selection of rat lines differing in 5-HT-1A receptor sensitivity: behavioral and functional correlates. Psychiatr Genet 6(3):107–117. https://doi.org/10.1097/00041444-199623000-00002

182. Knapp DJ, Overstreet DH, Crews FT (1998) Brain 5-HT(1A) receptor autoradiography and hypothermic responses in rats bred for differences in 8-OH-DPAT sensitivity. Brain

Res 782(1-2):1–10. https://doi.org/10.1016/S0006-8993(97)01127-X

183. Schijven D, Sousa VC, Roelofs J, Olivier B, Olivier JDA (2014) Serotonin 1A receptors and sexual behavior in a genetic model of depression. Pharmacol Biochem Behav 121:82–87. https://doi.org/10.1016/j.pbb.2013.12.012

184. Overstreet DH, Wegener G (2013) The flinders sensitive line rat model of depression-25 years and still producing. Pharmacol Rev 65(1):143–155. https://doi.org/10.1124/pr.111.005397

185. El Khoury A, Gruber SHM, Mørk A, Mathé AA (2006) Adult life behavioral consequences of early maternal separation are alleviated by escitalopram treatment in a rat model of depression. Prog Neuropsychopharmacol Biol Psychiatry 30(3):535–540. https://doi.org/10.1016/j.pnpbp.2005.11.011

186. Eriksson TM, Delagrange P, Spedding M, Popoli M, Mathé AA, Ögren SO, Svenningsson P (2012) Emotional memory impairments in a genetic rat model of depression: involvement of 5-HT/MEK/arc signaling in restoration. Mol Psychiatry 17(2):173–184. https://doi.org/10.1038/mp.2010.131

187. Burns-Cusato M, Scordalakes EM, Rissman EF (2004) Of mice and missing data: what we know (and need to learn) about male sexual behavior. Physiol Behav 83(2):217–232. https://doi.org/10.1016/j.physbeh.2004.08.015

188. Esquivel-Franco DC, Olivier B, Waldinger MD, Gutiérrez-Ospina G, Olivier JDA (2018) Tramadol's inhibitory effects on sexual behavior: pharmacological studies in serotonin transporter knockout rats. Front Pharmacol 9:676. https://doi.org/10.3389/fphar.2018.00676

189. Kennedy SH, Rizvi S (2009) Sexual dysfunction, depression, and the impact of antidepressants. J Clin Psychopharmacol 29(2):157–164. https://doi.org/10.1097/JCP.0b013e31819c76e9

190. Meyer JH (2007) Imaging the serotonin transporter during major depressive disorder and antidepressant treatment. J Psychiatry Neurosci 32(2):86–102. https://doi.org/10.1016/S1180-4882(07)50013-2

191. Homberg JR, Olivier JDA, Smits BMG, Mul JD, Mudde J, Verheul M, Nieuwenhuizen OFM, Cools AR, Ronken E, Cremers T, Schoffelmeer ANM, Ellenbroek BA, Cuppen E (2007) Characterization of the serotonin transporter knockout rat: a selective change in the functioning of the serotonergic system. Neuroscience 146(4):1662–1676. https://doi.org/10.1016/j.neuroscience.2007.03.030

192. Matuszcyk JV, Larsson K, Eriksson E (1998) The selective serotonin reuptake inhibitor fluoxetine reduces sexual motivation in male rats. Pharmacol Biochem Behav 60(2):527–532. https://doi.org/10.1016/S0091-3057(98)00010-0

193. Geng H, Peng D, Huang Y, Tang D, Gao J, Zhang Y, Zhang X (2019) Changes in sexual performance and biochemical characterisation of functional neural regions: a study in serotonin transporter knockout male rats. Andrologia 51(7):e13291. https://doi.org/10.1111/and.13291

194. Olivier JDA, Olivier B (2020) Antidepressants and sexual dysfunction: translational aspects. Curr Sex Health Rep 11(156–166):121–140

195. de Jong TR, Pattij T, Veening JG, Dederen PJWC, Waldinger MD, Cools AR, Olivier B (2005) Citalopram combined with WAY 100635 inhibits ejaculation and ejaculation-related Fos immunoreactivity. Eur J Pharmacol 509:49–59

196. Looney C, Thor KB, Ricca D, Marson L (2005) Differential effects of simultaneous or sequential administration of paroxetine and WAY-100,635 on ejaculatory behavior. Pharmacol Biochem Behav 82:427–433

197. Sniecikowska J, Newman-Tancredi A, Kolaczkowski M (2019) From receptor selectivity to functional selectivity: the rise of biased Agonism in 5-HT1A receptor drug discovery. Curr Top Med Chem 19(26):2393–2420. https://doi.org/10.2174/1568026619666190911122040

198. Garcia-Garcia AL, Newman-Tancredi A, Leonardo ED (2014) P5-HT1A receptors in mood and anxiety: recent insights into autoreceptor versus heteroreceptor function. Psychopharmacology 231(4):623–636. https://doi.org/10.1007/s00213-013-3389-x

199. Hennies HH, Friderichs E, Schneider J (1988) Receptor binding, analgesic and antitussive potency of tramadol and other selected opioids. Arzneimittelforschung 38(7):877–880

200. Frink MC, Hennies HH, Englberger W, Haurand M, Wilffert B (1996) Influence of tramadol on neurotransmitter systems of the rat brain. Arzneimittelforschung 46(11):1029–1036

201. Bar-Or D, Salottolo KM, Orlando A, Winkler JV (2012) A randomized double-blind, placebo-controlled multicenter study to evaluate the efficacy and safety of two doses of the

tramadol orally disintegrating tablet for the treatment of premature ejaculation within less than 2 minutes. Eur Urol 61(4): 736–743. https://doi.org/10.1016/j.eururo.2011.08.039

202. Adinolfi A, Zelli S, Leo D, Carbone C, Mus L, Illiano P, Alleva E, Gainetdinov RR, Adriani W (2019) Behavioral characterization of DAT-KO rats and evidence of asocial-like phenotypes in DAT-HET rats: the potential involvement of norepinephrine system. Behav Brain Res 359:516–527. https://doi.org/10.1016/j.bbr.2018.11.028

203. Cinque S, Zoratto F, Poleggi A, Leo D, Cerniglia L, Cimino S, Tambelli R, Alleva E, Gainetdinov RR, Laviola G, Adriani W (2018) Behavioral phenotyping of dopamine transporter knockout rats: compulsive traits, motor stereotypies, and anhedonia. Front. Psychiatry 9:43. https://doi.org/10.3389/fpsyt.2018.00043

204. Leo D, Sukhanov I, Zoratto F, Illiano P, Caffino L, Sanna F, Messa G, Emanuele M, Esposito A, Dorofeikova M, Budygin EA, Mus L, Efimova E, Niello M, Espinoza S, Sotnikova TD, Hoener MC, Laviola G, Fumagalli F, Adriani W, Gainetdinov RR (2018) Pronounced hyperactivity, cognitive dysfunctions, and BDNF dysregulation in dopamine transporter knock-out rats. J Neurosci 38(8):1959–1972. https://doi.org/10.1523/JNEUROSCI.1931-17.2018

205. Sanna F, Bratzu J, Serra MP, Leo D, Quartu M, Boi M, Espinoza S, Gainetdinov RR, Melis MR, Argiolas A (2020) Altered sexual behavior in dopamine transporter (DAT) knockout male rats: a behavioral, Neurochemical and Intracerebral Microdialysis Study. Front Behav Neurosci 14:58. https://doi.org/10.3389/fnbeh.2020.00058

206. Frohmader KS, Pitchers KK, Balfour ME, Coolen LM (2010) Mixing pleasures: review of the effects of drugs on sex behavior in humans and animal models. Horm Behav 58(1):149–162. https://doi.org/10.1016/j.yhbeh.2009.11.009

207. Pfaus JG, Wilkins MF, DiPietro N, Benibgui M, Toledano R, Rowe A, Couch MC (2010) Inhibitory and disinhibitory effects of psychomotor stimulants and depressants on the sexual behavior of male and female rats. Horm Behav 58(1):163–176. https://doi.org/10.1016/j.yhbeh.2009.10.004

208. Nathan SG (1986) The epidemiology of the DSM-III psychosexual dysfunctions. J Sex Marital Ther 12(4):267–281. https://doi.org/10.1080/00926238608415413

209. Jannini EA, Lenzi A (2005) Ejaculatory disorders: epidemiology and current approaches to definition, classification and subtyping. World J Urol 23(2):68–75. https://doi.org/10.1007/s00345-004-0486-9

210. Perelman MA, Rowland DL (2006) Retarded ejaculation. World J Urol 24(6):645–652. https://doi.org/10.1007/s00345-006-0127-6

211. Di Sante S, Mollaioli D, Gravina GL, Ciocca G, Limoncin E, Carosa E, Lenzi A, Jannini EA (2016) Epidemiology of delayed ejaculation. Transl Androl Urol 5(4):541–548. https://doi.org/10.21037/tau.2016.05.10

212. Olivier B, Van Oorschot R, Waldinger MD (1998) Serotonin, serotonergic receptors, selective serotonin reuptake inhibitors and sexual behaviour. Int Clin Psychopharmacol 13(Suppl 6):S9–S14. https://doi.org/10.1097/00004850-199807006-00003

213. Abdel-Hamid IA, Ali OI (2018) Delayed ejaculation: pathophysiology, diagnosis, and treatment. World J Mens Health 36(1): 22–40. https://doi.org/10.5534/wjmh.17051

214. Chen J (2016) The pathophysiology of delayed ejaculation. Transl Androl Urol 5(4):549–562. https://doi.org/10.21037/tau.2016.05.03

215. Otani T (2019) Clinical review of ejaculatory dysfunction. Reprod Med Biol 18(4): 331–343. https://doi.org/10.1002/rmb2.12289

216. Piche K, Mann U, Patel P (2020) Treatment of delayed ejaculation. Curr Sex Heal Rep 36(1):22–40. https://doi.org/10.1007/s11930-020-00287-z

217. Sharp T, Barnes NM (2020) Central 5-HT receptors and their function; present and future. Neuropharmacology 177:108155. https://doi.org/10.1016/j.neuropharm.2020.108155

218. De Deurwaerdère P, Bharatiya R, Chagraoui A, Di Giovanni G (2020) Constitutive activity of 5-HT receptors: factual analysis. Neuropharmacology 168:107967. https://doi.org/10.1016/j.neuropharm.2020.107967

219. de Almeida Kiguti LR, Pacheco TL, Antunes E, de G Kempinas W (2020) Lorcaserin administration has pro-ejaculatory effects in rats via 5-HT2C receptors activation: a putative pharmacologic strategy to delayed ejaculation? J Sex Med 17(6): 1060–1071. https://doi.org/10.1016/j.jsxm.2020.02.027

220. Olivier JDA, Franco DCE, Waldinger MD, Olivier B (2017) Sexual dysfunction, depression and antidepressants: a translational approach. In: Olivier B (ed) Sexual dysfunction. IntechOpen, London, pp 59–76

221. Oosting RS, Chan JSW, Olivier B, Banerjee P (2016) Vilazodone does not inhibit sexual behavior in male rats in contrast to paroxetine: a role for 5-HT1A receptors? Neuropharmacology 107:271–277. https://doi.org/10.1016/j.neuropharm.2016.03.045

222. Olivier B, Mos J (1991) Animal psychobiology. In: Archer S, Hansen T (eds) Behavioural biology: neuroendocrine axis. Lawrence Erlbaum, (Hillsdale, New Jersey), pp 207–227

223. Anderson EE (1936) Consistency of tests of copulatory frequency in the male albino rat. J Comp Psychol 21(3):447–459. https://doi.org/10.1037/h0054857

224. Beach FA (1938) Techniques useful in studying the sex behavior of the rat. J Comp Psychol 26(2):355–359. https://doi.org/10.1037/h0062437

225. Beach FA (1955) Characteristics of masculine sex drive. In: Nebraska symposium on motivation. University of Nebraska Press, Lincoln, pp 1–32

226. Portillo W, Díaz NF, Cabrera EA, Fernández-Guasti A, Paredes RG (2006) Comparative analysis of immunoreactive cells for androgen receptors and oestrogen receptor α in copulating and non-copulating male rats. J Neuroendocrinol 18(3):168–176. https://doi.org/10.1111/j.1365-2826.2005.01401.x

227. Portillo W, Díaz NF, Retana-Márquez S, Paredes RG (2006) Olfactory, partner preference and Fos expression in the vomeronasal projection pathway of sexually sluggish male rats. Physiol Behav 88(4-5):389–397. https://doi.org/10.1016/j.physbeh.2006.04.023

228. De Gasperín-Estrada GP, Camacho FJ, Paredes RG (2008) Olfactory discrimination and incentive value of non copulating and sexually sluggish male rats. Physiol Behav 93(4-5):742–747. https://doi.org/10.1016/j.physbeh.2007.11.027

229. Antonio-Cabrera E, Paredes RG (2012) Effects of chronic estradiol or testosterone treatment upon sexual behavior in sexually sluggish male rats. Pharmacol Biochem Behav 101(3):336–341. https://doi.org/10.1016/j.pbb.2012.01.021

230. Shimomi Y, Kondo Y (2020) Blunt olfaction in sexually sluggish male rats. Exp Anim 69(4):441–447. https://doi.org/10.1538/expanim.19-0161

231. Xia JD, Chen J, Sun HJ, Zhou LH, Zhu GQ, Chen Y, Dai YT (2017) Centrally mediated ejaculatory response via sympathetic outflow in rats: role of N-methyl-D-aspartic acid receptors in paraventricular nucleus. Andrology 5(1):153–159. https://doi.org/10.1111/andr.12274

232. Xia JD, Chen J, Yang BB, Sun HJ, Zhu GQ, Dai YT, Yang J, Wang ZJ (2018) Differences in sympathetic nervous system activity and NMDA receptor levels within the hypothalamic paraventricular nucleus in rats with differential ejaculatory behavior. Asian J Androl 20(4):355–359. https://doi.org/10.4103/aja.aja_4_18

233. Canseco-Alba A, Rodríguez-Manzo G (2013) Anandamide transforms noncopulating rats into sexually active animals. J Sex Med 10(3):686–693. https://doi.org/10.1111/j.1743-6109.2012.02890.x

234. Canseco-Alba A, Rodríguez-Manzo G (2014) Low anandamide doses facilitate male rat sexual behaviour through the activation of CB1 receptors. Psychopharmacology 231(20):4071–4080. https://doi.org/10.1007/s00213-014-3547-9

235. Rodríguez-Manzo G, Canseco-Alba A (2015) Biphasic effects of anandamide on behavioural responses: emphasis on copulatory behaviour. Behav Pharmacol 26(6):607–615. https://doi.org/10.1097/FBP.0000000000000154

236. Giuliani D, Ferrari F (1996) Differential behavioral response to dopamine D2 agonists by sexually naive, sexually active, and sexually inactive male rats. Behav Neurosci 110(4):802–808. https://doi.org/10.1037/0735-7044.110.4.802

237. Giuliani D, Ottani A, Ferrari F (2002) Influence of sildenafil on copulatory behaviour in sluggish or normal ejaculator male rats: a central dopamine mediated effect? Neuropharmacology 42(4):562–567. https://doi.org/10.1016/S0028-3908(01)00195-2

238. Trejo-Sánchez I, Pérez-Monter C, Huerta-Pacheco S, Gutiérrez-Ospina G (2020) Male ejaculatory Endophenotypes: revealing internal inconsistencies of the concept in heterosexual copulating rats. Front Behav Neurosci 14:90. https://doi.org/10.3389/fnbeh.2020.00090

239. Timmermans PJA (1978) Social behaviour in the rat. PhD thesis, University of Nijmegen

INDEX

Yong-Ku Kim and Meysam Amidfar (eds.), *Translational Research Methods for Major Depressive
Disorder*, Neuromethods, vol. 179, https://doi.org/10.1007/978-1-0716-2083-0,
© The Author(s), under exclusive license to Springer Science+Business Media, LLC, part of Springer Nature 2022

Printed in the United States
by Baker & Taylor Publisher Services